Aeronautical Applications *of* Non-destructive Testing

Abbas Fahr, Ph.D.

Senior Research Officer
Aerospace Structures, Materials and Manufacturing
National Research Council Canada

with editorial assistance from
William Wallace, Ph.D.

DES*tech* Publications, Inc.

Aeronautical Applications of Non-destructive Testing

DEStech Publications, Inc.
439 North Duke Street
Lancaster, Pennsylvania 17602 U.S.A.

Printed in the United States of America
10 9 8 7 6 5 4 3 2 1

Main entry under title:
  Aeronautical Applications of Non-destructive Testing

A DEStech Publications book
Bibliography: p.
Includes index p. 469

Library of Congress Control Number: 2013955473
ISBN No. 978-1-60595-120-1

HOW TO ORDER THIS BOOK

BY PHONE: 877-500-4337 or 717-290-1660, 9AM–5PM Eastern Time

BY FAX: 717-509-6100

BY MAIL: Order Department
DEStech Publications, Inc.
439 North Duke Street
Lancaster, PA 17602, U.S.A.

BY CREDIT CARD: American Express, VISA, MasterCard, Discover

BY WWW SITE: http://www.destechpub.com

Contents

List of Acronyms

AC	Alternating Current
AD	Airworthiness Directive
AE	Acoustic Emission
AFRL	Air Force Research Laboratory
AIA	Aerospace Industry Association
ANOVA	Analysis of Variance
ASIP	Aircraft Structural Integrity Program
ASME	American Society of Mechanical Engineers
ASNT	American Society for Non-destructive Testing
ASTM	American Society for Testing and Materials
ATA	Air Transport Association
ATESS	Aerospace Telecommunications Engineering Support Squadron
AU	Acousto-Ultrasonic
AUP	Acousto-Ultrasonic Parameter
AWI	American Welding Institute
BDP	Backscatter Depth Profilometry
BVID	Barely Visible Impact Damage
CBM	Condition-Based Maintenance
CCD	Charged Coupled Device
CF	Canadian Forces
CFRP	Carbon Fiber Reinforced Plastic
CGSB	Canadian General Standards Board
CT	Computed Tomography
DAIS	D-sight Aircraft Inspection System
DC	Direct Current
DFM	Deterministic Fracture Mechanics
DT	Damage Tolerance
DND	Department of National Defence
DSP	Digital Signal Processing

EASA	European Aviation Safety Agency
ECI	Eddy Current Inspection
ECT	Eddy Current Testing
EDM	Electron Discharge Machining
EM	Electro-Magnetic
EMAT	Electro-Magnetic Acoustic Transducers
ENSIP	Engine Structural Integrity Program
EW	Explosive Welding
FAA	Federal Aviation Authority
FML	Fiber Metal Laminates
FN	False Negative
FP	False Positive
FPI	Fluorescent Penetrant Inspection
FRP	Fiber Reinforced Plastic
FSH	Full Scale Height
FSW	Friction Stir Welding
GMR	Giant Magneto-Resistive
HOLSIP	Holistic Structural Integrity Program
HMV	Heavy Maintenance Visit
IACS	International Annealed Copper Standard
IAR	Institute for Aerospace Research
IQI	Image Quality Indicator
IR	Infra-Red
ISO	International Standardization Organization
JSF	Joint Strike Fighter
LFW	Linear Friction Welding
LG	Laboratory Grown
LOI	Lift-off Intersection
LPI	Liquid Penetrant Inspection
MANOVA	Multivariate Analysis of Variance
MAPOD	Model-Assisted Probability of Detection
MIA	Mechanical Impedance Analysis
MIL-HDBK	Military Handbook
MLE	Maximum Likelihood Estimators
MOI	Magneto-Optical Imaging
MPI	Magnetic Particle Inspection
MSD	Multi-Site Damage
MTF	Modulation Transfer Function
NAB	Nickel Aluminum Bronze
NASA	National Aeronautic and Space Agency
NDE	Non-Destructive Evaluation
NDI	Non-Destructive Inspection
NDT	Non-Destructive Testing

NRC	National Research Council
NRCC	National Research Council Canada
NRCan	Natural Resources Canada
NT	Neural Network
NTM	Non-destructive Testing Manual
OEP	Original Equipment Manufacturer
OFE	Oxygen Free Electron
OBUS	Obliquely Backscattered Ultrasonic Signal
PE	Pulse Echo
PEC	Pulsed Eddy Current
PEEK	Poly-Ether Ether Ketone
PHM	Prognostic Health Management
POD	Probability of Detection
POI	Probability of Indication
PPS	Poly-Phenylene Sulphide
PR	Pattern Recognition
PT	Pulse Thermography
PVC	Poly-Vinyl Chloride
RF	Repetitive Frequency
RFC	Retirement for Cause
RFEC	Remote Field Eddy Current
R&D	Research and Development
RGX	Reversed Geometry X-ray
RIM	Range Interval Method
ROC	Relative Operating Characteristic
RT	Radiographic Testing
SCC	Stress Corrosion Crack
SEM	Scanning Electron Microscope
SHM	Structural Health Monitoring
SII	Safe Inspection Interval
SSIP	Supplemental Structural Integrity Program
SSRMS	Space Station Remote Manipulator System
STD	Standard
SWF	Stress Wave Factor
TBC	Thermal Barrier Coating
TGO	Thermally Grown Oxide
TN	True Negative
TOF	Time of Flight
TOT	Time of Travel
TP	True Positive
TT	Through Transmission
UNS	Unified Numbering System
UT	Ultrasonic Testing

US	United States
USA	United States of America
USAF	US Air Force

Foreword

When I was asked to serve as review editor for this publication, I took no time to accept. I was responsible for creating the NDT program at the NRC Institute for Aerospace Research (IAR) in 1982, and Dr. Fahr was the first experienced researcher to be hired who had a background in non-destructive techniques. The invitation to edit this publication has allowed me to look back at what has been achieved over the past 30 years and to decide whether my early conviction was really warranted.

The reality today is that aircraft cannot be allowed to operate with defects in the materials, structures, and systems that keep these vehicles airborne and moving safely toward their destinations. They must also be able to take off and land safely, and to negotiate the runways and landing strips that they operate from. These are not always smooth, clean, flat surfaces. Civilian aircraft runways may have ice, snow, or slush coverings while military runways may be rough, un-paved, and dusty. Civilian and military aircraft alike will, from time to time, be exposed to rain, hail, fog, mist, and, in coastal regions, to salt spray. Any one of these negative environmental factors can cause damage to aircraft and to their ancillary equipment that could spread to become major risk factors. Fatigue and corrosion are particularly troublesome because they can affect hidden areas of an aircraft where they will initiate cracks that can grow rapidly to assume danger-ous proportions. Cracks will create regions of high stress concentration which will increase as they grow and the size of the remaining ligaments that are left to carry the loads will decrease. In the case of corrosion, load bearing metals will be damaged by thinning as the metal is converted to corrosion products, such as oxides or hydroxides. Stress in the remaining load carrying metal will increase. Compounding this problem is the fact that moisture often gets trapped between sheets of overlapping metals, as in lap-joints, where the corrosion is largely hid-den from view. But as the corrosion products grow, they expand and force the metal to bulge outward, only to be restrained by rivets or other fasteners holding the sheets together. But the metal between the fasteners is less constrained and is forced out to become plastically deformed, hence, permanently changing shape.

Local yield strengths will be exceeded. Stresses at the points of constraint that are around the rivet holes, will become excessively high and cracks will soon form and grow out into the surrounding metal. If unchecked or removed, a growing crack will ultimately lead to catastrophic rupture of the entire structure, and, potentially, the loss of the entire aircraft and injury or death to passengers and crew. NDT methods must be available to detect and assess the size and severity of these forms of damage. The present publication describes several NDT methods that are capable of dealing with the many different materials and forms of structural detail found in typical aircraft structure.

A further impetus for the development of NDT has been the gradual change in design philosophy for aircraft structures over the past 30 or more years. In the 1960s and early 70s, aircraft design was based largely on the principle that cracks should be avoided at all costs and parts were retired at a time when the probability of a crack being present was extremely low. These were generally referred to as Safe-life or Fail-safe philosophies. They have been described in more rigorous terms in Chapter 2 of this publication. Today there is much more acceptance of the fact that cracks are almost inevitable, and rather than trying to avoid them completely, engineers are now trying to live with them in a way that doesn't compromise safety and which may actually improve it. Thus it is accepted that crack-like defects will be present in virtually all aircraft structures and the urgent requirement now is to be able to detect them before they become a serious threat. This introduces a vital requirement for NDT. It must be possible to detect the presence of cracks with extremely high reliability as well as very high sensitivity. Recognizing the possibility that an extremely small crack might be missed during one inspection, the probability of detecting the growing crack on the next inspection must be high. The frequency of inspection must be such that crack detection is assured well before the growing crack can reach a critical size. This is the field of damage tolerance that requires deep understanding of fracture mechanics, crack growth kinetics, and quantitative NDT. This publication describes these in the last chapter on NDT Reliability.

Perhaps the obvious solution to fatigue cracking and corrosion is to build aircraft from materials that don't suffer these problems. Of course, aircraft designers and manufacturers are doing this by relying more and more on fiber reinforced composites. High strength fibers can carry loads far in excess of most metals; the fibers act as natural crack arrest features and provide outstanding fatigue strength while the thermosetting and thermoplastic resins that hold the fibers and fabrics together don't corrode. It sounds good, but they introduce a whole new set of problems, such as sensitivity to high and low energy impacts, delamination between plies, environmental sensitivity to moisture and other aircraft fluids, and ingress of moisture when honeycomb sandwich structures are used. These problems have provided impetus for the development of a whole new set of NDT procedures that are capable of detecting and quantifying these types of damage in composite aircraft structures. These, too, are described in this publication, in Chapter 11, with examples of applications provided.

These are frightening scenarios, and it is not surprising that so much focus has been directed towards the development of NDT methods over the past 30 or more years. In this publication, Dr. Fahr has described most of the standard NDT methods that are available for the inspection and characterization of aircraft materials and structures today. He has described the principles of each method, the benefits and limitations of each, and he has provided examples of how they can be applied in practice. Most, if not all, of the examples are from investigations carried out at IAR-NRC Aerospace over the last 25 years. The problems were almost all brought into the laboratory by various aircraft operators, thus, they were real-life, practical problems rather than academic exercises.

Having read all 13 Chapters several times, I am now confident that my decision nearly 30 years ago was fully justified. Dr. Fahr has led the development of the IAR NDT laboratory very well. With his team, he has assembled an extensive suite of NDT equipment and technologies that have been adopted, adapted, modified, and applied to a wide range of real engineering problems brought into the laboratory, mostly by external clients. It is fair to say that NRC Aerospace today has one of the finest collections of NDT equipment available in North America, along with the knowledge and expertise needed to use it in the interests of aircraft and public safety. Equally impressive is the fact that when one scans through the extensive lists of references collected at the end of each chapter, one has to be impressed by the number of authors who came to NRC as summer students, graduate students, guest workers, and even NRC employees who knew very little of NDT when they arrived. They have been educated, trained, and mentored, largely by Dr. Fahr, and many of them now have moved on to become leading practitioners of NDT in Canada and around the world.

I feel that the NRC, with the help and guidance of Dr. Fahr, has made a major and hopefully lasting contribution to NDT in Canada, and one that will remain at the leading edge continuing to play a vital role in promoting the safe and cost-effective operation of both civilian and military aircraft in Canada and around the world.

Dr. William Wallace, B.Sc. Tech.,
PhD, C.Eng., D.Eng., FASM, FCASI.
FIMMM. Researcher Emeritus,
Structures and Materials Performance
Laboratory, NRC Aerospace

Preface

Non-destructive testing (NDT) plays an important role in the safety of aircraft and other critical structures, yet the subject is not well-understood or recognized by many engineers. This book has been prepared with the aim of providing information on NDT technologies available for aeronautical applications in a simple and concise manner, such that it will be useful to readers of various backgrounds. It is intended to serve as a general educational or reference source for students and engineers in aeronautics or related fields who may require NDT as a part of their education or for their project activities. The book describes NDT methods in general, their applications for the evaluation and inspection of aerospace materials or components, the types of information they provide, and how such information is used to improve aircraft safety.

This book contains 13 chapters starting with an introduction to the NDT field and a glossary of terms related to the subject. A review of aircraft design, manufacture and maintenance methodologies as well as their NDT requirements, is provided in chapter 2. In chapter 3, a description of the way NDT is carried out and how the results are analyzed is given, which explains the need for NDT standards, references, and calibrations, as well as qualified personnel to perform the tests. Chapters 4 through 10 deal with the different NDT methods used in the aeronautical field describing the basic principles of the techniques, necessary instrumentation, procedures available for their use, a few examples of applications, and a summary of their capabilities and limitations.

Technologies covered in these chapters include: optical and enhanced optical methods; liquid penetrant, replication, and magnetic particle inspections; electromagnetic and eddy current approaches; acoustic and ultrasonic procedures; passive and active infrared thermal imaging; and x-rays and other radiographic techniques. In addition, there are three chapters which deal with important subjects related to aircraft inspection. Chapter 11 describes NDT of composite and hybrid materials, which are extensively being used in modern air-vehicles, while chapter 12 is concerned with aging aircraft issues. In these chapters, impact damage and environmental degradation of composites, corrosion and fatigue cracking of metallic parts, and the NDT methods available to detect them are explained. Finally, chapter 13 is dedicated to NDT reliability and the ways the probability of detection can be measured and employed to establish inspection intervals for the continued safe use of critical aircraft components or structures.

Acknowledgments

The non-destructive testing (NDT) technical information and application examples provided in this book are mostly from the research and technology development projects carried out at the Structures and Materials Performance Laboratory (SMPL) of the NRC Institute for Aerospace Research (currently NRC Aerospace). Many SMPL staff members, guest workers, and students have contributed to the work and we acknowledge their contributions with gratitude. In particular, the author would like to thank the following colleagues who have assisted directly in the preparation of the present publication:

- Mr. Mike Brothers for performing NDT tests and providing inspection images and data.
- Mr. Marc Genest for the technical material and test results on infrared thermography techniques.
- Mr. Muzibur Khan for NDT reliability analysis and for assisting with the formatting of this document.
- Dr. Zheng Liu for performing NDT signal and image processing and providing the associated examples.
- Dr. Catalin Mandache for the technical information on electromagnetic techniques and writing portions of the related chapter.
- Dr. William Wallace for editing the entire book manuscript and providing valuable comments.

The financial support for the work described in this publication was provided by the NRC and its government and industrial partners including the Canadian Department of National Defence.

Dr. Abbas Fahr, B.Sc, M.Tech, Ph.D.
Senior Research Officer Aerospace-
Structures and Materials Performance
National Research Council Canada

CHAPTER 1

Introduction

Non-destructive testing (NDT) plays an important role in the safety of aircraft components and structures, yet the subject is relatively obscure in the minds of many students and professionals engaged in the aeronautical field. As the term implies, NDT refers to testing methods that use special equipment and approaches to examine an object without harming it or affecting its future usefulness. NDT is mainly used to detect and characterize hidden defects or damage in critical components or structures, but it also has applications in dimensional measurements and evaluation of some material properties.

NDT is a very broad and interdisciplinary field that finds applications in many industries where failure of critical parts could endanger public safety. It is utilized to ensure the structural integrity of aircraft as well as nuclear reactors, high-speed trains, ships, off-shore platforms, bridges, cranes, gas or oil pipelines, and many other structures. NDT is employed during manufacturing as a quality control tool to ensure that the materials and finished products are free of internal defects or that detected abnormalities are below a size that would be considered potentially harmful. It is also used during service to identify cracks, corrosion, or other forms of degradation and damage that may cause failure of components or systems, or interfere with their proper functioning. In addition, NDT is utilized as a research tool in the laboratory to study certain material properties, internal features, or manufacturing processes. It is also employed during mechanical or environmental testing of specimens or full-scale structures to monitor material deformation, deterioration, crack propagation, and failure.

1.1 PURPOSE

The main purpose of this book is to provide information on NDT technologies available for aeronautical applications in a simple and concise manner such that it will be useful to readers of various backgrounds. It is intended to serve as a general educational or reference source for people who require NDT for their project activities as well as for students studying in aeronautical or related fields. The book describes the principles of the different NDT methods used in aerospace along with examples of common procedures, instruments, and applications to aircraft materials or components. The examples are intended to illustrate the type of information NDT provides and the way the results are related to the test part. While the main focus is on the aeronautical applications of NDT, the methods described can be applied in many other fields; however, the test procedures and instrumentations may be different.

1.2 DEFINITIONS

The American Society for Testing and Materials (ASTM) defines NDT as the following:

> *NDT is the development and application of technical methods to examine materials or components in ways that do not impair future usefulness and serviceability in order to detect, locate, measure and evaluate discontinuities, defects and other imperfections; assess integrity, properties and composition; and measure geometrical characteristics.*

In this book, the material or component that is being examined is referred to as the "test object" and the discontinuity, defect, imperfection, property, or dimension that is being examined is referred to as the "test target."

The terms non-destructive testing, non-destructive inspection (NDI), and non-destructive evaluation (NDE) are used synonymously. While NDT is the most common term, NDI is widely employed in the aerospace industry, and NDE is exclusive to tests that provide "quantitative" measurements. The term "quantitative" refers to the measurement of the "test target" characteristics or metrics (e.g., defect type, shape, size and location, or property value and component dimensions). Generally, when a test is carried out to determine the presence of defects or change of properties/dimensions, either NDT or NDI is used. However, when both the detection and characterization of defects or the measurement of dimensions, properties, or integrity of materials are of interest, NDE is employed. In this publication, for consistency, only the term NDT will be used regardless of the purpose of the test.

A "discontinuity" simply refers to any change or difference in the material's local physical characteristics, such as an abrupt change of density (e.g., air pockets, gaps, inclusions, etc.) or change of boundaries (sharp edges, corners, layers, grains, etc.) On a smaller scale, every material contains inherent discontinuities. When a component fails to perform its intended function due to the presence of a discontinuity, then the discontinuity is considered to be a "defect." Therefore, defect is a judgment call that is based on the criteria set for a specific component and a specific function or application. For example, a crack as small as 1 mm in the rotating components of aircraft engines is considered a serious defect while cracks as large as 10 mm on non-structural parts of the aircraft may be tolerated.

In addition to "defect," terms such as "flaw" and "damage" are also used to describe unacceptable discontinuities. While the word flaw is mostly used for manufacturing defects, damage refers to the service-induced discontinuities. The most important service-induced damage types in aircraft are cracks originating at stress concentration sites, such as bolt holes, corrosion of metallic surfaces due to moisture, delamination, and debonding of composite and bonded structures as results of foreign object impact or mechanical overloading. Terms such as "imperfections" and "abnormalities" are also used to describe material discontinuities that are considered defects if they exceed a predefined level. Words such as "degradation" and "deterioration" are utilized to describe adverse changes in both surface and bulk properties properties of materials. If a given material property degrades or deteriorates out of a range required for a specific application, then the material is considered to be defective.

The accept and reject criteria for the discontinuity sizes or property values that need to be detected or measured by NDT are usually defined using analytical and experimental tests on small samples, representative parts, or full-scale components. NDT professionals use a variety of methods to identify and characterize discontinuities and make accept or reject decisions based on predefined criteria that are provided to them.

1.3 AERONAUTICAL NDT

Many different NDT techniques are available for application to aircraft materials, components, and structures both during manufacturing and in service. An aircraft structure is regarded as a built-up assembly of individual components that are made of either a single material, such as aluminum, or a variety of materials including metals, composites, and hybrid materials. During manufacturing, NDT is used for quality control of the initial materials and the assembled components or structures to identify any defects, abnormalities, or imperfections. In service, NDT is employed to find any damage, degradation, or deterioration of the critical parts of the aircraft.

1.3.1 NDT METHODS

All NDT methods use some form of energy (e.g., light, heat, radiation, electromagnetism, mechanical force, or acoustic waves) to obtain information about the test object. The widely accepted and qualified methods for the inspection of aircraft materials, components, and structures are optical methods (including visual and aided-visual), radiography, acoustic and ultrasonic, electromagnetic, magnetic particles, and dye penetrant techniques. These are sometimes referred to as "conventional" NDT methods. Infrared thermography and laser shearography are also finding increasing aeronautical applications in recent years; however, they are not yet widely used for aircraft inspections. Several variants of a method may exist depending on the type of energy source, the way the tests are carried out, the approach used to process data, or the manner the results are presented. For example, radiography can be carried out using a variety of sources of radiation, such as x-rays, gamma-rays, or neutron particles, and processed or presented in many ways using films, digital pictures, or 3-dimensional computer tomography images. Similarly, acoustic and ultrasonic methods range from simple tap-testing and audio analysis to automated ultrasonic scanning and 2- or 3-dimensional display of the interior of the part.

In all the above-mentioned methods, the energy transmitted by the NDT device is affected by the material discontinuities or properties; thus, the measured response can be related directly to the target being tested. There are also a few methods that infer information about material conditions indirectly. For example, a localized surface deformation seen by some optical methods could be related to internal corrosion among other things. These indirect methods may be used occasionally for specific applications, but they often require verification by conventional NDT or other tests.

Each NDT method has certain capabilities and limitations and often more than one method is required to obtain the necessary information about the test target. For example, a dye penetrant inspection is very easy and cost-effective but can only identify discontinuities that are open to an accessible surface. On the other hand, electromagnetic or ultrasonic methods are sensitive to both internal and surface-connected defects, but they are more time-consuming and difficult to perform. Thus, the choice of the technique is dependent on many factors including inspection requirements and condition of the object as well as the capabilities and limitations of the NDT methods, access to the test site, ease of application, and cost of inspections.

1.3.2 NDT PERFORMANCE MEASURES

The performance of an NDT method can be described in many different ways; a few terms that are most commonly used are defined below in general and simple words.

- *Sensitivity* is an indication of how small of a defect can be detected or a change of dimension/property that can be measured.
- *Resolution* is a measure of how easily closely-located defects can be separated or small differences in properties/dimension can be measured.

- *Accuracy* is an indication of how precisely a defect can be detected or a change of property or dimension can be measured.
- *Repeatability* is a measure of how frequently the same detection or measurement results are obtained if the test is repeated.
- *Reliability* is a measure of how many times the technique provides the correct answer.
- *Applicability* is related to the practical use and usefulness of the tests and includes factors such as ease of application, cost of the tests, and benefit of results to achieve the test goals.

Details of these terms and their measurement metrics may differ in the context of the different techniques and the nature of the target being tested. In the ensuing technical chapters on NDT methods, more detailed descriptions of the important performance measures related to each technique will be provided.

Most NDT methods are based on comparison of the test object with a reference piece that has known characteristics, such as certain properties, calibrated dimensions, or specific discontinuities. NDT instruments and procedures are first calibrated with respect to established reference pieces, and the tests are carried out using standard practices that are developed for specific applications. NDT personnel who conduct aircraft inspections require a certain level of proficiency in the prescribed methods and certification by an authorized certifying agency. These are all important in obtaining valid and consistent results and will be covered in this book.

1.4 ORGANIZATION OF THE CHAPTERS

The first part of the book includes 10 chapters starting with this introduction to the NDT field and the definition of the important terms related to the subject. In Chapter 2, a short review of aircraft design, manufacture, and maintenance methodologies as well as the NDT requirements of each approach is provided. Chapter 3 describes the way NDT is carried out and the results are analyzed where topics such as NDT standards and references as well as the need for calibrations and the qualified personnel are explained. Chapters 4 to 10 deal with different NDT technologies used in the aeronautical field. In these chapters, first the basic principle of each technique is described along with typical instruments and procedures. Then, a few application examples on different materials or components is provided to illustrate the type of information each method can provide and how such information can be related to the state of the test object. Finally, a summary of the capabilities and limitations of the method is given. Technologies covered in these chapters include optical and enhanced optical methods, liquid penetrant, replication and magnetic particle inspections, electromagnetic or eddy current techniques, acoustic and ultrasonic methods, x-rays and other radiographic procedures, and finally infrared thermal imaging techniques. The following resources are extensively used throughout this document as primary NDT information sources:

- Non-destructive Testing Handbooks and Standards published by the American Society for Non-destructive Testing [1].
- Metals Handbook, Non-destructive Evaluation and Quality Control, published by the ASM International [2].
- NDE Resource Center of Iowa State University [3].
- Aeronautical Applications of Non-destructive Testing [4].

The second part of the book contains three more chapters. Chapter 11 deals with NDT of composite and hybrid materials that are increasingly being used in modern aircraft to replace aluminum alloys. In Chapter 12, aging aircraft issues related to corrosion and cracks in airframe aluminum structures as well as damage to engine parts and their NDT methods are described. Finally, Chapter 13 is dedicated to the reliability of NDT methods and the ways the probability of detection and confidence can be quantified and employed to establish inspection intervals for the continued safe use of aircraft components. In addition to the author's earlier publications referred to here, published literature and internet sources are used to obtain NDT material, figures, or photographs that were necessary for a comprehensive explanation of the subject. In such cases, however, reference is made to the original sources for further reading. The information gathered from the referenced sources available in the public domain may be copyright protected and are used in this book for educational and training purposes only.

1.5   CHAPTER REFERENCES

[1]   Non-destructive Testing Handbooks and Standards published by the American Society for Non-destructive Testing including; *Electromagnetic Testing*, 3rd Edition, Vol., 5, 2004 and *Ultrasonic Testing*, 2nd Edition, Vol. 7, 1991.

[2]   *Metals Handbook, Non-destructive Evaluation and Quality Control*, published by the ASM International, 9th Edition, Vol. 17, Metals Park, Ohio, USA, 44073, 1989.

[3]   NDE Resource Center of Iowa State University, Iowa, USA , available for public on the internet: http://www.ndt-ed.org/EducationResources/.

[4]   Aeronautical Applications of Non-destructive Testing, A. Fahr, National Research Council Canada report No: AN-SMPL- 2012-0108, September 2012.

CHAPTER 2

NDT in Aircraft Design, Manufacture, and Operation

NDT plays a major role in all stages of aircraft life from the initial concept and technology development, to the design and production of prototypes and the actual aircraft, to the operation of the fleet until the retirement of the last aircraft. The NDT objectives and tasks vary for the different phases depending on the initial design philosophy, materials and configurations employed in the construction, the operational conditions of aircraft, and maintenance approaches used.

In this chapter, first the aircraft structural design methodologies and their NDT requirements are briefly described. Then, a brief description of the development and construction of modern aircraft and the role of NDT in different stages of manufacturing are provided. Finally, the use of NDT during the operation and life management of aircraft is mentioned using a few examples applied to aging military aircraft.

2.1 AIRCRAFT DESIGN

Aircraft structural design has evolved over the years. In the early designs, the emphasis was on strength and the ability to withstand static loads, but soon it became clear that fatigue due to cyclic loads plays an equally prominent role in structural failures. The "safe-life" design philosophy was introduced in order to take cyclic loads into account. The safe-life design was followed by further improvements with the use of the "fail-safe" and "damage tolerance" approaches. Currently, all these approaches are used in the design of structural components of aircraft depending on their operational goals and applications. In addition, many other criteria and considerations are taken into account, and some of the important ones are illustrated in Figure 2.1 [1]. "Inspectability" or the ability to inspect the

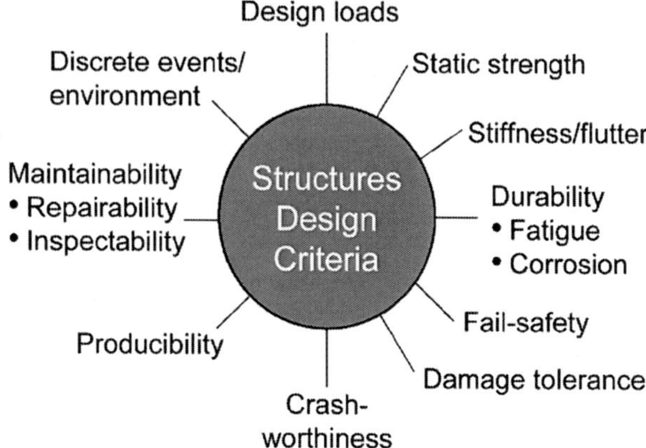

Figure 2.1 Typical considerations used in the aircraft structural design [1].

component during manufacturing and in service is one of the prime considerations of many that is relevant to the topic being covered in the present publication and will be discussed in this chapter.

2.1.1 SAFE-LIFE

The "safe-life" design philosophy was developed immediately after the Second World War in an effort to take into account cyclic stresses which led to fatigue failure of some early aircraft (e.g., Comet, B-47). Initially, these failures were attributed to poor static design but as the nature of the cyclic fatigue was identified, it became clear that structures designed based on static loads would not necessarily reach their design life under cyclic loading conditions [4]. This led to the introduction of the safe-life approach which considers both static and cyclic loads.

The safe-life of an aircraft is basically defined as the number of flight hours or cycles (a take-off and landing is considered one cycle) during which there is a low probability that the strength will degrade below its design value due to fatigue. A safety factor is incorporated to take into account variability in the quality of manufactured components and assemblies as well as the operational and environmental variability of the aircraft fleet as a whole and the statistical nature of the damage accumulation. Figure 2.2 illustrates a typical failure probability distribution as a function of the life span in cycles. In the safe-life approach, components are designed so that the probability of occurrence of cracks is 1 in 1000, that is equivalent to the lower bound of three standard deviations (-3σ) on the failure distribution curve.

The safe-life design involves rigorous fatigue testing of test coupons, representative load-bearing components, or sub-assemblies and full-size primary

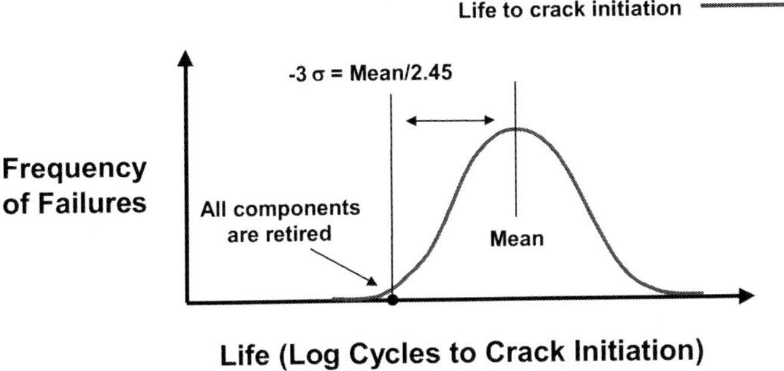

Figure 2.2 Safe-life approach for life management of fatigue-limited aircraft components.

structures for two to four times the number of flight hours or cycles expected during the actual life of the aircraft. This safety factor is to take into account the unknowns and assumptions of the design process and the operational and environmental variables of the aircraft fleet as a whole. However, this approach assumes that materials used in the production of aircraft structural components are homogenous and free of inherent defects because the safe-life is established by testing undamaged components. If a part is damaged during service, the established safe-life period is no longer valid and the part must be repaired to the original condition or replaced all together. The choice between repair and replacement is based on the availability and cost of the replacement part in relation to the existence, ease of use, adequacy, effectiveness, and cost of repairs.

The safe-life approach requires inspections of the manufactured materials and components to ensure their flaw-free conditions, but, in principle, it does not rely on NDT during aircraft operation. However, occasionally unanticipated cracks still occur before the end of the "design safe-life" creating safety problems and necessitating inspections. Such cracks may originate from manufacturing defects, which are below the detection limit of the NDT methods and escape the initial inspections, or subtle service-induced degradations, such as localized corrosion, wear, fretting, or impact damage. These, in combination with changes in operational environment and loading spectra, could contribute to premature nucleation of cracks. Although the overall design in fatigue situations may be acceptable, unforeseen and undetected flaws or random service damage can cause premature cracks or failures. Therefore, aircraft or components designed using this approach are often inspected by NDT in response to failures or accidents during operation.

In the safe-life approach, in addition to periodic visual checks of the critical components in-service NDT may be needed based on the service history of the aircraft. In-service NDT is carried out during overhauls to avoid premature failures. In this design, a large number of components are discarded at the end of their "design safe-life" regardless of their condition. Many of the discarded parts may still have considerable useful life remaining due to the safety factors considered. However, the safe-life approach has the advantage that the maintenance requirements are kept to a minimum, while the usage of components in service without inspection is maximized. Overall, this approach is overly conservative, because by default 999 out of 1000 components are retired with significant life remaining. Furthermore, the method is costly since all parts must be replaced at the end of their design-life, often at the same time, and the availability of spare parts may be a problem for old aircraft [3].

2.1.2   *F_{AIL}-S_{AFE}*

The "fail-safe" design is an extension to the safe-life approach and is used to minimize any unanticipated component failures. In the fail-safe design the principal structural components must retain their required residual strength for a period of usage after partial or full failure of a single element until the failed element is repaired or replaced [2]. The fail-safe design is achieved in different ways, such as using structural configurations with multiple load paths that will distribute the applied load to other members in the event of failure of individual elements and adding features to stop crack propagation.

In the fail-safe design a single component failure does not lead to immediate loss of the entire structure. For example, two skin layers (i.e., doublers) are used around windows, doors, access panels, and other cut-outs as well as at splices, frames, and longerons to reduce stress concentrations and to serve as tear-stoppers. With the use of the fail-safe design along with the safe-life, the usable structural life becomes much greater than the safe-life alone; however, the multiple load paths result in weight penalties.

Similar to the safe-life, the purpose of NDT in the fail-safe design is to detect defects of the manufactured parts during production and damage to any one of the load paths during service operation. The in-service inspections are based on predictive models that identify sites requiring periodic inspections, and the service history of the aircraft that identifies components with potential problems. In-service inspections are required for the identified sites to ensure that the aircraft operation and safety are not compromised. The fail-safe design philosophy is generally effective in allowing sufficient opportunities for timely detection of structural damage. However, failure modes cannot always be predicted with sufficient accuracy to ensure that structural damage would be accessible and obvious to NDT inspectors or be within safety limits.

In practice, however, a number of unanticipated failures of fail-safe structures due to unexpected fatigue damage occurred, which resulted in the expectations to find cracks that were difficult to detect using NDT methods [5]. This prompted the aerospace industry and airworthiness authorities to focus on the adequacy of inspections that would allow timely detection of structural damage before it became critical and led to the development of damage tolerance philosophy.

2.1.3 DAMAGE TOLERANCE

This design philosophy recognizes the potential for discontinuities and imperfections to exist in all critical components, which are the inherent result of the materials microstructure, processing, manufacturing, and usage. The U.S. Aircraft Structural Integrity Program (ASIP) defines damage tolerance (DT) as the attribute that permits the structure to retain its required residual strength for a period of unrepaired usage after it has sustained prescribed levels of fatigue, corrosion, and accidental damage, such as unstable propagation of initial flaws, fatigue cracks, or impact from a discrete source [2].

The DT approach is concerned with the propagation of cracks and uses fracture mechanics to establish the ability of the structure to operate safely for a specified period of time until the crack is detected during routine inspections. It is intended to ensure that if serious damage occurs within the design service goal of the airplane, the remaining structure can withstand reasonable loads without failure or excessive structural deformation until the damage is detected [5]. This is achieved by fail-safe design features including multiple load paths and the use of crack stoppers to control the rate of crack growth and provide adequate residual strength.

Figure 2.3 schematically illustrates the damage tolerance structural design principles. This design requires that should damage occur to the principal structural elements, the materials and stress levels must provide a controlled slow rate of crack propagation combined with high residual strength. In addition, the arrangement of design details must ensure a high probability that a damage or failure in any critical structural elements will be detected before the strength of the structure is reduced below the specified fail-safe level. This regulatory fail-safe

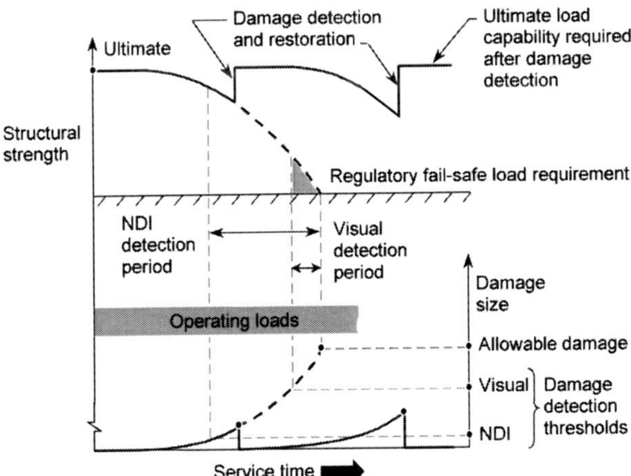

Figure 2.3 Schematic of the requirements of a damage tolerant structure [1].

limit is the necessary strength required to withstand the operational loading conditions, thereby allowing opportunities for timely detection of damage and replacement or repair of the failed element. In other words, the residual strength of the structure with detectable damage must always be higher than the regulatory fail-safe requirement of the structure and the operating load must remain below the specified limit.

For all safety-related structures, damage sizes that must be detected are determined analytically and verified by means of mechanical tests of the individual components as well as complete aircraft sections. The results of such tests are used to establish if in-service inspections are needed and also to determine the time or the number of cycles until the first inspection as well as the interval for subsequent inspections.

The US Air Force (USAF) uses the DT approach in the design and maintenance of the USAF aircraft based on the MIL-STD-1783 [6]. The U.S. Federal Aviation Authority (FAA) has also adopted the DT approach for most civilian transport aircraft [7, 8]. However, the initial design goal for commercial civilian aircraft is to have the inspection interval equal to or longer than the desired life of the aircraft. This is due to the fact that the cost and time involved for detailed NDT may prohibit economic operation of commercial aircraft. In addition, many of the airframe principal structural components are not easily accessible to allow economical routine in-service inspections.

Due to the above constraints along with the limitations of the in-service NDT methods to inspect large and complex geometries in a timely and economical manner, the majority of the principal airframe structures are designed and manufactured in such a way that they require minimal inspections during their service life. Only a limited number of safety-critical components will require periodic inspections, mostly by visual techniques, and any visible abnormalities will be verified by more elaborate NDT techniques. With this philosophy, NDT is integrated into aircraft design, manufacturing, and maintenance.

In the design stage, the inspectability of the critical components is considered. During the manufacturing phase, like the other approaches, NDT is used as a quality control tool to detect inherent material discontinuities or processing defects of individual components. The flaw sizes that must be detected in the manufactured parts are much smaller than the requirements of service inspections. Thus, automated NDT methods are often employed for production inspection in order to avoid the variability associated with the human operators and to improve the inspection reliability while increasing the through-put. In-service inspections are carried out during routine checks and at overhauls. NDT methods must be able to identify any damage before it reaches the allowable limit (Figure 2.3). The flaw sizes that must be detected by NDT during manufacturing and in-service are determined by analysis and tests and verified by demonstration of inspection methods. The latter provides information on the capability of the NDT approaches and is often established statistically using probability of detection (POD) and confidence analysis. The NDT POD data is then used to determine the frequency of in-service inspections or the safe inspection interval (SII). The way SII is determined will be described later in this chapter.

Damage tolerance design is required by the FAA for most civil transport aircraft unless it entails such complications that an effective DT structure cannot be achieved within the limitations of geometry and a good design practice [9]. Under these circumstances, a safe-life design that complies with the fatigue evaluation requirements is allowed. A typical example of a structure that is not conducive to damage tolerance design is the landing gear and its attachments that still use safe-life. Helicopters and some small aircraft (e.g., private and acrobatic) still use safe-life design.

Similar to the ASIP, the U.S. Engine Structural Integrity Program (ENSIP) has developed guidelines for damage tolerance application to engine systems [6]. ENSIP defines damage tolerance as the ability of the engine to resist failure due to the presence of initial flaws, cracks, or damage for a specified period of unrepaired use. NDT is used here also at the design and manufacturing phases of the materials and components as well as during service.

As described in *Damage Tolerance for High Energy Turbine Engine Rotors* [9], the FAA considers damage tolerance for high energy turbine engine rotating components as an element of the life management process. As with airframes, the DT assessment of engines addresses the potential existence of component imperfections through the incorporation of fracture-resistant design, fracture mechanics, process control, and NDT. The intent of damage tolerance assessment of engines is to supplement the existing safe-life methodology and is not intended to allow life-expired parts to remain in service beyond the limits established by the safe-life approach nor does it allow damaged parts to return to service.

2.1.3.1 Safe Inspection Interval (SII)

As mentioned above and explained in [10], [11], and [12], the damage tolerance approach assumes that the fracture-critical locations of a component or structure contain a flaw of a size just below the detection limit of the non-destructive inspection techniques. The flaw is assumed to grow during service in a manner that can be predicted by linear elastic fracture mechanics or other acceptable methods, until a predetermined dysfunction limit is reached beyond which the risk of failure due to rapid growth of the defect becomes excessive. The dysfunction flaw size is usually less than the critical size. The rate of growth and the dysfunction sizes are established analytically based on best estimates of service loads and material properties. The time or number of cycles required for the assumed flaw to grow from the initial size to its dysfunction size is then used to define a safe inspection interval (SII).

Figure 2.4(a) shows a typical crack growth curve and (b) the repeated inspections using the DT approach. As illustrated in this figure, at the end of one SII, the component is inspected and if no crack is detected, the part remains in service for another SII period. This process is repeated until a crack is found. In this manner, components are retired on an individual basis when their condition dictates.

In order to calculate SII and implement the DT approach, it is imperative to determine the detection limits of the candidate NDT methods and to ensure that the selected NDT approach does not miss a dysfunction crack size under any circumstances. Thus, it is necessary to establish the capability of the NDT methods in terms of sensitivity and reliability. As mentioned above, NDT capability is best determined by POD and confidence analysis. POD is a measure of both the sensitivity and reliability of NDT method and is used in damage tolerance life management of aircraft parts or structures. The NDT reliability is the subject of a separate chapter later in this publication where the different ways of establishing POD are explained.

Briefly, NDT POD and confidence data can be obtained experimentally by statistical assessments of the inspection procedures or by using a combination of analytical and empirical means. Flaw sizes detectable at 90% POD with 95%

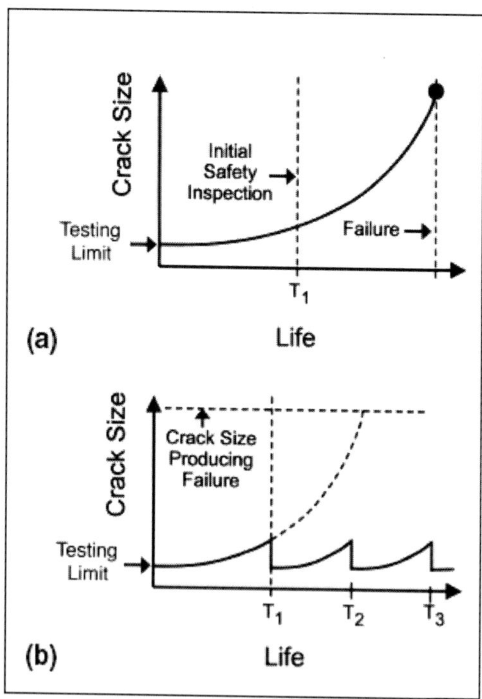

Figure 2.4 (a) shows a typical crack growth curve and (b) repeated inspections under the damage tolerance approach [25].

confidence are often defined as the NDT limits and used to calculate the safe inspection intervals. The smaller the flaw that can be detected with a higher POD and confidence, the less frequent the inspection needs would be (i.e., longer SII), which is more economical and attractive to the aircraft operators.

In addition to NDT limits, the SII calculations require information on material properties and damage growth rate for the component being inspected. Such data is obtained through vigorous fracture mechanics analysis and experimental tests. Examples are provided in references [25] and [26] for fatigue cracks in metallic parts. In this case, the cyclic stress intensity factor (Δk) ahead of the crack tip is given by:

$$\Delta k = \Delta\sigma\sqrt{\lambda\pi a} \qquad (2\text{-}1)$$

where λ is a factor that depends on the crack shape, structure, and gradient of stresses, and $\Delta\sigma$ is the stress amplitude. Assuming that stable fatigue crack growth conditions prevail, the crack growth rate is given:

$$da/dN = C\,(\Delta k)^{m} \qquad (2\text{-}2)$$

where C and m are experimentally-determined material constants. Substituting for Δk from equation (2-1) in equation (2-2) and integrating, the number of fatigue cycles (N) required to grow a crack from the assumed initial size a_i to the dysfunction size a_d is given by:

$$N = [1/C(\Delta\sigma\sqrt{\pi})^{m}].\int (\lambda a)^{-m/2}\, da \qquad (2\text{-}3)$$

A safety factor of two is often considered, and, thus, the SII is chosen to be half of N, or:

$$SII = \tfrac{1}{2}[1/C(\Delta\sigma\sqrt{\pi})^{m}].\int (\lambda a)^{-m/2}\, da \qquad (2\text{-}4)$$

Either deterministic fracture mechanics (DFM) or probabilistic fracture mechanics (PFM) may be employed. For DFM, the worst case values of m, C, Ds, a_i, and a_d are used to calculate the SII. However, it has been suggested that the worst case conditions for each and every variable in equation (2-4) may never occur in real life. Thus, the SII values obtained using the DFM can be overly conservative (i.e., shorter SII or more inspections). In the PFM approach, m, C, Ds, a_i, and a_d are treated as random variables with a statistical distribution. Therefore, rather than calculating a single SII value and using a safety factor based on good engineering judgement, a range of SII is calculated and appropriate values are selected to maintain a sufficiently low, but economically-feasible, failure probability. Thus, the SII values determined using the PFM approach are usually longer (fewer inspections) and more attractive from the practical stand point.

2.1.4 *HOLISTIC APPROACH*

In recent years, a more holistic structural integrity process (HOLSIP) has been developed [13, 14]. HOLSIP is based on the fundamental idea that failure modes or mechanisms are interconnected. It is a physics-based structural integrity approach that considers all time-dependent and time-related mechanisms of degradation, including fatigue, corrosion, creep, wear, and any combination of these. This framework assesses the structural integrity of aircraft components throughout their entire life cycle and considers all factors, including corrosion and fretting, that are absent in the above-mentioned life cycle management paradigms currently in use. With the present approaches, aircraft are assumed to be corrosion-free; thus, any such damage that is discovered to be above a certain limit must be removed and the structure repaired immediately [14]. This results in aircraft down-time and increased maintenance costs.

Like damage tolerance, HOLSIP assumes that initial discontinuities are present in materials and as-manufactured parts. In addition, it considers that the discontinuities will respond to the environment and extrinsic forces during the entire life of the aircraft. Thus, in this approach, extensive effort is expended to understand and characterize all degradation and failure processes that a structure may experience. The life of aircraft structural components is established by considering the individual and combined effects of overload, fatigue, environmental, and accidental damage. NDT is integrated into the HOLSIP process in order to assure structural integrity. Similar to the damage tolerance approach, prediction of inspection intervals is made by using analytical and experimental methods that take into account the detection threshold for the specific degradation mechanism and the POD for the method of detection.

As mentioned earlier, the damage tolerance approach is concerned with propagating discontinuities, such as cracks, and only requires detection by NDT when the discontinuity is approaching the maximum allowable limits. With the HOLSIP, the NDT requirements may include the detection and measurement of all discontinuities, including cracks, corrosion, fretting, impact damage, and other degradation types. This poses new challenges for NDT as hidden corrosion and fretting of metals are often more difficult to detect and characterize.

While cracks are two-dimensional and can be measured in terms of surface length or area, corrosion is three-dimensional and is usually characterized by the amount of material lost (e.g., reduction in thickness). This metric is not easily measured as corrosion could occur in different forms (e.g., pitting, exfoliation, cracking, etc.) creating non-uniform surfaces on a macro scale and breakdown of the microstructure. Also, NDT characterization of damage and degradation in composite materials used in the construction of modern aircraft is very difficult. Nevertheless, significant work has been carried out in recent years on both NDT of composite materials and components as well as NDT of corrosion in aluminum airframe structures that will be described later in this book.

2.2 NDT DURING AIRCRAFT LIFE CYCLE

As described in reference [15] and mentioned earlier, most modern transport aircraft are designed based on damage tolerance and employ a variety of materials, structural configurations, and joining technologies. In addition to the traditional aluminum alloys that still constitute a significant portion of airframe structures, fiber-reinforced plastics, such as carbon fiber epoxy laminates, Kevlar laminates, honeycomb-sandwiched panels, foam-core panels, and hybrid fiber metal laminates, are increasingly being used. Titanium alloys are also used in smaller quantities in airframes while steel and nickel-based alloys are employed in landing gear and engines. These materials are often coated with polymers, metals, or ceramics to provide protection against corrosion, erosion, or high temperature degradation. Individual aircraft components are put together using a variety of fabrication and assembly techniques to form complex structures, such as fuselages, wings, vertical stabilizers, engine mounts, or the entire engine. Airframe parts may be riveted, fastened, bolted, welded, or bonded together to create the structural elements while engine components are closely fitted together or bolted to make the power plant assemblies.

2.2.1 DEVELOPMENT AND MANUFACTURING

During the aircraft technology development stage, new materials, processes, fabrication, and test methods are developed and examined with respect to their suitability. These often require the development of new NDT approaches and equipment. The role of NDT during this phase includes inspection of test samples and individual parts to ensure their quality after fabrication and to determine the damage profile during or after dynamic tests. During this phase, both conventional and newer NDT methods are evaluated to establish their capabilities and limitations for the type of materials and components that are used and the defects or damage types that are encountered. At the end of this phase, materials, processes, and construction methods are defined and selected along with the discontinuity sizes that must be detected and the appropriate NDT methods that need to be used.

At the design stage, it is important to consider issues such as ease of access to the critical locations that require inspections, availability of NDT equipment, ability to inspect the assembled structures during service, and the capability of the inspection methods to identify defects larger than the established limits easily, reliably, and economically. Thus, there is a need for close interactions between the aircraft design, manufacturing, and NDT personnel. NDT professionals can contribute in terms of defining and developing suitable inspection methods or equipment for use during the development, manufacturing, and operational phases of the aircraft life.

At the aircraft development stage, the necessary quality levels, such as the tolerable defect sizes within the scope of the design are defined, and NDT tasks and objectives are specified accordingly. During this phase, NDT is used for testing coupons and prototype components to ensure their quality before dynamic tests are performed as well as to map and characterize damage after the tests.

During the aircraft production phase, NDT is used to perform quality control inspections of the materials and individual components in an efficient and cost-effective manner. Automated inspection methods are often used for this purpose. Also, in-service NDT methods and applicable standards are defined and developed for the inspection of the critical parts or locations that require periodic tests. The procedures are applied on the actual aircraft for verification and standardized inspection instructions are prepared. The applicable equipment, probes, calibration standards, and reference pieces are also identified. Issues, such as ease of use, cost, and commercial availability of equipment, are considered along with their capability to identify the required defects larger than the specified limits. Descriptive Non-destructive Testing Manuals (NTM) for the implementation of the inspections and the interpretation of results are prepared and provided to the aircraft operators or maintenance facilities. The purpose of NTM is to enable inspectors at different sites with a range of abilities to apply inspections more accurately and efficiently.

2.2.2 OPERATION

During the operation of aircraft, NDT is used to inspect the specified areas in accordance with the instructions and manuals provided by the original equipment manufacturer (OEM). For civilian airplanes certified after the introduction of damage tolerance regulations in 1978, inspections are carried out based on the Air Transport Association Maintenance Program Development Guidelines given in MSG-3 [16]. This document provides procedures for developing the initial in-service inspection program for three principal sources of aircraft structural damage: fatigue, corrosion, and accidental damage. The in-service inspection of new aircraft is initially directed towards detecting accidental damage and corrosion. As the fleet matures, the risk of fatigue cracking increases, and the initial inspection program is reassessed. If the initial inspection program is found to be inadequate for detecting fatigue cracks or other forms of damage, supplementary inspection programs are developed.

2.2.2.1 PERIODIC INSPECTIONS

Usually, aircraft manufacturers provide the operators with the initial inspection program that includes a list of structures or components requiring routine checks, the types of inspection to be performed, and signs or indications to look for. Generally, there are four inspection levels for transport aircraft, referred to as *"A-check," "B-check," "C-check,"* and *"D-check"* [17]. The A- and B-checks involve visual inspections and manual checks, but the C- and D-checks that are carried out during aircraft overhauls at a maintenance base also include other NDT methods. These four different levels are briefly described below.

A-check is performed approximately every month or 500 flight hours. This check is usually done overnight at an airport gate. The actual occurrence of this check varies by aircraft type, the cycle count, or the number of hours flown since the last check. The occurrence can be delayed by the airline if certain predetermined conditions are met.

B-check is performed approximately every three months. This check is also usually done overnight at an airport gate. A similar occurrence schedule applies to the B-check as to the A-check and B-checks may be incorporated into successive A-checks

C-check is performed approximately every 12–18 months or a specific number of actual flight hours as defined by the manufacturer. This maintenance check puts the aircraft out of service and is usually done in a hangar at a maintenance base. The schedule of occurrence has many factors and components and varies by aircraft category and type.

D-check is the most comprehensive check for the airplane and is also known as a Heavy Maintenance Visit (HMV). This check occurs approximately every 4–5 years depending on the aircraft and its usage. In the D-check, certain parts of the aircraft are disassembled for more thorough inspections that may require NDT techniques, such as liquid penetrant and magnetic particle tests, radiography, ultrasonic, eddy current, and others in addition to visual inspections. This requires significant time and must be performed at a maintenance base.

Visual and manual inspections are performed on the critical airframe structures or systems to determine their condition and to look for signs of deterioration, distortion, failure, accidental damage, apparent defects or dents, excessive wear, unsecured installations or attachments, improper operation, movements or rattling, leaks, burns, corrosion, and other abnormalities. Airframe structures that require visual inspections and manual checks include fuselage and hull, cabin and cockpit, engine and nacelle, landing gears, wing sections and movable control surfaces, empennage, propellers, and communication and navigation systems [17]. More detailed inspections are carried out on critical parts, such as engine mounts, exhaust stacks, cowling, landing gear wheels, internal spars, and rib attachments, and control surfaces for signs of fatigue cracks, excessive wear, impact damage or dents, and severe corrosion.

Critical engine components require periodic visual inspections by boroscopes (optical devises with long tubes to reach hidden sites) through ports that are available for this purpose. At engine overhauls that involve complete disassembly of components, detailed liquid penetrant inspections are performed on individual parts. Other NDT methods, such as eddy current, ultrasonic, or radiography, may also be used at this stage to identify and size any hidden damage that may exist.

Publications issued by the aircraft manufacturers and airworthiness authorities are the sources of information for guiding operators in the maintenance inspection of their aircraft fleet and engines. Information on the NDT techniques and standards to be used, the frequency and extent of inspections, and tools or equipment necessary are provided in the manufacturer's maintenance manuals, overhaul manuals, structural repair manuals, and service bulletins.

When unsafe conditions, such as the presence of cracks, severe corrosion, or other types of damage, are reported by an aircraft operator, the aircraft manufacturers, and/or airworthiness authorities, such as the FAA or Transport Canada, notify all operators of the same or similar aircraft specifying the details of the problem through publications such as Airworthiness Directives (AD). The AD's provide information about the affected aircraft model, engine or propeller type, serial number, etc. as well as the description of the problem experienced and the necessary corrective actions and the compliance time or period.

2.2.2.2   SUPPLEMENTAL INSPECTIONS

As aircraft age, the potential for severe corrosion and fatigue cracking increases, and many more of the principal components of primary structures will require frequent inspections. For civilian aircraft, the FAA has developed a number of programs to address the aging aircraft structural issues including [18]:

- Supplemental Structural Inspection Program (SSIP)
- Corrosion Prevention and Control Programs
- Structural Modification Programs
- Repair Assessment Programs
- Instructions for Continued Airworthiness
- Damage tolerance-based inspections and procedures.

The SSIP was one of the original aging aircraft initiatives that followed the Aloha Airlines accident in 1988. This program was developed in response to multi-site fatigue cracking of airframe structures in order to maintain the structural integrity and airworthiness of the airplanes that approach the manufacturer's original fatigue design life. The FAA issued AD for structural re-evaluation by the manufacturers that identified certain structural details where fatigue damage was likely to occur. Supplemental structural inspections were then mandated to detect fatigue cracking, and repair if necessary, to ensure continued airworthiness of transport airplanes.

The SSIP initially addressed a number of aging aircraft and made recommendations regarding supplemental structural inspections, repairs, and modifications [19]. Currently, updated FAA's Airworthiness Directives are available through the internet for every aircraft type and make that has experienced unforeseen problems and requires supplementary inspections of structural details [20].

The FAA Aging Airplane Safety Rule [18] mandated in the US in March 2005 imposes statutory requirements for certain transport airplanes to undergo supplemental inspections and review records after their 14th year in service and at specified intervals after that. Also, the rule imposes a requirement to include supplemental inspections by specified deadlines in the maintenance programs for certain airplanes. To date, the first five of the above-mentioned FAA programs have been incorporated into the maintenance schedules for most large transport category airplanes. The last element dealing with damage tolerance supplementary inspections apply only to the US-registered turbine-powered transport aircraft with 30 or more seats or >75,000 lbs payload. Presently, as of 2010, this program does not apply to other aircraft types, such as commuters with less than 30 seats, airplanes designed using safe-life, or structural components like landing gear certified as a safe-life design.

In Canada, currently Transport Canada is following the FAA and the European Aviation Safety Board (EASA) rulemaking activities in order to develop regulatory initiatives to harmonize and implement the FAA and EASA rules. Meanwhile, Transport Canada monitors the safety of aging Canadian aircraft through approved company maintenance inspections as well as random inspections by Transport Canada inspectors and engineers as aircraft undergo major overhauls [19].

Theoretically, once a civilian aircraft reaches its design life, it must be retired. However, with the increasing cost of new aircraft and engines, some operators find it more economical to refurbish and maintain their existing fleet than to purchase new aircraft. This is particularly true for aircraft operated by military agencies and cash-strapped countries.

2.2.3 AGING AIRCRAFT

Military aircraft are operated under different, and often more severe, flight conditions and environments than those experienced in civil aviation. Thus, their maintenance inspections are more rigorous and often more frequent. In addition, many military aircraft are kept in service beyond their initial design goals, and a number of approaches have been developed to achieve safety of the life-expired critical structural components (both engine and airframe). These approaches are based on damage tolerance and rely on periodic inspections to maintain aircraft operational safety [21]. Some are briefly described in the following paragraphs.

2.2.3.1 RETIREMENT-FOR-CAUSE

The retirement-for-cause (RFC) approach is used by the US and some other air forces to extend the life of gas turbine critical rotating parts [21]–[23]. The RFC replaces the "safe-life" operation of components where the retirement period is when 1 in 1000 components would have developed a detectable crack. As mentioned before, by definition, the remaining 999 components should have useable life remaining. This methodology was first developed and implemented for the U.S. Air Force F100 engine's rotor components. It was estimated that the use of RFC would result in considerable life-cycle cost savings for that engine system in the US fleet [23].

The RFC requires periodic inspections to assess whether or not detectable damage exists. Those components with no detectable damage are returned to service until the next inspection. This process is repeated until damage is found, and at that time the part is either repaired to the original condition or retired and discarded [23]. This approach allows parts with damage to be identified and taken out of the service before they can cause an incident and parts without detectable damage to be used to their full potential. For increased safety, inspection intervals are selected such that if the damage is missed in one inspection there will be one or more additional chances to detect it in subsequent inspections. The RFC approach requires data on inspection reliability that is established experimentally (e.g., [12], [25], [26]) assisted with modeling [24]. The inspections are scheduled based on flight hours or cycles. The cost and time associated with the inspections as compared with the availability and cost of replacement parts are factors considered in the use of the RFC approach.

This approach has been applied in the US for the life extension of the major rotating parts of aircraft engines, such as the F100 since 1984 [22], [23]. The Canadian Forces have also used the approach for the J85-CAN 40 engine compressor discs and spacers as well as the Nene 10 engine impellers based on the POD and safe-inspection interval analysis described in references [26] and [27].

2.2.3.2 SAFETY-BY-INSPECTION

While the retirement-for-cause approach extends the useful life of the critical components of aircraft engines, the "safety-by-inspection" approach enables the effective management of the airframe major structures beyond the established fatigue life. This approach is also based on damage tolerance and uses periodic inspections to achieve safety. For example, the Canadian Forces CP 130 and CC 140 aircraft wing structures are maintained using this approach by implementing periodic eddy current inspections to identify fatigue cracks that may occur in the bolt holes of lap splice joints. Similar to the RFC approach, inspection limits of eddy current procedures used at the Department of National Defense (DND) for identifying bolt hole cracks in lap joints are established by using probability of detection analysis as described in Chapter 13 of this book.

2.2.3.3　CONDITION-BASED MAINTENANCE

The condition-based maintenance or CBM is another approach that has been developed in an attempt to achieve structural safety beyond the original design goals [28]. The CBM is based on the "Structural Health Monitoring (SHM)" approach where embedded sensors and on-board tools are employed to obtain information on the condition of the structure for prognostic and diagnostic purposes. These sensors provide aircraft condition data (temperature, vibration, cycles, time, damage site, size, etc.) that are analyzed in combination with environmental factors (loads, humidity, usage profiles, etc.) in order to anticipate when component failure will occur.

In this approach, scheduled inspections are supplemented by continuous monitoring to allow predictive and proactive maintenance before component failure. CBM requires not only the reliable detection of damage by the installed NDT sensors but also their characterization in terms of size, type, location, etc. This is significantly more information than would be required under the other maintenance approaches mentioned before. Damage characterization is needed in order to make an accurate assessment of the severity of the detected abnormalities so that proper corrective actions can be taken before the damage reaches a critical size and component failure occurs. This approach relies on quantitative NDT.

2.2.3.4　PROGNOSTICS HEALTH MANAGEMENT

The "Prognostic Health Management (PHM)" is an extension of the CBM approach for aircraft maintenance that includes an additional process. In other words, in addition to the assessment of the current conditions (diagnostics) and the prediction of future conditions (prognostics) of the structure, it involves a maintenance optimization and planning process (i.e., health management) [29], [30]. Each module's typical interactions are shown in Figure 2.5 [29]. The goal of PHM is to optimize inspection and maintenance actions based on risk and cost. The three main processes involved in PHM are described below:

> *Diagnostic* processes involve using a variety of sensors to monitor a component's physical and structural conditions and to detect any faults or damage that may be present. The fundamental modes of degradation or damage to structural components are fatigue cracking, corrosion, creep, fretting, delamination, and disbonding. The NDT sensors are expected to provide a high degree of fault detection and fault isolation capability with very low

false alarm rates in order to determine the state of the component in terms of performing its functions.

Prognostic processes require the ability to predict the future state of a system based on the current condition and the expected usage. This process uses material properties and load prediction data along with an appropriate life prediction methodology (such as HOLSIP) and a damage progression modeling program to predict the remaining useful life of the component and its future performance.

Health Management processes involve making intelligent, informed, and appropriate decisions about maintenance actions based on diagnostic/ prognostic information, operational logistics, requirements, available resources, and repair methodologies.

Like CBM, this approach will replace the scheduled-based maintenance with a prediction-based maintenance to avoid failures. This is expected to enhance aircraft readiness and maintenance cost-effectiveness. This approach requires many sensors to be integrated into the aircraft design for real-time in-situ monitoring, maintenance, and operational considerations as required in PHM. Although modern aircraft already use significant number of sensors to monitor many of the critical systems by sensing loads, strain, pressure, temperature, etc., the use of permanently fixed NDT sensors for augmentation or replacement of conventional NDT inspections of aircraft structures and engines is still in its infancy. New aircraft development projects, such as the Joint Strike Fighter (JSF), are placing more emphasis on PHM, which requires embedded NDT sensors and on-board NDT systems to provide quantitative information needed for the successful implementation of this approach.

The incorporation of NDT sensors and devices in aircraft structures requires careful consideration of issues associated with the additional weight, power needs, communication with the central control system, sensor malfunction, and

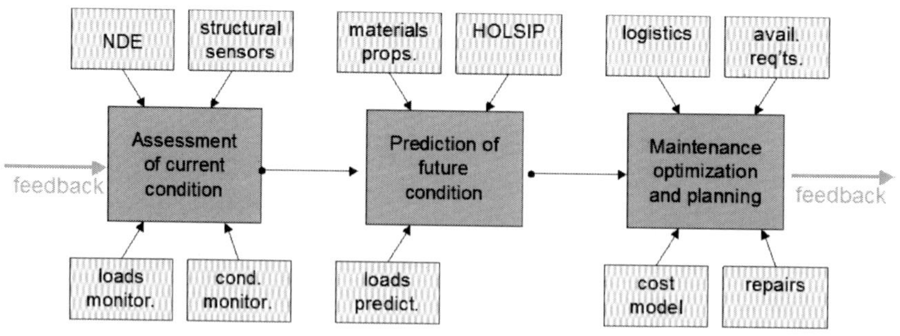

Figure 2.5 A block diagram of a typical PHM approach [28].

accessibility for replacement or repairs as well as the reliability of results and means of validation. A literature review on *Advanced Sensor Technology for Aircraft* [29], [30] provides information on NDT sensor technology for airframe and engine health monitoring applications. In these publications, important characteristics, capabilities, and issues related to NDT sensors for PHM applications have been identified. There are numerous recent PHM activities that involve experimental studies of a wide range of applications, such as monitoring of bonded composite repairs (e.g., [31], [32]) and in-flight systems (e.g., [33]).

2.3 CHAPTER REFERENCES

[1] The impact of mandated aging air plane programs on jet transport air plane scheduled inspection programs, A. Akdeniz, Aircraft Engineering and Aerospace Technology, Vol. 3, No. 1, pp 4–16, 2001.

[2] Aircraft Structural Integrity Program General Guidelines, US Department of Defence Handbook, MIL-HDBK-1530A (USAF), 2002.

[3] The Aging of Engines: An Operator's Perspective, A.K. Koul, J-P. Immarigeon, W. Beres, P. Au, A. Fahr and W. Wallace, the RTO AVT Lecture Series on Aging Engines, Avionics, Subsystems and Helicopters, published as RTO EN-14, 2000.

[4] From Safe-Life to Fracture Mechanics, G. Redmod, Asia-Pacific Conference on NDT, Brisbane, Australia, 17–21 Sept. 2001.

[5] Damage Tolerance and Fatigue Evaluation of Structure, FAA Advisory Circular no. 25.571-1D, 2011.

[6] Engine Structural Integrity Program General Guidelines, US Department of Defence Handbook, MIL-HDBK-1783B, 2002.

[7] Damage Tolerance Assessment of Repairs to Pressurized Fuselages, FAA Advisory Circular no. 120-73, 2000.

[8] Damage Tolerance: Facts and Fiction, U.G. Goranson, 14th Plantema Memorial Lecture, Presented at the 17th Symposium of the International Committee on Aeronautical Fatigue, Stockholm, Sweden, June 09, 1993.

[9] Damage Tolerance for High Energy Turbine Engine Rotors, FAA Advisory Circular 33.14-1, 2001.

[10] Damage-tolerance-based Life Prediction of Aeroengine Compressor Discs: I. A deterministic Fracture Mechanics Approach, A.K. Koul, N.C. Bellinger and A. Fahr, International Journal of Fatigue, pp 379–387, September 1990.

[11] Damage-tolerance-based Life Prediction of Aeroengine Compressor Discs: II. A Probabilistic Fracture Mechanics Approach, A.K. Koul, N.C. Bellinger and G. Gould, International Journal of Fatigue, pp 388–397, September 1990.

[12] Importance of Sensitivity and Reliability of NDI Techniques on Damage-tolerance-based Life Predication of Turbine Discs, A.K. Koul, A. Fahr, G. Gould and N.C. Bellinger, NATO Advisory Group for Aerospace Research and Development (AGARD) Report No. 768, 1988.

[13] HOLSIP Structural Integrity Process, www.holsip.com.

[14] Holistic Structural Integrity Process, A New Methodology for Aircraft Life Assessment, www.nrc-cnrc.gc.ca/eng/news/iar/2005/08/04/aircraft-assessment.html, displayed in 2010.

[15] NDT in Aerospace-State of Art, G. Tober, D. Schiller, Airbus, Bremen, Germany, www.ndt.net, displayed in 2010.

[16] Air Transport Association Manufacturer/ Operator Scheduled Maintenance Development Guidelines MSG-3, Revision 2003-01, 2003.

[17] FAA Aviation Maintenance Technician Handbook, Chapter 8, http://www. faa.gov/library/manuals/aircraft/amt_handbook/media/FAA-8083-30_ Ch08.pdf, displayed in 2010.

[18] FAA Aging Airplane Safety Final Rule, Federal Register Vol. 70, No. 21, mandated in the US on March 4, 2005

[19] A Report of the Ageing Airplane Rulemaking & Harmonization Initiatives Working Group, Transport Canada Civil Aviation Final Report, July 27, 2007.

[20] http://rgl.faa.gov/Regulatory_and_Guidance_Library/rgAD.nsf/, displayed in 2010.

[21] Aging Aircraft NDE; Capabilities, Challenges and Opportunities, E.A. Lindgren, J.S. Knopp, J.C. Aldrin, G.J. Steffes, and C.F. Buynak, US Air Force Research Laboratory Report AFRL-ML-WP-2006-497, October 2006.

[22] Concept Definition: Retirement for Cause of F100 Rotor Components, J. A. Harris Jr, D. L. Sims, C. G. Annis Jr AFWAL-TR-80-4118, September 1980.

[23] Retirement for Cause as an Alternative Means of Managing Component Lives, Peter J. Bonacuse, www.grc.nasa.gov, displayed in 2010.

[24] The Role of Model Based Inversion, R.B. Thompson, ASNT—back to basics, July 2008, www.asnt.org/publications/materialseval/basics, displayed in 2010.

[25] Influence of Sensitivity and Reliability of NDI Techniques on Damage Tolerance Based Life Prediction of Turbine Discs, A.K. Koul, A. Fahr, N.C. Bellinger, and G.M. Gould, NRC Report: IAR-LTR-SMPL-1988-000630, March 1988.

[26] The Sensitivity and Reliability of NDI Techniques for Gas Turbine Component Inspection and Life Prediction, A. Fahr, and D. Forsyth, NRC Report: IAR-SMPL-1993-0007, 1993.

[27] NDI Techniques for Damage Tolerance-Based Life Prediction of Aero-Engine Turbine Disks, A. Fahr, D.S. Forsyth, M. Bullock, and W. Wallace, NRC Report: IAR -LTR-ST-1961, February 1994.

[28] Condition Based Maintenance for Army Aviation, J.H. Pillsbury, BENET Business Library, www.bnet.com, displayed in 2010.

[29] Advanced Sensor Technology for Aircraft–Project Report for Tasks 1–3, D. S. Forsyth, Z. Liu, M. Genest, A. Fahr, and N. Mrad, NRC Report: IAR-LTR-SMPL-2005-0064, Oct. 13, 2005.

[30] NDE for Engine Components and Engine Health Monitoring Applications—Literature Review, Genest, M., Fahr, A. and Mrad, N, NRC Report: IAR-LM-SMPL-2007-0036, 19 February 2007.

[31] Structural Health Monitoring (SHM) of Bonded Composite Repairs, M. Martinez, M. Genest, M. Brothers, F. Habib, NRC report number; CPR-SMPL-2011-0216, 2011.

[32] Structural Health Monitoring of Bonded Composite Repairs Based on Acoustic Lamb Waves, M. Martinez, M. Brothers, and F. Habib, NRC Report: IAR-JA-SMPL-2011-0250, 2011.

[33] A Feasibility Study on an In-Flight Damage Detection System, M. Martinez, D. Backman, G. Li, and T. Benak, NRC Report: IAR-CPR-SMPL-2009-0175, 2009.

CHAPTER 3

NDT Process

3.1 BACKGROUND

Generally, NDT is used in the aeronautical field mainly to detect and characterize defects or damage, measure dimensions, or obtain information about certain material properties. As mentioned earlier, NDT is carried out by examination of the unknown test object using an energy source and comparing the object's response with that of a known reference piece.

The energy may be provided by a light or laser source, acoustic or ultrasonic waves, mechanical vibration or static loads, radiation or nuclear particles, heat, or electromagnetic fields. Some NDT methods like acoustic and ultrasonic techniques require a coupling medium between the energy source and the test part. Others, like liquid penetrant and magnetic particle inspections or replication, rely on specific enhancing fluids, powders, or compounds and defined processes to bring up discontinuities. The target's response to the input energy may be in the form of visual traces in "optical methods," minute local deformation of a materials surface in "acoustic and ultrasonic techniques," minor variations in the materials electrical or magnetic behavior for the "electromagnetic methods," changes in the object surface temperature for "thermography," or differences in the recorded radiation in the case of "radiography." These are the basis for the different NDT techniques.

Depending on the NDT method, different devices, media, sensors, or instruments are used to view, capture, or sense the target's response. In the case of optical methods, enhancing devices, such as magnifying glasses, ultraviolet lights, optical instruments, and cameras, may be used to view traces of the target. Piezoelectric or similar-type sensors are employed in conventional acoustic and ultrasonic methods to convert the surface displacement into electrical signals. For electromagnetic methods, coils are usually used to sense the target's effect on the electromagnetic field. Thermography relies on infrared cameras to measure minute variations of the object's surface temperature while films, photographic papers, or specialized radiation-sensitive plates are employed to record the radiation beams or particles passed through the target in radiographic techniques.

In newer and less-conventional NDT methods, a combination of the different approaches is used. For example, in "shearography" either mechanical or thermal energy is applied to excite the target that in turn creates out-of-plane deformation of the object face while optical devices are used to view the minute local displacements of the surface. Similarly, in laser ultrasonic, optical laser beams are employed to create ultrasonic waves in the material and also to detect the surface displacements caused by the echoes returned from the target within the object. These techniques are generally more complicated and costly than the well-established conventional methods.

The detected signals, or captured images, are viewed and interpreted by the NDT practitioners or professionals either directly or after performing additional "signal and image processing." Regardless of the techniques employed, the application of NDT in aerospace and other industries relies on well-trained personnel as well as the use of standard practices, reference pieces, and careful calibrations in order to obtain accurate, consistent, and repeatable results.

3.2 NDT PERSONNEL

The development and validation of NDT techniques, sensors, instruments, and scanning devices or systems, and analysis and interpretation of results and their potential applications are carried out by NDT professionals, research scientists, and engineers in research and development (R&D) laboratories. Such developments require a good knowledge of physics, engineering physics, signal and image processing, and other related disciplines, such as materials science and fracture mechanics, statistics, electronics, and mechanics of automated systems. It involves theoretical analysis and modeling of the NDT energy source, transfer and propagation in materials or components, and interaction with the target. The development of specific NDT procedures for applications on complex components or structures of aircraft is also carried out by NDT professionals in close interactions with NDT practitioners who apply the methods in the field. Once the NDT methods, the associated equipment, specific applications, and standard practices are developed they are applied in the field by certified NDT practitioners or technicians.

Certified NDT practitioners define and perform tests in order to determine the condition of materials or components and establish their suitability for the intended application. They analyze and interpret the test results and make judgments based on preset criteria that are defined analytically and/or experimentally.

In practice, often calibration and reference standards of known characteristics are used to set accept and reject criteria. Personnel conducting field NDT tests to ensure the integrity or safety of critical components or structures, such as aircraft, nuclear reactors, trains, pipelines, etc., require rigorous training and certification in the appropriate NDT methods related to specific applications. Certification of NDT practitioners is provided by an authorized certifying agency.

In North America, the American Society for Non-destructive Testing (ASNT) plays a prominent role in the training and certification of NDT practitioners as well as in developing and maintaining recommended NDT practices and guidelines. ASNT has a large catalogue of NDT education and reference materials, including several Non-destructive Testing Handbooks, which are major reference sources on specific methods (e.g., [1]).

In Canada, currently the Natural Resources Canada (NRCan) manages the certification of individuals performing field NDT. The NDT Certifying Agency of NRCan is an independent body that certifies individuals according to the National Standard of Canada—CAN/CGSB-48.9712-2006 "Qualification and Certification of Non-Destructive Testing Personnel." This standard is developed and maintained by the Canadian General Standards Board (CGSB). The certification complies with the requirements of the International Organization for Standardization ISO 9712:2005 and European Standard EN 473:2000.

There are several standards for the qualification and certification of aircraft NDT technicians including:

- NAS 410—Certification & Qualification of Non-destructive Test Personnel issued by the US Aerospace Industries Association (AIA).
- ATA Specification 105—Guidelines for Training and Qualifying Personnel in Non-Destructive Testing Methods issued by the US Air Transport Association (ATA)
- MIL-STD-410—Non-destructive Testing Personnel Qualification and Certification issued by the US Air Force.

For civil aircraft inspection, Transport Canada CAR 571 Schedule I—Personnel Certification for NDT—requires certification based on the CAN/CGSB 48.9712-95, or MIL-Std-410, or ATA Specification 105 at Level 2 or 3 in all conventional NDT methods. These standards establish the requirements for the qualification and certification of personnel performing NDT in the aerospace manufacturing, service, maintenance, and overhaul industries.

3.3 STANDARD PRACTICES

NDT standard practices are described in traceable and numbered documents prepared by an organized group of international NDT experts under private or public standardizing agencies, such as the American Society for Testing and Materials (ASTM), ASNT, CGSB, or the European Committee for Standardization. Such documents contain technical specifications, criteria, or requirements for NDT methods, procedures, and practices. NDT standards describe how to perform a particular test in order to achieve accurate results consistently. These standards are based on the R&D work carried out by NDT professionals, scientists, and practitioners worldwide. NDT standards are periodically revised to incorporate technical advances made in the field.

The Annual Book of the ASTM Volume 03.03 on Non-destructive Testing is an important source for NDT standards, guides, codes, and recommendations. This document is revised annually and covers most NDT methods, including ultrasonic, electromagnetic, liquid penetrants, magnetic particle, radiography, acoustic emission, thermography, and shearography. The NDT standards are handled by the ASTM Committee E-7 and developed by several sub-committees comprising of international experts in specific NDT methods or applications. To date, nearly 130 standard references, guides, test methods, and practices covering most NDT techniques and applications have been developed by the ASTM. A partial list of the ASTM general and aerospace-related NDT standard practices is provided in the Appendices at the end of this chapter. Most of the ASTM standard practices are of a general nature; however, there are also some standards that are specific to a particular procedure, material, or application area.

A typical ASTM NDT Standard Practice has a specific designation number (e.g., E2533-09) and a title (e.g., Standard Guide for NDT of Polymer Matrix Composites Used in Aerospace Applications). It first describes the scope of the test and provides the related ASTM and other referenced documents as well as the terminology used in the standard and their definition. The standard then describes the test practice along with the requirements, such as the facility and personnel qualifications, standard references that may be needed for calibrations of equipment and procedures, necessary equipment and their performance needs, and setup and adjustments of equipment. The standard also describes the specifics of the test, such as the sensor location with respect to the part and any adjustments that may be needed, or the test object physical condition and its preparation for the test, the type, extent, and timing of examination, etc. It also mentions the significance of the test with respect to the purpose, analysis of results in relation to pre-determined accept/reject criteria, and other important considerations related to the test. Finally, the standard provides guidelines for reporting and documenting of the outcome. In the ensuing chapters that cover the different NDT methods, important ASTM Standards related to the specific techniques or applications are provided.

Other professional societies, such as the American Society of Mechanical Engineers (ASME) and the American Welding Society (AWS), also develop, maintain, and distribute NDT codes and standards, some of which are applicable to aeronautics. At the international level, some countries have their own standards while others follow the ASTM guidelines. The International Organization for Standardization (ISO) has a Technical Committee on NDT (TC 135) that deals with NDT quality standardization worldwide. The ISO range of standards includes a glossary of terms, methods of testing, and performance specifications for equipment and apparatus related to NDT. A partial list of ISO standards that are related to NDT applications in aeronautics is also provided in the Appendices of this chapter.

3.4 REFERENCE STANDARDS

Generally, NDT reference standards are test pieces with specific characteristics (e.g., specific properties, uniform geometries, calibrated dimensions, known discontinuities) that are used for calibration of NDT instruments and procedures. They are used as a guide for adjusting instrument controls to measure certain properties, to reveal the presence of discontinuities that may be considered as harmful defects, or for determining indications that are insignificant. They are essential for establishing repeatability and consistency of the NDT tests. A variety of traceable NDT reference standards are designed and made for specific techniques, test procedures, or applications and are commercially available. They are made of specific materials with specific properties in specific shapes and dimensions and may contain discontinuities, such as flat-bottom holes, side-drilled holes, slots, notches, foreign material inserts, and other geometric features. A few examples of commercial reference standards used in NDT are provided in Figures 3.1 to 3.3, and their descriptions are given in the related captions.

Such reference standards are used for the calibration of NDT instruments and procedures before the actual tests are carried out. The NDT Standard Practices specify the appropriate reference standard to be used for a specific test. Once it is established that the NDT test (the instrument and the procedure) is capable of clearly and accurately indicating a known target (e.g., a certain size discontinuity in a reference standard) it is expected that the same test will be able to indicate similar targets in the actual part and provide identical indications if the test is

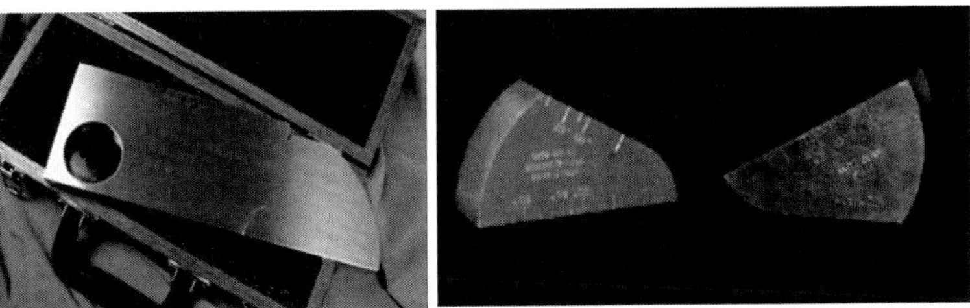

Figure 3.1 Reference standards for calibration of ultrasonic instruments, (left) the ASTM E164 and (right) the IIW miniature. These references are commercially available in a selection of materials [1].

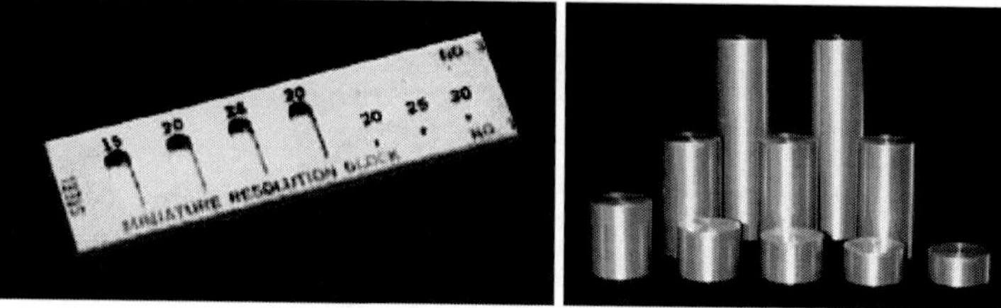

Figure 3.2 Ultrasonic reference standards for evaluating the near-surface resolution (left) and distance and area (right) [2].

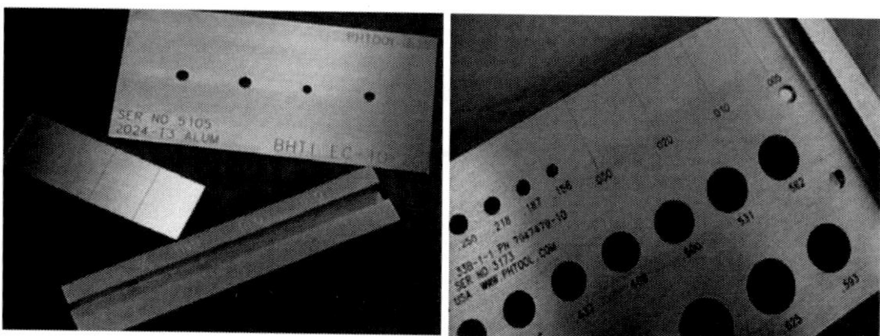

Figure 3.3 Eddy current reference standards containing EDM notches in flat areas and holes [1].

repeated. In this way, it is ensured that the needed information about the test object can be obtained in a repeatable and consistent manner. The use of standard procedures, calibrations, and references allows conformance to a uniform testing specification regardless of the instrument employed, the time the test is carried out, or the location of the test. The only remaining variables are uncontrollable factors, such as the inspector skills and judgment, and these will be described in more detail under a separate chapter on NDT reliability.

3.5 NON-STANDARD REFERENCE PIECES

In addition to the above-mentioned reference standards that are official, uniform, traceable, and commercially available to NDT users, the aircraft manufacturers or operators and the NDT test facilities or laboratories often develop reference pieces for their own or clients use. Such reference pieces usually resemble the actual part in terms of material, design, and geometry but contain embedded defects or damage similar to those expected in the actual part. If real defects cannot be introduced, artificial defects simulating real ones are used.

In metals, drilled holes of various diameters and depths are employed to simulate discontinuities, such as voids, cavities, and pores, while narrow slots of various sizes and locations are used to represent cracks. Inclusions may be replicated by introducing small pieces of different foreign materials into the reference samples. Metal loss due to corrosion is mimicked by removing the material to create a step-wedge with steps indicating different levels of the remaining thicknesses.

In composites, artificial delamination and disbonding are often produced by embedding two layers of Teflon tapes sealed at the edges to entrap air. Often paper, nylon, or plastic materials used during construction of composite structures are placed in reference pieces to represent foreign material inclusions. Lead tapes or other materials of different sizes, shapes, and characteristics attached to the flaw-free areas of the actual test part also provide an easy way of generating reference indications for comparative assessment of the inspection results.

Non-standard reference pieces are used to help in the analysis and interpretation of the NDT outcome as well as for obtaining information about the test object. These tasks are carried out by comparison of the response signals from the unknown targets with those of the known reference pieces of the same material and geometrical features. The ASTM has standards for manufacturing reference pieces (e.g., [3]). In section 3.10.3 of the Appendices to this chapter, several non-standard metallic and composite reference pieces have been described that provide the readers with ideas about the ways such references can be made.

3.6 CALIBRATIONS

Calibration of an NDT instrument involves testing and/or adjusting the characteristics, settings, or controls to ensure that the instrument performs as intended for the specific application. The initial calibrations are carried out by the manufacturer but may have to be repeated periodically to ensure that they are still valid and effective. Such calibrations are performed using guidelines provided by the equipment manufacturers. In addition, a "user calibration" by the operator is often necessary when the equipment is employed in different applications or under different conditions.

User calibration may be needed after changing hardware (e.g., change of sensors); when the purpose of the measurement is changed (e.g., from property measurement to defect detection); after changes to instrument settings (e.g., changes in frequency, gain, threshold, filter, etc.); or when the test material, environment, and conditions are different from the ones used at the earlier calibration. This is to verify that the instrument still conforms with the initial manufacturer's calibrations. For methods, such as LPI, MPI, or x-radiography, that use specific fluids, powders, or recording media to visualize the test results, in addition to instrument calibrations, the employed media also involve calibrations. For example, liquid penetrants or x-ray films come in different sensitivity levels that are required to be checked periodically for their accuracy and consistency. User calibrations of instruments are carried out based on the guidelines provided by the manufacturers using NDT Standard Practices and Standard References.

NDT procedures are also calibrated before performing the actual tests to ensure that the sensitivity, resolution, accuracy, and other performance characteristics are as required for the specific application. The calibration of NDT procedures are also carried out using NDT Standard Practices and Standard References.

When a proper reference sample is not available, lead or Teflon tapes of different sizes are affixed on the surface of the object as artificial discontinuities or references to set the threshold levels and accept/reject limits.

3.7 NDT MEASUREMENTS

NDT measurements are carried out for different purposes and the way such tests are performed varies with the test component being examined, the method or procedure being used, and the outcome that is required. Therefore, it is difficult to describe an NDT measurement in a simple general manner. In a typical inspection, NDT traces or response signals of a test piece are measured from which the needed information about the object (e.g., presence of discontinuities) is obtained. In practice, this is achieved by comparing the response of the test object with that of a calibration or reference sample that resembles the object and has known properties or defects. Defective areas are identified when the response signal exceeds the threshold limits. Inspection of an aircraft part may involve the following steps:

1. Identification and selection of the appropriate techniques, standard practices, or recommended guidelines for the specific test to be carried out.
2. Identification and selection of the necessary equipment, peripherals, and calibration standards for the test.
3. Provision of the qualified personnel for the specific test and application.
4. Equipment calibration and setup based on the provided guidelines.
5. Preparation of the test object or the inspection site based on the appropriate guidelines or standard practices.
6. Selection of the test conditions (e.g., temperature, lighting) and test parameters (e.g., equipment sensitivity level) based on the recommended guidelines.
7. Tests on known calibration standards based on the guidelines or standard practices and equipment adjustments to achieve the required performance on specific reference targets.
8. Tests on actual parts based on the approved procedures or standard practices and analysis of results as compared with the results from the reference targets.
9. Decision making based on predefined threshold levels and accept/reject criteria.
10. Repetition of the tests to confirm consistency and repeatability if necessary.
11. Documentation of results.

3.7.1 EXAMPLE

The example provided below shows a typical inspection aimed at detecting cracks emanating from bolt holes of airframe metallic structures [4]. The eddy current method is identified for this application using either manual inspection or a rotary probe that is inserted into the holes by the inspector. In this case, technicians are required to be qualified by the Canadian General Standards Board (CGSB) certification in eddy current techniques at levels II or III.

The provided inspection document [4] identifies the required equipment and sensors for the tests and gives the necessary reference documents where the details of the equipment setup are specified. It also identifies the reference standards made of different metals, including aluminum, titanium, and stainless steel, that are needed for calibrations. The identified reference standards contain holes of different dimensions with artificial cracks of various sizes that are used to calibrate the equipment and the inspection process. Also, the preparation of the inspection area for cleanliness, acceptable paint conditions, and temperature range over which the calibration and tests are to be carried out are provided.

The tests involve first equipment calibration based on the provided approach for both manual and rotary eddy current scanning inspections and comparisons with examples of screen presentations or signals for each case. After the calibrations and the inspector's satisfaction that the instruments are performing as required, the inspector proceeds with the tests on the actual bolt holes in aircraft structures based on the guidelines provided in advance of the tests. The manual inspections are performed starting at the near-surface with a 360° scan of the sensor inside of the hole and repeat of the scan in small increments until the entire hole is covered. A similar approach is used when a rotary scanner is employed. Examples of the eddy current signals from normal conditions and those with defects are provided to assist the inspector in his decision-making process. The procedure requires further investigation of any indications other than those of the normal conditions.

In the ensuing chapters, more specific information on the different NDT techniques and their applications for the detection of discontinuities or measurement of dimensions and properties will be provided. Also, details of the analysis and interpretation of NDT response signals or images that may vary for the different methods will be explained. In the following section, the common aspects of the NDT signal/image processing routines are described.

3.8 NDT SIGNAL AND IMAGE PROCESSING

The majority of NDT methods use electronic equipment to obtain signal responses from the test component and display them in analog or digital formats for viewing and interpretation by the inspector. Initial analog signal processing operations include filtering, circuitry matching, and smoothing that are performed through appropriate electronic components within the NDT instrument before the response signal is displayed. From the analog output signal, basic characteristics, such as amplitude, duration, time-shift, etc., can be obtained in real time and related to the target. The analog output can be digitized either within the instrument or by an external digitizer. Most modern NDT instruments contain both analog and digital circuitry enabling the real-time and digital processing of the response signals.

Digital signal processing (DSP) is basically performing numerical calculations on the signals either through specific electronic chips installed in the measurement device or by using specialized software packages and computer post-processing. A wide variety of commercial signal processing software packages, such as Matlab™, are available for this purpose.

During a test, a large volume of data is generated that must be analyzed and presented in a user-friendly manner for easy interpretation by the inspector. The embedded DSP chips are capable of performing millions of numerical calculations in a very short time enabling real-time display of processed data. However, they usually provide limited processing capabilities that are pre-selected by the instrument manufacturer for specific procedures or applications. DSP chips are integrated into all modern NDT instruments that provide digital output for post-processing using computers and specialized or generic software packages. The generic programs, such as Matlab™, have a wider range of capabilities and are mostly employed in research laboratories during technique development.

NDT signal and image processing is a vast subject that requires a dedicated chapter. However, since the focus of this book is on NDT applications, only the commonly-used signal/image processing terms and methods are described here. When necessary, the basic mathematical functions related to the described methods are also provided. Technique-specific signal treatment procedures are explained in the chapters dealing with the different NDT methods.

3.8.1 *NDT Signals*

In the majority of NDT methods, the output signals are expressed in terms of amplitude (in volts) as a function of time (in seconds) or in "time-domain." The time-domain display is the most logical and intuitive way of presenting the NDT output and is used in methods such as acoustic and ultrasonic, thermography, shearography, pulsed-eddy current, etc. Often, a time gate is used to isolate and analyze a portion of the output signal that corresponds to the target response.

Another way of displaying the NDT response is to show the amplitude variation as a function of frequency, or in "frequency-domain." Figure 3.4 provides a time-dependent signal and the corresponding frequency spectra. This signal is from an acoustic test and the time-domain response contains numerous overlapping echoes that are not easily recognizable to the naked eye. However, in the frequency-domain, the response signal becomes more evident making the interpretation easier.

In electromagnetic (EM) methods, such as the conventional eddy current, the instrument output may be presented in terms of the two EM components, e.g.,

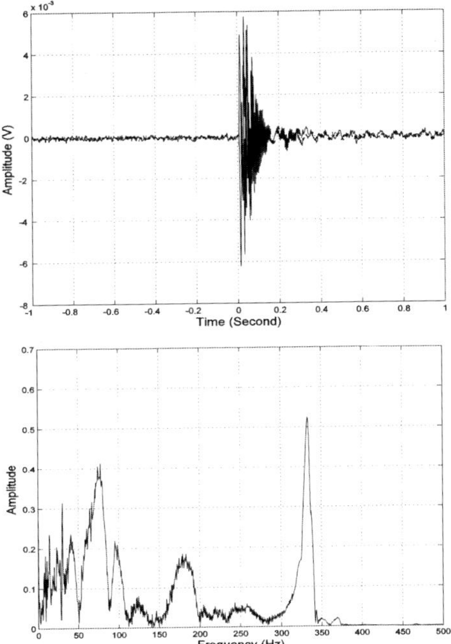

Figure 3.4 Typical time-domain (top) and the corresponding frequency-domain (bottom) signals from a sonic method.

electrical resistance and magnetic inductance. This type of presentation is referred to as an impedance plane diagram (e.g., Figure 3.5), and its processing is different from those of the time-dependent signals. This section deals only with the time-dependent NDT signals; however, in the chapter dealing with electromagnetic techniques more detailed analysis of hysteresis signals is provided.

3.8.2 *Signal-to-noise Ratio*

The signal-to-noise ratio (S/N) is a measure of the strength of the target signal (e.g., discontinuity signal) as compared to the background noise (including noise due to material, equipment, and environment). The ability of the NDT approach to identify the target or "detectability" is directly related to the S/N and thus high S/N values are desirable.

A time-dependent signal may be random or deterministic, periodic or transient, analog or digital. NDT signals are deterministic, in other words, they can be produced with specific characteristics and repeated, while the background noise is considered as random. When a signal is repeated at a regular time interval, it is called periodic while a transient signal appears and then disappears within a short period. For example, ultrasonic signals are periodic while acoustic emission

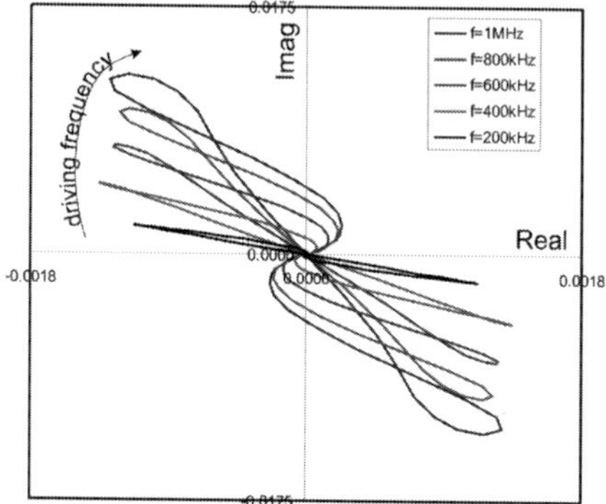

Figure 3.5 Typical impedance plane diagram used to present eddy current signal output.

signals are transient. An analog signal is a continuous function of time while a digital signal is a discrete sequence of numbers.

Mathematically, an acquired one-dimensional NDT signal in time domain, $s(t)$, is usually a combination of a true signal, $x(t)$, and noise, $w(t)$, and can be expressed as:

$$s(t) = x(t) + w(t) \qquad (3\text{-}1)$$

The operations to remove the noise include local or global averaging, filtering, and signal fitting [5]. Similarly, a two-dimensional image, $s(i,j)$, can be expressed as [6]:

$$s(i, j) = x(i, j) * h(i, j) + w(i, j) \qquad (3\text{-}2)$$

where $w(i,j)$ is the noise, $h(i,j)$ represents the function that deteriorates the original NDT image, $x(i,j)$, and * represents a 2D convolution of the two functions. Inversely, a deconvolution refers to the process of separating the two functions $x(i,j)$ and $h(i,j)$. Reference [6] provides further information on these processes.

3.8.3   *FOURIER ANALYSIS*

The Fourier analysis is a mathematical operation that converts a signal from a time-domain to a frequency-domain where its constituent frequencies can be identified. As described in reference [7], a periodic analog signal, $s(t)$, mathematically can be expressed as:

$$s(t) = s(t - T_0) \qquad (3\text{-}3)$$

where t is the time variable and T_0 is the constant repetition period. The $s(t)$ can be written as a Fourier series:

$$s(t) = \sum_{n=1}^{\infty} A_n \cos(2n\pi f_o t + \varphi_n) \tag{3-4}$$

where A_n and φ_n are the amplitude and phase values of the nth harmonic ($n = 1.2,3,.....$), respectively and the term $2n\pi f_o t$ denotes the frequency content of the periodic signal that is made of discrete individual frequencies, nf_0.

The spectral representation of a transient signal is given by its Fourier integral:

$$S(f) = \int_{-\infty}^{+\infty} s(t) e^{-i2\pi ft} dt \tag{3-5}$$

where f is the variable frequency and $S(f)$ is a *continuous Fourier transform*. $s(t)$ can be reconstructed from $S(f)$ by the inverse Fourier transformation:

$$s(t) = \int_{-\infty}^{+\infty} S(f) e^{i2\pi ft} df \tag{3-6}$$

A digital signal, $s(n)$, is a sequence of numbers defined for every integer n that has a discrete time and amplitude. Digitization of a signal is carried out at a specific sampling rate or interval producing a sampled signal, $s[n]$, that is the multiplication of $s(t)$ by the sampling function, $d(t)$ [9]:

$$d(t) = \sum_{n=-\infty}^{\infty} \delta(t - nT) \tag{3-7}$$

where T is the sampling interval and δ is the discrete delta function. The Fourier transform of a sampled signal is also a sampled function expressed by:

$$D(f) = 1/T \sum_{k=-\infty}^{\infty} \delta(f - kF) \tag{3-8}$$

where $F = 1/T$. The $\delta(t - nT)$ term is used to define a discrete sequence $s[n]$ as a weighted sum of discrete impulse functions:

$$s[n] = \sum_{n=-\infty}^{\infty} s(t)\delta(t - nT) \tag{3-9}$$

where $s[n]$ denotes the sampled signal. The Fourier transform of $s[n]$ can be written as the convolution of the Fourier transforms of $s(t)$ and $\delta(t - nT)$:

$$S[k] = S(f) * D(f) = 1/T \sum_{k=-\infty}^{\infty} S(f - kF) \tag{3-10}$$

It must be noted that the convolution is a more complicated operation than a simple multiplication and the discrete Fourier transform $S[k]$ is a periodic replication of $S(f)$ with a frequency separation $F=1/T$.

3.8.4 GATED SIGNALS

The output of an NDT system may contain several signals from which the target signal can be separated with the use of a gate. For time-dependent signals, the spectrum of a gated signal is the convolution of the Fourier transform of the time-domain signal and the gate function. The gated spectrum can be very different from the non-gated signal in the frequency-domain. The most commonly used gate function is the square or rectangular function. For the time-gate of $t_o-T < t < t_o+T$, the Fourier spectrum is expressed by:

$$G(f) = 2T(\frac{\sin 2\pi fT}{2\pi fT})e^{i2\pi ft} \qquad (3\text{-}11)$$

Choosing a long gate (i.e., when T is large) will result in a narrow $G(f)$ and negligible spectral distortion. In contrast, if the gate is too short, distortion will occur in the resulting spectrum.

3.8.5 SIGNAL AVERAGING

Digital signal averaging is commonly used to improve the received NDT signals. The average, or mean value of a series of signals, is a weighted sum of all V individual values and is expressed by [7]:

$$\overline{s} = \sum_{i=1}^{V} P_i s_i(t) \qquad (3\text{-}12)$$

where P_i is the probability of occurrence of the particular s_i. For repetitive stationary signals, P_i is equal to $1/V$.

3.8.6 WAVELET TRANSFORMATION

In wavelet transformation, the original signal is presented as a series of functions spanned over both the time and frequency domains so that the signal features can be easily accessed. When the wavelets are discretely sampled, the term "discrete wavelet transform" is used. In this analysis, the NDT signal is analyzed in a "multi-resolution" vector space. Reference [8] provides a detailed description of the multi-resolution analysis theory and its application in wavelet transformation.

Like the Fourier analysis, the Wavelet transformation can be performed using commercial signal processing software packages such as Matlab™.

3.8.7 TIME-FREQUENCY ANALYSIS

The analysis of NDT signals is usually done either in the time-domain or in the frequency-domain separately. The combined time-frequency analysis is another approach for transient signals varying in time. There are several different ways to formulate a time-frequency analysis, and details are available in reference [9]. The time-frequency analysis provides a three-dimensional representation of signals in time-frequency-amplitude space (see Figure 3.6). Often, the projection of the three-dimensional display is shown in a time-frequency plane with color representing the amplitude.

3.8.8 STATISTICAL PATTERN RECOGNITION

Pattern recognition (PR) generally refers to the process of evaluating NDT test results in order to automatically recognize and classify them into groups. There are different PR approaches as described in [10]; one is the statistical pattern recognition. This approach uses statistics and mathematics to analyze and compare unknown signals with known ones that have been previously stored in the computer and to group them into separate classes on the basis of their similarity. The process generally involves training and testing of the PR system on known data before it is applied to unknown signals.

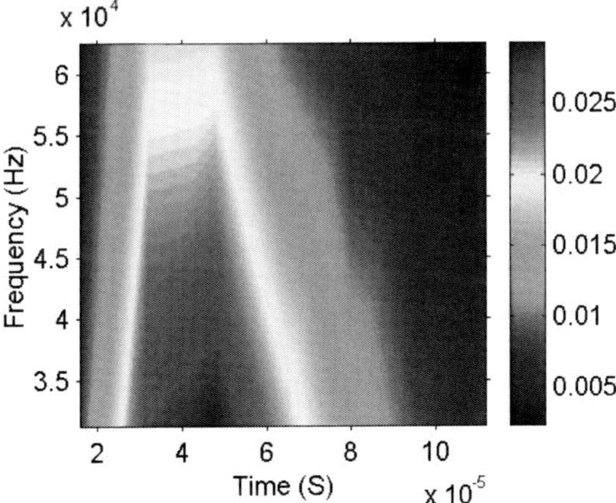

Figure 3.6 A two-dimensional presentation of time-frequency-amplitude signals from a pulsed eddy current inspection. Amplitude is represented by different colors.

The first step in a PR is the data acquisition and transformation using one or several of the methods described above. Then useful signal features are selected, using statistical ranking algorithms, based on the ability of the individual features to discriminate between the different classes. One such algorithm is known as "principal component analysis," which examines the relationships between the features by calculating the correlation matrix of all features. This matrix is then used to find and keep the features that have a high correlation factor or "principal components" [11]. In this way, the PR system is trained to partition input signals into groups in a feature space. Once the training and testing is completed, the PR system is used to analyze and classify the unknown signals to one of the predefined groups based on the similarity of the measured features. This approach has been applied for the classification of NDT signals as well as images resulting from inspection of aircraft parts; two examples are provided in references [11] and [12].

3.8.9 NEURAL NETWORK

A neural network (NT) operation essentially mimics the biological nervous system to process information and is used in NDT for pattern recognition purposes. The main difference between the neural network and statistical pattern recognition is that the NT is able to learn complex and non-linear input-output relationships by sequential training. A network can be viewed as a large number of parallel processing elements with many interconnections. NT models use organizational principles in a network of weighted graphs in which the nodes are artificial neurons and directed edges are connections between neuron outputs and inputs. Multilayer neural networks are usually used for classifying NDT signals. Further information on NT theory is provided in reference [11] while reference [13] gives an example of its application in NDT.

3.8.10 DATA FUSION

As expectations from the NDT to identify and characterize a target with a high accuracy, precision, and reliability increase, any single method may not be able to achieve the objectives of the test. The use of two or more complimentary techniques and fusion of the resulting data may potentially help to achieve the test objectives. The aim of data fusion (DF) is to assist NDT inspectors to better identify and characterize the target being tested. This is done by gathering data from multiple NDT methods, processing of the information in an efficient way, and assisting in the decision making process in order to improve the overall performance.

In reference [14], a review of commonly-used data fusion frameworks is provided together with important factors that need to be considered during the development of an effective DF problem-solving strategy. As explained in this reference, there are several DF approaches and algorithms available; some are applied for the analysis of multi-mode NDT data from aircraft components as described in references [15], [16], and [17].

3.8.11 BASIC SCANNING AND PRESENTATION MODES

3.8.11.1 A-SCAN

A-scan refers to the situation where the sensing device is on a single location with respect to the object and the signal is presented in terms of amplitude (voltage) versus time (seconds) as illustrated in Figure 3.7(a). As mentioned above, a time-domain A-scan signal may be gated to identify the target or the free surfaces of the object (e.g., front and rear surfaces). The A-scan characteristics, such as amplitude or time, may be related to the target dimension or location. A-scan is the basis for all other time-based presentation modes.

3.8.11.2 B-SCAN

B-scanning refers to the linear movement of the sensing device across the test object and display of the target signal as a function of the sensor location as illustrated in Figure 3.7(b). The B-scan is mostly used in acoustic and ultrasonic methods to measure the time-laps between the sensor and the target. It basically provides a two-dimensional slice view of the cross-section of the object.

3.8.11.3 C-SCAN

C-scanning refers to the movement of the sensing device over the test area. A C-scan display refers to the two-dimensional image of a signal feature as a function of the position of the sensing device as shown in Figure 3.7 (c). When signal amplitudes are plotted as a function of the sensor location, amplitude C-scans are generated. An amplitude C-scan basically displays the gated segment of the test object as seen by the NDT sensor. Figure 3.8 shows the screen presentation of a commercial ultrasonic instrument that shows A-, B-, and C-scan displays of a typical ultrasonic test on a composite reference sample.

3.8.11.4 D-SCAN

D-scan is a term primarily used in ultrasonic weld inspection to display the cross-sectional view of the weld in its longitudinal direction and will be described in the chapter dealing with ultrasonic testing.

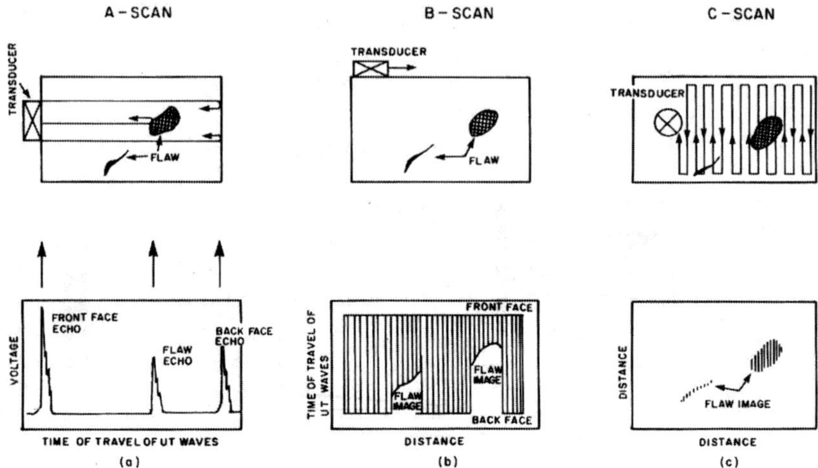

Figure 3.7 A schematic presentation of different scanning and presentation modes. (a): A-scan, (b): B-scan, and (c): C-scan.

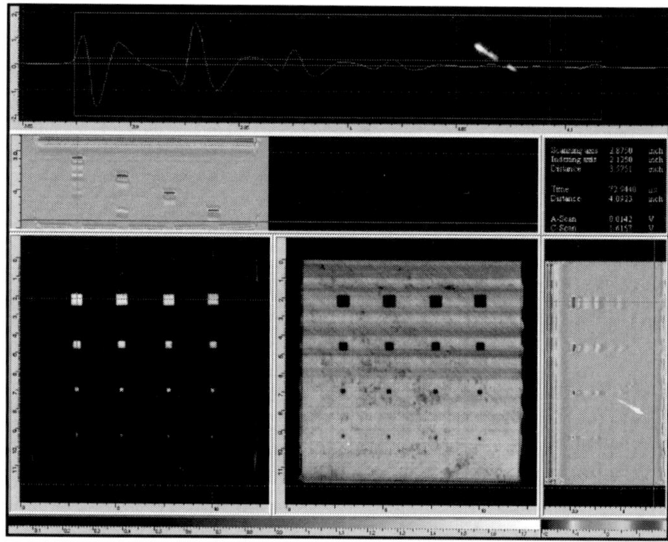

Figure 3.8 A composite reference sample with artificial defects of different sizes and locations tested using ultrasonic method. (Top): A-scan display of the time-domain signal corresponding to the point on the top-left corner defect. (Middle-left): B-scan display of the signal time for a single scan line along the top-row defects showing their cross-sectional locations. (Bottom-left and middle): C-scan displays of the entire panel showing planar view of the internal defects. (Bottom-right): B-scan display of the signal time for a single scan line along left-column defects showing their cross-sectional locations.

3.8.11.5 P-SCAN

P-scan is another term used in ultrasonic weld inspection and refers to a presentation that combines B-, C-, and D-scans in a single image creating a complete view of the discontinuities. This will also be described later in the ultrasonic chapter.

3.8.11.6 POLAR SCAN

A polar scan provides the same information as the C-scan and is used for circular-shaped components. It involves rotating the part around the central axis of the part on a turn-table while indexing the sensor by a small increment along its diameter after every revolution.

3.8.11.7 T-SCAN

T-scan, or time-of-flight C-scan, presents layer-by-layer views of the cross-section of the test object.

3.8.12 IMAGE PROCESSING

The majority of modern NDT instruments provide two-dimensional images of the inspection results in digital format (e.g., C-scans, x-ray images, thermal maps, etc.). With image processing and enhancement, subtle features can be seen better making it easier to analyze and interpret the results. Some image processing functions are common to all NDT images (e.g., contrast and edge enhancements, background subtraction, scaling, palette adjustments, filtering, etc.). Others are technique-dependent (e.g., time or frequency analysis, time-dependent thermal mapping, data-fusion, tomographic reconstruction, etc.). This section only deals with the common processing routines and the method-specific approaches are described separately with the description of the different NDT techniques when necessary.

In most NDT methods, such as ultrasonic, eddy current, thermography, and shearography, the sensor output signals are acquired in digital format from which two dimensional images are constructed. In some methods, such as conventional radiography and optical photography, two-dimensional images are recorded directly on films, papers, or specific devices and digitized later. Once a digital image is available, image processing can be performed to enhance the quality. Image enhancement is done to facilitate viewing, identify minute features or subtle changes, and measure the features or changes more accurately. A number of algorithms are used for NDT image enhancement and most of the current commercial instruments come with relevant signal/image enhancement hardware or software packages.

Generally, a digital NDT image consists of a large number of pixels, each presented in different shades of gray or in different colors. Each gray shade or color corresponds to a value of the output signal parameter, such as signal amplitude in volts. In a typical inspection, after the NDT instrument has first been calibrated, tests are carried out on the unknown samples and either gray-scale or color images are recorded. An important consideration is that the human eye normally has the capability of discriminating about a 2% change in brightness and a better ability to recognize a color difference than shades of gray. Thus, color images are usually used in NDT in order to provide easier interpretation and more accurate results. Various image processing methods are then applied to improve the visibility of features of interest. One example is presented below to demonstrate the effects of the different image enhancement methods.

In the ultrasonic gray-scale C-scan image presented below, the total loss of signal or zero is represented by black, 100% intensity is represented by white, and different shades of gray in between correspond to various signal amplitude values. Although all the embedded Teflon inserts are identified in the gray-scale image and the planner size can be measured, this image cannot specify their depth locations (Figure 3.9, left). However, when the same data is presented in color, more information, including the depth location of the inserts, becomes evident. In this case, the different colors indicate various depth levels within the sample based on prior calibrations (Figure 3.9). The color image also shows variations in the background red color due to the sample's geometrical waviness that were not visible in the gray-scale C-scan. The geometrical noise is filtered out by further image enhancements as shown in Figure 3.10. The process also has improved the detectability showing small inherent discontinuities in addition to the inserts.

Finally, it is important to mention that any image enhancement carried out on the NDT results must always include the references too in order to avoid inadvertent changes in the threshold setting that may occur as a result of additional processing. Also, NDT images must always include the original scale pallet as well as the threshold level and the maximum size of acceptable defects.

3.9 NDT MODELING

As NDT methods use an energy source to obtain a signal response from the test object, physics-based models may be employed to simulate the interaction of the energy with the test piece and the target. Such simulations may be able to either predict the NDT outcome (e.g., response signal) or help to characterize the target. For prediction of the NDT response signals, inputs on material properties, component geometry, and the NDT system as well as the target characteristics are needed. On the other hand, by using the NDT signals along with information on the component and the measurement system, one may be able to characterize the target. These are often termed as forward and inverse problems.

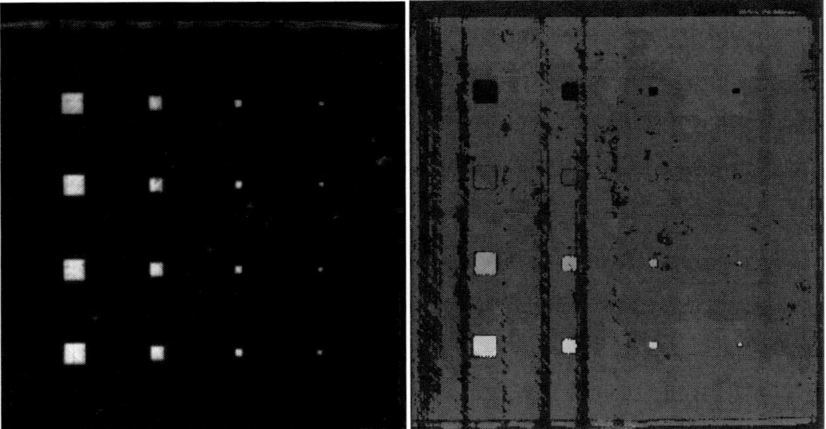

Figure 3.9 Ultrasonic C-scan images of a composite reference piece with embedded Teflon inserts of various sizes located at different depth in gray and in color. Both images show the size and planar location of the inserts but the color image also shows the depth location by the different colors. The dark red lines are caused by geometrical waviness of the panel.

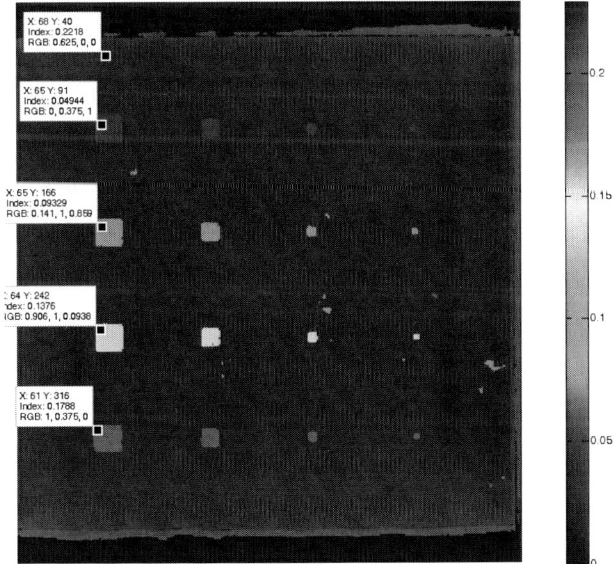

Figure 3.10 Same as Figure 3.9 color C-scan after additional signal gating to remove the geometrical waviness effects. Note that the inherent manufacturing defects are clearer here. The color scale helps to establish at what depth the inserts or defects are located.

3.9.1 *Forward Problem*

Analytical or numerical models are used to predict the sensor response of a given geometry, discontinuity configuration, or test condition in order to understand or characterize the physical processes involved in the NDT measurements. For example, forward modeling is commonly used in the design of new sensors or for system optimization when the optimal input signal parameters are needed to achieve the desirable output response.

Numerical models are often employed when analytical solutions are difficult or not possible. These models attempt to predict the output signal for certain known parameters of the input signal and the object. A numerical modeling process generally involves the following steps:

- Definition of the object geometry.
- Selection of appropriate equations related to physics of the process.
- Selection of the material property data and boundary conditions.
- Selection of a mesh configuration and solving equations.
- Post-processing and visualization of the solution.

3.9.2 *Inverse Problems*

In the case of inverse problems, the output response function is available in the form of measured data or signals and the object characteristics are estimated. Inverse problems are usually solved numerically, and, due to the availability of computational resources, such solutions are becoming more useful. Estimation of material properties or discontinuities may be possible using inverse approaches if both the inspection input and output signals are available.

Overall, modeling is a useful tool for understanding and optimizating the NDT process as well as for the development of new sensors and procedures or optimization and interpretation of experimental results. However, modeling requires highly knowledgeable and skilled personnel as the choice of wrong inputs or models may provide invalid or erroneous results. NDT modeling should not be used as substitute but as a means of complimenting the experimental measurements.

3.10 EXAMPLES OF NON-STANDARD REFERENCE PIECES

3.10.1 *Metallic Samples with Cracks*

Reference [18] describes a methodology used to create fatigue cracks in samples of different geometries that are employed for calibration or analysis of the NDT instruments or procedures. This methodology generally involves the following steps:

1. Machine samples with the required geometry and extra material that can be slotted. The thickness and size of the extra material are tailored to develop the appropriate crack length versus crack depth ratio.
2. Using Electron Discharge Machining (EDM) with a wire or shim electrode; introduce a starter slot typically 1 mm deep.
3. Using a load frame with cyclic loading capability to apply fatigue loading to grow a crack from the starter EDM slot. Appropriate load levels based on the material's fatigue crack growth properties must be selected. Crack growth predication software packages, such as AFGROW [19], can aid.
4. Track the crack development and growth using an aided visual inspection, such as a traveling microscope, borescope, vernier callipers, and cameras. Liquid penetrant, ultrasonic, or eddy current methods can be used to detect or measure the cracks.
5. Once the desired crack size is achieved, machine away the extra material feature to obtain the final detail size.
6. Re-install the coupon in the load frame and cycle for a few extra hundred cycles to help open the crack after potential smearing of crack edges due to machining.
7. Perform the final measurement of crack size under load while the sample is still in the load frame. Replication of the crack surface lengths can also be performed using silicone rubber.
8. Perform additional manufacture and assembly of the sample to obtain the required configuration.

A few examples are provided below.

3.10.1.1 FATIGUE CRACKS INSIDE FASTENER HOLES

Samples with fatigue cracks inside fastener holes were needed for the evaluation of the capability of an eddy current approach used in the inspection of aircraft wing box structures, such as skin splice, spar-to-web and skin-to-spar joints, as described in reference [18]. Such joints were simulated by manufacturing simple assemblies of two-layer coupons, either "thick" or "thin," with a centrally located bolt hole. The coupons were representative of the target structures in terms of material, geometry, surface condition, thickness, hole size, and crack characteristics.

To create fatigue cracks of various sizes, undersized straight-through non-countersunk pilot holes were drilled in aluminum panels and then counter bored to create fins, which were then EDM slotted using an electrode shim for creating corner or mid-bore cracks. The drawing of the panel from which coupons were subsequently machined is shown in Figure 3.11. Panels were fatigue loaded in a load frame, as shown in Figure 3.12. The crack shapes and sizes for the four hole

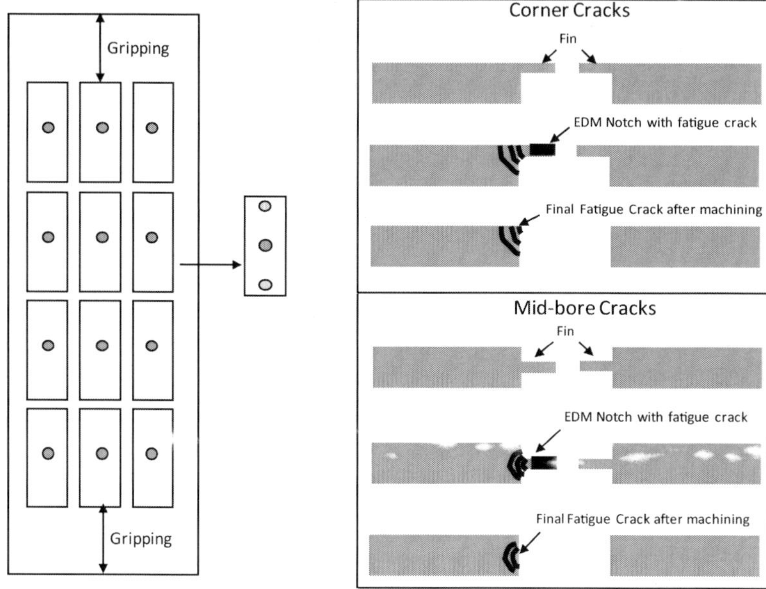

Figure 3.11 Schematic of a panel with several hole sites (left) and an extracted reference coupon (center). The approaches used to generate corner cracks (top right) and mid-bore cracks (bottom right).

Figure 3.12 Test panel installed in a load frame and a video camera used to monitor crack growth.

configurations, in other words, corner crack in thin material, corner crack in thick material, mid-bore crack in thick material, and number of fatigue cycles required to grow the EDM slots to the required crack sizes were estimated using models in the AFGROW crack growth prediction software program [19].

Once the cracks were developed, the fins were machined out and final crack sizes measured by traveling microscope under load as well as by a replica technique. The panels were cut into coupons for further assembly. Depending on the required configuration, the inspection sets were assembled and placed in specially designed boxes for inspections to generate data necessary for reliability assessment, as shown in Figure 3.13.

3.10.1.2 Fatigue Cracks under Countersunk Fasteners

For the case of countersunk holes, a similar strategy to that presented earlier can be used to develop cracks inside the countersunk hole in different locations as shown in Figure 3.14.

3.10.1.3 Fatigue Cracks on Free Surfaces

In this work, fatigue cracks on free surfaces of aluminum plates were needed for the evaluation of NDT methods. To create surface cracks, an aluminum plate was

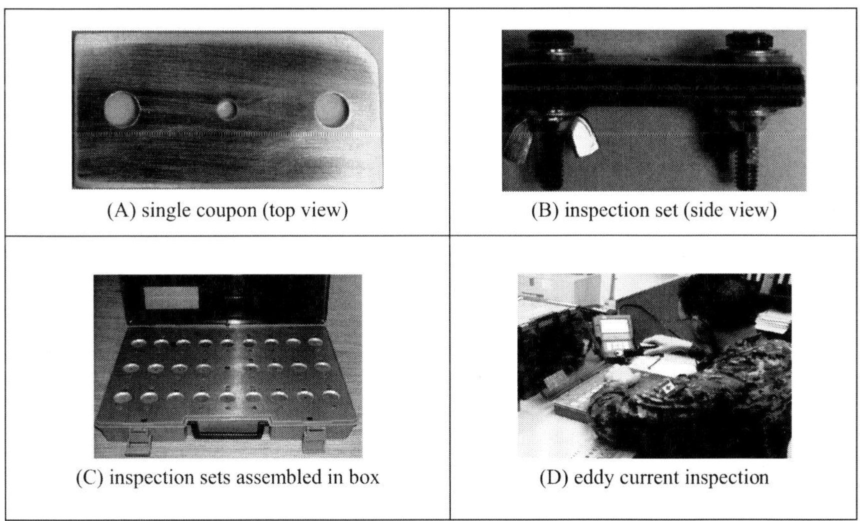

(A) single coupon (top view)	(B) inspection set (side view)
(C) inspection sets assembled in box	(D) eddy current inspection

Figure 3.13 Examples of: (A) single coupon (top view), (B) inspection set (side view), (C) inspection sets assembled in box, and (D) an inspector performing a bolt hole eddy current inspection.

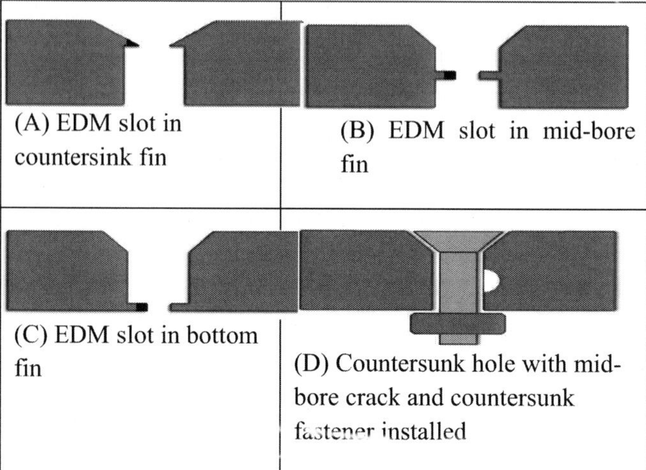

(A) EDM slot in countersink fin	(B) EDM slot in mid-bore fin
(C) EDM slot in bottom fin	(D) Countersunk hole with mid-bore crack and countersunk fastener installed

Figure 3.14 Strategies for locating fatigue cracks in countersunk holes at the: (A) countersink; (B) mid-bore; and (C) bottom surface. (D) A schematic of a final machined hole with mid-bore fatigue crack and countersunk fastener installed.

machined with a raised rib in which a single EDM slot was cut as illustrated in Figure 3.15 (left). The specimen was then loaded in fatigue until a surface crack grew from the EDM slot into the Al plate. An optical traveling microscope was utilized to track growth of the crack to the desired size.

The rib with EDM slot was subsequently machined away leaving the surface crack in the aluminum plate. A secondary machining process was necessary on the opposite side of the specimen in order to eliminate bending that occurred following the machining of the primary surface. To obtain cracks of different surface lengths, aspect ratios, and locations within the same coupon, numerous ribs of different thicknesses and EDM slots were used as shown in Figure 3.15 (right).

3.10.2 METALLIC SAMPLES WITH MATERIAL LOSS

The sample seen in Figure 3.16 was made to assess the detection and measurement capability of some NDT methods (thermography and ultrasonic) as a function of the discontinuity size, separation, and depth location in aluminum plates. The discontinuities here are areas of removed material that simulate the material loss due to corrosion. The specimen shown is a 174 mm × 200 mm aluminum plate of 3 m thickness containing 4 columns of 25 mm wide rectangular removed material regions. These discontinuities are positioned into four columns of the same depth. The material loss in each of the four columns is 85%, 66.7%, 35%, and 20% of the total thickness.

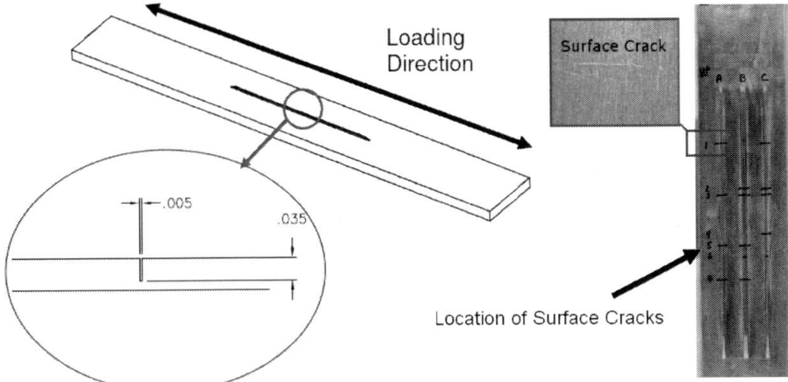

Figure 3.15 Surface crack specimens showing a single rib with single crack at left and multiple ribs with multiple cracks at right.

Figure 3.16 A photograph of a reference aluminum sample with areas of removed material.

3.10.3 COMPOSITE REFERENCE PIECES

Similar to metallic parts, reference panels are needed to demonstrate the capabilities of the NDT techniques in the detection of defects or damage in composites (e.g., cavities, inclusions, delamination, impact damage, etc.). Two examples of composite reference specimens made of solid laminates and honeycomb sandwich materials are provided below.

3.10.3.1 COMPOSITE SOLID LAMINATE SAMPLES

The following example is a curved carbon fiber solid laminate that has 10 plies and contains 25 discontinuities (Teflon inserts) of different sizes embedded at different depths as illustrated schematically in Figure 3.17. The panel simulates aircraft parts made of composite solid laminates and was made to assess the detection and

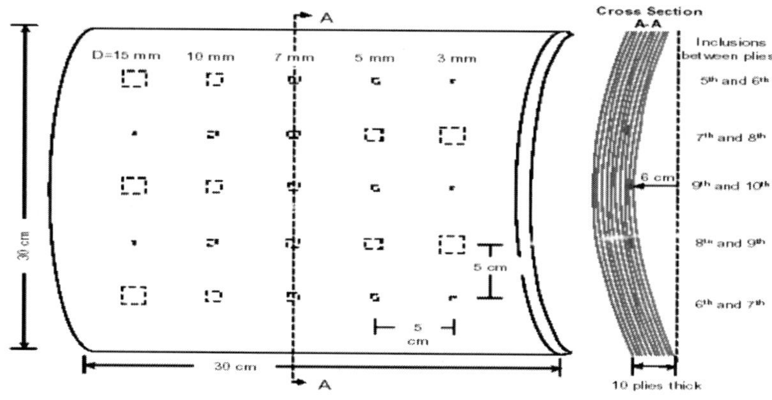

Figure 3.17 A composite solid laminate reference panel with embedded artificial defects.

measurement capability of NDT methods (e.g., ultrasonic and thermography) for delamination-like discontinuities.

3.10.3.2 HONEYCOMB SANDWICHED PANELS

The honeycomb sandwich panels are made of graphite-epoxy-skins bonded to aluminum honeycomb core using a commercial aerospace adhesive. The panels are similar in materials and design to composite structures of the CF-18 aircraft. The panels are made of two variants of aluminum honeycomb core. One is 15.875 mm thick with larger cell size or lower core density while the other is 50 mm thick and has a smaller cell size or higher core density. A photograph of the reference panels is shown in Figure 3.18 along with the x-ray images indicating the honeycomb cell size and density. These panels contain several simulated defects and their descriptions, sizes, and location are provided in Figure 3.19. Figure 3.20 shows another honeycomb reference panel with artificial defects that are identified and described in Table 3.1 Description of artificial defects embedded in the panels shown in Figure 3.20.

Figure 3.18 (top): Photograph of two honeycomb sandwich panels representing the CF-18 rudder material and design. (middle and bottom): X-ray images of the panels showing the honeycomb cell size and core density differences.

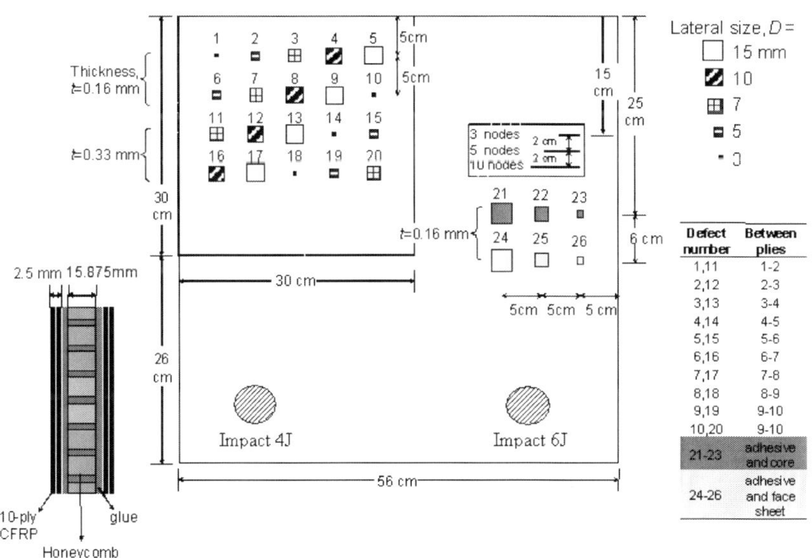

Defect number	Between plies
1,11	1-2
2,12	2-3
3,13	3-4
4,14	4-5
5,15	5-6
6,16	6-7
7,17	7-8
8,18	8-9
9,19	9-10
10,20	9-10
21-23	adhesive and core
24-26	adhesive and face sheet

Figure 3.19 A honeycomb composite reference panel with different artificial defects (identified by numbers 1–26) and two impact damaged sites.

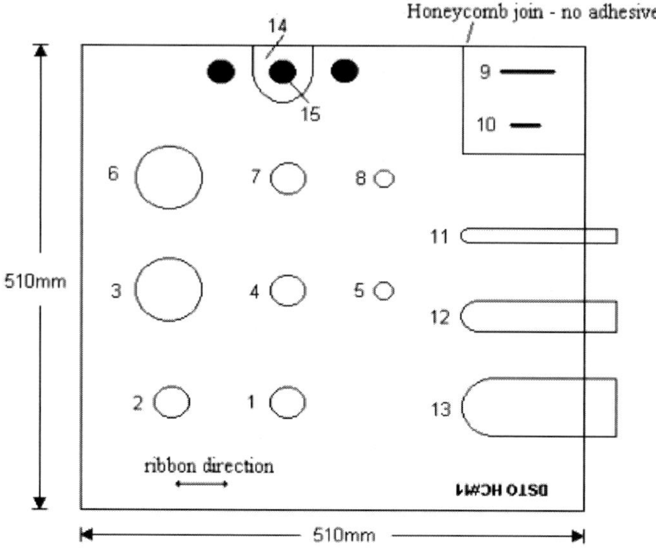

Figure 3.20 A honeycomb reference panels containing artificial defects of different types, sizes, and locations. The description of the defects is provided in Table 3.1

Table 3.1 Description of artificial defects embedded in the panels shown in Figure 3.20.

	Simulated Skin Disbonds
1	Double layer 25 mm diameter Teflon between skin and adhesive
2	Single layer 25 mm diameter Teflon between skin and adhesive
6–8	ER200 (skin) 50, 25, 12 mm φ
11–13	0.08 mm thick Teflon shim

	Simulated Weak Fillet Bond
3–5	ER200 (honeycomb) 50, 25, 12 mm φ
14	ER200 (honeycomb)
15	25 mm φ stub bonded to skin and pulled off

	Simulated Node Disbonds
9	5 node bonds separated
10	3 node bonds separated

3.11 APPENDICES

3.11.1 A PARTIAL LIST OF ASTM GENERAL AND AEROSPACE-RELATED NDT STANDARDS

General NDT
* E2533-09, Standard Guide for Non-destructive Testing of Polymer Matrix Composites Used in Aerospace Applications
* E1316, Standard Terminology for Non-destructive Examinations.
* F1469, Standard Guide for Conducting a Repeatability and Reproducibility Study on Test Equipment for Non-destructive Testing.

NDT Agencies
* E543, Standard Practice for Agencies Performing Non-destructive Testing.
* E1212, Standard Practice for Quality Control Systems for Non-destructive Testing Agencies.
* E1359, Standard Guide for Evaluating Capabilities of Non-destructive Testing Agencies.

Radiography (X and Gamma)
* E746, Standard Test Method for Determining Relative Image Quality Response of Industrial Radiographic Film.
* E747, Standard Practice for Design, Manufacture, and Material Grouping Classification of Wire Image Quality Indicators (IQI) used for Radiology.
* E1254, Standard Guide for Storage of Radiographs and Unexposed Industrial Radiographic Films.
* E1255, Standard Practice for Radioscopy E1316, Standard Terminology for Non-destructive Examinations.
* E1320, Standard Reference Radiographs for Titanium Castings.
* E1742, Standard Practice for Radiographic Examination.
* E142, Standard Method for Controlling Quality of Radiographic Testing.
* E155, Standard Reference Radiographs for Inspection of Aluminum and Magnesium Castings.
* E2662, Standard Practice for Radiologic Examination of Flat Panel Composites and Sandwich Core Materials Used in Aerospace Applications.
* E94, Standard Guide for Radiographic Examination.
* E192, Standard Reference Radiographs for Investment Steel Castings of Aerospace Applications.
* E215, Standard Practice for Standardizing Equipment for Electromagnetic Examination of Seamless Aluminum-Alloy Tube.
* E242, Standard Reference Radiographs for Appearances of Radiographic Images as Certain Parameters Are Changed.
* E272, Standard Reference Radiographs for High-Strength Copper-Base and Nickel-Copper Alloy Castings.

- E310, Standard Reference Radiographs for Tin Bronze Castings.
- E390, Standard Reference Radiographs for Steel Fusion Welds.
- E431, Standard Guide to Interpretation of Radiographs of Semiconductors and Related Devices.
- E505, Standard Reference Radiographs for Inspection of Aluminum and Magnesium Die Castings.
- E592, Standard Guide to Obtainable ASTM Equivalent Penetrameter Sensitivity for Radiography of Steel Plates 1/4 to 2 in. (6 to 51 mm) Thick with X-Rays and 1 to 6 in. (25 to 152 mm) Thick with Cobalt-60.
- E689, Standard Reference Radiographs for Ductile Iron Castings.
- E801, Standard Practice for Controlling Quality of Radiological Examination of Electronic Devices.
- E802, Standard Reference Radiographs for Gray Iron Castings Up to 4 1/2 in. (114 mm) in Thickness.
- E999, Standard Guide for Controlling the Quality of Industrial Radiographic Film Processing.
- E1025, Standard Practice for Design, Manufacture, and Material Grouping Classification of Hole-Type Image Quality Indicators (IQI) used for Radiology.
- E1030, Standard Test Method for Radiographic Examination of Metallic Castings.
- E1032, Standard Test Method for Radiographic Examination of Weldments.
- E1114, Standard Test Method for Determining the Focal Size of Iridium-192 Industrial Radiographic Sources.
- E1161, Standard Test Method for Radiologic Examination of Semiconductors and Electronic Components.
- E1165, Standard Test Method for Measurement of Focal Spots of Industrial X-Ray Tubes by Pinhole Imaging.
- E1390, Standard Guide for Illuminators Used for Viewing Industrial Radiographs.
- E1411, Standard Practice for Qualification of Radioscopic Systems.
- E1416, Standard Test Method for Radioscopic Examination of Weldments.
- E1475, Standard Guide for Data Fields for Computerized Transfer of Digital Radiological Test Data.
- E1647, Standard Practice for Determining Contrast Sensitivity in Radioscopy.
- E1648, Standard Reference Radiographs for Examination of Aluminum Fusion Welds.
- E1734, Standard Practice for Radioscopic Examination of Castings.
- E1735, Standard Test Method for Determining Relative Image Quality of Industrial Radiographic Film Exposed to X-Radiation from 4 to 25 MV.
- E1815, Standard Test Method for Classification of Film Systems for Industrial Radiography.

- E1817, Standard Practice for Controlling Quality of Radiological Examination by Using Representative Quality Indicators (RQIs).
- E1936, Standard Reference Radiograph for Evaluating the Performance of Radiographic Digitization Systems.
- E1955, Standard Radiographic Examination for Soundness of Welds in Steel by Comparison to Graded ASTM E390 Reference Radiographs.
- E2002, Standard Practice for Determining Total Image Unsharpness in Radiology.
- E2104, Standard Practice for Radiographic Examination of Advanced Aero and Turbine Materials and Components.
- F629, Standard Practice for Radiography of Cast Metallic Surgical Implants.
- E446, Standard Reference Radiographs for Steel Castings Up to 2 in. (51 mm) in Thickness.

Neutron Radiography
- E545, Standard Test Method for Determining Image Quality in Direct Thermal Neutron Radiographic Examination.
- E748, Standard Practices for Thermal Neutron Radiography of Materials.
- E803, Standard Test Method for Determining the L/D Ratio of Neutron Radiography Beams.
- E1453, Standard Guide for Storage of Media That Contains Analog or Digital Radioscopic Data.
- E1496, Standard Test Method for Neutron Radiographic Dimensional Measurements.
- E2003, Standard Practice for Fabrication of the Neutron Radiographic Beam Purity Indicators.
- E2023, Standard Practice for Fabrication of Neutron Radiographic Sensitivity Indicators.

Computed Radiography/ Tomography
- E2007, Standard Guide for Computed Radiography.
- E2033, Standard Practice for Computed Radiology (Photostimulable Luminescence Method).
- E1441, Standard Guide for Computed Tomography (CT) Imaging.
- E1570, Standard Practice for Computed Tomographic (CT) Examination.
- E1672, Standard Guide for Computed Tomography (CT) System Selection.
- E1695, Standard Test Method for Measurement of Computed Tomography (CT) System Performance.
- E1814, Standard Practice for Computed Tomographic (CT) Examination of Castings.
- E1935, Standard Test Method for Calibrating and Measuring CT Density.

- E2033, Standard Practice for Computed Radiology (Photostimulable Luminescence Method).

Compton Scattering
- E1931, Standard Guide for X-Ray Compton Scatter Tomography.

Eddy current
- B244, Standard Test Method for Measurement of Thickness of Anodic Coatings on Aluminum and of Other Nonconductive Coatings on Nonmagnetic Basis Metals with Eddy-Current Instruments.
- B499, Measurement of Coating Thickness by the Magnetic Method: Nonmagnetic Coatings on Magnetic Base Metals.
- B457, Standard Test Method for Measurement of Impedance of Anodic Coatings on Aluminum.
- E243, Standard Practice for Electromagnetic (Eddy-Current) Examination of Copper and Copper-Alloy Tubes.
- E309, Standard Practice for Eddy-Current Examination of Steel Tubular Products Using Magnetic Saturation.
- E376, Standard Practice for Measuring Coating Thickness by Magnetic-Field or Eddy-Current (Electromagnetic) Test Methods.
- E426, Standard Practice for Electromagnetic (Eddy-Current) Examination of Seamless and Welded Tubular Products, Austenitic Stainless Steel and Similar Alloys.
- E566, Standard Practice for Electromagnetic (Eddy-Current) Sorting of Ferrous Metals.
- E571, Standard Practice for Electromagnetic (Eddy-Current) Examination of Nickel and Nickel Alloy Tubular Products.
- E690, Standard Practice for In Situ Electromagnetic (Eddy-Current) Examination of Nonmagnetic Heat Exchanger Tubes.
- E703, Standard Practice for Electromagnetic (Eddy-Current) Sorting of Nonferrous Metals.
- E1033, Standard Practice for Electromagnetic (Eddy-Current) Examination of Type F-Continuously Welded (CW) Ferromagnetic Pipe and Tubing Above the Curie Temperature.
- E1312, Standard Practice for Electromagnetic (Eddy-Current) Examination of Ferromagnetic Cylindrical Bar Product Above the Curie Temperature.
- E1571, Standard Practice for Electromagnetic Examination of Ferromagnetic Steel Wire Rope.
- E1606, Standard Practice for Electromagnetic (Eddy-Current) Examination of Copper Redraw Rod for Electrical Purposes.
- E1629, Standard Practice for Determining the Impedance of Absolute Eddy-Current Probes.

- E2096, Standard Practice for In Situ Examination of Ferromagnetic Heat-Exchanger Tubes Using Remote Field Testing.
- E570, Standard Practice for Flux Leakage Examination of Ferromagnetic Steel Tubular Products.

Electrical conductivity
- E1004, Standard Test Method for Determining Electrical Conductivity Using the Electromagnetic (Eddy-Current) Method.
- E977, Standard Practice for Thermoelectric Sorting of Electrically Conductive Materials.

Ultrasonic
- E127, Standard Practice for Fabricating and Checking Aluminum Alloy Ultrasonic Standard Reference Blocks.
- E428, Standard Practice for Fabrication and Control of Steel Reference Blocks Used in Ultrasonic Examination.
- E664, Standard Practice for the Measurement of the Apparent Attenuation of Longitudinal Ultrasonic Waves by Immersion Method.
- E2580, Standard Practice for Ultrasonic Testing of Flat Panel Composites and Sandwich Core Materials Used in Aerospace Applications.
- B594, Standard Practice for Ultrasonic Inspection of Aluminum-Alloy Wrought Products for Aerospace Applications.
- E114, Standard Practice for Ultrasonic Pulse-Echo Straight-Beam Contact Testing.
- E2373, Standard Practice for Use of the Ultrasonic Time of Flight Diffraction (TOFD) Technique.
- E1315, Standard Practice for Ultrasonic Examination of Steel with Convex Cylindrically Curved Entry Surfaces.
- E587, Standard Practice for Ultrasonic Angle-Beam Contact Testing.
- E164, Standard Practice for Ultrasonic Contact Examination of Weldments.
- E213, Standard Practice for Ultrasonic Examination of Metal Pipe and Tubing.
- E214, Standard Practice for Immersed Ultrasonic Examination by the Reflection Method Using Pulsed Longitudinal Waves.
- E273, Standard Practice for Ultrasonic Examination of the Weld Zone of Welded Pipe and Tubing.
- E317, Standard Practice for Evaluating Performance Characteristics of Ultrasonic Pulse- Echo Examination Instruments and Systems Without the Use of Electronic Measurement Instruments.
- E494, Standard Practice for Measuring Ultrasonic Velocity in Materials.
- E587, Standard Practice for Ultrasonic Angle-Beam Examination by the Contact Method.

- E588, Standard Practice for Detection of Large Inclusions in Bearing Quality Steel by the Ultrasonic Method.
- E797, Standard Practice for Measuring Thickness by Manual Ultrasonic Pulse-Echo Contact Method.
- E1001, Standard Practice for Detection and Evaluation of Discontinuities by the Immersed Pulse-Echo Ultrasonic Method Using Longitudinal Waves.
- E1065, Standard Guide for Evaluating Characteristics of Ultrasonic Search Units.
- E1158, Standard Guide for Material Selection and Fabrication of Reference Blocks for the Pulsed Longitudinal Wave Ultrasonic Examination of Metal and Metal Alloy Production Material.
- E1324, Standard Guide for Measuring Some Electronic Characteristics of Ultrasonic Examination Instruments.
- E1454, Standard Guide for Data Fields for Computerized Transfer of Digital Ultrasonic Testing Data.
- E1774, Standard Guide for Electromagnetic Acoustic Transducers (EMATs).
- E1816, Standard Practice for Ultrasonic Examinations Using Electromagnetic Acoustic Transducer (EMAT) Techniques.
- E1901, Standard Guide for Detection and Evaluation of Discontinuities by Contact Pulse-Echo Straight-Beam Ultrasonic Methods.
- E1961, Standard Practice for Mechanized Ultrasonic Examination of Girth Welds Using Zonal Discrimination with Focused Search Units.
- E1962, Standard Test Method for Ultrasonic Surface Examinations Using Electromagnetic Acoustic Transducer (EMAT) Techniques.
- E2001, Standard Guide for Resonant Ultrasound Spectroscopy for Defect Detection in Both Metallic and Non-Metallic Parts.
- E2192, Standard Guide for Planar Flaw Height Sizing by Ultrasonics.
- E2223, Standard Practice for Examination of Seamless, Gas-Filled, Steel Pressure Vessels Using Angle Beam Ultrasonic.
- F1512, Standard Practice for Ultrasonic C-Scan Bond Evaluation of Sputtering Target-Backing Plate Assemblies.

Acoustic Emission
- E650, Standard Guide for Mounting Piezoelectric Acoustic Emission Sensors.
- E976, Standard Guide for Determining the Reproducibility of Acoustic Emission Sensor Response.
- E2661/ E2661M-10, Standard Practice for Acoustic Emission Examination of Plate-like and Flat Panel Composite Structures Used in Aerospace Applications.
- E569, Standard Practice for Acoustic Emission Monitoring of Structures during Controlled Stimulation.

- E749, Standard Practice for Acoustic Emission Monitoring During Continuous Welding.
- E750, Standard Practice for Characterizing Acoustic Emission Instrumentation.
- E751, Standard Practice for Acoustic Emission Monitoring During Resistance Spot-Welding.
- E1067, Standard Practice for Acoustic Emission Examination of Fiberglass Reinforced Plastic Resin (FRP) Tanks/Vessels.
- E1106, Standard Method for Primary Calibration of Acoustic Emission Sensors.
- E1118, Standard Practice for Acoustic Emission Examination of Reinforced Thermosetting Resin Pipe (RTRP).
- E1139, Standard Practice for Continuous Monitoring of Acoustic Emission from Metal Pressure Boundaries.
- E1419, Standard Test Method for Examination of Seamless, Gas- Filled, Pressure Vessels Using Acoustic Emission.
- E1781, Standard Practice for Secondary Calibration of Acoustic Emission Sensors.
- E1888, Standard Test Method for Acoustic Emission testing of Pressurized Containers Made of Fiberglass Reinforced Plastic with Balsa Wood Cores.
- E1930, Standard Test Method for Examination of Liquid Filled Atmospheric and Low Pressure Metal Storage Tanks Using Acoustic Emission.
- E1932, Standard Guide for Acoustic Emission Examination of Small Parts.
- E2075, Standard Practice for Verifying the Consistency of AE-Sensor Response Using an Acrylic Rod.
- E2076, Standard Test Method for Examination of Fiberglass Reinforced Plastic Fan Blades Using Acoustic Emission.
- E2191, Standard Test Method for Examination of Gas-Filled Filament-Wound Composite Pressure Vessels Using Acoustic Emission.
- F914, Standard Test Method for Acoustic Emission for Insulated Aerial Personnel Devices.
- F1430, Standard Test Method for Acoustic Emission Testing of Insulated Aerial Personnel Devices with Supplemental Load Handling Attachments.

Acousto-ultrasonic
- E1495, Standard Guide for Acousto-Ultrasonic Assessment of Composites, Laminates, and Bonded Joints.
- E1736, Standard Practice for Acousto-Ultrasonic Assessment of Filament-Wound Pressure Vessels.

Liquid Penetrant
- E1208, Standard Test Method for Fluorescent Liquid Penetrant Examination Using the Lipophilic Post-Emulsification Process.

- E1209, Standard Test Method for Fluorescent Liquid Penetrant Examination Using the Water-Washable Process.
- E1210, Standard Test Method for Fluorescent Liquid Penetrant Examination Using the Hydrophilic Post-Emulsification Process. ASTM E1417 - 05e1 Standard Practice for Liquid Penetrant Testing.
- E165, Standard Test Method for Liquid Penetrant Examination.
- E433, Standard Reference Photographs for Liquid Penetrant Inspection.
- E1135, Standard Test Method for Comparing the Brightness of Fluorescent Penetrants.
- E1219, Standard Test Method for Fluorescent Liquid Penetrant Examination Using the Solvent-Removable Process.
- E1220, Standard Test Method for Visible Penetrant Examination Using the Solvent-Removable Process.
- E1417, Standard Practice for Liquid Penetrant Examination.
- E1418, Standard Test Method for Visible Penetrant Examination Using the Water-Washable Process.

Magnetic Particles
- E709, Standard Guide for Magnetic Particle Testing.
- E125, Standard Reference Photographs for Magnetic Particle Indications on Ferrous Castings.
- E709, Standard Guide for Magnetic Particle Examination.
- E1444, Standard Practice for Magnetic Particle Examination.

Thermography
- E2582, Standard Practice for Infrared Flash Thermography of Composite Panels and Repair Patches Used in Aerospace Applications.
- E1213, Standard Test Method for Minimum Resolvable Temperature Difference for Thermal Imaging Systems.
- E1311, Standard Test Method for Minimum Detectable Temperature Difference for Thermal Imaging Systems.
- E1543, Standard Test Method for Noise Equivalent Temperature Difference of Thermal Imaging Systems.
- E1862, Standard Test Methods for Measuring and Compensating for Reflected Temperature Using Infrared Imaging Radiometers.
- E1897, Standard Test Methods for Measuring and Compensating for Transmittance of an Attenuating Medium Using Infrared Imaging.
- E1933, Standard Test Methods for Measuring and Compensating for Emissivity Using Infrared Imaging Radiometers.
- E1934, Standard Guide for Examining Electrical and Mechanical Equipment with Infrared Thermography.

Shearography
- E2581, Standard Practice for Shearography of Polymer Matrix Composites, Sandwich Core Materials and Filament-Wound Pressure Vessels in Aerospace Applications.

Optical
- E408, Standard Test Methods for Total Normal Emittance of Surfaces Using Inspection-Meter Techniques.

Leak testing
- E908, Standard Practice for Calibrating Gaseous Reference Leaks.
- E427, Standard Practice for Testing for Leaks Using the Halogen Leak Detector (Alkali-Ion Diode).
- E432, Standard Guide for Selection of a Leak Testing Method.
- E479, Standard Guide for Preparation of a Leak Testing Specification.
- E493, Standard Test Methods for Leaks Using the Mass Spectrometer Leak Detector in the Inside-Out Testing Mode.
- E498, Standard Test Methods for Leaks Using the Mass Spectrometer Leak Detector or Residual Gas Analyzer in the Tracer Probe Mode.
- E499, Standard Test Methods for Leaks Using the Mass Spectrometer Leak Detector in the Detector Probe Mode.
- E515, Standard Test Method for Leaks Using Bubble Emission Techniques.
- E1002, Standard Test Method for Leaks Using Ultrasonics.
- E1003, Standard Test Method for Hydrostatic Leak Testing.
- E1066, Standard Test Method for Ammonia Colorimetric Leak Testing.
- E1211, Standard Practice for Leak Detection and Location Using Surface-Mounted Acoustic Emission Sensors.
- E1603, Standard Test Methods for Leakage Measurement Using the Mass Spectrometer Leak Detector or Residual Gas Analyzer in the Hood Mode.
- E2024, Standard Test Methods for Atmospheric Leaks Using a Thermal Conductivity Leak Detector.

3.11.2 A PARTIAL LIST OF ISO NDT STANDARDS

- ISO 10375: Non-destructive testing—Ultrasonic inspection—Characterization of search unit and sound field.
- ISO/DIS 12709: Non-destructive testing—Ultrasonic inspection—Detection and evaluation of discontinuities by the immersed pulse-echo ultrasonic method.
- ISO/DIS 12710: Non-destructive testing—Ultrasonic inspection—Evaluation of electronic characteristics of instruments.

- ISO/FDIS 12713: Non-destructive testing—Acoustic emission inspection—Primary calibration of transducers.
- ISO/DIS 12714: Non-destructive testing—Secondary calibration of acoustic emission transducers.
- ISO/DIS 12715: Non-destructive testing—Ultrasonic inspection—Reference blocks and test procedures for the characterization of contact search unit beam profiles.

3.12 CHAPTER REFERENCES

[1]   ASTM Standard, Non-destructive Testing, Volume 03.03, October 2010.

[2]   http://www.ndt-ed.org/GeneralResources/Standards/standards.htm.

[3]   E127-10 Standard Practice for Fabricating and Checking Aluminum Alloy Ultrasonic Standard Reference Blocks Volume 03.03, Issued October 2010.

[4]   Canadian Forces Eddy Current Inspection no. GEN-74-E REV 2, September 2003.

[5]   Health Monitoring of Aerospace Structures: Smart Sensor Technologies and Signal Processing, W. Staszewski, C. Boller, and G. Tomlinson, John Wiley & Son Ltd., England, 2004.

[6]   Advanced Image Processing Methods for Ultrasonic NDE Research, C. H. Chen, World Conference on Non-destructive Testing, Montreal, QC, Canada, 2004.

[7]   Non-destructive Testing Handbook, 2nd Edition, Volume 7, Ultrasonic Testing, Published by the American Society for Non-destructive Testing, 1991.

[8]   A theory for Multi-resolution Signal Decomposition: The Wavelet Representation, S. G. Mallat, IEEE Trans. Pattern Anal. Mach. Intell., Vol. 11, No. 7, pp. 674–693, July 1989.

[9]   The Wigner Distribution: A Tool for Time-Frequency Signal Analysis - Part 2: Discrete-Time Signals, T.A, C.M. Classen, and W.F.G. Mecklenbrauker, Philips Journal of Research, Vol. 35, pp. 276–350, 1980.

[10] Pattern Recognition and Neural Networks, B. D. Ripley, Cambridge University Press, UK, 2007.

[11] Automated Measurement of Corrosion Damage Using Edge of Light Technique, Z. Liu, D.S. Forsyth, A. Marincak, and P. Vesely, American Institute of Physics, Vol. 23, pp. 1347–1354, 2003.

[12] Automated Pulsed Eddy Current Method for Detection and Classification of Hidden Corrosion, M.S. Safizadeh, Z. Liu, C. Mandache, D. S. Forsyth, and A. Fahr, Proceedings of the 6th International Workshop on Advances in Signal Processing for NDE and Materials, Quebec City, Quebec, Canada, August 2005.

[13] Investigations on Classifying Pulsed Eddy Current Signal with a Neural Network, Z. Liu, D. S. Forsyth, B. A. Lepine, I. Hammad, and B. Farahbakhsh, Insight: Non-destructive Testing and Condition Monitoring, Vol. 45, No. 9, pp. 608–614, September 2003.

[14] https://www.e-education.psu.edu/drupal6/files/sgam/FUSION%20902341. pdf

[15] The Role of Data Fusion in NDE for Aging Aircraft, D. S. Forsyth and J. P. Komorowski, SPIE Conference on Aging Aircraft, Airports, and Aerospace Hardware, Newport Beach, SPIE, Vol. 3994, 2000.

[16] A Data Fusion Scheme for Quantitative Image Analysis by Using Locally Weighted Regression and Dempster-Shafer Theory, Z. Liu, D. S. Forsyth, M. S. Safizadeh, and A. Fahr, IEEE Transactions on Instrumentation and Measurement, Vol. 57, No. 11, pp. 2554–2560, November 2008.

[17] Combining Multiple Non-destructive Inspection Images with a Generalized Additive Model, Z. Liu, P. Ramuhalli, M. S. Safizadeh, and D. S. Forsyth, Measurement Science and Technology, Vol. 19, No. 085701, pp. 1–8, August 2008.

[18] Artificial Seeding of Fatigue Cracks in NDI Reference Coupons, M. Yanishevsky, M. Martinez, C. Mandache, M. Khan, A. Fahr, and D. Backman, The British Institute of Non-Destructive Testing Insight Journal, June 2010.

[19] AFGROW Version 4.10.13.0, 02/03/2006, http://www.afgrow.net/downloads/ pdownload.aspx.

CHAPTER 4

Visual Inspection and Optical Methods

4.1 VISUAL INSPECTION

Visual inspection is the most widely used approach for detecting surface abnormalities and is universally practiced by aircraft maintenance people. Often the visual inspections are aided by the use of light sources and optical devices, such as magnifying glasses, borescopes, fiberscopes, video-scopes, and a variety of other optical imaging and image enhancement tools. Live video imaging is also used for viewing of the test site from a distance and is useful for monitoring time-dependent processes. A few examples of these devices and tools are described in this chapter.

Visual and aided visual tests are suitable for first-line inspections of aircraft to avoid the long downtime that is usually needed for conventional NDT. With visual inspections, however, only surface abnormalities, such as contaminations, surface-open cracks, impact dents, paint chip-offs, corrosion blisters, and cracked seals, may be seen. These could be indicative of underlining problems that would necessitate further examinations using other NDT methods. While fast, economical, and commonly practiced, the visual inspections rely on the inspector's acuity, experience, and judgment.

4.1.1 BORESCOPES

A borescope is a long tubular device that illuminates and magnifies the test site allowing visual inspection of hard-to-reach surfaces inside an aircraft airframe, structure, or engine. The tube is either rigid or flexible and comes in different lengths that provide optical connection between the viewing end and the tip of the borescope. A

Figure 4.1 Aided visual inspection of (top) an aircraft engine, (middle) a helicopter tail rotor blade, and (bottom) a portion of the Canadarm composite arm boom.

variety of commercial borescopes are available and some examples are mentioned below. They are used during aircraft manufacturing to verify the proper placement and fit of seals, bonds, gaskets, and subassemblies as well as during maintenance to detect corrosion, cracks, or other internal damage. They are commercially available in a wide variety of standard and customized designs with different focusing, illumination, magnification, working lengths, and field of view characteristics.

4.1.1.1 RIGID BORESCOPES

These devices use a rigid tube with a light source and a series of relay lenses to provide illumination and magnification of the inspection site. The tip can be either a fixed- or swing-type. They are easy to use but are limited to applications where access is through a straight-line path. Rigid borescopes come in different sizes

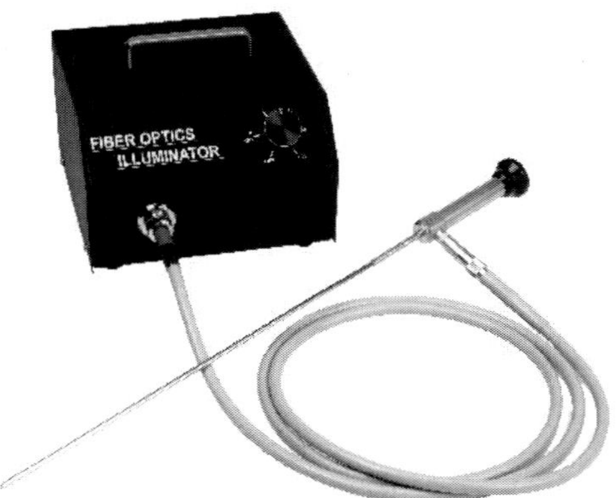

Figure 4.2 A typical commercial rigid borescope [1].

ranging from 0.15 to 30 m in length and 0.9 to 70 mm in diameter. Magnification is usually in the order of 4× but could be powered up to 50×. The viewing field often ranges from 10 to 90 degrees and the viewing head is either fixed or rotational. The depth of focus is usually fixed and may be quite large in most borescopes.

4.1.1.2 FLEXIBLE FIBERSCOPES

Fiberscopes are flexible instruments that use a non-coherent fiber bundle to transfer the illumination from the light source to the inspection area and a coherent fiber bundle to transmit the viewing image back to the inspector's eye. They often come with different interchangeable viewing heads or objective lenses and a remote control for articulation of the viewing head. Most fiberscopes have adjustable focusing capability that with the inter-changeable and articulated distal tips provide various fields of view at different directions during the inspection. Fiberscopes usually can be attached to digital imaging equipment for recording, play back, and analysis. Fiberscopes are available in lengths up to 12 m and diameters in the range of 1.4 to 13 mm.

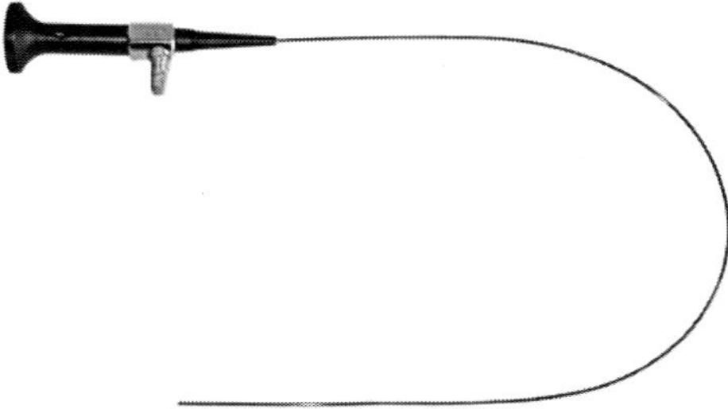

Figure 4.3 A typical commercial fiberscope [2].

Figure 4.4 A commercial videoscope with a flexible viewing tube [3].

4.1.1.3 VIDEOSCOPES

Videoscopes use a flexible tube with wiring to carry the image signal from a charge-coupled device (CCD) sensor at the distal tip to a video monitor. The CCD sensor consists of thousands of light-sensitive elements that convert the optical image obtained by the objective lens to electrical signals that form each picture pixel. Like the borescopes and fiberscopes, the resolution of videoscopes depends on the object-to-lens distance and the field of view as these affect the magnification and the area of coverage. Generally, videoscopes produce higher resolution images than the fiberscopes. Videoscopes are advantageous in terms of reducing picture distortion and eye fatigue. Also, image recording, enhancements, and integration with automated systems are possible with videoscopes.

4.1.2 APPLICATION EXAMPLES

A few examples of the application of visual inspection using borescopes in aerospace are presented in Figure 4.5 to Figure 4.7 and related descriptions are provided in their captions.

4.1.3 CAPABILITIES AND LIMITATIONS

Overall, visual and aided visual inspections have the following capabilities and limitations.

Capabilities:
- Fast and economical for viewing surfaces for contamination, corrosion, cracking, and other abnormalities.
- Suitable for hard-to-reach areas.
- Reduce costly tear-downs.
- Provide instantaneous views or images of the inspection site.

Limitations:

- Only surface abnormalities can be seen.
- Dependent on the inspector acuity, experience, and interpretation skills.
- Limited area of coverage.
- Sub-surface defects cannot be detected.

A number of enhanced optical techniques are available for viewing large components or small specimen surfaces, and a few that have shown potential applications in aerospace NDT are mentioned below.

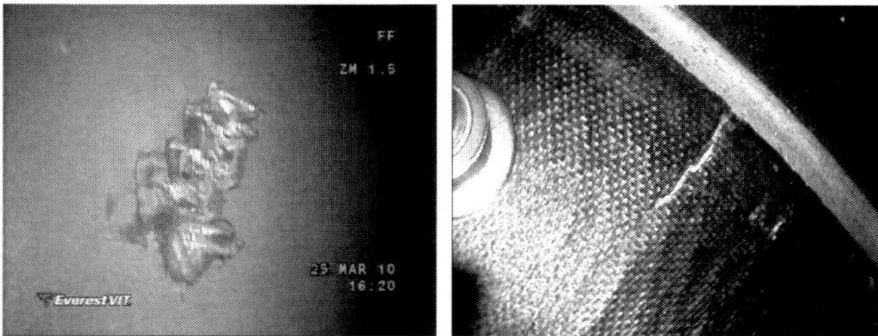

Figure 4.5 Borescope images of surface damage on a helicopter blade (left) and a crack in an aerospace composite tube (right).

Figure 4.6 A view of the interior of a space component (a segment of Canadarm) made of CFRP.

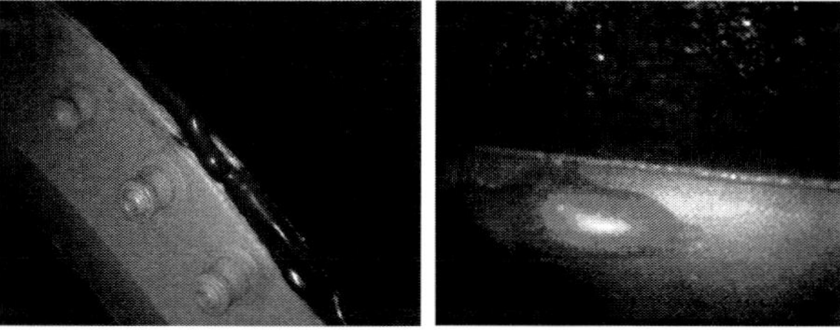

Figure 4.7 Borescope views of the interior of a space component showing insufficient sealant (left) and excessive sealant (right) between composite structure and metallic flange.

4.2 EDGE OF LIGHT (EOL)

4.2.1 PRINCIPLES

The EOL™ technique was developed and patented at the NRC Institute for Aerospace Research. It consists of a narrow light source and a detector held at a constant distance from the target [4]. A light source behind a slit is used to produce a rectangular band of light on the surface of the object to be scanned. In the main zone of the band, the intensity is almost constant, but at the edges it drops off rapidly with distance due to diffraction effects. The detector of the EOL™ scanner is set up to operate in the middle of this "edge of light" zone of rapidly changing intensity as seen in Figure 4.8. The detector is held at a constant position with respect to the light source and captures the light reflected off the surface. The intensity of the reflected light is constant when the surface is smooth. However, any out-of-plane surface distortions appearing in the edge of light zone change the angle at which the light is reflected from the surface. This alters the position of the "edge of light" band in relation to the detector changing the intensity of light captured by the detector.

EOL scanner performance can be optimized for a particular application by expanding or compressing the edge zone, varying the angle of illumination, or changing the viewing angle. The scanner travels along the surface of interest providing a high contrast map of the surface topography. A commercial portable EOL hand scanning device is shown in Figure 4.9.

4.2.2 APPLICATION EXAMPLES

EOL may be used to detect cracks and corrosion in metallic aircraft parts and impact damage in composite components. The plastic zone at the crack tip in some metals like aluminum creates sufficient surface deformation to produce the intensity shifts on the EOL scanned image. Figure 4.10 shows an EOL image of cracks emanating from fastener holes in an aircraft aluminum lap-joint specimen. Under optimal experimental conditions, the EOL technique may provide better crack detectability than the widely-used liquid penetrant and magnetic particle inspections. This was demonstrated in a study carried out on aircraft engine compressor discs containing low cycle fatigue cracks in their bolt holes [5, 6, 7].

In addition to cracks, corrosion of aluminum lap joints may also be detected by the EOL technique. The EOL lap-joint corrosion detection is possible when there is a surface bulging between the fasteners that hold the Al plates together. The bulging is usually created due to the underlying corrosion products that have a higher volume than the original aluminum. This deformation is known as "pillowing" and the EOL technique has the ability to enhance the visual appearance of this effect. Figure 4.11 is an EOL image of an aircraft lap-joint specimen showing surface pillowing due to corrosion of faying Al surfaces. In composites, barely visible impact damage (BVID) produced by foreign object strikes can be

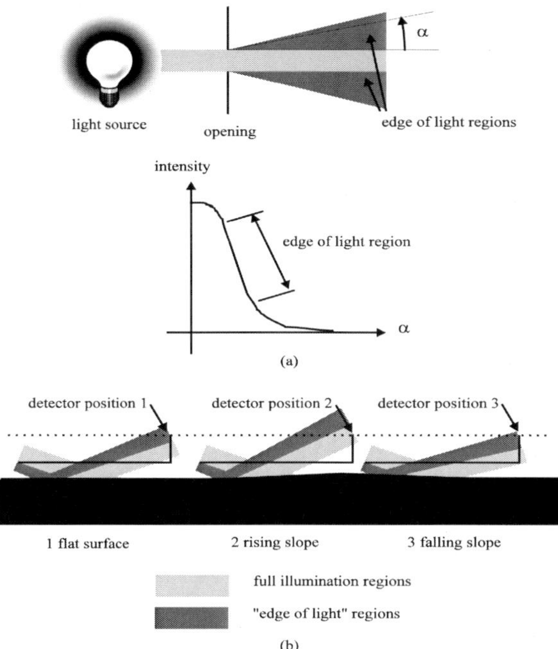

Figure 4.8 Schematic presentation of the principles of the edge of light (EOL) enhanced optical technique showing (a) light diffraction and the "edge of light" zone and a (b) change of the inspection surface slope that causes the edge of light zone to move with respect to the detector.

Figure 4.9 A prototype EOL hand scanner unit (image courtesy of RD Tech/Olympus Inc.).

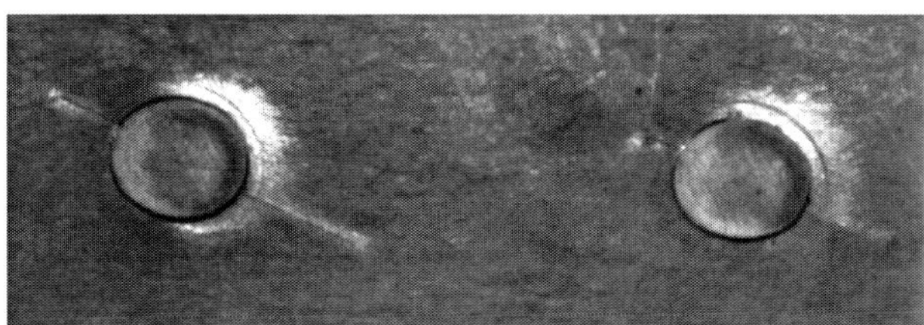

Figure 4.10 An EOL image of an aircraft outer skin showing cracks extending from the rivet holes. The rivet heads are about 7 mm in diameter.

identified by the EOL as shown in Figure 4.12. Such damage may not be visible to the human eye without the EOL enhancement.

4.2.3   *CAPABILITIES AND LIMITATIONS*

Capabilities:
*   A fast, economical, and easy method for viewing surface distortions.
*   Provides high contrast views of surface cracks, corrosion, BVD, and other surface abnormalities.
*   Suitable for first line inspection of large areas to identify surface features that may be indicative of subsurface damage.

Limitations:
*   Only sensitive to surface deformation.
*   Does not interrogate flaws directly, it only looks at the surface distortions that may be created by defects.
*   Internal damage that does not produce surface deformation cannot be detected.
*   Normal surface distortions, due to manufacturing or substructures, could be mistaken for defects, thus causing false calls.
*   Verification by other NDT methods is often necessary.
*   Standards do not exist for this approach, and the method is not universally accepted as a NDT method.

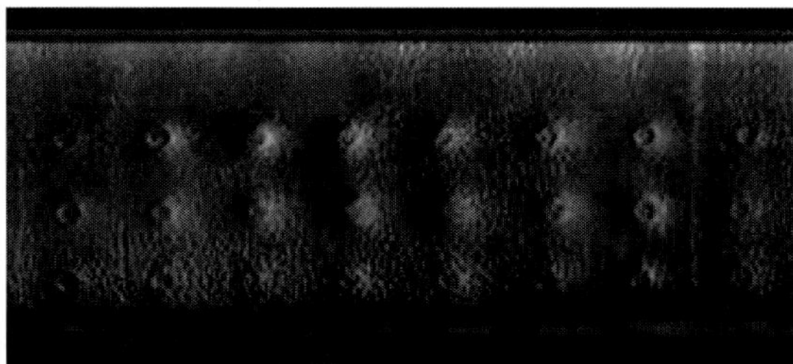

Figure 4.11 An EOL image of an aircraft lap splice joint showing pillowing deformation due to corrosion around rivet holes in most of the right half of the image [4].

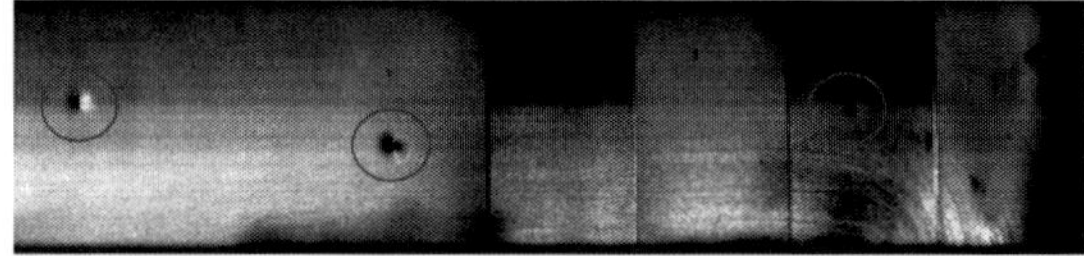

Figure 4.12 An EOL inspection of a section of a helicopter tail rotor blade, showing the enhancement of impact damage in the circled sections.

4.3 DOUBLE PASS RETROREFLECTION OR D-SIGHT

4.3.1 Principles

The D-sight technique is a double-pass retroreflection optical method that has been developed by Diffracto Ltd in Canada. Like the edge of light, it is a real-time, enhanced visual technique for viewing small out-of-plane surface distortions, such as dents, pillowing, and other deformations [8]. The D-Sight optical set-up consists of a light source, a camera, a retroreflective screen, and the test specimen (Figure 4.13). The surface of the test article must be reflective in order to return the light from a standard divergent light source. If the surface is not adequately reflective, a thin layer of fluid, such as water, can be applied to make it reflective. Alternatively, a thin transparent plastic foil may be placed vacuum-tight on the surface [9]. The light reflected off the surface of the test article strikes the retroreflective screen which consists of numerous half-silvered glass beads, typically 60 μm in diameter. The screen returns a diverging light cone instead of a single ray at the same angle onto the surface of the test article that creates the D-sight effect. The light is reflected again by the specimen and collected by a camera placed slightly off-set from the light source.

When the specimen is perfectly flat, the camera sees the specimen with a uniform light intensity over the surface. An out-of-plane surface distortion, on the other hand, will result in local intensity differences caused by the returning light from the screen. The light intensity returned by the screen glass beads is dependent on the viewing angle of the light cone relative to the incoming light ray; thus,

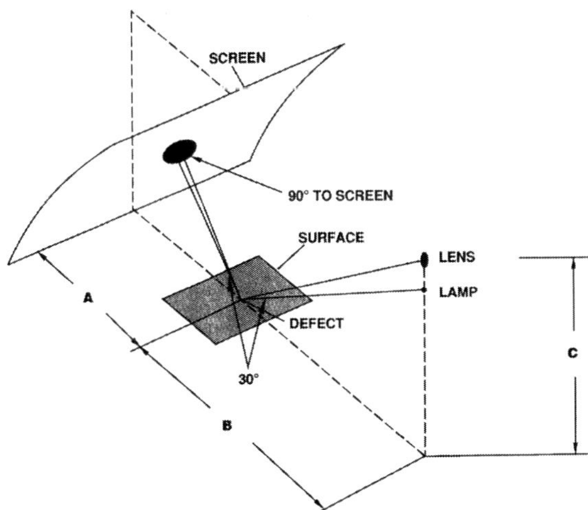

Figure 4.13 A schematic presentation of the principles of the D-sight enhanced optical technique.

the surface distortion will appear brighter on one side and darker on the opposite side of the surface distortion, depending on the local slope of the surface. The process converts variations in surface slope to changes in light intensity.

A commercial D-sight Aircraft Inspection System (DAIS) is available (Figure 4.14) that provides a digitized image of the inspection surface along with the position of possible defects. This device is portable and can be used both in the field or laboratory environment. Depending on the size of the inspection hood, areas close to a square meter can be inspected with reasonable sensitivity. This commercial hardware along with the related inspection database software allows quick surveys of large areas to be performed as a "triage," so that slower and more expensive NDT methods can be used to focus on suspect areas.

4.3.2 *APPLICATION EXAMPLES*

The D-sight is a simple viewing of the surface reflections at an oblique angle that enhances minute surface distortions. The approach can be used to inspect relatively large surface areas (flat or moderately curved) for surface distortions that may be indicative of underlying internal defects. This method may be used for inspection of accessible surfaces of aircraft structures, such as aluminum lap-joints, laminated composite parts, or sandwich components [10]. As an example, in Figure 4.15 the D-sight image of a lap-joint specimen is shown that indicates pillowing due to internal corrosion. An example of a DAIS survey of a complete aircraft vertical stabilizer is presented in Figure 4.16. This image shows the normal surface roughness associated with the manufacturing process of the composite skin along with a darker suspect area that requires further assessment by other NDT methods.

4.3.3 *CAPABILITIES AND LIMITATIONS*

Capabilities:
- A fast, economical, and easy method for viewing surface distortions caused by corrosion or impact.
- Useful for first line inspection of large areas to identify surface abnormalities.

Limitations:
- Only sensitive to surface deformation.
- Does not interrogate flaws directly, it only looks at the surface distortions that may be created by defects.
- Flaws that do not produce surface deformation cannot be detected.

- Surface distortions that are not created by defects could be mistaken for flaws, thus causing false calls.
- Verification by other NDT methods is usually necessary.
- Non-reflective or rough surfaces require application of a high-lighter (fluid or a thin transparent plastic).
- Standards do not exist for this approach, and it is not universally accepted as a NDT method.

Figure 4.14 The DAIS 500, with the inspection head in the background and pendant on the right (image courtesy of Diffracto Ltd.).

Figure 4.15 An example of a D-sight image from the inspection of a lap-splice joint with internal corrosion.

Figure 4.16 D-sight inspection of one side of an aircraft vertical stabilizer.

4.4 LASER SHEAROGRAPHY

4.4.1 PRINCIPLES

Laser shearography is based on speckle pattern shearing interferometry that measures the out-of-plane deformation gradient of an object under load to detect anomalies [11]. Figure 4.17 schematically illustrates the principles of the laser shearography approach. In this method, diffused light from expanded laser sources (e.g., laser diodes) illuminates the surface of the test part while a shearing interferometer (such as a Michelson interferometer) is used to laterally shear the return beam from the object, with respect to a reference beam, and create a sheared speckle image of the object. A CCD camera (CCD stands for charge-coupled device) monitors the generated light speckles.

Under either mechanical or thermal load, the inherent features of the structure and the internal discontinuities create variations in the specimen's local stiffness (or thermal behavior if thermal loading is used) that result in a different speckle distribution that is recorded by the CCD camera. By subtracting the original speckles obtained under no load from the new ones that are created under load, a fringe pattern will be generated that can be related to the out-of-plane deformation of the object surface. The fringe patterns are processed using specialized signal processing methods to provide images that are easier to interpret by the inspectors. Shearography equipment manufacturers use different ways to present the results; some examples are presented in Figure 4.18.

Laser shearography is a relatively new NDT method that has unique applications in aerospace. The ASTM has a standard practice (E-2581-7) for the use of shearography on polymer matrix composites, sandwich-core materials, and filament-wound pressure vessels employed in aerospace applications [12].

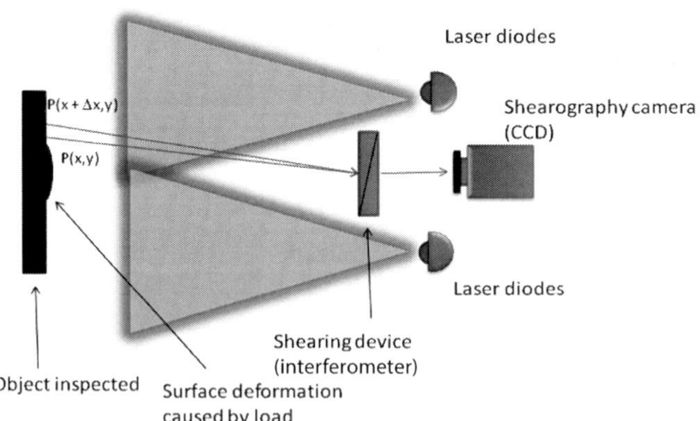

Figure 4.17 Schematic diagram showing the principles of the laser shearography. P(x, y) is the reference beam and P(x+Δx, y) is the return beam from the object.

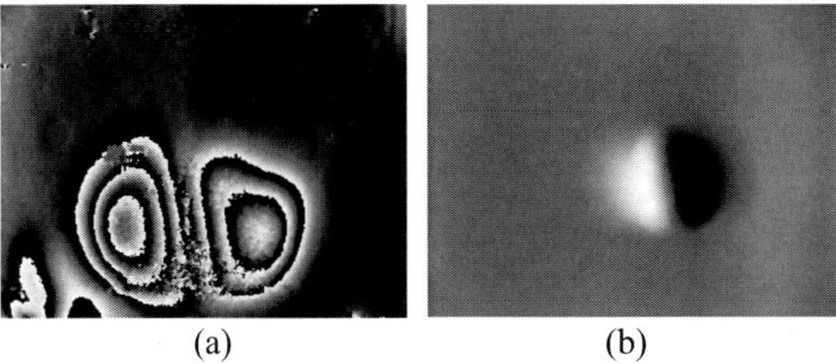

(a)   (b)

Figure 4.18 Examples of images created by shearography. (a) A sample with a disbond showing fringe pattern of shearography test. (b) Same as (a) after further signal processing using unwrapped phase map.

4.4.2 APPLICATION EXAMPLES

This approach was applied for the inspection of composite sandwich parts made of either aluminum or CFRP skins and Nomex® honeycomb core with the purpose of determining if the method was capable of identifying and sizing impact damage typical of those that occur during service [13], [14]. In this study, a commercial system equipped with both thermal and vacuum loading was used. A photograph of the shearography system employed is shown in Figure 4.19. The test results were compared with those of ultrasonic pulse-echo and through-transmission as well as pulse thermography to establish the capabilities and limitations of the laser shearography as compared to the more established procedures. Details of the comparison are provided later in Chapter 11 of this book.

Briefly, tests on the aluminum-Nomex® honeycomb panel indicates that both thermal and vacuum shearography were capable of identifying the impact damage [13]. However, thermal shearography indications were smaller and less visible than those of vacuum shearography as shown in Figure 4.20. The size difference was associated with the sensitivity of each approach to the different sections of the damage. Like pulse thermography and pulse-echo ultrasonic, thermal shearography was only capable of identifying the debonding of the top skin from the core as thermal waves dissipate quickly with the depth of penetration. In contrast, with vacuum shearography it was possible to detect the skin debonding as well as the core damage.

Since vacuum shearography worked better in identifying the entire impact damage on sandwich constructions [13], this approach was employed on a large structure simulating a portion of an aircraft fuselage. The structure was made of CFRP skins and Nomex® honeycomb core. The structure was impacted at several locations taking into consideration the types and energy levels of different damage types a typical fuselage structure encounters during its lifetime. The damage types included tool drops, hail impact, tire debris on runway, etc. The impact energies imparted on the specimen varied from 5 J to 80 J, with impact head size varying from 12.7 mm (0.5 in.) to 89 mm (3.5 in.) in diameter [14].

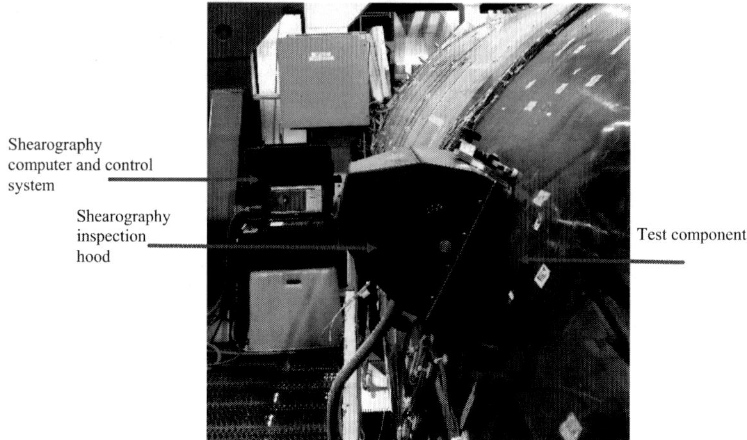

Figure 4.19 Photograph of a shearography experimental setup used to inspect the composite barrel.

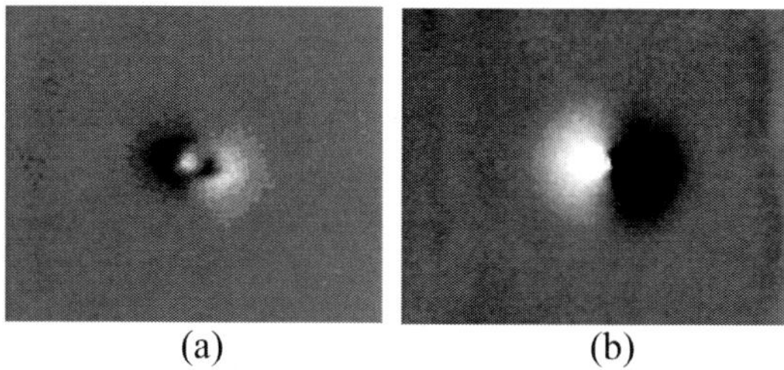

Figure 4.20 (a): Impact damage detected by thermal shearography. (b) The same impact damage detected by vacuum shearography.

A typical image of an impact-damaged area obtained by vacuum shearography and the corresponding line-profile through the middle of the damaged-zone are shown in Figure 4.21. Since the shearography indications do not clearly show the damage border, the line-profile was used to identify the extremities of the damage, as indicated by the small circles in Figure 4.21. From profiles, such as this one, the damage size could be estimated. The damage sizes obtained by vacuum shearography were consistently larger than those obtained by other NDT methods (e.g., pulsed thermography and pulse-echo ultrasonic). This was expected based on the previous study which showed that vacuum shearography indications correlate well with the overall damage, including the core crushing, while pulse thermography and pulse-echo ultrasonic traces relate only to the skin damage.

Overall, damage sizing using a shearography image without line-profiling can be subjective. Thus, a commonly accepted criterion (such as the line-profile used here) is necessary to identify the edge of the detected discontinuities automatically using an appropriate algorithm in order to obtain consistent size information.

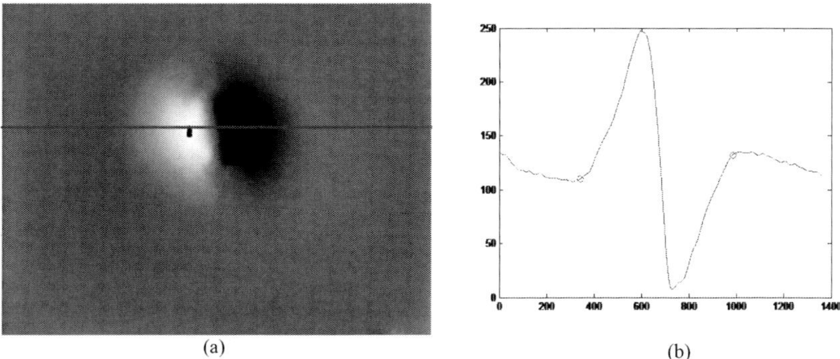

(a)   (b)

Figure 4.21 Shearography image of impact damage in a honeycomb composite parts and the related line-profile identifying the indication border. The measured damage size corresponding to the distance between the two circles is about 0.6 in. or 15 mm.

4.4.3 CAPABILITIES AND LIMITATIONS

Capabilities:
- A fast approach for identifying impact damage on honeycomb sandwich components.
- Access to only one side of the test object is needed.
- Damage size can be estimated by using vacuum shearography.
- Core damage in honeycomb structures may be detectable.

Limitations:
- Relies on localized load application to create surface deformation at discontinuity sites.
- Discontinuities that are very close to the surface can be identified while those located deep inside the material cannot be detected.
- Does not work well on very stiff materials.
- Discontinuity edges cannot be easily established from shearography indications.
- Size estimates are not precise.

4.5 CHAPTER REFERENCES

[1]　http://www.endoscopy4you.com

[2]　Boroscope It Inc., Manitoba, Canada, http://borescopeit.org.

[3]　http://www.olympus-global.com

[4]   The Edge of Light Enhanced Optical NDI Technique, D.S. Forsyth, J.P. Komorowski, A. Marincak, R.W. Gould, Canadian Aeronautics and Space Journal, Vol. 43, No.4, pp. 231–235, December 1997.

[5]   POD Assessment of NDI Procedures, Results of a Round Robin Test, A. Fahr, D.S. Forsyth, M. Bullock and W. Wallace, Proceedings of Review of Progress in Quantitative NDE, Edited by: D.O. Thompson and D.E. Chimenti, Plenum Press, New York, Vol.14B, pp 2391–8, 1994.

[6]   The Sensitivity and Reliability of NDI Techniques for Gas Turbine Component inspection and Life Prediction, D.S. Forsyth and A. Fahr, NRC Report: LTR-ST-2055, July 1996.

[7]   POD Assessment using Real Aircraft Engine Components, A. Fahr and D.S. Forsyth, Proceeding of the Review of Progress in Quantitative NDE, Edited by: D.O. Thompson and D.E. Chimenti, Plenum Press, New York, Vol. 17B, pp 2005–12, 1997.

[8]   A Technique for Rapid Impact Damage Detection with Implication for Composite Aircraft Structures, J.P. Komorowski, D. L. Simpson, R.W. Gould, Composites, Vol. 21, Issue 2, pp. 169–173, March 1990.

[9]   Method for Preparing Solid Surfaces for Inspection, R.W. Gould, J.P. Komorowski, United States Patent, 5,569,342, Oct. 29, 1996.

[10] Synergy between Advanced Composites and New NDI Methods, J.P. Komorowski, R.W. Gould, and D.L. Simpson, Advanced Performance Materials, Vol. 5, No. 1-2, January 1998, pp. 137–151, January 1998.

[11] Digital Shearography: Theory and Application of Digital Speckle Pattern Shearing Interferometry, W. Steinchen and L. Yang L, SPIE—The International Society for Optical Engineering, Bellingham, Washington, 2003.

[12] ASTM Standard Practice E-2581-7: Shearography of Polymer Matrix Composites, Sandwich Core Materials and Filament-Wound Pressure Vessels in Aerospace Applications, http://www.astm.org/Standard/index.shtml.

[13] NDE of Composite Structures using Ultrasonic, Thermography and Laser Shearography, M. Genest, M. Brothers, R. LeBlanc, and A. Fahr., Proceedings of SAMPE Conference, Seattle, WA, May 17–20, 2010.

[14] Pulsed Thermography and Laser Shearography for Damage Growth Monitoring, M Genest, R.S. Rutledge, R. Kothari, K. Mourtazov, and C. Turgeon, Proceedings of the International Workshop on Smart Materials, Structures & NDT in Aerospace, 2–4, Montreal, Quebec, Canada, November 2011.

CHAPTER 5

Liquid Penetrant, Replication, and Magnetic Particle Methods

5.1 LIQUID PENETRANT INSPECTION (LPI)

5.1.1 *PRINCIPLES*

The basic steps of a liquid penetrant inspection process are schematically illustrated in Figure 5.1 [1]. First, the inspection surface is thoroughly cleaned. Then, a bright-colored fluid (penetrant) is applied to the surface and allowed to seep into surface-breaking discontinuities. After that, the penetrant is removed from the surface leaving some residuals inside the surface-breaking discontinuities. Finally, the surface is examined under an appropriate lighting condition to reveal discontinuities. In some LPI processes, a developer (e.g., a fine powder dust) is applied to draw the penetrant residuals out of flaws in order to enhance their visibility [2].

The LPI process is only suited for the detection of surface-breaking discontinuities in smooth and non-porous surfaces. Cracks, pores, laminations, laps, seams, and cavities in a variety of metallic and non-metallic components can be identified using the LPI method. It is a fast and economical approach that is used extensively in the inspection of aircraft parts, particularly engine components, both after production and in service. A number of ASTM Standards are available for the liquid penetrant testing where standard practices and requirements for the different penetrant and developer types or applications are provided in detail [1–11].

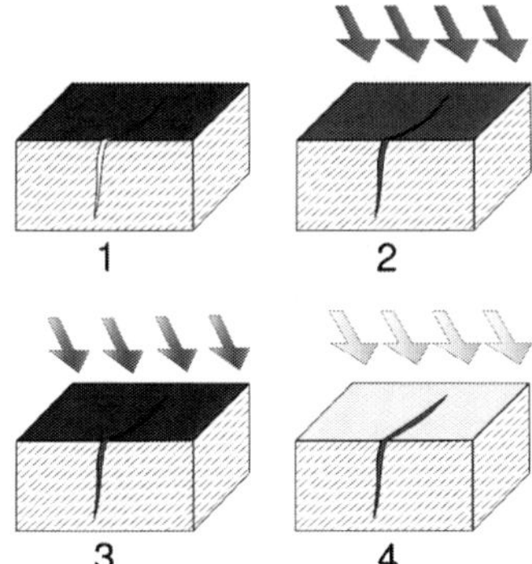

Figure 5.1 A simplified LPI process involving: 1—surface cleaning, 2—penetrant application, 3—removal of excess penetrant, and 4—development and viewing [1].

5.1.2 LPI Process

5.1.2.1 Surface Preparation

All surfaces to be inspected are first prepared to ensure that they are clean, dry, and free of dirt, oil, grease, and other contaminants that could prevent the penetrant from entering discontinuities. If surfaces are covered by paint, sealants, corrosion products, or coated with chemicals, these must be removed to allow the penetrant to flow freely into the discontinuities. One or more cleaning methods, such as solvent-cleaning, vapour-degreasing, ultrasonic vibration, or aqueous-based, may be needed. When these methods are not fully effective, etching, mechanical cleaning, grit blasting, or the use of chemicals may be necessary. After cleaning, the surfaces are dried to remove the residuals of the cleaning agents from defects. Drying may be done by forced air circulation in an oven or using a radiant heat source. The latter is used in automated LPI systems.

5.1.2.2 Penetrant Application

The penetrant fluid is applied by spraying, dipping, brushing, or other methods to cover the entire inspection surface and let fluid seep into the surface-open flaws. The temperature of the component, penetrant, and test environment is controlled to prevent early drying of the penetrant. A minimum dwell time of 10 minutes is needed for normal ambient temperatures (~25°C). For lower or higher temperatures (or humidity levels), a different dwell time may be necessary. The recommended dwell time for the different penetrant types is provided by their suppliers.

There are two types of penetrants: fluorescent (type I) and visible (type II).

Type I Penetrants are usually green in color and fluoresce brilliantly under an ultraviolet light. Their sensitivity varies and is classified into five levels:

- Ultralow: Level ½
- Low: Level 1
- Medium: Level 2
- High: Level 3
- Ultrahigh: Level 4

Penetrants with the higher sensitivities are also more costly. When fluorescent penetrants are used, the technique is sometimes referred to as fluorescent penetrant inspection (FPI). Fluorescent penetrants are generally used in production inspections where the entire part can be immersed in the penetrant tank to cover all surfaces at once. They are also available in spray cans for in-service localized inspections.

Type II Penetrants are usually red and produce vivid indications under white light that are visible to the naked eye. The sensitivity range for this type of penetrants is limited. Visible penetrants are also available in spray cans for in-service inspection of small areas.

The choice of the penetrant depends on the surface condition of the test piece, the required sensitivity level, and the available processing facilities. The method of application may vary from one product to another, and manufacturers often provide directions for their product use that must be followed carefully.

5.1.2.3 Removal of Excess Penetrant

Excess penetrant is removed from the surfaces either by water-washing or using a suitable solvent, depending on the type of penetrant used. It is important that the penetrant is properly and adequately removed from the surfaces but not from the flaws. Over-washing of the surfaces may result in missing the flaws, and under-cleaning of the excess penetrant could create false indications.

Water-washable penetrants may be removed with a manual wipe, water-spray, or an air agitated immersion wash. The water temperature and pressure are kept at the levels recommended by the penetrant suppliers. An automated water spray may be employed at prescribed pressures and temperatures to speed up the process.

Solvent-removable penetrants are removed by first wiping the excess penetrants with a clean, lint-free, dry cloth and the process is repeated with a solvent-dampened cloth. In this case, the component surface is not flushed or the cloth is not saturated with solvent to avoid over-removal of the penetrant from discontinuities.

If a developer is required, the component is drained of excess water before the application of the developer. It is important that the penetrant residuals inside defects are not completely evaporated. The component is then either air-dried or placed in an oven at a specified temperature and drying time depending on the developer type.

5.1.2.4 DEVELOPMENT

The process of development involves the application of either a dry or a liquid developer to the inspection surface in order to draw the penetrant out of the surface-breaking flaws. This additional step is to enhance the visibility. A few Type I penetrants that are qualified to MIL-I-25135 could be used without a developer in some applications. The majority of Type I and all Type II penetrants require the use of a developing agent.

There are different types of developers including:

- dry powder
- aqueous (water-soluble or suspendable)
- non-aqueous (solvent-soluble or suspendable)

Dry powder developers are used with fluorescent penetrants (Type I). They are applied by spraying the part with the powder dust in a chamber or spray booth. After a prescribed dwell time, the excess developer is removed by light tapping or air blow-off at low pressures (< 5 psi). The dry developer dwell time may be from 10 minutes to 4 hours.

Aqueous water-soluble developers are used only with Type I penetrants, but aqueous water-suspendable developers are used with both fluorescent and visible penetrants. However, they are not recommended for use with water-washable penetrants. The water-suspendable developers are supplied in a dry powder form, which is then dispersed in water in recommended concentrations. The aqueous developers can be applied after water rinsing of the components by spraying, flowing, or immersing. The component is then either air-dried or oven-dried for a required time.

Non-aqueous developers are used with both Type I and Type II penetrants. They are applied by spraying the component surface with a thin, uniform coat. The uniformity and thickness of the developer coating is important for both types of penetrants. The dwell time could vary from 10 minutes to 1 hour. Non-aqueous developers are the most sensitive of all the developers as the solvent contributes to the penetrant absorption mechanism. In the case of very deep or tight cracks, the dry powder or water-soluble developers may not be able to reach the penetrant residuals inside the crack but a non-aqueous solvent-based developer can be dissolved into the entrapped penetrant, decreasing the viscosity while increasing the volume that together help to bring up the penetrant to the surface.

Immediately after the developer is applied, the penetrant trace is almost the same as the surface-size of the flaw, but it will gradually increase with time lapse due to the absorption or spread of the penetrant in the developer. Thus, for size measurements, it is important to inspect the surface immediately after the development process is complete.

5.1.2.5 INSPECTION

Finally, the surface is examined under a suitable lighting condition for indications of the bleeding of the penetrant. For visible penetrants, ambient white light is adequate, but for fluorescent penetrants, a dark room and an ultraviolet light are necessary. In the latter case, the inspector's vision must be allowed to adapt to the darkness of the inspection booth. Also, the ultraviolet (black light) intensity must meet the specified requirements of the particular standard used. The indications found during inspections are evaluated in accordance with the specified accept/reject criteria. If the inspector is uncertain about some indications, they can be verified by wiping the suspect traces with a dampened cloth and repeating the development process. If no indication appears, the original traces may be considered to be false. Finally, an estimate of the surface size of the flaws can be made from the penetrant traces.

5.1.3 LPI METHODS

There are four basic LPI methods depending on the penetrant type and removal approach:

5.1.3.1 METHOD "A" — WATER-WASHABLE PENETRANTS

In this case, the penetrant fluid is of a type that can be removed by water rinsing after sufficient dwell time. The water-washable types can be applied easily and quickly, but they require very careful removal of the excess penetrant to avoid over-washing. Method "A" can be used for both small and large pieces as a fast and economical approach. However, shallow flaws may be missed as a result of the penetrant being washed out. Also, very tight cracks may not be detected due to the low sensitivity of the penetrant fluid.

5.1.3.2 METHOD "B" — LIPOPHILIC POST-EMULSIFIABLE PENETRANTS

This type of penetrant is insoluble in water by itself and is made to be selectively removed from the surface by the use of a separate emulsifier. The emulsifier differs from a solvent or cleaner in that it does not remove the penetrant, but when applied and left for a specific time, it diffuses into the excess penetrant and makes it water-washable. This allows the excess penetrant to be rinsed off the surface of the test piece with water. Method "B" is more time-consuming than method "A" but has less risk of over-washing and is more suitable for detecting small and shallow flaws. However, emulsification is an additional process, which increases the time and the cost of inspections.

5.1.3.3 METHOD "C" — SOLVENT-REMOVABLE PENETRANTS

This type of penetrant is only soluble in a solvent, and, thus, the excess penetrant on the surface of the part can be removed by simply wiping the surface with a clean, lint-free cloth that is lightly moistened with an appropriate solvent. This approach is most often used for in-service inspection of localized areas of aircraft structures or components that can be easily accessed. It is more sensitive than the water-washable method but requires additional time for careful removal of the penetrant.

5.1.3.4 METHOD "D" — HYDROPHILIC POST-EMULSIFIABLE PENETRANTS

This approach is the same as method "B" except that the emulsifier is water-based and acts to displace the excess penetrant on the surface of the part by detergent action. Hydrophilic emulsifiers are slower-acting than the lipophilic type and provide a better control of the cleaning process. Method "D" also requires a pre-rinse step by water spray before the application of the emulsifier to remove the excessive penetrant from the surface leaving only a very thin uniform film that is removed by the emulsifier. The liquid penetrant manufacturers often recommend nominal concentrations and emulsification times for specific penetrants and emulsifiers that they supply. Method "D" is more sensitive than the lipophilic approach. It is widely used at aircraft engine maintenance facilities that require a large number of parts to go through the individual LPI processes at the same time.

5.1.4 EQUIPMENT

LPI instruments vary from portable aerosol spray cans to fully automated stationary facilities. Sprays in small containers or cans are generally used for in-situ inspection of localized critical areas of aircraft. They are commercially available in kits that include the necessary penetrants, cleaners, developers, and accessories, along with instructions for their use. An example is provided in Figure 5.2. On the other hand, modular stationary systems are mostly used in laboratories and aircraft manufacturing or maintenance facilities. They typically consist of several dip tanks for penetrant, emulsifier, and developer application; a drain and dwell area; a washing area; a drying oven; and a viewing area or inspection booth. Figure 5.3 shows a typical stationary LPI facility.

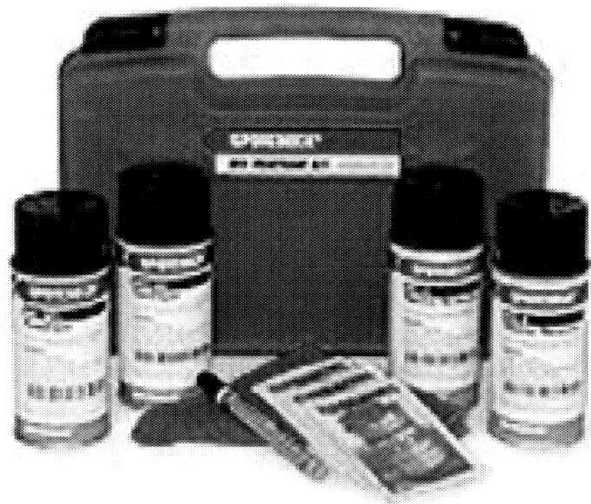

Figure 5.2 A commercial LPI kit from Magnaflux.

Figure 5.3 A typical LPI station (courtesy of the Canadian Institute for NDE).

Automated processing systems are similar to stationery ones except that the penetrant application, washing, and drying processes are done by a machine instead of a human. However, the actual inspection under ultraviolet light and interpretation of the indications are carried out by a qualified person. Such facilities are used for inspection of large numbers of parts in production or maintenance facilities.

Overall, the LPI method is very process and operator dependent. As such, the sensitivity and reliability of this approach may vary significantly with the type of penetrants, developers, process, and facilities employed as well as the operator's skill and judgment. Further discussion on this topic is provided later in the chapter dealing with the subject of NDT reliability. In summary, the following factors are important to the effectiveness of the LPI method:

* Smoothness and cleanliness of the surface to be inspected.
* Size and cleanliness of the surface-breaking discontinuities.

- The use of a proper liquid penetrant, developer, and process.
- The ability of the inspector to carry out the test and interpret the results.

Finally, it must be noted that the chemicals used in the LPI approach may pose safety problems, such as toxicity or flammability, and they must be handled properly. Also, proper disposal of the waste chemicals is very important and must be considered as a part of the overall cost.

5.1.5 *APPLICATION EXAMPLES*

The following pictures provide a few examples of the application of LPI to engineering components [1]. Description of the components is provided in the respective caption.

5.1.6 *CAPABILITIES AND LIMITATIONS*

Capabilities:
- Simple, fast, and inexpensive compared to other NDT methods.
- Applicable in production and maintenance environments as well as in the field.
- Provides visible indications of surface-breaking discontinuities.
- Appropriate for the detection of surface-breaking cracks.
- An estimate of the surface size of cracks can be made.
- Large surface areas or large numbers of parts can be inspected.
- Applicable on a wide variety of materials and component geometries.
- Limited training is required

Limitations:
- Internal discontinuities cannot be detected.
- Surface must be cleaned.
- Rough or porous surfaces cannot be inspected.
- Not suitable for very shallow or wide defects.
- Cannot give flaw depth information.
- Post-inspection cleaning is necessary to remove chemicals.
- Requires multiple processes under controlled conditions.
- Chemicals used pose safety problems (e.g., toxicity, flammability) if not handled properly.
- Proper disposal of the waste chemicals is essential.
- Reliability may be an issue especially for small defects.

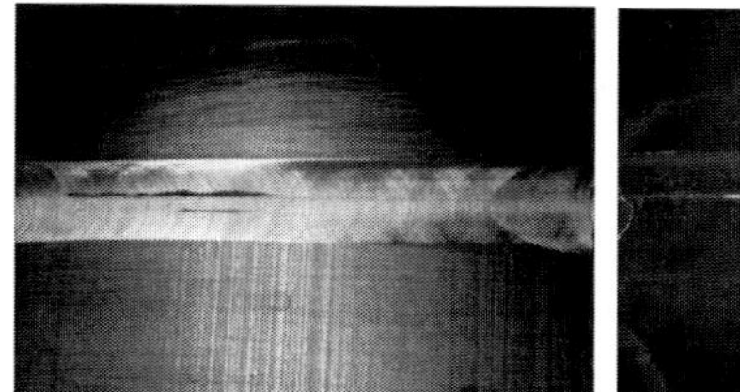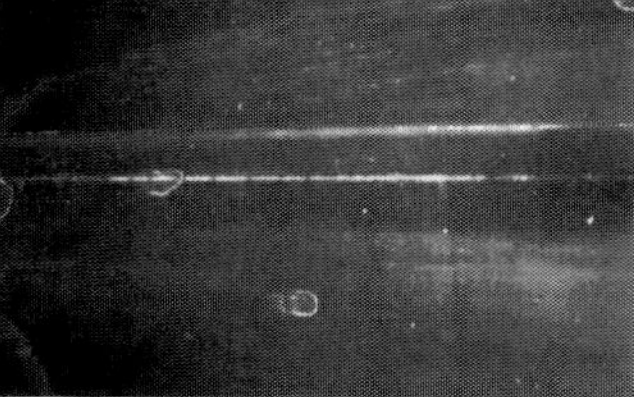

Figure 5.4 Liquid penetrant inspection of a weld using a dye penetrant (left) and a fluorescent penetrant (right) showing surface-breaking cracks.

Figure 5.5 Fluorescent penetrant inspection (FPI) of a friction stir weld showing manufacturing voids and pores.

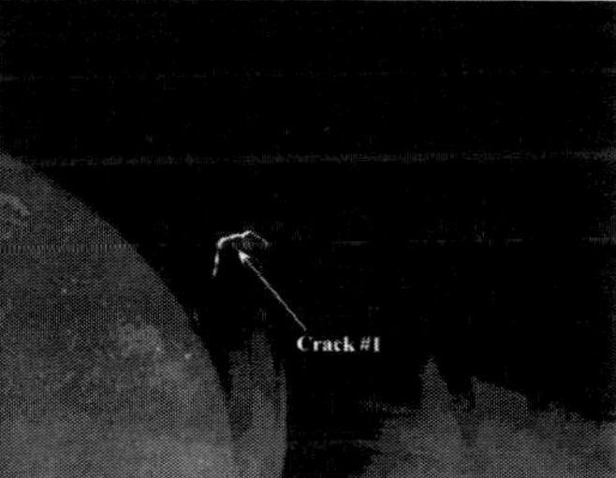

Figure 5.6 Fluorescent penetrant inspection (FPI) of an aircraft engine mandrel (left) and a spacer (right) showing service-induced cracks.

5.2 REPLICATION

5.2.1 PRINCIPLES

In this technique, an exact replica of the surface is produced for examination under an optical microscope in the laboratory. This technique is occasionally used as a substitute for optical techniques in situations where direct viewing of the inspection site is not possible or easy due to limited access (e.g., thread roots, bolt holes) or geometry (too small for NDT or too large for microscopy). Replication is employed as a means of viewing component surfaces at high magnifications for defects, surface finish, or microstructure. It is also employed for the verification of NDT results and measurement of the defects surface size.

In this method, special viscous compounds that easily fill surface cavities and harden after a time period are used to create a negative copy of the surface topography. Replication of the surface features is widely used in archaeology and has evolved from plaster casts of dinosaur footprints to fast-setting dimensionally stable silicon-based replicas of defects in aircraft structures.

5.2.2   REPLICATION MATERIALS

Cellulose acetate film is a replicating material that softens in acetone. It is applied to a surface which has been wetted with acetone and stripped off when dried for examination under a microscope. Thicknesses of 20–35 μm are used for replication of fine surfaces and thicker sheets (up to100 μm) are employed for replicating rough surfaces. Surface-breaking discontinuities a few microns in size can be detected. More recent materials have the capability of replicating features on the scale of nanometers.

Pressure-sensitive films produce fast and inexpensive replicas by rubbing or burnishing a piece of foam-backed film to make it conform to the sample surface. They are primarily used for rough surfaces (e.g., grit-blasted metals) and are useful for inspecting sheet products and vertical or inverted surfaces.

Vinyl polysiloxane, originally developed for dental impressions, flows easily and has a short curing time. It also has good dimensional stability and provides a resolution in the order of one micron. The two-part material is mixed and applied to the surface by hand and provides a cost-effective solution for replication of large surfaces. Recently, two-part silicone rubbers have been developed specifically for high resolution replication. These compounds, which are typically applied with a dispensing gun (Figure 5.7), can reproduce features down to 0.1 μm or below. Some can be used over a wide temperature range and even can be applied under water for corrosion, wear, and damage detection.

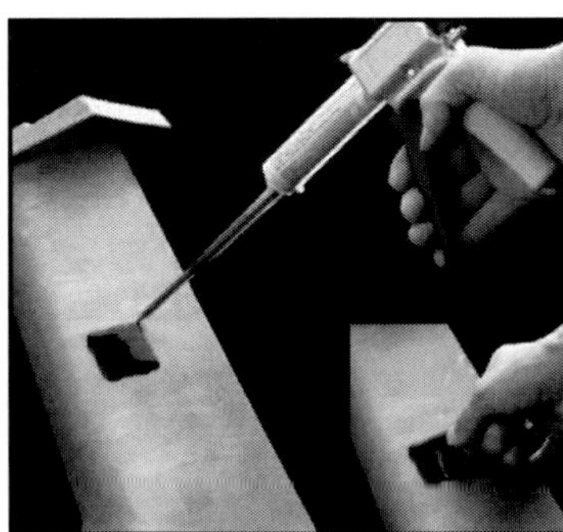

Figure 5.7 Surface replication using a commercial dispensing gun and silicon rubber [14].

5.2.3 APPLICATION EXAMPLES

The aerospace applications of the replication method include measurement of the surface size of cracks, dents, scratches, and corrosion pillowing as well as field metallography. For the latter, a replica of the test site (e.g., fractured surface) is produced for detailed examination under microscope in the laboratory. The approach is widely used to produce a permanent record of crack growth or corrosion progression. Figure 5.8 illustrates replicas of a corner crack that originated at the edge of a bolt-hole in an aluminum sample. In this figure, the surface length of the crack, as seen on the top surface of the sample and inside the bolt-hole, is shown. The difference in the surface length indicates that the crack is in the shape of a quarter of an ellipse. The scanning electron micrograph of the fracture surface of the same sample is illustrated in Figure 5.9 that provides the actual crack profile and verifies the replica finding.

Figure 5.10 shows the first and second replicas of an aircraft lap-joint component containing corrosion. The second replica is a replica of the first one, and, therefore, directly resembles the original subject. It illustrates the rivets and fasteners of the lap-joint, as well as the corrosion pillowing deformation between them.

Replication is also used for field metallography when the part cannot be removed or sectioned for laboratory tests. The ASTM E 1351-01(2006) Standard

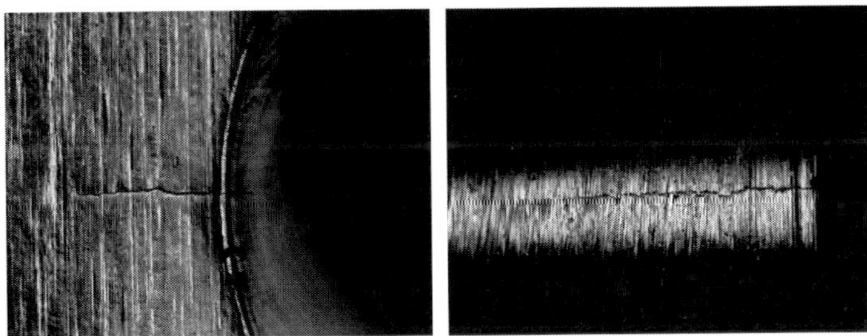

Figure 5.8 Replica of a bolt-hole fatigue crack in an aluminum sample as seen on the surface of the sample (left) and inside the bolt hole (right).

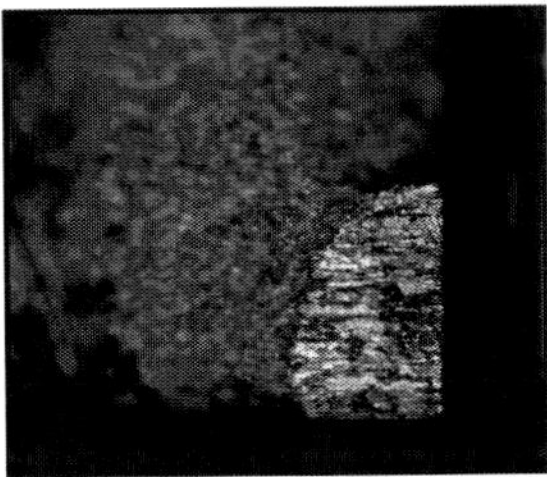

Figure 5.9 Micrograph of the above crack obtained after pry-opening and examination under a scanning electron microscope (SEM).

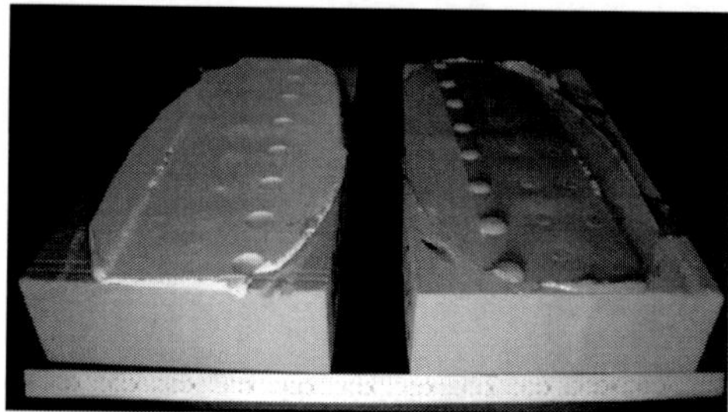

Figure 5.10 A replica of a section of an aircraft aluminum lap joint (left) and a replica from the first replica (right) which provides a record matching the actual test specimen. The replicas are bonded to foam blocks for ease of handling.

provides the details of the approach. Figure 5.11 shows the microstructure of a steel sample obtained by optical microscopy of the actual sample and its replica. The two micrographs are taken from different regions of the same sample.

5.2.4 *CAPABILITIES AND LIMITATIONS*

Capabilities:
- A simple and inexpensive approach to produce replicas of surface discontinuities.
- Provides a permanent trace of surface features for microscopy and size measurements.
- Can be applied in the laboratory or in the field if the test site is accessible.
- Applicable to different solid materials.
- Replicas can be archived for future reference.
- Limited training is needed.

Limitations:
- Replicas must be examined by optical or scanning electron microscopy.
- The inspection surface must be accessible.
- Limited to surface discontinuities.
- Applicable to small area inspections.
- Time consuming and labor intensive.

5.3   MAGNETIC PARTICLE INSPECTION (MPI)

5.3.1 *PRINCIPLES*

When a ferromagnetic material is magnetized, the presence of surface or near-surface discontinuities that are nearly perpendicular to the direction of the magnetic field will cause a magnetic leakage at the surface of the part as illustrated

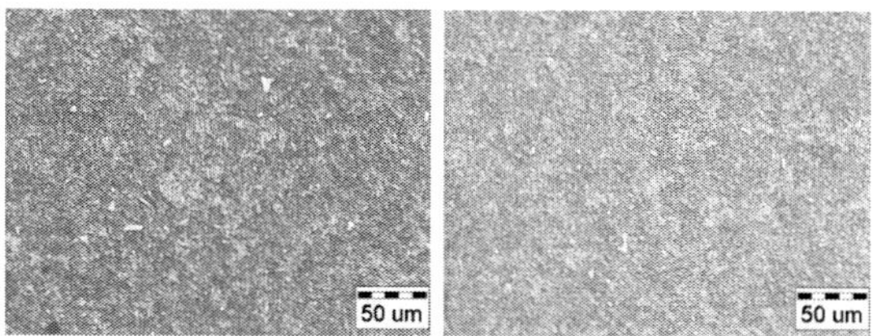

Figure 5.11 Microstructure of a steel sample (left) and replica of the same material of a different site (right) [14].

schematically in Figure 5.12 [1]. A small quantity of fine magnetic powder is applied over the surface of the part while it is still under the influence of the magnetic field. The magnetic particles gather and are held where the leakage occurs forming an outline of the discontinuity. In this way, the location of the discontinuity is identified and its general shape is traced [12].

Ferromagnetic materials, including alloys that contain iron, nickel, or cobalt, can be inspected using the MPI method for surface- and near-surface discontinuities. MPI is used during manufacturing processes for inspection of castings, forgings, extrusions, welds, etc. Parts such as metal billets, tubes, bars, and shafts can be inspected. It is also used during service to inspect engine or gear box components that are ferromagnetic. A number of ASTM Standards are available for the MPI and some are provided in references [15] to [17]. Generally, an MPI process consists of the following three steps:

- First, a suitable magnetic flux is established in the test object by magnetization.
- Then, magnetic particles, either in a dry-powder form or in a liquid-suspension, are applied.
- Finally, the test object is examined under a suitable lighting condition.

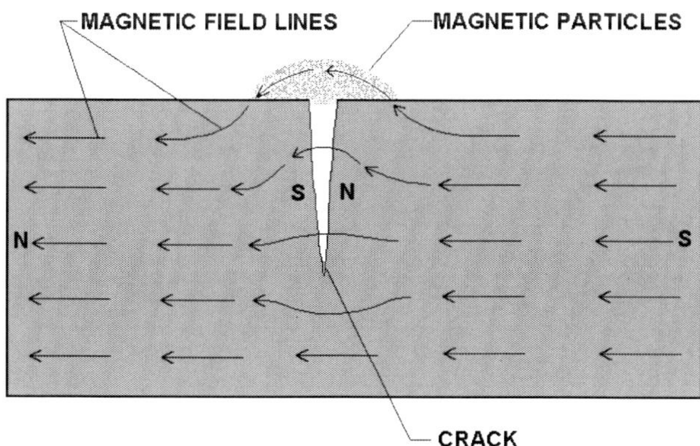

Figure 5.12 Schematic diagram showing the principles of magnetic particle inspection [1].

5.3.2 MAGNETIZATION METHODS AND DEVICES

The magnetic field can be applied either locally or to the entire part. The basic requirement is that the discontinuities must be between the two poles of the magnet and lie close to perpendicular to the line connecting the poles. Maximum magnetic leakage normally occurs when discontinuities are perpendicular to the direction of the magnetic field. Since discontinuities may exist in various unknown directions, each part must be magnetized in at least two directions approximately at right angles to each other. The generated magnetic field may be longitudinal, transverse, circular, or multidirectional. Depending on the component shape and whether or not all or only part of the part requires inspection, different magnetizing devices can be used to achieve the appropriate magnetic field. The devices may be made of permanent magnets, electromagnets, flexible cables carrying an electrical current, hand-held prods, or bench magnetizing units. The most commonly used devices are described below.

5.3.2.1 YOKES

Yokes are portable hand-held devices made of either permanent magnets or electromagnets (Figure 5.13). Permanent magnets are suitable for on-site inspections where electrical power is not available or where arcing is not permitted. They have a constant low magnetic strength and can be separated from the part easily. They are most effective for small area inspections but have very limited magnetization depth.

Electromagnetic yokes consist of a coil wound around a U-shaped iron core that can be turned into a magnet by introducing an electric current using an on-off switch. Both direct and alternating currents can be used. The DC current enables easy control of the flux density and the magnetization depth providing different depths of penetration. The use of high flux densities enables one to detect deeper subsurface flaws. On the other hand, the magnetic field introduced by an AC current is mainly at the surface and, therefore, provides better sensitivity for

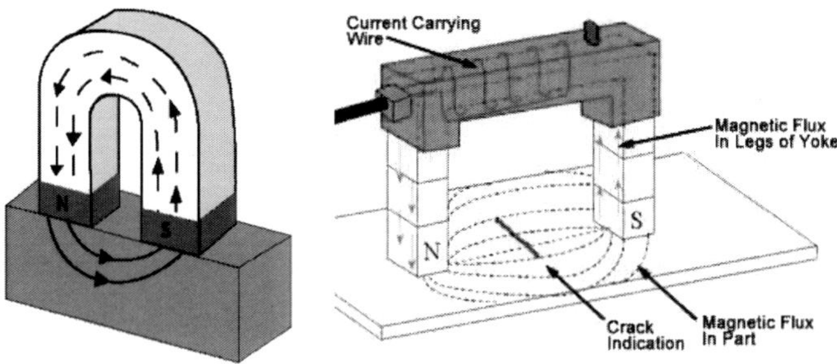

Figure 5.13 MPI using a permanent magnetic yoke (left) and an electromagnetic yoke (right) [1].

surface-breaking discontinuities over large areas. The legs of the yoke can be made adjustable to accommodate complex parts and their magnetic strength can be varied or adjusted to suit the coverage area. They can be easily removed from the part once the current is switched off. For these reasons, electromagnetic yokes are preferable to permanent magnets and are widely used.

5.3.2.2　Coils

Single- or multiple-loop coils are used for longitudinal magnetization of long cylindrical parts such as rods, tubes, and bars (Figure 5.14). The flux density in the coil is proportional to the product of the current, I, and the number of turns in the coil, N. Therefore, the magnetic strength of such devices can be varied by changing either I or N. Generally, portable coils provide high magnetizing forces by using a higher number of turns to compensate for their lower current flow, while bench equipment use capacitor discharge units that provide high currents for a very short duration. Coils are used for inspection of transverse cracks in shafts, spindles, and rods.

5.3.2.3　Conductors

For inspection of tubes or ring-shaped parts, it is sometimes easier to use a separate conductor placed inside of the tube or the ring to carry the magnetizing current rather than the part itself. Central conductors produce a magnetic field perpendicular to the current path that is uniform along the length of the conductor with the flux density inversely proportional to the distance from the center of the conductor. The current is applied to the conductor by clamping it between the contact heads of a bench MPI unit as seen in Figure 5.15. Cracks oriented in the same direction as the conductor can be detected. For transverse cracks that require longitudinal magnetization, the part is placed within a ring-coil of the bench unit.

Figure 5.14 Longitudinal magnetization using coils [18].

5.3.2.4 PROD CONTACT

For inspection of parts that are too bulky to be tested in a bench unit or for in-situ inspection of large structures, magnetization is done using prod contacts (Figure 5.16). In this approach, the current is passed through the part, or a portion of the part, producing magnetic fields between and around the contact points. Hand-held electrical prods are useful in confined spaces and are used in the inspection of welded structures. However, they suffer from two major disadvantages. First, arc strikes can occur at the prod contact points that can damage the objects' surface. This can be minimized by using aluminum or copper pads instead of solid copper rods and turning off the device before removing the prods. Second, because the particles must be applied when the current is on, the inspection becomes a two-man operation.

5.3.2.5 DIRECT CONTACT

For parts that have no openings, a circular magnetic field is produced by clamping the part between the contact heads of a bench unit that applies the current directly through the part. This approach is used to detect cracks oriented in the same direction as the line connecting the two contact points. For complete inspection of complex parts, it is necessary to attach clamps at several points on the

Figure 5.15 Magnetization of a part using a central conductor [15].

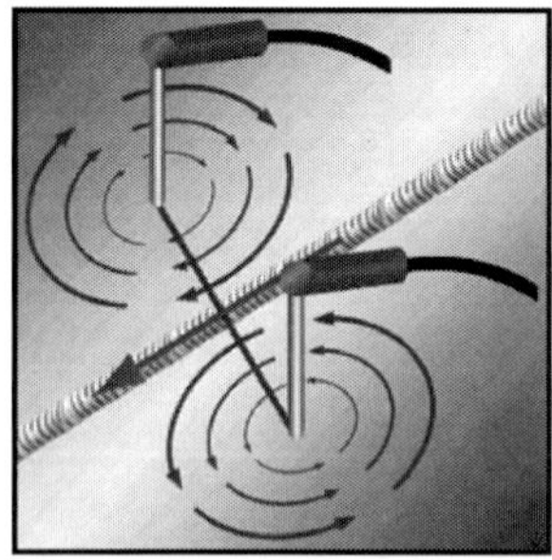

Figure 5.16 Magnetization using a pair of prods in contact with the part [1].

part. Direct contact bench units are fixed installations used to test large numbers of manufactured parts of various sizes. The electrical components of a hand-held prod are incorporated into a bench unit making testing more rapid, convenient, and efficient for a one-man operation (Figure 5.17).

5.3.3 TYPES OF MAGNETIC PARTICLES

The particles used in MPI are basically finely-divided ferromagnetic materials, such as iron or iron oxide powders. These compounds are magnetized quickly due to their high magnetic permeability and are attracted to the sites of the flux leakage. At the same time, they have the ability to avoid being attracted to each other. The size and shape of the particles are important in obtaining consistent results. The particles are applied over the inspection surface either as dry powder or in a liquid carrier, such as water or light oil. The use of a liquid is to enable the particles to flow over the surface and to migrate to the flux leakage sites faster and easier. The particles are often coated with a color (black, grey, red, or yellow) or a fluorescent material that enables them to be seen easily under the ambient light or an ultraviolet (UV) lamp. Part surfaces may be coated with a thin layer of white paint to increase the contrast between the background and the black or colored magnetic particles.

5.3.3.1 DRY PARTICLES

Dry particles are applied by spraying or dusting the inspection surface with two or three short bursts of the powder (1/2 second) to help improve particle mobility. They are best suited for use with portable equipment during local magnetization of large objects. In this case, while the magnetizing force is still applied, the excess particles are carefully blown off the surface with a few gentle puffs of dry air. It is important that the force of the air is strong enough to remove the excess particles but not too strong to dislodge particles held by a magnetic flux leakage field. They are used when inspections have to be carried out at very low or

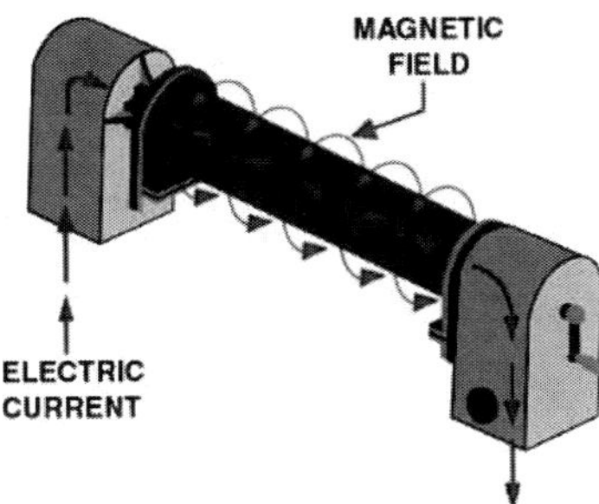

Figure 5.17 Magnetization by direct contact in a bench MPI unit [1].

very high temperatures. Dry magnetic powders are well suited for the inspection of rough surfaces, hot surfaces, or where contamination is a concern. Surface-open, near-surface, or relatively deep flaws can be detected. The particles can be removed from the part easily but cannot be reused. However, dry particles are difficult to apply for overhead inspection positions and require a proper safe-breathing apparatus.

5.3.3.2 WET PARTICLES

Magnetic particles suspended in a fluid, or wet particles, are much smaller than the dry particles and provide better detection sensitivity. Wet particles are mostly used indoors or in sheltered areas where the temperature and lighting conditions can be controlled. Water-suspended particles cannot be used at freezing temperatures or on corrosion-prone parts. While petroleum-suspended particles are more suitable in such applications and provide lower viscosity, they are highly flammable and produce odorous fumes that require proper air circulation and ventilation. They are most commonly used in production or maintenance facilities with horizontal bench units. The large bench units retain and re-circulate the fluid for continuous inspections. Wet particles are also available in spray cans for use with electromagnetic yokes in the field.

With wet particles, all surfaces of the component can be quickly and easily covered with a uniform liquid coat. Also, the fluid carrier provides a better mobility of the particles for a longer time than the dry ones, allowing more particles to float to the magnetic leakage fields. Thus, the wet particles are more suited for detecting very small discontinuities on smooth surfaces. On rough surfaces, however, wet particles that are much smaller than dry powders can settle in surface valleys and lose mobility, becoming less effective than the dry particles under the same conditions. With the wet particles, surface discontinuities will produce sharp indications while subsurface flaws will have less defined indications, losing their definition with increase of the flaw depth.

5.3.3.3 MAGNETIC RUBBER

In addition to the above approaches, occasionally magnetic rubber compounds are used for inspection of hard-to-reach areas, such as threads on the inside diameter of holes. They can reveal much smaller cracks than those possible with the conventional MPI. In this approach, an uncured liquid rubber containing suspended magnetic particles is applied to the inspection area on a magnetized component. The rubber is allowed to set while the magnetic field is maintained in the parts. The rubber cast is then removed and examined for the evidence of discontinuities. The moulding can be retained as a permanent record of the inspection. The same equipment used in the dry MPI may be used with the magnetic rubber approach, but the direct current yoke is most commonly employed. The drawback of the magnetic rubber is that it requires much longer inspection times.

5.3.4 MPI Process

An MPI process involves surface preparation, magnetization, and application of magnetic particles followed by viewing and interpretation. Magnetic particles are applied either before or after magnetization of the part depending on the method used.

Surface preparation involves first cleaning the inspection site to ensure that it is free of oil, grease, moisture, loose dust, corrosion, or other contaminants that could keep particles from moving freely. This can be accomplished by using detergents, organic solvents, or mechanical means. The latter approach is only used when there is either a non-conductive coating or the discontinuities are masked by the coating. Parts that have been previously magnetized require demagnetization before conducting an MPI test.

Magnetic particle inspection of ferromagnetic components can be carried out under continuous magnetization, short pulsed magnetization, or residual magnetic field. Continuous magnetization is employed for most applications to provide high magnetic field strength. It is used when performing multi-directional MPI.

The sequence of particle application when using dry or wet particles with the continuous magnetization is different. For dry particles, it is necessary to apply the magnetic field prior to the application of the particles and to terminate after the powder is applied and the excess is blown off. This is because the dry particles lose most of their mobility when they contact the surface. In the case of wet particles, magnetization can be done either continuously or in short pulses (0.5 sec. each). When using continuous magnetization, the part is covered with wet particles while it is being magnetized using a sustained or prolonged current flow and the indications are viewed and interpreted at the same time.

When short magnetic pulses are used, the part is covered with an abundant amount of wet particles terminating the supply immediately before turning off the current flow. The viewing and interpretation are done at this stage. In the case of residual magnetization, magnetic particles are applied after the magnetizing force has been discontinued. This approach is used only if the test material has high magnetic receptivity so that the residual leakage fields will be of sufficient strength to attract and hold the particle. The residual magnetization method is attractive for production applications and particularly for inspection of long tubular parts.

Viewing and interpretation are done under adequate white light or ultraviolet light depending on the type of the particles used to identify areas where the magnetic particles are clustered.

5.3.5 *APPLICATION EXAMPLES*

MPI is a simple, fast, and effective method of inspecting ferromagnetic metals for surface-breaking or near-surface flaws, and in particular cracks, even if the crack-opening is very narrow. However, if the cracks run parallel to the direction of the magnetic field, the interruption of the field and magnetic leakage will not take place. Therefore, if the crack direction is unknown, it is necessary to magnetise the part in at least two directions at about 90° to each other. Commercial portable (e.g., Figure 5.18) and fixed bench units (e.g., Figure 5.19) are available that allow component magnetization in one or more directions. In some multi-directional units, longitudinal, transverse, or circumferential magnetizations can be done in rapid succession, producing two balanced fields for detecting discontinuities of different orientations.

Aircraft parts made of ferromagnetic materials of any shape, removed from airframe or engines, pumps, landing gears, gear boxes, shafts, shock struts, etc., can be inspected using MPI to check for the presence of fatigue cracks. Some examples are provided in Figures 5.20 to 5.22 [1]. MPI is also widely used to detect cracks within the threads of bolts. In some cases, MPI can leave residual fields which can be removed by demagnetization.

Like the LPI method, the outcome of a magnetic particle inspection when carried out manually is very dependent on the process and the operator. As such, the sensitivity and reliability of this method may vary significantly with the type of

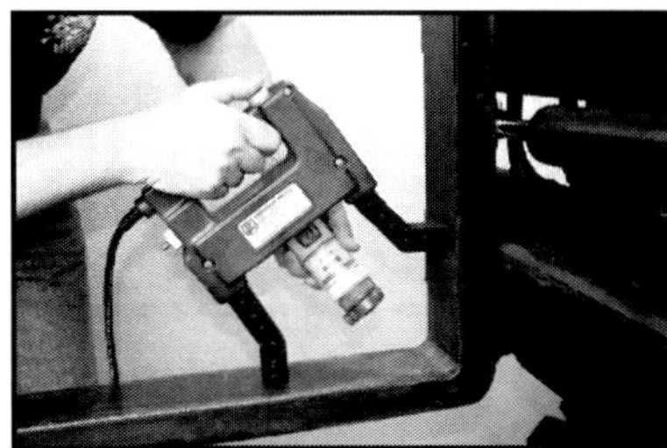

Figure 5.18 A typical commercial portable MPI unit with adjustable poles [15].

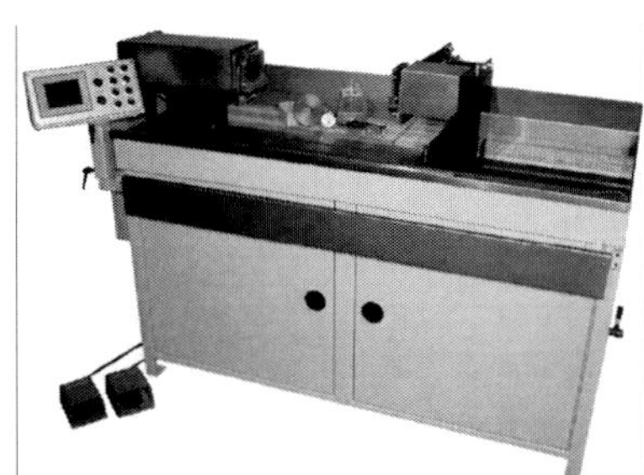

Figure 5.19 A commercial Magnaflux magnetic particle equipment with the capability of linear and circumferential magnetization [15].

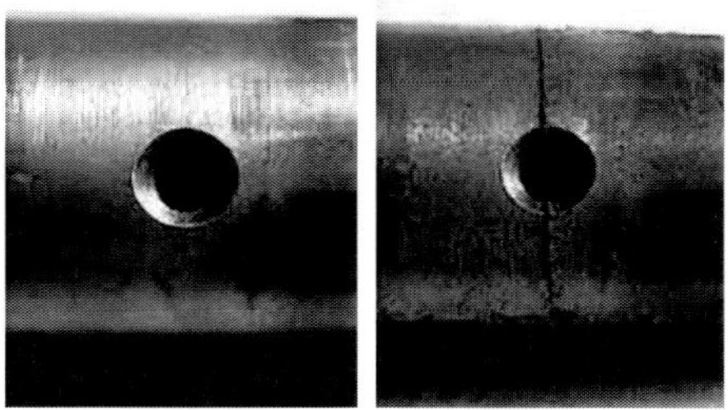

Figure 5.20 A metal part before (left) and after (right) MPI using red iron oxide powder indicating the presence of cracks emanating from a hole [1].

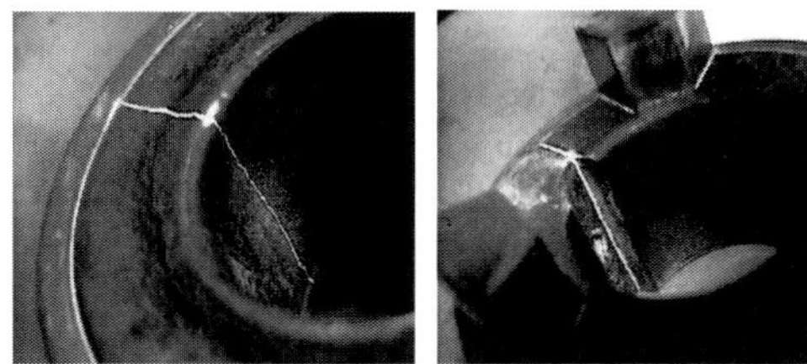

Figure 5.21 Wet fluorescent magnetic particle indication of cracks in a bearing (left) and in a sharp radius (right) [1].

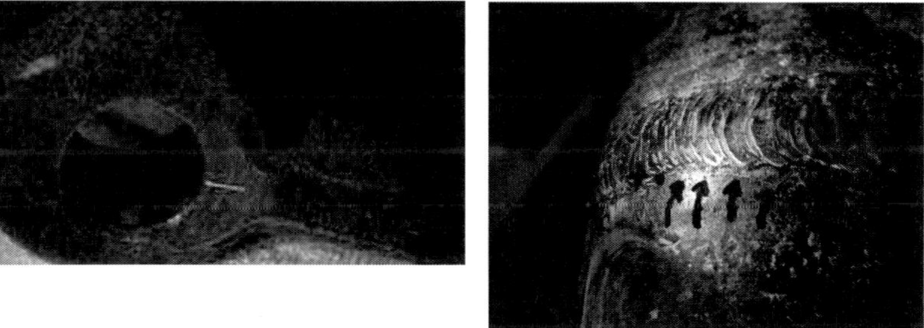

Figure 5.22 Magnetic particle indication of a crack in a casting (left) and in a pipe close to a welding site (right) [1].

particles, the MPI process, and devices used as well as the operator's skill and judgment. Further discussion on this topic is provided later in the chapter dealing with the subject of NDT reliability.

5.3.6 CAPABILITIES AND LIMITATIONS

Capabilities:
- MPI is a fast and inexpensive method of testing ferromagnetic materials.
- Suitable for surface and near-surface crack detection.

- It can reveal surface cracks even when they are filled with foreign materials.
- Large surface areas of complex parts can be inspected rapidly.
- Surface preparation is less critical than with penetrant inspections.
- Magnetic particle indications are produced directly on the surface of the part and form an image that is close to the discontinuity shape.
- Only limited training is required.

Limitations:
- MPI is exclusive to ferromagnetic materials.
- Requires relatively smooth surfaces (not appropriate for porous or very rough surfaces).
- Deep internal discontinuities cannot be detected.
- Flaw detection is dependent on the magnetization direction.
- Depth of flaws cannot be determined.
- Large parts may require excessive currents to be magnetized.
- Using high currents may cause local heating at the points of electrical contact.
- Demagnetization is sometimes necessary before and after inspections.
- Post-cleaning is usually necessary.
- Paints or other non-magnetic coverings adversely affect sensitivity.
- Magnetizing may be difficult for some shapes.
- Conventional MPI does not provide permanent records.

5.4   CHAPTER REFERENCES

[1]   Introduction to Non-destructive Testing, NDT Resource Centre, Center for Non-destructive Evaluation, Iowa State University, Ames, Iowa 50011, USA. http://www.ndt-ed.org/Resources/resources.

[2]   ASTM Standards E 165, Standard Test Method for Liquid Penetrant Examination, ASTM International, 100 Barr Harbour Drive, PO Box C700, West Conshohocken, PA 19428-2959, USA.

[3]   ASTM Standard E 1417, Standard Practice for Liquid Penetrant Examination, ASTM International, 100 Barr Harbour Drive, PO Box C700, West Conshohocken, PA 19428-2959, USA.

[4]   ASTM Standard E 1418, Standard Test Method for Visible Penetrant Examination Using the Water-Washable Process, ASTM International, 100 Barr Harbour Drive, PO Box C700, West Conshohocken, PA 19428-2959, USA.

[5] ASTM Standard E 1208, Standard Test Method for Fluorescent Liquid Penetrant Examination Using the Lipophilic Post-Emulsification Process,. ASTM International, 100 Barr Harbour Drive, PO Box C700, West Conshohocken, PA 19428-2959, USA.

[6] ASTM Standard E 1209, Standard Test Method for Fluorescent Liquid Penetrant Examination Using the Water-Washable Process, ASTM International, 100 Barr Harbour Drive, PO Box C700, West Conshohocken, PA 19428-2959, USA.

[7] ASTM Standard E 1210, Standard Test Method for Fluorescent Liquid Penetrant Examination Using the Hydrophilic Post-Emulsification Process, ASTM International, 100 Barr Harbour Drive, PO Box C700, West Conshohocken, PA 19428-2959, USA.

[8] ASTM Standard E 433, Standard Reference Photographs for Liquid Penetrant Inspection, ASTM International, 100 Barr Harbour Drive, PO Box C700, West Conshohocken, PA 19428-2959, USA.

[9] ASTM Standard E 1135, Standard Test Method for Comparing the Brightness of Fluorescent Penetrants, ASTM International, 100 Barr Harbour Drive, PO Box C700, West Conshohocken, PA 19428-2959, USA.

[10] ASTM Standard E 1219, Standard Test Method for Fluorescent Liquid Penetrant Examination Using the Solvent-Removable Process, ASTM International, 100 Barr Harbour Drive, PO Box C700, West Conshohocken, PA 19428-2959, USA.

[11] ASTM Standard E 1220, Standard Test Method for Visible Penetrant Examination Using the Solvent-Removable Process, ASTM International, 100 Barr Harbour Drive, PO Box C700, West Conshohocken, PA 19428-2959, USA.

[12] Metals Handbook, Non-destructive Evaluation and Quality Control, Ninth Edition, Volume 17, 1989.

[13] American Society for Non-destructive Testing. Non-destructive Testing Handbook, 2nd Edition, Volume 7, Ultrasonic Testing, 1991.

[14] SEM-EDS and XRD Laboratory, http://www2.arnes.si/~sgszmera1/html/examples/15.html

[15] ASTM Standard E 709-08 Standard Guide for Magnetic Particle Testing, ASTM International, 100 Barr Harbour Drive, PO Box C700, West Conshohocken, PA 19428-2959, USA.

[16] ASTM Standard E 125, Standard Reference Photographs for Magnetic Particle Indications on Ferrous Castings, ASTM International, 100 Barr Harbour Drive, PO Box C700, West Conshohocken, PA 19428-2959, USA.

[17] ASTM Standard E 1444, Standard Practice for Magnetic Particle Examination, ASTM International, 100 Barr Harbour Drive, PO Box C700, West Conshohocken, PA 19428-2959, USA.

[18] Magnetic Particle Testing, http://aem.eng.ua.edu/people/haque/Magnetic_files/frame.htm

CHAPTER 6

Electromagnetic Methods

Publications in electromagnetic techniques are in abundance; however, this chapter is primarily based on an NRCC report prepared by Catalin Mandache [1]. In general, electromagnetic techniques are applicable to conductive materials and components only. Except for visual methods, eddy current testing, which uses electromagnetic energy, accounts for nearly 70% of all inspections employed on aircraft, followed by ultrasonic methods and radiographic testing [2]. The American Society for Nondestructive Testing Handbook on Electromagnetic Testing [3] provides a comprehensive description of the theoretical principles and practical applications of the different electromagnetic methods. The majority of the algorithms and equations referred to in this chapter are based on this handbook, a textbook [4] and two other sources, references [5] and [6]; however, the focus here is on aircraft applications.

6.1 PRINCIPLES

Generally, electromagnetic NDT of conductive materials can be carried out in any one of the following three ways:

(i) By passive inspection, which is when an electromagnetic sensor detects the intrinsic magnetization of the test part in the absence of an applied field. This type of evaluation is performed mainly for material characterization and measurement of properties and residual stresses in ferromagnetic materials.

(ii) By static excitation, which is when a continuous magnetic field is applied by a permanent magnet or by a direct-current electromagnet. This type of excitation is found in the inspection of the magnetic materials and used in magnetic particle inspection (MPI) and magnetic flux leakage techniques.

111

(iii) By time-varying excitation, which is when a time dependent magnetic field is applied to a test piece. Eddy current inspection belongs to this category. However, there are other methods that also use a time-varying field as an excitation source, such as magnetic particle inspection described earlier, alternating-current field measurement, and magnetic Barkhausen noise measurement [7]. The Barkhausen effect (discovered by German physicist Heinrich Barkhausen in 1919) is the noise in the magnetic output of a ferromagnetic material when the magnetizing force applied to it is changed and is caused by rapid changes in the size of magnetic domains.

There are situations in which two or even all three types of electromagnetic excitations described above are applied concurrently.

In Figure 6.1, the principles of static; direct field;and time-dependent, alternating field excitations are shown. In Figure 6.1 (left), a permanent magnet introduces a flux through a ferromagnetic material; in the presence of a defect, the magnetic flux lines could leak outside of the test specimen where they can be detected as an indication of a material discontinuity. Leakage might be due to local saturation or existing boundaries. In Figure 6.1 (below), a time-varying current passes through a coil and creates an excitation field. When a conductive test piece is placed nearby, an eddy current flow is established within the piece creating a magnetic field that opposes the incident field. Discontinuities in the material result in a non-uniform magnetic field outside of the conductive part.

In order to underline the principles of electromagnetic NDT, one has to start from Maxwell's equations [3]. Ampere's law expressed by equation (6-1) states that a magnetic field, H (*ampere per meter*), is created by a current (indicated here by the current density, j (*ampere per square meter*), and a time-changing electric field, E (*volt per meter*):

$$\nabla \times \vec{H} = \vec{j} + \varepsilon \frac{\partial \vec{E}}{dt} \tag{6-1}$$

where ε is the electrical permittivity or dielectric constant (*farad per meter*). Similarly, according to Faraday's law, a time-varying magnetic field creates an electric field according to equation (6-2):

$$\nabla \times \vec{E} = -\mu \frac{\partial \vec{H}}{\partial t} \tag{6-2}$$

where μ is the magnetic permeability (*henry per meter*). Combining the above two equations and using Ohm's law, $j = \sigma E$, where σ is the electric conductivity, one obtains the following general field equation:

$$\nabla^2 \vec{H} - \mu\varepsilon \frac{\partial^2 \vec{H}}{\partial t^2} - \sigma\mu \frac{\partial \vec{H}}{\partial t} = 0 \tag{6-3}$$

The same equation stands for the electric field, E.

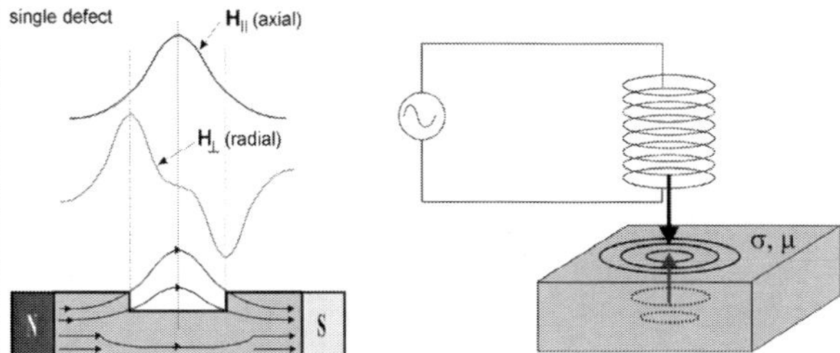

Figure 6.1 Basic principles of electromagnetic NDT. (Left): A static field flows inside a ferromagnetic part and the flux lines escape from the thinner section of the material. (Right): A time-varying source field induces an eddy current flow in the test piece, which in turn creates an opposing magnetic field to the excitation field.

When the magnetic field is produced by current displacement (curl $H = \varepsilon dE/dt$) and the propagation is without loss, the above equation is like a wave equation:

$$\nabla^2 \vec{H} - \mu\varepsilon \frac{\partial^2 \vec{H}}{\partial t^2} = 0 \tag{6-4}$$

This is the case for a low-loss, absorbing medium with small σ where the solution is a sinusoidal wave. On the other hand, when the magnetic field is produced by current conduction, the equation is of a diffusion type, and the solution is an exponential decay.

$$\nabla^2 \vec{H} - \sigma\mu \frac{\partial \vec{H}}{\partial t} = 0 \tag{6-5}$$

This is the case for conductive materials and the last equation represents the diffusion of the magnetic field created by the small-magnitude currents (or eddy currents) in a conductive material.

6.2   EDDY CURRENT TESTING (ECT)

Since this document is primarily concerned with aeronautical applications of NDT where most parts requiring inspections are non-magnetic and electrically conductive, eddy current testing will be the focus of this chapter. Principles of electromagnetic induction are used in eddy current inspection to detect surface and near surface discontinuities in electrically conductive materials. Eddy current testing is used both during manufacturing as well as in service to determine materials' properties, assess condition, measure dimensions, or identify defects. For example, a material's electrical conductivity or magnetic permeability can be determined and related to properties such as hardness. Also, thickness of non-conductive coatings on conductive metals, or thickness of non-magnetic metal cladding on magnetic metals, can be measured using eddy current techniques. Discontinuities, such as voids, cracks, seams, and gaps, or conditions such as dissimilar metals and change in heat-treatments, can be identified by ECT.

A large number of ASTM and other standards are available for eddy current testing; a few ASTM Standards that are aerospace-related are listed below:

- B244—Standard Test Method for Measurement of Thickness of Anodic Coatings on Aluminum and of Other Nonconductive Coatings on Nonmagnetic Basis Metals with Eddy-Current Instruments.
- B 499—Measurement of Coating Thickness by the Magnetic Method, Metals.
- B 457—Standard Test Method for Measurement of Impedance of Anodic Coatings on Aluminum.
- E 566—Standard Practice for Electromagnetic (Eddy-Current) Sorting of Ferrous Metals.
- E 1629—Standard Practice for Determining the Impedance of Absolute Eddy-Current Probes.
- E1004—Standard Test Method for Determining Electrical Conductivity Using the Electromagnetic (Eddy-Current) Method

Next to general visual inspections, eddy current testing is the most commonly used approach for NDT of aircraft during service. It uses a single probe, requires access to one side of the test piece only, and provides information on the component conditions, such as changes in heat-treatment, thickness loss due to corrosion, or presence of fatigue cracks. Eddy current testing can be carried out using different forms of excitations as schematically shown in Figure 6.2 including:

- Single frequency excitation or conventional eddy current.
- Mixed-frequency excitation or multi-frequency eddy current.
- Broad-band excitation or pulsed eddy current.

Despite many capabilities of the eddy current technique for inspection of electrically-conductive materials, the method suffers from the difficulty of discriminating between the different component and test variables that might simultaneously affect the response signal, making interpretation difficult. This challenge is due to the non-linear spatial and temporal distributions of the electromagnetic field in a conductive material. The test piece properties (e.g., electrical conductivity and magnetic permeability) and thickness as well as the probe characteristics, such as type, frequency, geometry, and separation, influence the response signal and complicate the identification and interpretation of discontinuities.

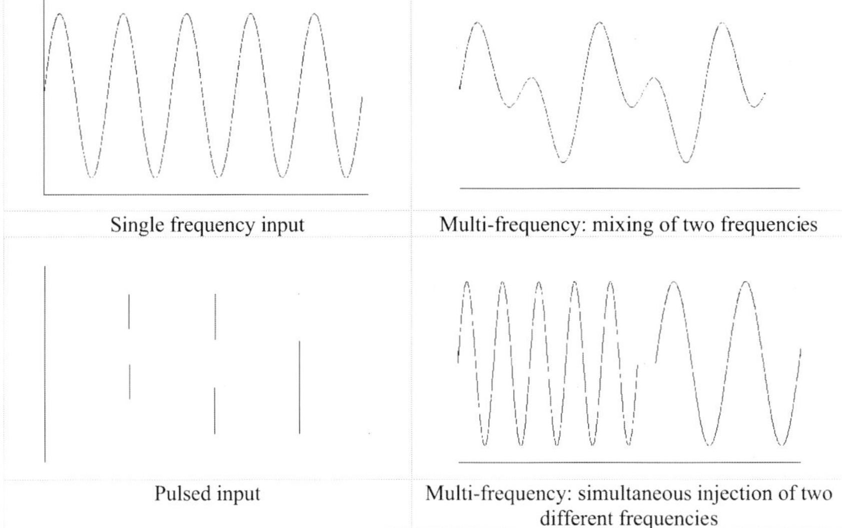

Figure 6.2 Different types of input excitation used in eddy current testing.

6.2.1 CONVENTIONAL EDDY CURRENT

In the conventional eddy current method, a coil (probe) carrying an alternating sinusoidal current (Figure 6.2, top left) is brought close to the test piece. In this way, an alternating magnetic field is induced in the material, which generates a small current (eddy current). This eddy current, in turn, generates its own magnetic field, which interacts with the magnetic field of the exciting probe. The presence of flaws causes interference with the passage of the eddy current and changes the impedance of the coil, which can be detected and measured. Instrument frequency, gain, and threshold as well as the probe's proximity to the test piece and the material discontinuities are the important parameters affecting the response signal.

In this approach, the response signal is a plot of the normalized inductive reactance of the probe as a function of its resistance. In this plane, the locus variation with the electrical conductivity of the material follows a curve as shown in Figure 6.3 (left). The direction of movement of the signal away from the conductivity curve could indicate the possible causes, such as probe lift-off or presence of a discontinuity. However, this representation is not easy to interpret and relies heavily on the experience and intuition of the inspector. By constraining the signal variation due to the lift-off change to the negative horizontal axis as schematically shown in Figure 6.3 (right), specimen discontinuities can be revealed more easily. Using this figure, the size and depth of a crack can be estimated from the amplitude and phase of the eddy current signals, respectively. Representation of eddy

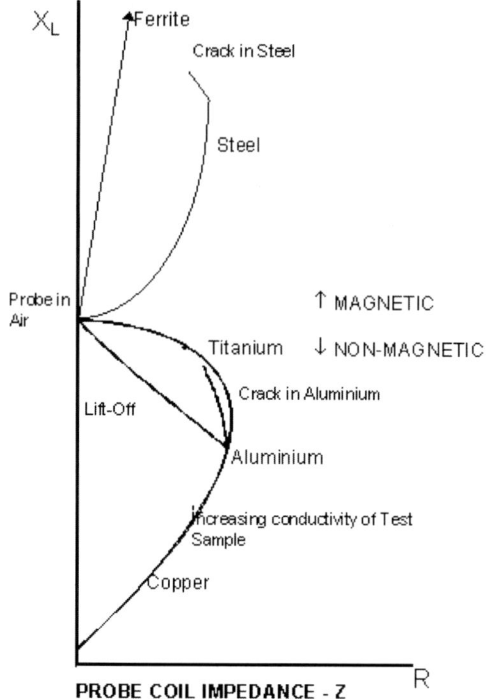

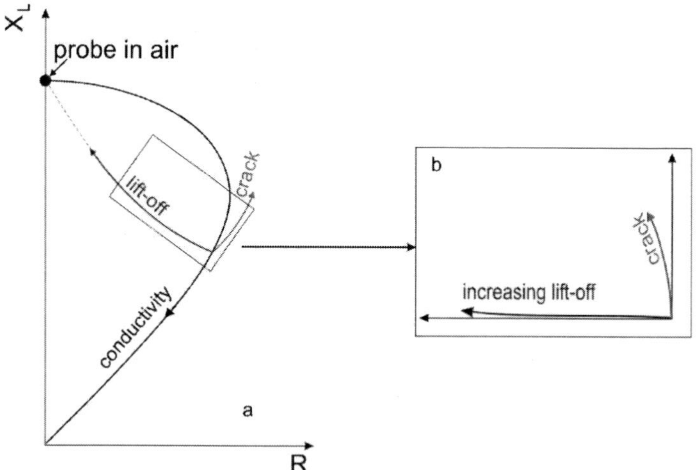

Figure 6.3 (Top): Impedance plane diagram of conventional eddy current for magnetic and non-magnetic metals. (Bottom left): Same as above but only for non-magnetic metals with air point on vertical axis. (Bottom right): Zoom-in on the operating point and rotation of the plane to keep the lift-off variation along the horizontal axis.

current signal response on the impedance plane diagrams similar to this figure make it possible to distinguish between various material discontinuities.

Single frequency eddy current techniques are the method of choice for most aircraft inspection problems, including detection of fatigue cracks in both engine and airframe components, identification of hidden corrosion in lap-joint structures, non-metallic coatings thickness measurement, etc. For crack detection, typically surface-open cracks 1 mm or smaller can be identified; however, the detection limit and reliability often vary depending on the component geometry, crack type, location, ease of access, and inspection conditions used. The chapter on NDT reliability discusses these issues in more detail. Hidden cracks can also be identified although the limit of detection is larger and increases as the distance from the surface increases.

For corrosion detection at frequencies of less than 10 kHz, approximately 10% material loss due to corrosion in the first layer of thin structures can be identified. The spatial resolution of the conventional eddy current technique is not high enough to identify pitting corrosion, but, rather, an average material thickness for the area underneath the probe is provided. The area of corrosion damage would need to be larger than the dimensions of the probe to be detected. Typical probe dimensions for the eddy current thickness loss measurements are in the order of 1 cm. At lower frequencies, eddy currents can penetrate into the second or perhaps third layers and detect corrosion. Later in this chapter, a few examples of inspection of actual aircraft parts by the conventional eddy current approach will be provided which will illustrate the types of eddy current scans obtained from aluminum lap-joints having corrosion sites between the layers.

6.2.2 MULTI-FREQUENCY EDDY CURRENT

Multi-frequency eddy current inspection can be implemented using conventional ECT equipment to find discontinuities located at different depth levels (e.g., cracks or corrosion in either first or second layers of lap-joints). This approach has a practical advantage in that it uses commercially available equipment similar to that of the conventional ECT method, which is familiar to many inspectors. However, setup and calibration for the multi-frequency inspection are time-consuming and complex. Multi-frequency eddy current tests can be performed using mixed frequency input signals (see Figure 6.2, top right) or by simultaneous injection of discrete but different frequency signals (see Figure 6.2, bottom right). Using a dual frequency mixing, corrosion greater than 10% in the second layer of two-layer structures can be found. However, the approach cannot determine the exact location of corrosion (i.e., on the bottom of the first layer or on the top of the second layer).

6.2.3 Pulsed Eddy Current

Besides the conventional and multi-frequency eddy current testing methods, which characteristically use sinusoidal currents passing through the excitation coil, there is another approach that uses rectangular or square waveforms to excite the probe (see Figure 6.2, bottom left). This approach is called "pulsed eddy current" (PEC) or "transient eddy current." Although both of these terms are employed in the literature, it should be noted that this approach uses a "pulsed" excitation signal while the response signal is in a "transient" form. Also, it must be mentioned that the sinusoidal input of the conventional eddy current method produces a single frequency signal while the rectangular or square wave excitation of the pulsed eddy current approach provides a broad-band frequency response.

The majority of commercial eddy current instruments are designed for sinusoidal excitation at a single frequency and occasionally at two or more frequencies. As mentioned above, eddy current inspection at a single-frequency suffers from the limited depth of penetration if a high frequency is used or a limited sensitivity if low frequencies are employed.

The use of the multi-frequency approach is aimed at achieving a reasonable sensitivity for a larger thickness range but compromising on the simplicity of the test. Pulsed excitation partially solves this problem by using a broadband excitation equivalent to a simultaneous injection of a multi-frequency signal. This increases the transient ECT response amplitude thus improving the signal-to-noise ratio for a wider range of depth levels. Also, in terms of presentation, the PEC signals are displayed in volts vs. time, analogous to ultrasonic A-scan, from which B- and C-scan images can be produced and are more intuitive than the conventional eddy current impedance diagrams.

6.3 EDDY CURRENT PROBES

6.3.1 Search Coils

The most commonly-employed types of magnetic field sensors in the NDT field are search coils. The type and shape of the search coils vary depending on the component material and shape as well as the purpose of the test. For example, when inspecting for cracks, the coils must be placed in such a way that the flow of the eddy currents is close to normal to the crack face in order to obtain the maximum response. However, when inspecting for thickness loss due to corrosion, the coil direction must be optimized based on the type of corrosion (i.e., exfoliation or pitting).

Eddy current sensors are available in a large variety of shapes and sizes (e.g., Figure 6.4, left) and also can be custom-designed for specific applications. Eddy current probes are classified by the configuration and mode of operation of the test coils [5]. Probe configuration generally refers to the way the coil or coils are "coupled" to the test area of interest. For example, in a bobbin probe, the coils are wound around a cylindrical-shaped ferrite core that is inserted into a pipe to inspect from the inside out. In contrast, an encircling probe is one that uses ring-shaped coils around the pipe to inspect from the outside in. The mode of operation refers to the way the coil or coils are wound and interfaced with the test equipment. The mode of operation of a probe generally falls into one of the following four categories [8].

6.3.1.1 ABSOLUTE PROBES

Absolute probes generally have a single test coil that is used to generate the eddy currents and sense changes in the electromagnetic field. When an alternating current (AC) is passed through the coil, it creates a changing electromagnetic field that generates eddy currents within the conductive material (Figure 6.4, right). This takes energy from the coil, which appears as an increase in the electrical resistance of the coil. The eddy currents generate their own magnetic field that opposes the magnetic field of the coil, thus changing the inductive reactance of the coil. By measuring the absolute change in impedance of the test coil, information about the test material is obtained.

Absolute coils are very versatile and can be used for measurement of conductivity, lift-off, and thickness as well as for flaw detection. Since absolute probes are sensitive to the material's conductivity and permeability, as well as lift-off and temperature, special care must be taken to minimize these variables when the probe is used for flaw detection. In some commercial absolute probes, a fixed "air-loaded" reference coil is added to compensate for temperature variations.

6.3.1.2 DIFFERENTIAL PROBES

These probes have two active coils usually wound in parallel but in opposite directions and with some distance from each other. During inspections when one coil is over a defect and the other is over the defect-free portion of the material, a differential signal is produced. Differential probes have the advantage of being relatively insensitive to slowly-changing specimen conditions, such as gradual

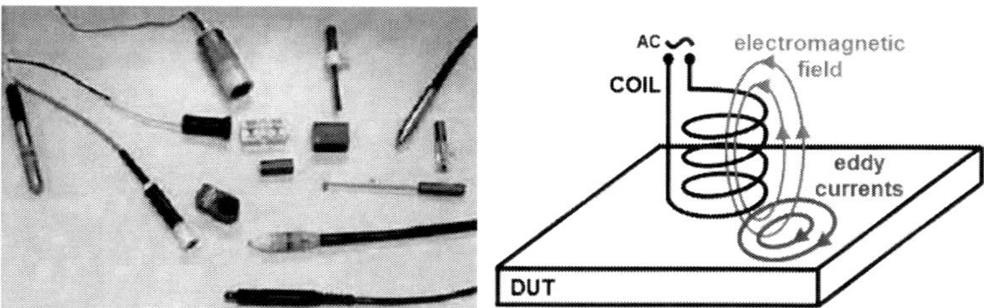

Figure 6.4 (Left): Different types of commercial eddy current probes. (Right): Basic principle of eddy current testing using a single coil.

dimensional variations, minor changes in lift-off, or temperature differences, and yet very sensitive to defects. Also, with this type of probe, wobbling does not affect the response signals significantly. However, signal interpretation may be difficult. For example, if the size of a crack is much larger than the space between the two coils, only the leading and trailing edges of the crack may be detected. The mid-length of the crack may remain undetected due to signal balance as both coils sense the field equally.

6.3.1.3 REFLECTION PROBES

Reflection probes also have two coils, but one coil is used to induce eddy currents and the other is used to sense changes in the test material. They are usually referred to as driver/pickup probes. In this type of probe, the driver and pickup coils can be separately optimized for their intended purpose. The driver coil can be made large enough to produce a strong electromagnetic field in the test material, while the pickup coil can be made sufficiently small to be sensitive to desired minute defects. However, often a compromise between the two coil sizes is necessary. For example, very small pick-up coils generally have higher frequency bands and are less sensitive to deeper flaws. In this type of probe, the coils may be placed in a concentric configuration as illustrated in Figure 6.5.

6.3.1.4 HYBRID PROBES

This type of probe comes in different designs. In one design, a driver coil surrounds two D-shaped sensing coils. In this configuration, the probe operates in the reflection mode while its sensing coil is a differential type that is very sensitive to surface cracks. Another design relies on a conventional coil to generate eddy currents in the material but uses a different sensing device to detect changes in the test material. Hybrid probes are usually specially designed for specific applications.

6.3.2 SOLID STATE SENSORS

Conventional eddy current methods suffer from limited depth of penetration and are only effective for the identification of surface or close-to-surface flaws. A

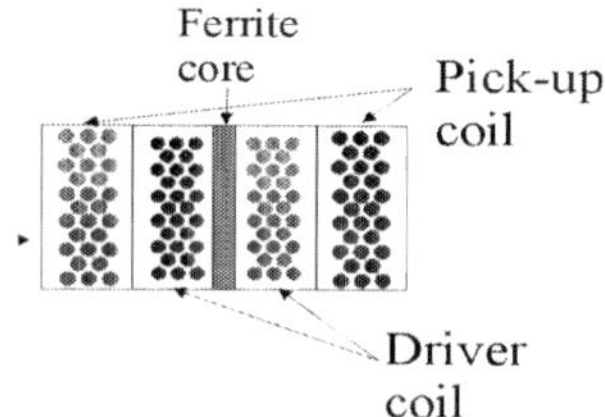

Figure 6.5 Cross-sectional illustration of a concentric probe design.

low excitation frequency is needed for deep penetration and is achieved by using large-diameter coils. However, a large coil results in poor sensing resolution. The use of solid-state magnetic field sensing devices instead of receiving coils can help to solve this problem [9].

These devices are manufactured in very small sizes allowing improved spatial resolution while providing good sensitivity to weak magnetic fields. Moreover, they can operate over a wide range of frequencies, starting from a few hertz, meaning that they are able to pick-up signals from discontinuities in deeper sections of the material that can be reached through low frequency input signals.

Solid-state sensors are particularly advantageous when used with pulsed eddy current methods that employ broadband excitations. This, in combination with the higher sensitivity of the solid-state sensors to low frequencies, allows inspection of thick-section components or ferromagnetic materials that otherwise would be difficult to inspect using ECT methods.

6.3.2.1 HALL SENSORS

The "Hall Effect" is the product of a voltage difference across an electrical conductor, transverse to an electric current in the conductor, and a magnetic field perpendicular to the current [10]. It was discovered by Edwin Hall in 1879. The Hall coefficient is defined as the ratio of the induced electric field to the product of the current density and the applied magnetic field. It is a characteristic of the material from which the conductor is made, and its value depends on the type, number, and properties of the charge carriers that constitute the current.

A Hall sensor is a four-terminal solid-state device that produces an output voltage proportional to the product of its input current, I_c, and the magnetic flux component that it detects, B. By holding its input current constant, the Hall sensor output is proportional to the magnetic flux. Under the influence of the Lorentz force, the electrons will move to one of the sides whose normal is mutually perpendicular to both B and I, while the positive charge carriers accumulate on the opposing side, as shown in Figure 6.6 and equation (6-6).

$$\vec{F}_{Lorentz} = q(\vec{v} \times \vec{B}) \tag{6-6}$$

The Lorentz force separates the positive and negative electric charges, q, that are entering with a velocity v in the magnetic field B. This separation of charges will induce a potential difference, which represents the output of the Hall sensor, as in equation (6-7).

$$V_{Hall} = k_H \cdot I \cdot B \cdot \sin \varphi \tag{6-7}$$

where φ is the angle of the magnetic field with the input current, while k_H is the open circuit sensitivity of the sensor. Hall sensors cover the magnetic field range from 1 G (gauss) to 10 kG (to give an indication of the order of magnitude—the Earth's magnetic field is about 0.5 G) and have a typical sensitivity of a few mV per G. These types of sensors are bipolar and have a linearity range of a few hundred G. Hall sensors are extensively used as magnetic field readers for non-destructive evaluation applications, such as Barkhausen noise and magnetic flux monitoring [11]. They are also employed for a multitude of other purposes, among which are: proximity sensors, current detectors, gaussmeters, in-air gap measurements, and ignition systems.

6.3.2.2 GMR Sensors

The giant magneto-resistive (GMR) effect was discovered in 1988 and is evident in thin-layer sandwiched structures of ferromagnetic and non-magnetic materials. The resistance of such a structure depends on the alignment of the magnetization vectors of the sandwich layers, being highest when these vectors are anti-parallel and lowest when they are aligned along the external magnetic field [12]. When the magnetization directions of the magnetic layers are aligned, the electrons can pass easily from one layer to another, decreasing the electric resistance. The electrons moving in the magnetic material are scattered as they attempt to transverse these domains; when the thin magnetic layers have parallel oriented magnetizations, the tunneling of the electrons from one layer to another is facilitated, decreasing the GMR's sensor resistance.

Similar to the Hall elements discussed above, the GMR sensors are also based on the Lorentz force. By applying a voltage along the length of the magneto-resistive layers, a current will flow and the resistance could be measured by reading the voltage. The application of a magnetic field will change the path of the

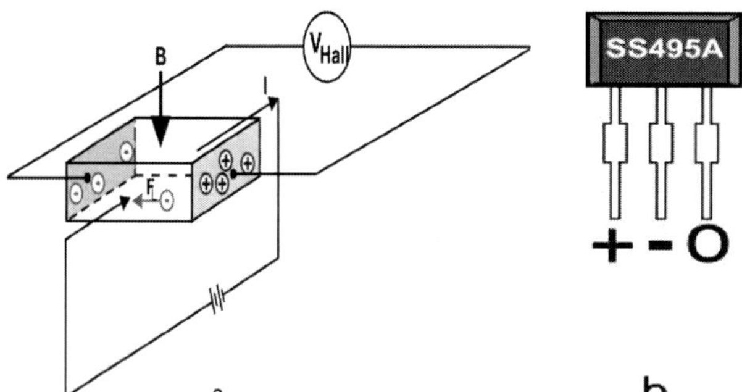

Figure 6.6 Schematic of a Hall Effect sensor. (a): The charge carriers move on opposite sides of the cell under the influence of the Lorentz force. (b): A commercial (Honeywell SS495) linear Hall Effect sensor.

charge carriers affecting the electrical resistance of the sensor. In comparison to the Hall sensors, the GMR sensors have higher sensitivity and higher temperature stability (up to 200°C). Also, while the Hall sensors are sensitive to a direction orthogonal to their chip, the GMR's are sensitive along an axis parallel to their length in the plane of the chip [11]. Simple GMR sensors are made of alloys of nickel and indium antimonide (InSb) [12]. They are nonlinear and their electrical resistance changes up to 2% when placed in a magnetic field.

GMR sensors are copper alloys sandwiched between magnetic layers that are in turn sandwiched between cobalt and iron alloys. They are linear within a broader range and exhibit a typical change in resistance of 15–20% when placed in a magnetic field (newer types of sensors, called "colossal magneto-resistive" might exhibit 70% change in resistance). GMR sensors are manufactured as single chips on silicon wafers using integrated circuit technology. The thin layers forming the GMR sensors have thicknesses from 0.5 to 2 nm. A schematic representation of a GMR sensor and its behavior in a magnetic field are shown in Figure 6.7.

GMR sensors have many advantages that are not found in other types of electromagnetic sensors, such as linear response, increased sensitivity with the input voltage increase, broadband frequency range, and ability to operate at practical temperatures. GMR sensors are used in a wide range of industrial applications, such as magnetic encoder detectors; magnetic ink detection, as used in printing banknotes; anti-lock braking systems; and NDT. The first real impact of GMR sensors came at the end of 1997 when IBM announced the use of this technology for manufacturing hard-disk read-heads [12].

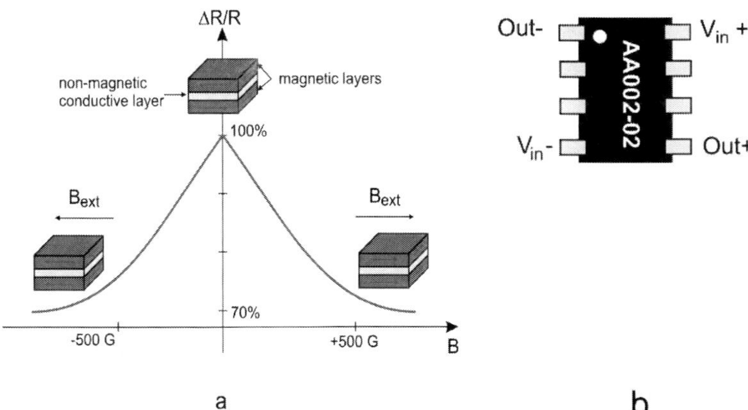

Figure 6.7 (a): Typical GMR sensor output along with magnetization orientations within its layers. (b): A commercial (NVE Corporation AA002-02) GMR sensor.

6.4   EDDY CURRENT MEASUREMENT

6.4.1   *FREQUENCY EFFECTS*

A typical eddy current probe widely used in aerospace applications consists of two concentric coils. The outer one induces the electromagnetic field in the conductive material, while the shielded inner coil picks up the induced field after being affected by the test piece. The coils cannot detect static magnetic fields, but they respond to magnetic field changes based on Faraday's law, expressed in equation (6-2). These changes affect the amplitude and phase of the eddy currents induced in the test piece. The voltage picked-up at the search coil, V_{coil}, is proportional to the coil's number of turns, N, the cross-sectional area, A, and the rate of change of the magnetic flux density, B, whose field lines intersect the area of the coil.

$$V_{coil} = NA\frac{dB}{dt} \tag{6-8}$$

For a sinusoidal excitation field, the magnetic induction, at any time, t, is given by the following relation:

$$B(t) = B_0 \cdot \sin(\omega \cdot t - \phi_0) \tag{6-9}$$

where $\omega = 2\pi f$, f is the frequency and ϕ_0 is the phase with respect to a reference. The picked-up voltage is, thus, proportional to the frequency of excitation:

$$V_{coil} = 2\pi k f \cdot N \cdot A \cdot B_0 \cdot \cos(2\pi f \cdot t - \phi_1) \tag{6-10}$$

where k is a coupling constant ($0<k<1$), and ϕ_1 is the phase measured with respect to the same reference as ϕ_0.

As indicated by the above equation, the coil sensitivity increases with increasing frequency, f, the number of turns, N, and the cross sectional area or size, A. However, the depth of penetration of the electromagnetic field decreases with increasing frequency and decreasing coil size. For detecting deeper flaws, a large driving coil and low excitation frequency are required along with a highly sensitive receiving coil. Inductive pick-up coils provide a trade-off between sensitivity and depth of penetration or between the coil size and operating frequency as they respond to the rate of change in the magnetic field. Detection of deep discontinuities by the eddy current technique is possible when a low excitation frequency and highly sensitive receivers are used. Also, Hall and magneto-resistive sensors that respond to the magnitude of the field rather than the rate of change are sensitive to low-frequency applications and deeper discontinuity detection.

6.4.2 SKIN DEPTH OF PENETRATION

A normal incidence magnetic field (assumed here as aligned with the z-axis) decreases in amplitude with depth increase inside the conductive material following an exponential law. Therefore, the magnetic induction at a depth z is given by:

$$B(z) = B(z = 0) \cdot \exp(-z / \delta) \tag{6-11}$$

where δ is the depth from the material's surface. The "skin depth of penetration" is defined as the depth at which the field value decreases to 37% of its value at the surface of the material. This skin depth, δ, is dependent on the magnetic permeability, μ, the electrical conductivity, σ, of the test piece, as well as the frequency, f, of the field as described by the following equation:

$$\delta = \frac{1}{\sqrt{\pi \cdot \mu \cdot \sigma \cdot f}} \tag{6-12}$$

Therefore, according to equation (6-12), for a specific material, the frequency of the field, f, alone defines the thickness range of the inspection: the lower the frequency of operation, the deeper the penetration of the magnetic field. According to equation (6-10), the flaw detection sensitivity increases with its operating frequency, but the electromagnetic field penetration decreases as indicated by equations (6-11) and (6-12).

6.4.3 PROBE LIFT-OFF

One of the most important factors affecting eddy current signals is the separation gap between the test piece and the probe, which is referred to as the probe-to-specimen "lift-off." In real testing situations, variations in the applied non-conductive coatings (e.g., paints) and geometrical changes of the specimen result in locally-varying lift-off effects. Lift-off elimination methods have been developed that help in the analysis and interpretation of eddy current signals for the identification of material discontinuities.

In conventional eddy current testing, the lift-off variation in the impedance plane diagram is eliminated by adjusting the phase of the signal such that the lift-off creates horizontal displacement of the loci. By doing so, the lift-off effects remain constant on the vertical axis enabling correlation of amplitude and phase variations with the materials discontinuities. In the pulsed eddy current method, there is a point on the response signal that is independent of lift-off, and this is described further in the following section.

6.4.4   Lift-off Intersection (LOI)

Pulsed eddy current testing uses voltage vs. time to display the signal response of the test component. As the probe lift-off changes, the voltage vs. time signal also changes as shown in Figure 6.8. However, the response signals mostly intersect at a certain spot that is called the "lift-off intersection" point or LOI in short. At the LOI point, the signal coordinates (i.e., voltage and time values) are independent of the lift-off [13]. The example shown in Figure 6.8 presents the pulsed eddy current signals obtained on a thick aluminum plate in conditions of varying probe-specimen distance using a reflection type probe. It can be seen that, regardless of the lift-off value, the transient responses pass through the same point on the voltage-time plane. The LOI, which is independent of lift-off, is used for identification of material discontinuities.

6.5   MEASUREMENT SYSTEMS

6.5.1   Conventional Eddy Current

The conventional and multi-frequency eddy current methods are widely used in the aerospace industry, and there are numerous commercial instruments of various sizes and capabilities available for a range of applications. The choice of equipment is dependent on the component material and geometry, as well as the

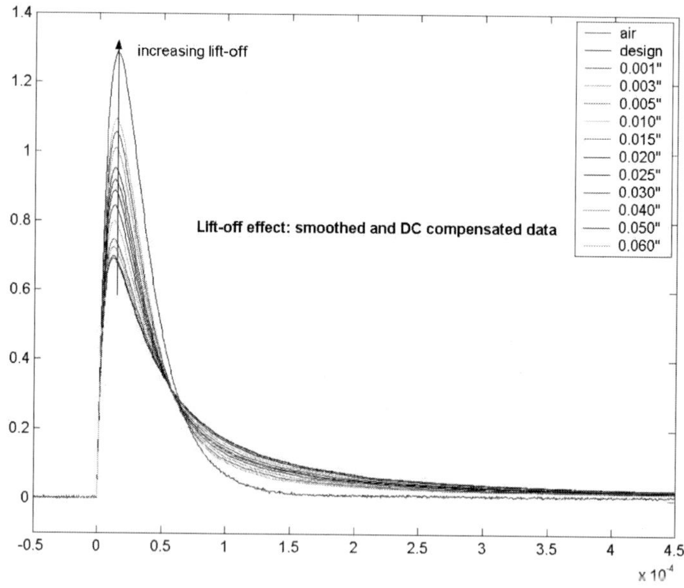

Figure 6.8 Experimentally-observed LOI point in an aluminum plate.

type of measurement or application being considered. The instruments range from simple hand-held conductivity meters to large multi-axis automated systems for the inspection of complex engine parts. Some examples are shown in Figure 6.9 and Figure 6.10.

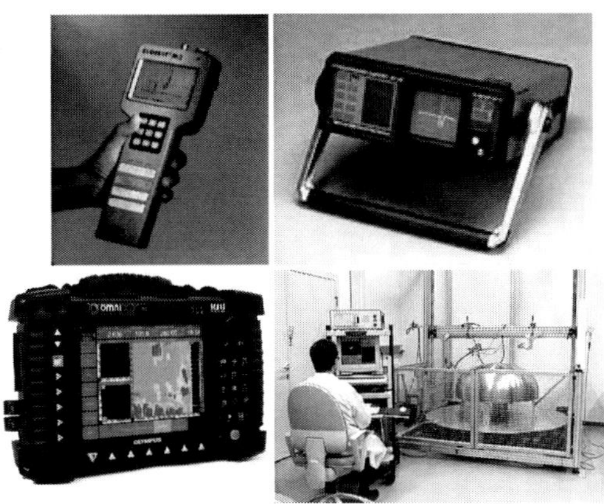

Figure 6.9 Examples of commercial eddy current instruments. (Top-left): Elotest M2. (Top-right): Elotest B1. (Bottom-left): Olympus (RD Tech) Omniscan. (Bottom-right): Astrium Automated Eddy Current System.

Figure 6.10 Home made eddy current scanning systems available at NRCC. (Top): Automated x-y scanning system for inspection of airframe parts. (Bottom): Automated multi-axis scanning system for inspection of aircraft engine components.

6.5.2 PULSED EDDY CURRENT MEASUREMENT EXAMPLES

Unlike conventional and multi-frequency eddy current testing that are carried out using specific commercial equipment, the pulsed eddy current measurements are often performed by universal instruments. The probe is excited by a pulse or rectangular impulse using a pulse and/or function generator, while the response signal is recorded and analyzed in the time-domain as with any digital oscilloscope. In this way, both time and frequency information can be extracted from the response signal.

Figure 6.11 shows a schematic illustration of a pulsed eddy current setup. Since the conventional single-frequency and multi-frequency eddy current technologies are relatively mature, the current research has concentrated mostly on the pulsed eddy current technique. The following sections briefly describe some of the work carried out by C. Mandache as provided in more details in reference [1]. The general objective of the work was to improve this new NDT approach and develop aerospace applications for PEC.

6.5.2.1 THICKNESS MEASUREMENT

Coating thickness measurement by the conventional eddy current method is well-developed, and there are ASTM standards for gauging non-conductive coatings on conductive and non-magnetic metals (ASTM-STD-B244) as well as for non-magnetic coatings on conductive and magnetic metals (ASTM-STD-B 499). These approaches rely on accurate determination of lift-off due to the coating and

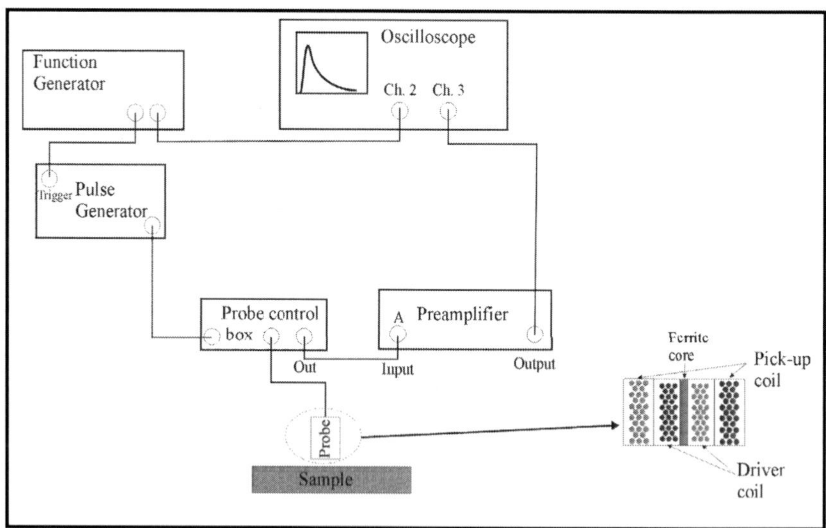

Figure 6.11 Schematic of a simple PEC experimental setup.

are used for the measurement of anodic coatings on aluminum as well as for the determination of paint thickness, etc.

An approach was developed at NRCC to establish the thickness of a conductive metal substrate using the pulsed eddy current approach and LOI feature of the PEC signals that is independent of lift-off [13]. A calibration curve for the specific probe and the test material of a constant conductivity is necessary so that the signals from the field inspections can be compared to the known calibration data as explained in the following example. Six brass shims of various thicknesses [i.e., 0.12 mm (0.005 in), 0.17 mm (0.007 in), 0.25 mm (0.010 in), 0.3 mm (0.012 in), 0.5 mm (0.020 in), and 0.67 mm (0.025 in)] were tested using the PEC technique, and their LOI points were determined as a function of thickness to generate the calibration curve. The LOI point for each sample was established by recording the transient signal when the probe was directly on the sample (no lift-off) and when a non-conductive coating was applied that created a lift-off of 0.37 mm (0.015 in).

As already seen in Figure 6.8, the display of PEC signals in amplitude (volts) vs. time (seconds) at different lift-off values provides the lift-off intersection point or LOI. By adjusting the coordinates, the LOI amplitude is brought to zero (i.e., zero-amplitude crossing), and the reference-subtracted LOI time for each case is established. This time is introduced into the raw data in order to find the LOI amplitude and plot the LOI vs. time (Figure 6.12). This figure can now be used as a calibration curve to determine the thickness of the unknown brass samples. Two Brass samples of unknown thickness were then tested, and their LOI amplitudes and times were determined. Using the calibration curve, the thickness of the unknown brass samples could be found by the interpolation or extrapolation of their LOI points on the calibration curve, as shown in Figure 6.12. A LOI

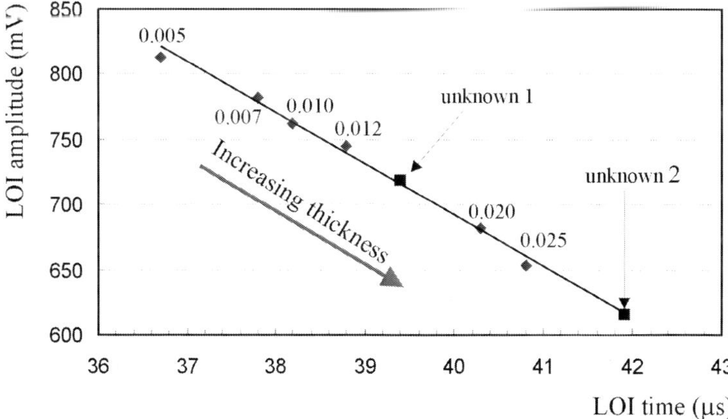

Figure 6.12 LOI-based calibration curve built in terms of sample thickness when the conductivity is kept constant (similar curve would be obtained if conductivity was varied and the thickness kept constant).

point has the same time and voltage coordinates, regardless of the layer of non-conductive coating.

From the calibration curve shown in the above figure it can be estimated that unknown brass #1 has a thickness of about 0.37 mm (0.015 in), while the thickness of unknown brass #2 is 0.75 mm (0.030 in). These values agreed well with the actual thicknesses of these samples measured by a caliper.

Similar curves could be obtained in terms of conductivity for a test piece that has a constant thickness and is used for establishing the conductivity of unknown samples of the same thickness. In this case, the calibration curve could be a plot of the LOI amplitude vs. time for samples of known thickness but different conductivities. If both conductivity and thickness change, it would be difficult to differentiate between the two effects from the PEC signals and the resulting LOI. Either a higher conductivity or a larger thickness will result in LOI of smaller amplitude and higher time. Note that calibration curves developed in this way will be dependent on the probe characteristics (i.e., geometry, frequency, etc.) as well as the material being tested.

6.5.2.2 HARMONIC LOI

Harmonic excitation refers to a condition when a series of periodic waves is applied instead of a single pulse or a single square-wave at a certain time interval. A square wave can be represented as a summation of sinusoidal waves as depicted in Figure 6.13. It has been shown that the lift-off point of intersection also exists for the harmonically-excited eddy current technique [14, 15]. For distinction from the LOI feature of the pulsed eddy current, the lift-off independent crossing point for the sinusoidal excitation eddy current approach is called the "harmonic LOI." In this case, the amplitude and the phase of the response signal, measured here

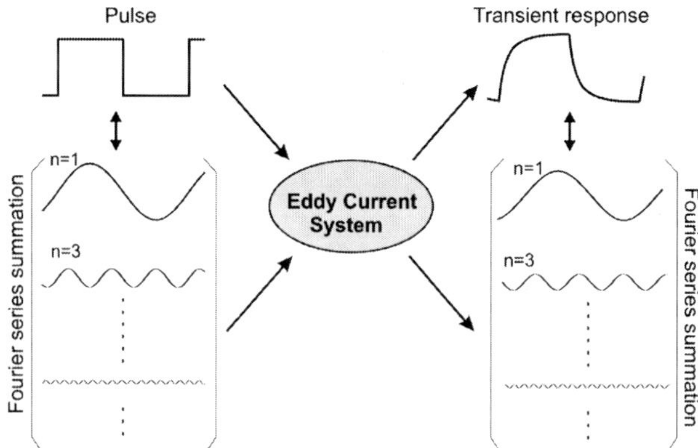

Figure 6.13 Schematic representation of the input and output of a pulsed eddy current excitation and its harmonic components.

with respect to the excitation signal, change in such a way that the sinusoidal waves always cross at the same point regardless of the probe-specimen lift-off similar to the effect shown before in Figure 6.8.

For a periodic square-wave signal of amplitude V_0 the equivalent Fourier series can be written as in equation (6-13), where n represents the harmonic order and t is the time coordinate. Note that only odd harmonics contribute in this Fourier series of reconstruction of individual positive square waves. Therefore, a fundamental frequency of $f_1 = 1/T$ in pulsed eddy current is equivalent to simultaneous sinusoidal excitation with a multitude of frequencies of: $f_n = nf_1$, where n = 1, 3, 5, ...

In this case, the input signal, $g(t)$, can be written as:

$$g(t) = \frac{V_o}{2} + \frac{2V_o}{\pi} \cdot \sum_{n=1,3,5,...} \frac{1}{n} \cdot \sin n\omega t \tag{6-13}$$

Here, the harmonic LOI time can be easily determined analytically if response signals corresponding to two separate lift-off values are known. In this case, the harmonic LOI point occurs due to a combination of both amplitude variation and phase shift [16]. Therefore, the intersection time for any two sinusoidal outputs, t_{LOI}, can be given by:

$$A_1 \cdot \sin(n\omega t_{LOI} - \varphi_1) = A_2 \cdot \sin(n\omega t_{LOI} - \varphi_2) \tag{6-14}$$

where A_1 and A_2 are the amplitudes and φ_1 and φ_2 are the phases of any two harmonic responses corresponding to different lift-off values. The time of intersection of any two sinusoidal signals is provided by solving equation (6-14). This LOI time is given below:

$$t_{LOI} = \frac{1}{n\omega} \cdot \tan^{-1}\left[\frac{A_1 \cdot \sin\varphi_1 - A_2 \cdot \sin\varphi_2}{A_1 \cdot \cos\varphi_1 - A_2 \cdot \cos\varphi_2}\right] \tag{6-15}$$

It is possible to generate a set of harmonic LOI by using equation (6-15) or experimentally by measuring the amplitude and phase data for different lift-off and frequency values.

In a study described in reference [16], it was shown that there is a good agreement between the experimental and analytical LOI values and the error is less than 1.5%. The plot shown in Experimental observation of the harmonic LOI feature—the input signal and four output signals corresponding to different lift-off values are shown. represents the experimental observation of the harmonic LOI that shows the output signals change in amplitude and in phase, but they all cross at the same location over the length of a half period when different lift-off is applied. Harmonic LOI is used to establish flaw depth as explained in the next section.

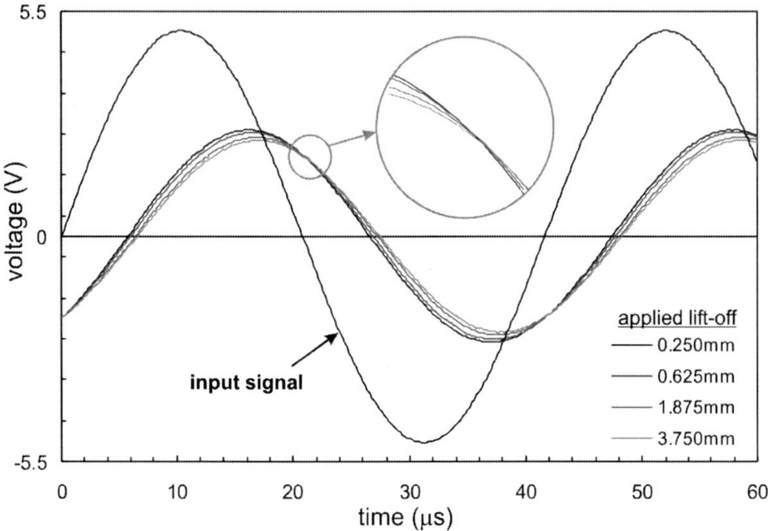

Figure 6.14 Experimental observation of the harmonic LOI feature—the input signal and four output signals corresponding to different lift-off values are shown.

6.5.2.3   FLAW DEPTH MEASUREMENT USING HARMONIC LOI

The example provided below describes a procedure that is used to measure the depth-location of a small flat-bottom hole drilled from the rear surface of an aluminum plate that simulates pitting corrosion on the inner surface of an aircraft skin. The distance from the front surface to the root of the hole is measured here. This value corresponds to the thickness of the material not affected by corrosion at that location. The methodology to measure the thickness is schematically illustrated in Figure 6.15. For this purpose, first the LOI time and amplitude as a function of lift-off are determined on a defect-free area of the specimen by simply moving the probe up and down from the test piece surface in order to simulate lift-off variation. Once the LOI location is established, there are two ways to determine the thickness value, as shown in Figure 6.15.

In one approach displayed on the left side of Figure 6.15, first the time location of the LOI is established and then the sample surface is scanned while only the amplitude within a small time-gate at the LOI is recorded to create a real-time C-scan image of the discontinuity. In this case, the image intensity (or color) corresponds to the LOI value on the time axis that is related to the distance of the root of the hole to the surface of the specimen. In the second approach illustrated on the right side of Figure 6.15, the sample is scanned while the signal amplitude and phase for the entire test piece are recorded with respect to a flaw-free location on the specimen (reference). In this case, after completion of the scan, the LOI time for each scan point is determined analytically by using equation (6-13), and the results are presented in C-scan format.

The two approaches yield similar but not quite identical results, as seen in Figure 6.15. The slight difference is due to the fact that in the approach shown on the left, the time set for the gate is constant, while in the approach displayed on the right, the time varies according to the point of intersection between individual signals and the reference signal. It should be noted that the time and voltage of the LOI point are constant when changing the lift-off, but they vary when inspecting for discontinuities. In the latter case, when a discontinuity is present, both the LOI time and amplitude change.

Similarly, in the following example, the harmonic LOI feature is used for the detection of cracks in the second layer of lap-joints [17, 18]. The sample here was a two-layer aluminum lap-joint specimen (Figure 6.16) that included an EDM notch of 6 mm in surface length, 1 mm in depth, and 0.125 mm in width located on the top surface of the second layer. The Al plates were 2 mm thick. Probe-specimen lift-off was simulated by applying Teflon tapes of different thicknesses

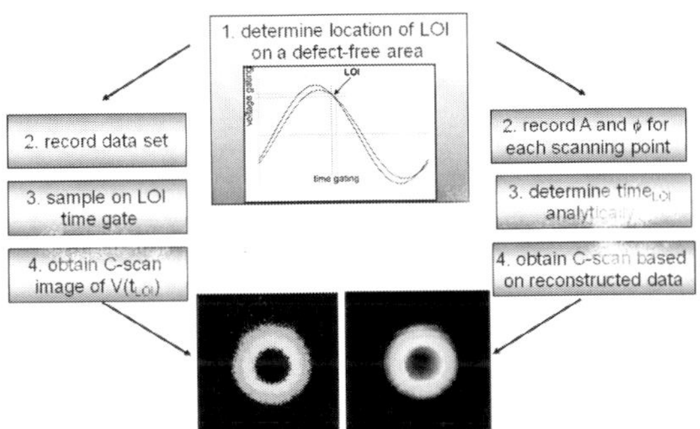

Figure 6.15 Two types of eddy current scanning procedure that are used with the harmonic LOI feature [18].

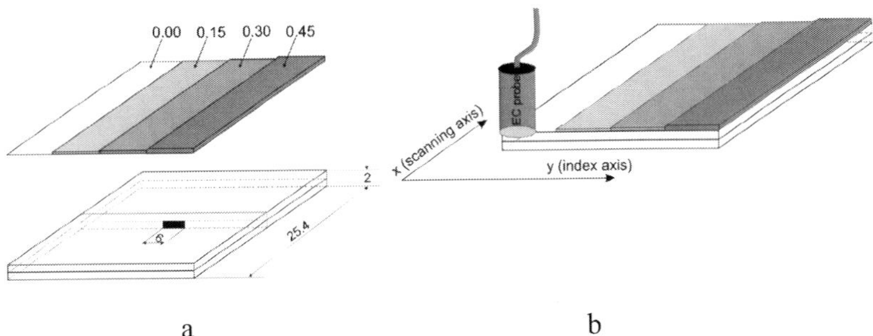

Figure 6.16 Detection of a second layer notch in a lap-joint specimen using the harmonic LOI principles (dimensions are in mm). (a): Schematic of notch location and applied lift-off. (b): Schematic illustration of scanning method [18].

on the sample covering equal areas of the specimen's top surface. In this way, four lift-off values were created corresponding to: 0.45 mm, 0.30 mm, 0.15 mm, and 0.00 mm (i.e., no lift-off). Pulsed eddy current inspections were performed by scanning the specimen with the EC probe in an automated x-y scanner with a scan resolution of 0.25 mm on both axes.

The scanning plane was perpendicular to the EDM slot face. The eddy current probe was a sliding type consisting of two concentric coils; the outer one was for the excitation of the electromagnetic energy while the shielded inner coil was used as the pick-up sensor. The input harmonic signal frequency was set at 7.5 kHz, which corresponded to a standard depth of penetration of 1.38 mm in aluminum. C-scan images were obtained by measuring and plotting the response signal amplitude, the signal phase, and the amplitude of the signal at the LOI time. As shown in Figure 6.17, the two amplitude C-scans clearly show the presence of the second-layer EDM notch, but the signal phase did not provide a clear indication of the discontinuity. While the discontinuity can be detected using amplitude C-scans, size estimation from these plots is not easy.

The same approach was also applied for the thickness change measurement, as schematically illustrated in Figure 6.18. Here, four sections of an Al plate were milled to different depths simulating thickness loss due to corrosion corresponding to: 0% (no-corrosion), 14%, 16%, and 35% of the total plate thickness.

Again, the same type of reflection probe used in the previous example was employed with a driving frequency of 10 kHz. The resulting C-scans are shown in (a) Signal amplitude, (b) signal phase, and (c) signal amplitude gated at the LOI time for the experiment illustrated in Figure 6.18.. Here, again, the phase image fails to clearly indicate any of the milled areas while the amplitude images show all three thinned sections. The measurement of the signal amplitude gated at the LOI time provides the best resolution and a good estimate of the thickness loss using the image intensity (or color) variations.

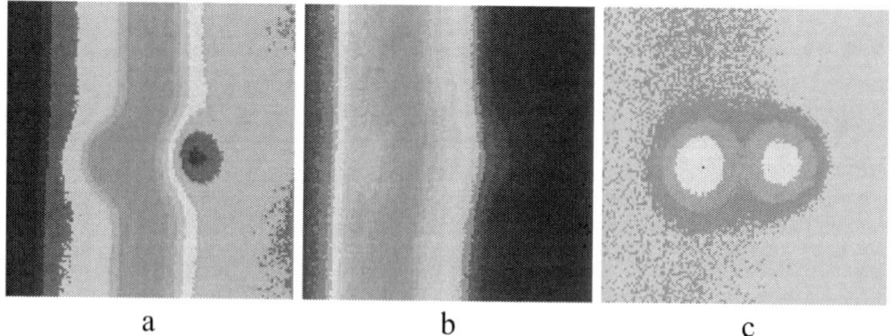

a   b   c

Figure 6.17 Eddy current C-scan images of an aluminum lap-joint specimen with a notch in the second layer corresponding to: (a) signal amplitude, (b): signal phase, and (c): the signal amplitude gated at the LOI time. The scanning approach is shown in Figure 6.16.

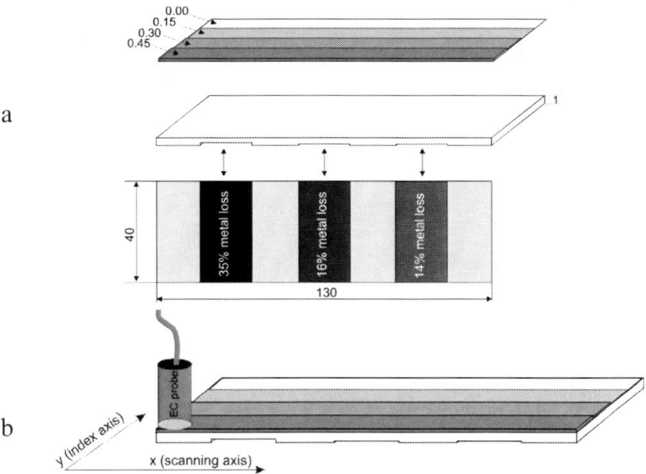

Figure 6.18 Thickness change sensing based on harmonic LOI principles: (a) schematic of the specimen and (b) scanning approach.

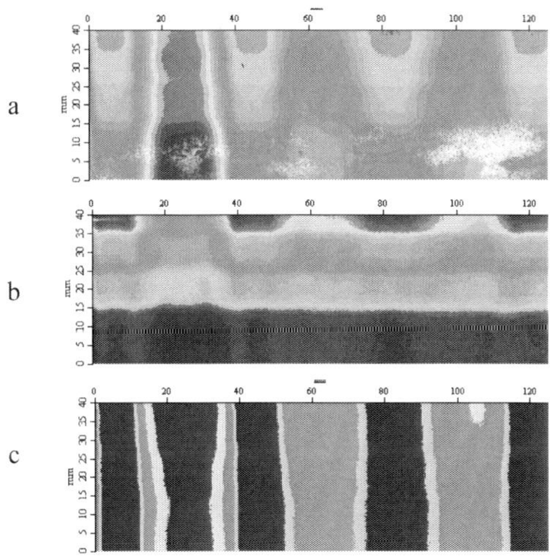

Figure 6.19 (a) Signal amplitude, (b) signal phase, and (c) signal amplitude gated at the LOI time for the experiment illustrated in Figure 6.18.

6.5.2.4 PULSED EDDY CURRENT MEASUREMENT OF LIFT-OFF

In eddy current inspections, the lift-off variation due to non-uniform paint thickness, probe tilt, or protruding rivets can cause undesirable effects and affect flaw detectability. On the other hand, controlled lift-off can provide useful information as demonstrated in the previous sections for the measurement of the thickness of non-conductive coatings on conductive materials [19] or gauging surface corrosion on metals.

With the pulsed eddy current technique, it is possible to detect and quantify the lift-off without changing the hardware or using complicated algorithms. Moreover, the LOI feature in pulsed eddy current response can be used for both the inspection of metallic materials as well as for the measurement of the thickness of any non-conductive coating applied on the metal surface.

It has been found (e.g., [14, 15, 19]) that the signal amplitude at the LOI time is lift-off independent, and, thus, any amplitude variations at the LOI point can be related to the material discontinuities. On the other hand, the time derivative of the signal at the same location is directly proportional to the thickness of the non-conductive layer between the probe and the metallic test piece, such as paint. In the following example, these findings are applied to determine metal-loss due to corrosion in aluminum lap-joint structures.

The example uses an aluminum plate with a milled-out area in the far-side of the plate to simulate metal-loss due to corrosion of the faying surface. The thickness of the Al plate was 1.025 mm, and the radius of the circular milled area was 10 mm, which corresponded to a material loss of 35%. Paint thickness variation was simulated by placing layers of Teflon tape on the top surface of the specimen to provide varying probe lift-off corresponding to 0.0 mm (bare metal), 0.115 mm (one layer of Teflon tape), and 0.230 mm (two layers of Teflon tape) as illustrated in Figure 6.20(a).

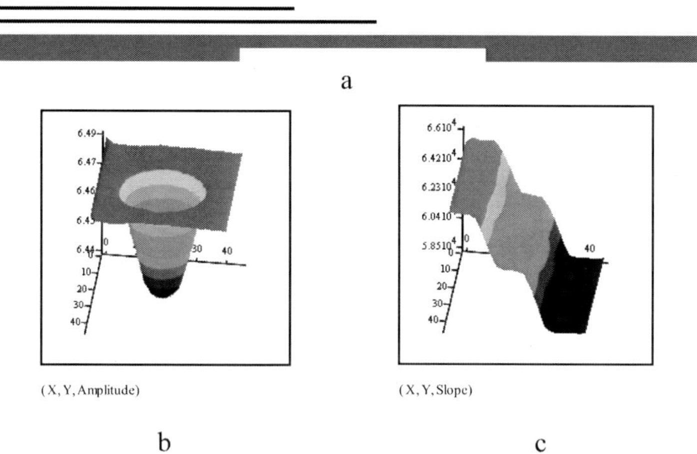

Figure 6.20 (a): Cross-sectional view of an Al specimen with a milled area simulating metal-loss due to corrosion. (b): Eddy current C-scan using the signal amplitude at the LOI point that reveals the milled-out area in the part. (c): Eddy current C-scan using the time derivative at the LOI point that shows the direct relation between LOI with the thickness of the non-conductive Teflon layer simulating paint thickness change [19].

As described earlier, the signal amplitude at the LOI point and its time-derivative were recorded while scanning the specimen in an automated eddy current system to generate the corresponding C-scan images as shown Figure 6.20(b) and (c). While the amplitude scan reveals the milled-out area clearly, its time derivative at the LOI point also indicates the changes due to the non-conductive Teflon layers covering portions of the milled-out area corresponding to lift-off variation (or paint thickness changes) [20].

6.5.2.5 PEC SIGNAL ENHANCEMENT USING A MU-METAL

In general, eddy current testing at high frequencies suffers from the limited depth of penetration of the electromagnetic field into the conductive test material. At low frequencies, the depth of penetration is better but the sensitivity is less; thus, small defects may not be detectable. As discussed earlier in this chapter, the pulsed eddy current approach uses a broadband excitation that is equivalent to performing conventional eddy current tests at a multitude of frequencies. However, PEC signal enhancement may be necessary to increase the magnitude of the response signal.

One way of increasing the response signal is by improving the signal-to-noise ratio. This may be achieved by backing the test piece with a thin layer of soft magnetic material, such as a mu-metal[1] sheet, that serves the purpose of amplifying the eddy current response. The mu-metal reflects the magnetic field reaching the backside of the test piece and, in effect, amplifies the small-amplitude response signals.

To visualize this concept, consider a conductive plate of thickness D that is subjected to eddy current testing on the top surface as illustrated in Figure 6.21. To inspect the far side of the specimen, it is necessary that the skin depth of penetration, δ, associated with the test is such that electromagnetic field of sufficient strength reaches the far side.

As described in section 6.4.2 of this chapter, in conventional eddy current testing, in order to achieve good sensitivity at the far side (either for flaw detection or plate thickness measurement), the frequency must be set so that the standard skin depth of penetration is less than three times the material thickness. When inspecting a single layer conductive part, the magnetic field B_0 incident at the normal angle to the part decays exponentially with depth, z, as given by the following equation:

$$B \propto B_o \cdot \exp(-z / \delta) \tag{6-16}$$

---

[1]Mu-metal is a nickel-iron alloy (approximately 77% nickel, 16% iron, 5% copper, and 2% chromium or molybdenum) that is notable for its high magnetic permeability. The high permeability makes mu-metal useful for shielding against static or low-frequency magnetic fields, which cannot be attenuated by other methods. The name came from the Greek letter mu (μ) which represents permeability.

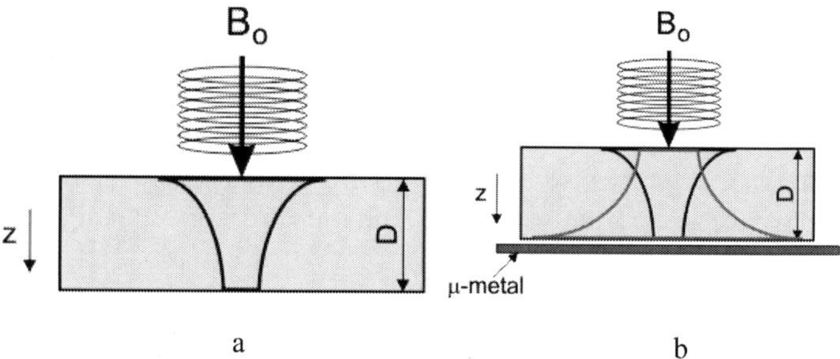

Figure 6.21 (a): In a conductive material, the excitation magnetic field attenuates with depth and (b) placing a ferromagnetic backing, the magnetic field reaching the backside of the test piece is amplified.

Placing a magnetic material with a relative permeability of μ under the test specimen of thickness D will result in an increase of the field density inside the test piece as depicted in Figure 6.21(b) and expressed in equation (6-17).

$$B' \propto \mu \cdot B_o \cdot \exp(-D/\delta) \cdot \exp(z/\delta) \qquad (6\text{-}17)$$

The addition of the magnetic substrate shortens the coil's magnetic field path while increasing its impedance. This, in combination with the main effect of the magnetic substrate due to the additional magnetic permeability, results in increased voltage across the sensing coil, thus, enhancing the response signal.

This approach may be applied to components that are made of a conductive top layer over a ferromagnetic substrate to improve the detection sensitivity to top layer metal-loss or for more accurate thickness measurements [13].

6.5.2.6   MU-METAL AND LIFT-OFF

Earlier, it was shown that the transient eddy current signals displayed in time-domain show a crossing point that is independent of the lift-off (i.e., LOI point). However, when pulsed eddy current signals are obtained on a 0.1 mm thick mu-metal sheet, there are no apparent crossing points independent of lift-off as shown in Figure 6.22 [13]. Here, increasing lift-off decreases the signal amplitude as it eventually approaches the signal recorded when the probe is in air away from any external magnetic field source. Similar to LOI, this approach also uses the crossing point of the pulsed eddy current signals to overcome the lift-off effects [16, 18, 19]. However, in this case, the coordinates of the lift-off crossing points are employed as signal features.

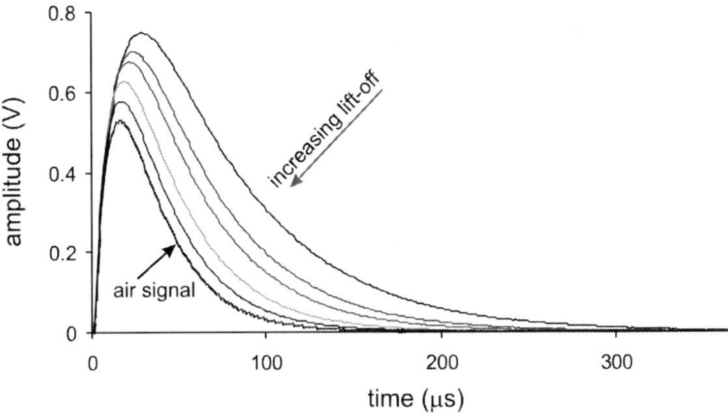

Figure 6.22 Transient responses for various lift-offs from a 0.1 mm thick mu-metal sheet. The air signal is recorded with the probe far away from any conductive material.

6.5.2.7 Mu-metal with Non-magnetic Materials

When a layer of conductive but non-magnetic material is placed on top of the mu-metal, it may be shown experimentally that the LOI feature still exists [20]. This LOI point typically occurs at an earlier time and has higher amplitude than the LOI obtained for a non-magnetic and non-conductive material. An example is shown in Figure 6.23, where the transient signals for a 0.64 mm brass plate of 26% IACS conductivity (brass is non-magnetic) are displayed with and without the 0.1 mm mu-metal backing for three lift-off values of: 0 mm, 0.25 mm, and 0.50 mm. These experimental results have important applications in practice; two examples are provided below:

1. **Detection of Metal Loss due to Corrosion in Inner Surfaces**: A ferromagnetic backing material may be used to increase the sensitivity and depth resolution of pulsed eddy current in the detection of metal-loss due to corrosion in the inner layers of aircraft lap-joints. To assess this application, a test piece was made of 2024 aluminum plate (1.6 mm thick, 50 mm wide, and 260 mm long) which contained two rows of 10 mm diameter flat-bottom circular holes in the far-side simulating metal-loss due to corrosion. The depth of the holes was 3%, 7%, 14%, 22%, and 30% of the plate thickness. Figure 6.24 is a schematic representation of this test piece. A 0.1 mm thick mu-metal sheet was placed under the first half of the specimen while it was scanned from the top surface using the pulsed eddy current method. The amplitude of the resulting signals was measured at the LOI point and is displayed in Figure 6.25. It is clear from this figure that, while the 30%, 22%, and 14% milled areas are detectable without the mu-metal, traces of all, including the 7% and 3% metal-losses, became more apparent with the

mu-metal backing. Thus, improvement in the detection sensitivity and resolution are achieved with the mu-metal backing.

2. **Measurement of Cladding Thickness on Ferromagnetic Substrates**:
   The thickness of a conductive non-magnetic cladding over a ferromagnetic substrate may be measured using the LOI feature. Such a measurement is barely possible with the conventional or pulsed eddy current methods if the substrate is not a magnetic material. Metal cladding provides improved wear resistance and corrosion protection; thus, it used in aerospace as well as many other industries.

In the work described in reference [20], experiments were carried out on a 3 mm thick mild steel (ferrous) plate covered with a 250 μm brass layer on two-thirds of the plate. Another brass layer of 50 μm was added to the upper half for a

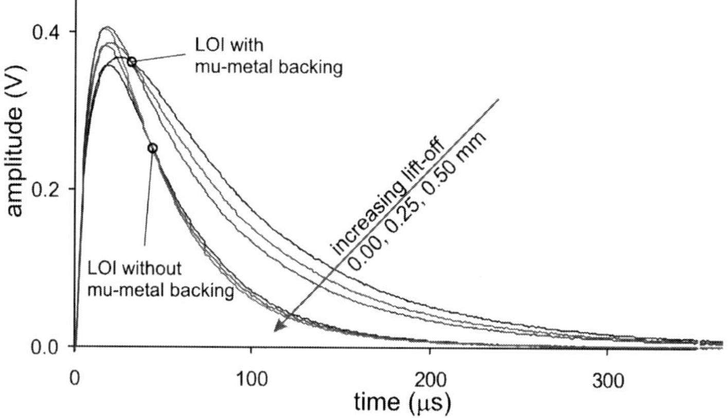

Figure 6.23 Eddy current signal amplitude vs. time measurements taken from a 0.64 mm thick brass specimen with and without mu-metal backing [13].

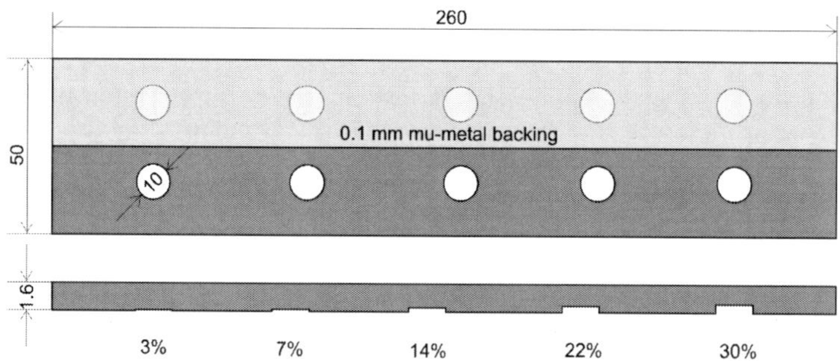

Figure 6.24 An aluminum specimen with flat-bottom holes simulating metal-loss due to corrosion. The top half of the specimen has a thin sheet of mu-metal backing. The cross-sectional thickness is exaggerated for improved visibility of the discontinuities [13].

total of 300 μm for the top one-third and leaving 250 μm thickness for the middle one-third. The lower one-third was left without a brass layer. The design of this specimen is illustrated in Figure 6.26. Thin brass layers on mild steel simulate conductive non-magnetic cladding over a magnetic material. To simulate different degrees of lift-off, the right one-third of the plate was covered with Teflon strips totalling 200 μm in thickness, the middle one-third with 100 μm, and the left one-third had no Teflon. With this specimen, nine different conditions were created.

Here again, pulsed eddy current tests were carried out, and the signal amplitude values were taken at the LOI point and plotted in the C-scan format. As previously shown in Figure 6.22, when a ferrous material is tested, the signal responses do not cross each other with varying lift-off. However, the signals cross each other when the specimen is covered with a conductive but non-magnetic layer (as seen in Figure 6.23). In this example, the amplitude and time coordinates of the LOI

Figure 6.25 Pulse eddy current scan result of the above specimen showing metal-loss traces. Dark gray area (below) without the mu-metal backing and light gray area (above) with the mu-metal backing [13].

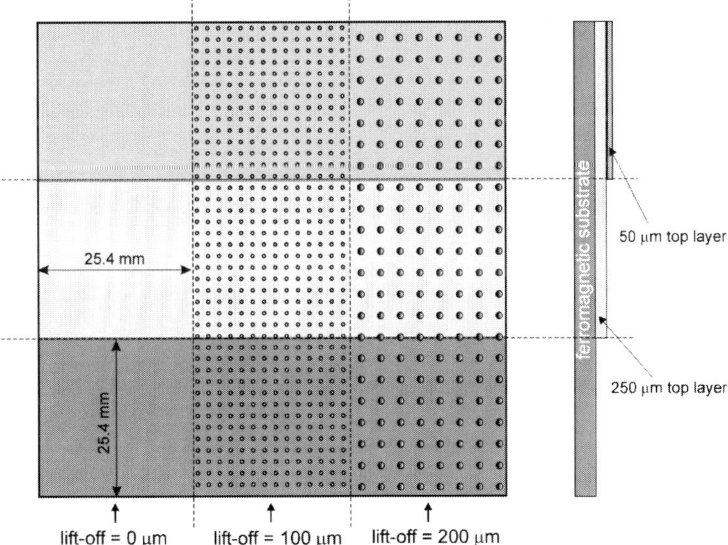

Figure 6.26 Test specimen consisting of a mild steel (ferromagnetic) substrate covered by 250 μm and 300 μm thick brass layers (conductive non-magnetic). Teflon tapes (non-conductive) are added to simulate lift-off of 200 μm, 100 μm, and 0 μm creating a matrix of nine test conditions. Overall area of the specimen: 76.2 mm × 76.2 mm [20].

point were used to measure the thickness of brass layers over the mild steel plate. The test results were obtained using a reflection-type probe with a pulse width of 500 μs.

As shown in Figure 6.27, the lift-off effect is substantially reduced for the top two sections where brass layers clearly produce a color (i.e., amplitude) distinction between the 300 μm and 250 μm brass strips. Using appropriate calibration pieces, the color can be related to the thickness for more quantitative measurements. For the bottom one-third of the test sample where there is no conductive brass strip, the LOI crossing point does not exist; thus, the lift-off effects are more visible as shown by the different shades of red (dark gray in figure).

6.6   APPLICATION EXAMPLES OF EDDY CURRENT TESTING

In this section, some examples of aerospace applications of eddy current techniques are provided. Generally, the aerospace applications of eddy current methods include inspection of the critical engine and airframe components for fatigue cracks as well as identification and measurement of inter-layer corrosion of lap-joints in fuselage or wing structures. These and a few other examples that are not directly related to aerospace are included here.

6.6.1 INSPECTION OF ENGINE PARTS

Fatigue cracks may develop in dovetail slots, bolt-holes, or other critical locations of engine parts. Eddy current techniques are used to detect such cracks. For the eddy current inspection of dovetail slots, often, custom-made sliding probes are

Figure 6.27 Pulsed eddy current surface scan of the above specimen showing the 300 μm and 250 μm brass layers (dark blue on top and lighter blue in middle, respectively) and the bare metal (red on bottom). Lift-off effects are visible only on the bare metal area [20].

employed. The work on the development of eddy current inspection techniques for the detection of fatigue cracks in T-56 turbine wheels provides a typical example [21]. This component had experienced several fatigue-induced in-service failures over the engine life. Cracks had been found near the base of the blade dovetail slots at a relatively sharp radius on the inner-most serration area. An eddy current inspection procedure was needed to detect such cracks. The requirement was to identify cracks greater than 0.75 mm in length.

A probe was developed to operate in a hand-held scan mode as shown in Figure 6.28. The probe was a reflection-differential type and the coil size and location were chosen to cover the area where cracks occur. A reference calibration specimen with a range of EDM sizes, shapes, and placements was fabricated to evaluate the probe's performance. The calibration specimen also included real fatigue cracks. The probe was tested on the reference specimen and found to be capable of detecting EDM notches as small as 0.5 mm when the notches were located away from the edges of the serration. However, when the notch was close to an edge, the sensitivity was reduced. Since cracks usually occurred in the inner portion of the slots, the reduced sensitivity at the edges was considered to be acceptable.

Using the calibration piece, an experimental comparison was carried out of the eddy current response from an EDM notch relative to a fatigue crack of the same length. The assessment showed 30 to 40% lower signal strength for the fatigue cracks as compared to the same size EDM notch. This was likely due to the difference in the gap width between the two faces of these discontinuities, which was in the order of 0.125 mm for EDM notches and almost zero for fatigue cracks. The larger the gap, the more resistance to the flow of the eddy currents it would be and, thus, the more difference at the pick up coil. This difference was taken into account when the signal intensity was used to obtain crack size information. In addition, it was noticed that the presence of contaminants, such as dirt or grease,

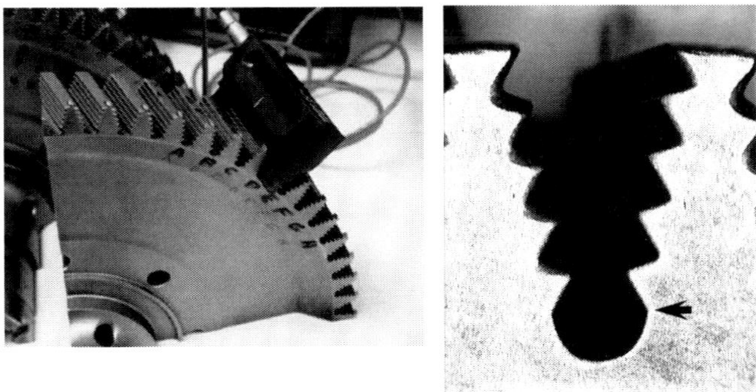

Figure 6.28 (Left): Eddy current inspection of dovetail slots in T-56 turbine discs using a sliding probe. (Right): A high magnification view of a dovetail slot and the critical region identified by an arrow.

in the slots significantly increased the noise level; thus, careful cleaning of the parts was necessary before the inspections of the actual parts. After the above-mentioned initial tests, service-exposed disks were fully inspected for cracks larger than 0.75 mm. Although the inspections produced signal indications, they were all smaller than those of the 0.75 mm target flaw size with a large margin. At the end of this work, a written inspection procedure was prepared and provided to the client along with the manufactured probe and the calibration piece.

The eddy current technique is widely used for the detection of fatigue cracks in the bolt-holes of engine parts. Many probes and instruments with a range of specifications and capabilities are commercially available for bolt-hole inspections using either manual or automated approaches. For crack detection in bolt-holes, spinning probes are often employed. Such probes use a device to rotate a pencil-shaped probe at a high speed with the coil on the side while it is inserted manually into the hole. This enables the coverage of the entire inner surface of the hole in one insertion. These devices may be incorporated into an automated scanning system, such as the one shown in Figure 6.10 developed for the bolt-hole inspection of J85-CAN-40 compressor discs.

This system consisted of a commercial eddy current instrument with a rotating probe device, an automated XYZ table, and an advanced pattern recognition software package. As described in more details in the chapter on NDT Reliability, with this system, it was possible to identify the presence of service-induced fatigue

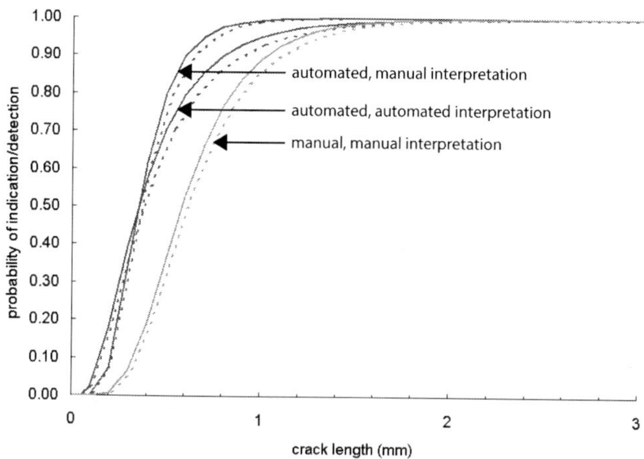

Figure 6.29 Comparison of the probability of detection of eddy current inspection of bolt-hole cracks in J85-CAN-40 compressor discs. (Top): Automated inspection and human identification of cracks. (Middle): Automated inspection and pattern recognition identification of cracks based on their signal characteristics. (Bottom): Manual inspection and human identification of cracks. Solid lines are the mean POD and dashed lines are the POD with 95% confidence.

cracks as small as 0.05 mm in the bolt-holes of J85-Can-40 compressor discs with a 90% probability of detection at 95% confidence (Figure 6.29) [22]. This figure also provides a comparison of the manual eddy current tests by a highly-qualified technician with the automated inspection using either pattern recognition analysis or human interpretation. The figure shows that manual tests in the laboratory environment are less reliable than inspections performed using an automated system. However, in this case, the use of pattern recognition to identify cracks from their signal characteristics did not improve the probability of detection when compared with the human interpretation of the same signals.

6.6.2 *INSPECTION OF AIRFRAME STRUCTURES*

Next to general visual examinations, eddy current techniques are the most commonly used methods in the inspection of airframe structures. These methods require access to one side only and can detect cracks as well as material thinning due to corrosion. However, finding cracks and corrosion in multi-layer structures and under installed fasteners is not an easy task. A multitude of variables may influence the eddy current signals in a non-distinctive way, thus, influencing their detection capability. Important variables include:

- **Component design** (single layer, multi-layer, thickness of the layers, thickness variation, with or without inter-layer sealant or adhesive, interior stiffeners, doublers, or substructures).
- **Component surface condition** (painted or unpainted, local deformation, cleanliness, etc.), **fastener material and geometry** (conductivity, magnetic properties, size, flat or protruding, etc.).
- **Probe type and geometry** (coil/sensor design, size, placement, etc.).
- **Probe-component coupling** (paint thickness variation, lift-off effects, etc.).
- **Flaws** (type, size, location, etc.).

Some examples of the work that deal with these issues are described in the following sections.

6.6.2.1 DETECTION OF CRACKS IN AIRFRAME STRUCTURE

Cracks emanating from fastener holes can be detected using the conventional eddy current approach when they extend beyond the fastener heads. However, cracks located in the inner layers of lap joints are very challenging to identify. Detection of deep discontinuities may be possible by using low excitation frequencies and highly sensitive receivers. As mentioned before, Hall and GMR probes respond to the magnitude of the field and are more sensitive to low-frequency/low-amplitude signals associated with deeper discontinuities.

Eddy current inspection for cracks originating at fastener holes is customarily performed by removal of the fasteners or rivets, followed by testing of the inside surfaces of the empty holes using an appropriate spinning probe. This approach is very time consuming considering the large number of fasteners used in airframes as it requires removal of the fasteners. On the other hand, the detection of cracks under installed fasteners is challenging due to the unknown crack orientation and the presence of the fasteners. The induced eddy currents flow into the lap-joint material as well as the metallic fasteners that are conductive and reduce the current density where the crack is located. The difficulty increases when the fasteners are magnetic (e.g., Alodine coated rivets, steel fasteners). The protrusion of the fastener heads often create additional problems and prevent the probe from scanning at reasonable speeds.

Although modern airframe structures are assembled using flush-head fasteners [14], a thin layer of Teflon tape must be placed on the row of fasteners to allow the free movement of the probe while creating a uniform lift-off. Without the Teflon tape, lift-off variations reduce the ability of the eddy current approach to identify cracks, especially at deeper layers.

In the work described in reference [14], three different probe designs were evaluated in terms of their ability to detect EDM notches in the second layer of a riveted panel as below:

(i)   A reflection-type probe with concentrically positioned driver and pick-up coils.

(ii) A sliding differential probe, where the driver and receiver were placed side-by-side with the distance in between optimized to the required detection size.

(iii) A linear-field absolute probe, which used a driver coil wound in such a way that a uniform magnetic field was generated in the direction parallel to the specimen surface while the pick-up coil was oriented in the normal direction to detect the magnetic field perpendicular to the specimen surface. This probe is referred to as a self-nulling probe.

Using the lift-off intersection point (LOI), the detectability of EDM slots in the fastener holes of the second layer was assessed for the three probe designs and a correlation of the results with the EDM sizes was obtained. As shown in Figure 6.30, the linear field probe performed the best amongst the three designs mentioned above, both in terms of detection and size estimation. When the eddy current signal amplitude was used instead of LOI, it was still possible to identify EDM slots, but the correlation of the amplitude with size was not quite linear as shown in Figure 6.31. This figure was obtained using the sliding probe. The study showed that the probe's sensitivity to magnetic field magnitude rather than its rate of change is important for increased detectability of flaws in the second layer of lap-joint structures.

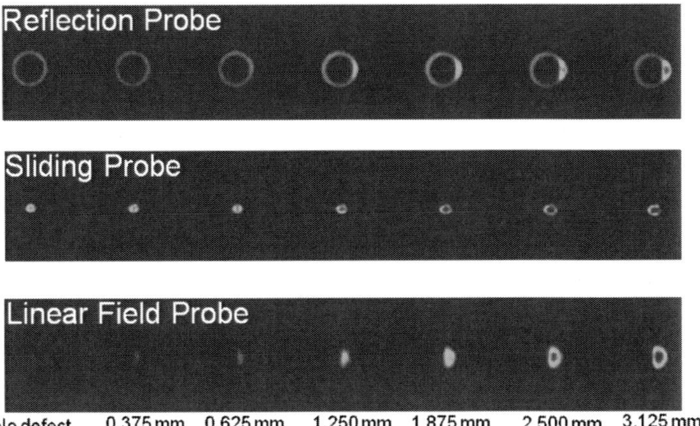

Figure 6.30 Pulse eddy current inspection results obtained using three different probe designs on a two-layer lap-joint specimen containing EDM notches in the fastener holes of the second layer. In all three scans, the lift-off point of intersection is monitored and mapped. The presence and size of EDM slots are represented by the intensity of the resulting traces.

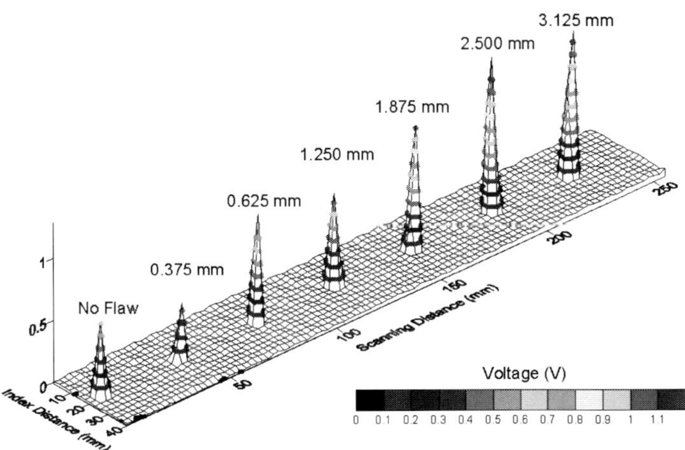

Figure 6.31 EDM notches in the second layer of a lap-joint specimen identified using pulsed eddy current method by scanning the sample with a sliding probe and monitoring the eddy current signal amplitude instead of LOI [14].

In a similar study explained in reference [23], it was shown that solid-state sensors are also capable of mapping the magnetic field directly rather than its rate of change, and their sensitivity is independent of the signal frequency. Thus, they have the ability to pick up a small magnetic field generated by a low current density that flows around the crack tip. In practice, a combination of pulsed eddy currents and solid-state detectors, such as GMR, are probably the most viable. This hypothesis was investigated using a homemade GMR probe consisting of two commercially-available elements, an Eddytech S/N 10005 coil as the driver and a NVE Corp. AAL002 solid-state device as the receiver [23]. The coil-sensor arrangement was such that the incident magnetic field was perpendicular to the specimen surface, but the sensing direction was aligned parallel to the surface (self-nulling). A homemade GMR probe design and shape are shown in Figure 6.32. Because of this arrangement, GMR probes are not sensitive to the excitation field but are sensitive to any field distortions along the surface of the inspection part.

In the work described in reference [11], the above-mentioned probe was tested on a three-layer lap-joint panel with three rows of flat-head Alodine rivets. EDM notches of three different lengths: 4.8 mm, 3.2 mm, and 1.6 mm, and a constant opening of 0.175 mm and depth of 1 mm were present in the rivet-holes of all three layers. Each layer was 1 mm thick; thus, the EDM slots were cut through both surfaces of the affected layer. A schematic diagram of the panel is shown in Figure 6.33. In the first layer, due to the countersinking, the visible surface-length of the EDM slots was shorter by 0.36 mm (i.e., fastener dead radius), but they extended to the same radial length from the hole as those in the second and third layers. However, the shortest EDM slots were hidden under the fastener heads, regardless of their location in the panel.

Two separate scans were performed, one with the sensitive axis of the GMR sensor along the length of the EDM slots (x-axis) and the other normal to the length (y-axis). The magnetic field in the plane parallel to the specimen's surface was obtained by the vector addition of the two orthogonal components, as indicated in equation (6-18).

$$B_{surface} = (B_x^2 + B_y^2)^{1/2} \tag{6-18}$$

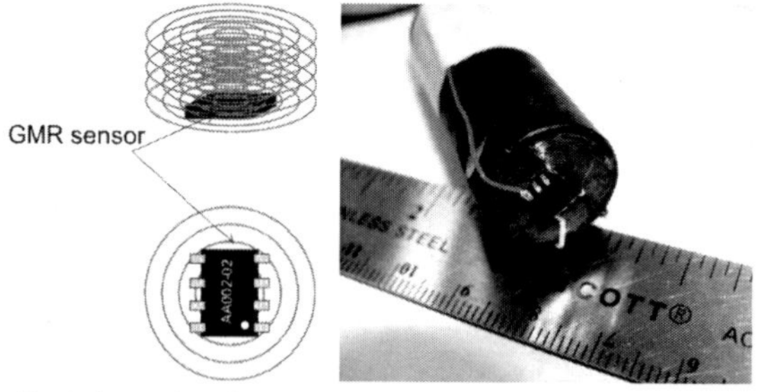

Figure 6.32 (Left): Schematic diagram of a GMR solid state probe design. (Right): A homemade GMR probe consisting of a pancake-shaped excitation coil and a solid-state sensing element [23].

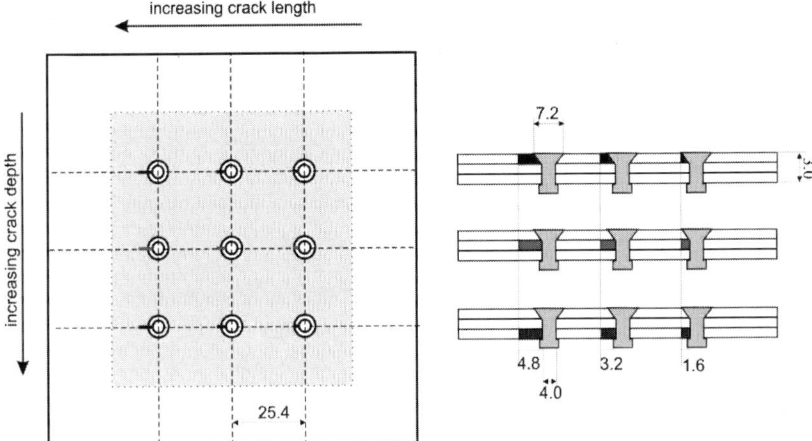

Figure 6.33 Schematic representation of a three-layer lap-joint specimen containing EDM notches of three different sizes in the vicinity of fasteners in all the three layers [11].

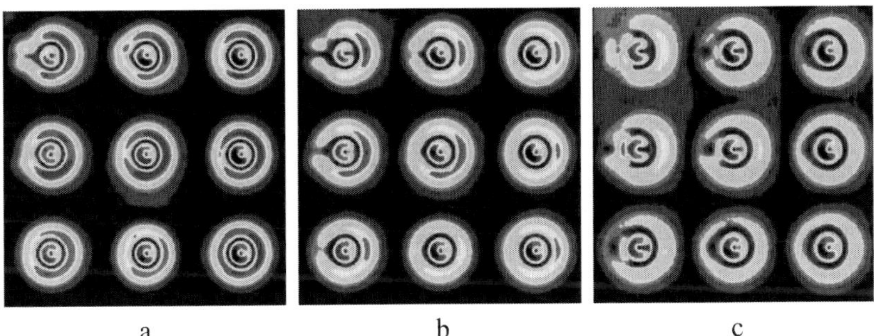

Figure 6.34 C-scan images of the above-mentioned specimen showing the magnetic field sensed by the GMR sensor at different time gates. (a): after 60 µs, (b): after 100 µs, and (c): after 180 µs [11]. Distortion of the outer rings indicates the presence of defects and the extent of distortion is indicative of the defect size.

This value was measured for each scanning point and plotted as a function of the probe location to generate C-scan images shown in Figure 6.34. The C-scans here represent the magnetic field amplitude on the surface recorded by the GMR sensor at three different time gates. The longer the time, the higher the diffusion of the eddy current into the sample (i.e., penetration into deeper layers).

As seen in the C-scans, the EDM slots produce different degrees of distortion in the sensed magnetic field depending on the size as well as the depth location. While at the 60 µs time-gate, most of the larger EDM slots and those that are closer to the surface are identifiable. At the longer time-gate of 100 µs, deeper and smaller notches become more evident until at 180 µs when all the EDM notches, even those smaller than the fastener radius in the third layer, are identifiable. The resulting traces indicate that with this approach, it is possible to make an estimate of the discontinuity size. However, like many other NDT approaches, size estimation is by comparison of the C-scan images with those of reference samples containing known discontinuities. The part used above is an example of a reference sample that could be used when inspecting for fastener-hole cracks in actual three-layer lap-joint structures of aircraft fuselage.

Although the above results were obtained when the GMR probe was excited with square waveforms for pulsed eddy current inspection of a multi-layer specimen, it can also be used with sinusoidal excitation for conventional eddy current testing. The latter approach would be more suitable for inspection of the first layer, very thin multi-layers, or low-conductivity materials where the low depth of penetration is not an issue, but the high sensitivity and resolution of a GMR sensor is needed.

6.6.2.2 DETECTION OF CORROSION IN AIRFRAME STRUCTURES

When assessing the condition of multi-layer aircraft structures using NDT methods, there is often a requirement to identify hidden corrosion in the affected layers in addition to fatigue cracks. The NDT methods for aircraft corrosion inspection need to be sensitive to thickness loss of at least 10% or more in each layer while being simple, fast, quantitative, and suitable for large-area inspections. Eddy current testing is considered to be the most suitable as compared to the other methods that often present serious limitations. For example, ultrasonic testing is only applicable for the inspection of the upper skin, since the ultrasonic waves do not propagate through air gaps that usually exist between the layers. Radiography, on the other hand, requires access to both the outer and inner surfaces for source and film positioning. Thermography and shearography are also ineffective for the detection of corrosion in deeper sections of metallic parts.

Eddy current methods need only single-side access to the inspection site, are relatively easy to implement in the field, do not require coupling, and can provide quantitative information. In chapter 12, "NDT of Corrosion in Aluminum Aircraft Structures," the application of eddy current methods for corrosion and crack detection in airframe lap-joint components has been described in more detail. In this chapter, specifics of the eddy current approaches when used for corrosion measurement in actual aircraft fuselage lap-joint structures are provided.

The conventional single-frequency eddy current tests are generally dependent on the probe frequency, and the results lack information necessary to completely characterize corrosion or loss of material thickness. Generally, high frequency scans provide indication of corrosion in the first layer; while, at lower frequencies, second layer corrosion can be mapped.

Figure 6.35 shows a schematic diagram and the actual picture of a two-layer lap-joint sample removed from a retired aircraft fuselage that contained hidden natural corrosion. The sample consisted of two 1.27 mm thick 2024-T3 Al sheets attached to a stringer with a row of fasteners along the middle of the lap-joint. Figure 6.36 shows conventional eddy current inspection results obtained using three different probe frequencies. The images provided here all identify corrosion sites mostly toward the two ends of the sample, but each frequency provides a different indication that cannot be easily related to the depth location of corrosion (first or second layer). Also, the extent of material loss due to corrosion cannot be easily determined.

Multi-frequency eddy current measurements can combine the results to provide a more accurate measure of the condition of a component. Some commercial instruments feature frequency mixing functions that are shown [24] to be useful in reducing the effects of inter-plate gap variations when inspecting for second layer corrosion. However, the multi-frequency approach is still incapable

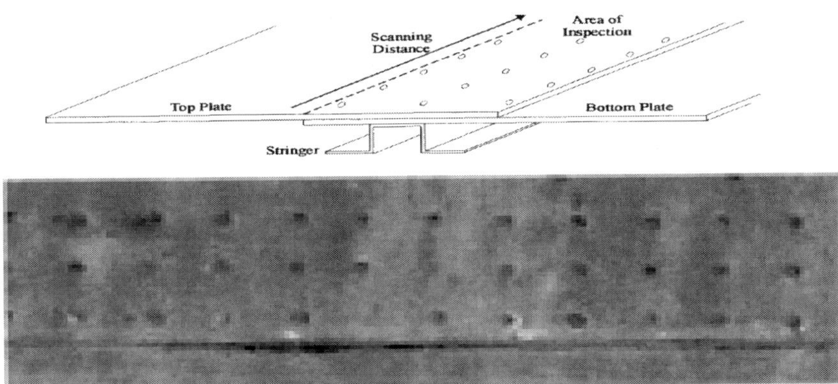

Figure 6.35 (Top): Schematic of an aircraft lap-splice joint. (Bottom): Photograph of a segment of fuselage lap-joint with natural corrosion [26].

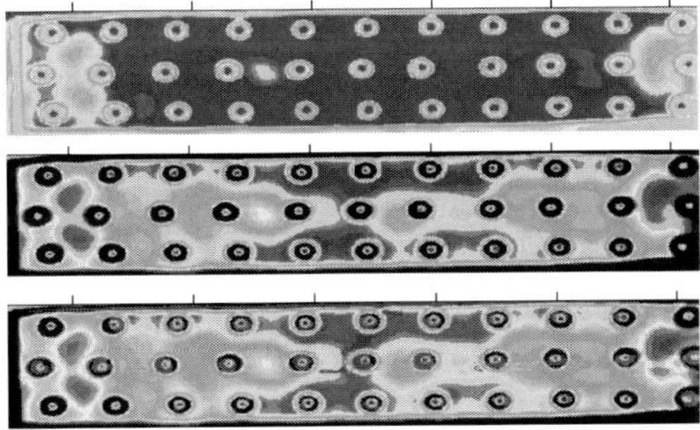

Figure 6.36 Inspection of a two-layer aircraft lap-joint using conventional eddy currents at different frequencies. (Top): 6 MHz. (Middle): 10 MHz. (Bottom): 12 MHz. Inter-layer corrosion is identified at both ends, however, the depth location and extent of corrosion cannot be determined [26].

of quantitative measurements. Swept frequency measurements using impedance analyzers are reported [25] to perform better for quantitative corrosion characterization, especially when they are interpreted with theoretical models.

These techniques, however, are much too laborious for practical applications and do not lend themselves to scanning approaches. In contrast, the pulsed eddy current technique uses a broad frequency spectrum that contains depth information and can be done using conventional C-scanning. In practice, the voltage vs. time signal of the pulsed eddy current is broadened and delayed as it travels deeper into a dispersive material. Therefore, anomalies that are closer to the surface will affect the eddy current response earlier in time than deep flaws. Figure 6.37 shows the result of a pulsed eddy current inspection on a similar sample to that described above carried out using a sliding probe with the receiver coil located co-axially within the drive coil. The outer diameter of the drive coil was 14 mm and that of pick-up coil was 6.35 mm. The diameter of the ferrite-core was 1.6 mm. The probe was designed to operate with a center frequency of 12 kHz. An inter-layer corrosion site is identified indicating thickness changes of 3 to 10% occurring between the first and second plates when compared with the reference scale shown below the image. The reference scale is based on calibration with respect to samples of known thickness and the same material examined using the identical test parameters.

6.6.3 INSPECTION OF FRICTION STIR WELDS

Friction stir welding (FSW) is a solid-state joining process that is being considered as an alternative to conventional mechanical fastening or adhesive bonding. It uses the heat generated from a friction tool to plasticize and bond metals in a highly controllable and repeatable manner [27, 28]. Since its invention by the Welding Institute in 1991, FSW has gained popularity in the aerospace, automotive, railway, and naval industries. The technique uses a rotating tool, as shown schematically in Figure 6.38, whose pin and shoulder generate enough heat to plasticize and mix the material without melting it. Unlike conventional welding, welds created in this way have a fine-grained structure with no entrapped oxides or gas porosity. A single tool can be used for welding materials in lengths up

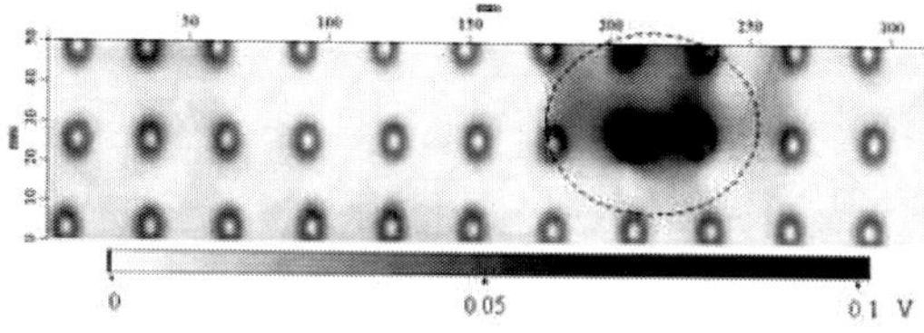

Figure 6.37 Pulsed eddy current inspection results of a two-layer lap-joint with inter-layer corrosion.

to 1000 m; however, the travel and pin rotational speed, its down force, diameter, and geometry, have to be optimized for specific situations. Lower costs, reduced weight, increased strength, and faster manufacturing contribute to the attractiveness of FSW for aerospace applications. The process is being considered as an alternative to bonding and riveting for the manufacture of large fuselage components.

Validation of the desired weld quality is done by both destructive and nondestructive testing. The types of discontinuity that may exist in friction stir welds are different from those of conventional welded joints. Kissing bonds, voids, lack of penetration, conductivity changes, and inconsistent weld depth are typical defects that require identification by NDT. There are challenges in NDT of FSW after manufacturing, especially due to the different discontinuity types and the random orientation of the defects that may exist. Also, there is inadequate knowledge of the types of damage that components with FSW may experience in service.

The initial work at Institute for Aerospace Research (IAR) described here concentrate on the NDT of FSW for the identification of manufacturing defects. The results may provide some indication of the possible testing challenges that may occur later when the friction stir welded components have been used in service. Typical discontinuities found during manufacturing of friction stir welded parts include:

- **Voids** that are subsurface volumetric air pockets with no material.
- **Worm holes** are identified as voids that are aligned in the direction of the weld.
- **Lack-of-penetration** is a term used to describe a gap between adjacent surfaces that should have been filled with the material.
- **Kissing bond** is a condition where the welded materials are physically mixed but there is no metallurgical bonding between the adjacent surfaces.

In addition to these discontinuities that must be detected by NDT, the electrical conductivity of the friction stir-welded joint must be measured as it provides information on properties, such as fatigue resistance, residual stresses, and hardness. An example of conductivity mapping is provided later in this chapter.

Of the above, lack-of-penetration is the most common type of manufacturing defect. Thus, the work described in [29] deals with the eddy current detection of this type of discontinuity in FSW. In this study, 400 mm × 200 mm coupons of Al 2024-T3 were machined from a 2.56-mm thick plate, placed on a steel-backing anvil, clamped along the two long edges, and friction stir welded. The welding tool was made of H13 steel and consisted of a cylindrical threaded retractable pin having a diameter of 6.3 mm and a smooth concave shoulder of 19 mm in diameter. Schematics of the FSW tool, its cross-section, and the welding process are shown in Figure 6.38.

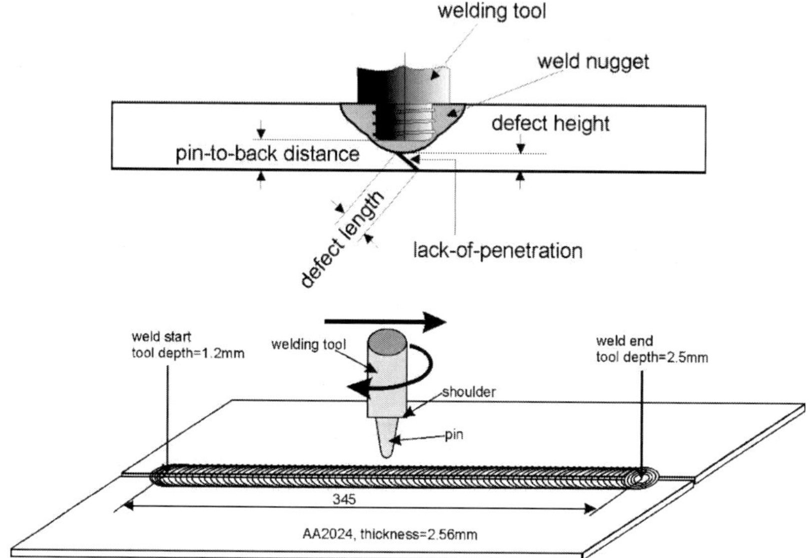

Figure 6.38 Schematics of a friction stir weld and the welding process [27].

The tool uses a retractable pin, whose length varies linearly from the beginning to the end of the weld, while keeping the shoulder penetration constant at 0.2 mm. During the welding process, the tool (i.e., pin and shoulder) penetration increases linearly from 1.2 mm at the weld beginning to 2.5 mm at the weld end. This introduces a variable weld depth and, consequently, lack-of-penetration of decreasing depth as the welding progresses. The welding path was parallel to the rolling direction of the plates. The welding was carried out at a rotational speed of 1000 RPM, a traveling speed of 10 mm/s, and a tilt angle of 2°.

Conventional eddy current tests were performed by using a 400 kHz sinusoidal waveform as the excitation signal using an absolute probe-coil. The probe was connected to a commercial instrument that was interfaced with a computer. The scans were performed from the far-side of the weld, which contained the discontinuities.

In Figure 6.39, the results of this inspection are shown in two sections, identified with the coordinate along the weld: 0 mm for the beginning of the weld and 348 mm for its end. The change of color is due to the change in the coil impedance induced by the test piece. Minute variations in the distance between the probe and specimen along the weld length create lift-off effects; thus, the results presented in Figure 6.39 are after lift-off compensation at the acquisition time using previously-described procedures. Moreover, the raw data is post-processed by normalization in order to display discontinuities more clearly.

The lack-of-penetration is seen as an almost horizontal line from the start of the weld up to about 168 mm (narrow red, yellow, and light blue lines). After this

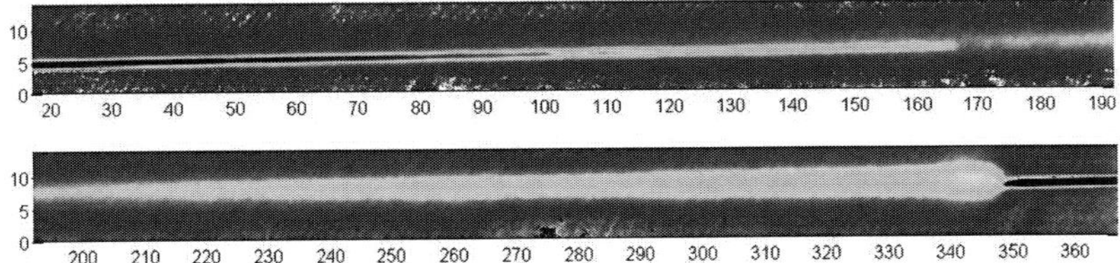

Figure 6.39 Conventional eddy current results of the inspection of a FSW with lack-of-penetration of decreasing depth (from left to right up to 168 mm), well-welded area (168 to 348 mm), and lack of weld (after 348 mm) [29].

point until close to the end of the weld (wider blue trace from 168 to 348 mm coordinates), the two plates appear to be well-joined. The eddy current response corresponding to larger and continuous lack-of-penetration (red line) appears as a long crack along the length of the weld.

Destructive testing was carried out to verify the NDT results. The test involved cross-sectioning and metallographic examination of samples taken from along the weld length. The results confirmed that the lack-of-penetration continued up to about 168 mm from the beginning (the thin red line in of the upper eddy current scan of Figure 6.39), followed by a well-welded segment (wider blue trace), and, finally, an un-welded gap toward the end of the weld (red line in the lower ECT scan). The change of color in the wider blue indications of the welded areas may be due to the change in the electrical conductivity in the absence of defects. A comparison of the ECT scans and metallographic results indicated that eddy current testing is capable of detecting lack of penetration as little as 200 μm in depth. These results were also verified with liquid penetrant testing [29].

6.6.4 Inspection of Thermal Barrier Coatings (TBC)

The hot-section components of aircraft engines are often coated with a thermal barrier layer to protect them from high-temperature degradation in order to extend their service lives. A TBC system consists of a ceramic topcoat and a metallic bond-coat sprayed or vapor deposited onto the metal substrate. TBC failure is often associated with the formation of thick layers of mixed oxides at elevated temperatures that cause internal stresses leading to the final disbonding or spallation of the TBC. Depending on the bond-coat composition, thermally-grown oxides (TGO) may consist of chromium oxides or chromia, nickel oxides, spinel ($MgAl_2O_4$), or alumina (Al_2O_3). The TGO plays an important role in bonding the outer ceramic layer to the bond coat. However, it is essential that the TGO is a thin layer of alpha alumina rather than a thick layer of spinel. The latter, due to the mismatch of the thermal expansion with the original bond-coat and topcoat, may result in micro-cracking and eventual separation of the coating [1]. NDT is needed to identify extensive formation of spinel before spallation occurs.

In the study described in reference [30], two sets of circular coupons were manufactured using a plasma spray technique as typically used in the TBC coating of aircraft engine parts. In both sets, the metallic substrate was Hastalloy-X (50 mm diameter and 2 mm thickness) and the ceramic top-coat was Metco 204NS (7wt% yttria-stabilized zirconia). The thickness of the ceramic topcoat was 0.25–0.30 mm. Two types of bond-coat powders were used in the fabrication of the specimens. In the specimens identified by "G," a pre-alloyed NiCoCrAlY bond-coat powder was used. The elemental composition of the powder in weight percent was: 23Co, 18Cr, 13Al, 0.2Y, and the balance Ni. The specimens identified by "P" were made of a powder with 2.5Co, 17.5Cr, 5.5Al, 0.5Y, in weight percent of each element and the balance Ni. This powder had a larger particle size distribution and a greater amount of segregation than the "G" powder resulting in a less uniform coating and poorer quality specimens.

Some specimens were exposed to temperatures up to 1250°C, which was higher than the typical engine operating temperatures, to produce oxidation in practical times and also to simulate temperature spikes of the engine or hot spots in the components. The specimens were provided in the as-sprayed and thermally-exposed conditions. Pictures of the specimens are shown in Figure 6.40. Small Al tapes were attached on all specimens for identification of their orientation during the tests and also for the calibration of inspection techniques.

These studies, described in references [31] and [32], indicated that the degradation of the bond coat would produce magnetic oxides, which can be discriminated on impedance plane diagrams of eddy current signals, as illustrated in Figure 6.41. Thus, eddy current tests were performed along with other NDT inspections (e.g., ultrasonic and thermography) to establish the potential of each approach. Conventional eddy current measurements were carried out using an absolute coil (diameter 3 mm) with a ferrite core (diameter 1 mm) at a frequency of 50 kHz. The scanned area was 35 mm × 35 mm, while the scanning increment was set at 0.2 mm.

The eddy current amplitude C-scan images obtained on the "G" and "P" TBC samples for the as-manufactured and heat-treated conditions are shown in Figure 6.42 and Figure 6.43. These images are after normalization with respect

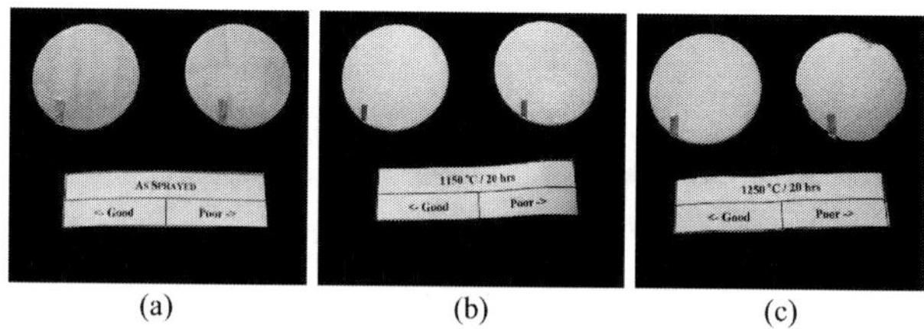

Figure 6.40 Pictures of the TBC coupons investigated. (a): as-sprayed, (b): exposed to 1150°C for 20 hours, and (c): exposed to 1250°C for 20 hours. For each picture the left coupon is made of a "good" bond-coat powder identified by "G" and the one to the right is made of a "poor" bond-coat powder identified by "P."

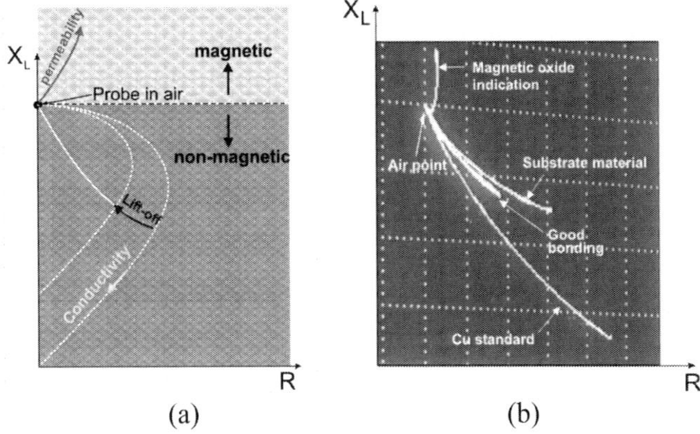

Figure 6.41　(a): Schematic representation of the impedance plane diagram in eddy current testing. (b): Actual eddy current impedance plane diagram responses from the inspection of a TBC coupon.

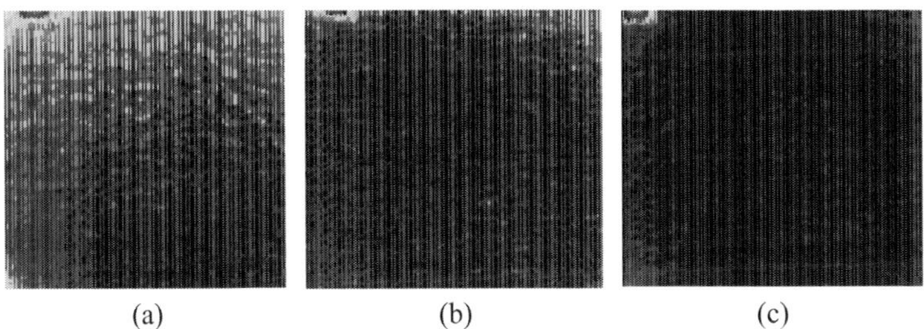

Figure 6.42　Eddy current amplitude images of sample "G." (a): as-sprayed, (b): exposed to 1150°C for 20 hours, and (c): exposed to 1250°C for 20 hours.

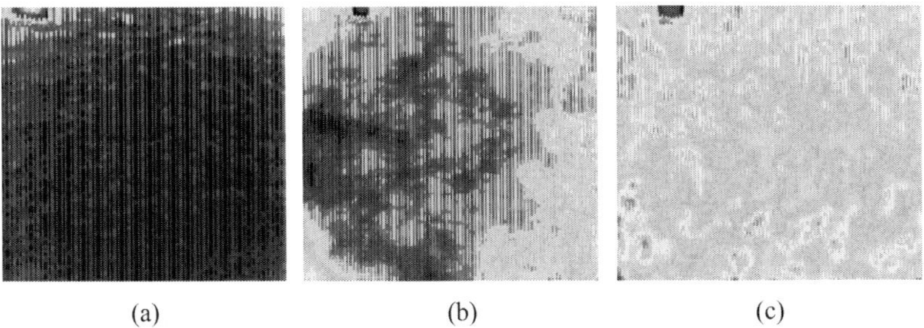

Figure 6.43　Eddy current amplitude images of sample "P." (a): as-sprayed, (b): exposed to 1150°C for 20 hours, and (c): exposed to 1250°C for 20 hours.

to the ECT response produced by a small aluminum reference tape attached to all samples. The rectangular indication seen in the images is due to the reference Al tape. Because the electrical properties of Al are constant in all cases, its eddy current response should remain the same for all tests. Thus, the ECT signal responses of the samples were normalized with respect to their reference Al signal response.

The ECT C-scans clearly show differences between the as-sprayed samples and those exposed to high temperatures in the case of the "P" specimens that were made of a poor bond-coat powder. In contrast, the ECT results of the "G" samples that were made of a good bond-coat powder are very similar. Figure 6.44 shows SEM micrographs of the cross-section of both "G" and "P" samples that had been exposed to 1250°C for 20 hours. The formation of a rather uniform thin alumina layer at the original interface between the ceramic topcoat and bond-coat can be observed in sample "G." In this sample, only a small cluster of other oxides (i.e., spinel, chromia and nickel oxides, or mixed oxides) can be found in the interfacial region or within the bond-coat. In contrast, the microstructure of sample "P" showed a thick layer of mixed oxides at the interface through which cracking has occurred. Stresses associated with the volume change after the formation of oxides combined with thermal mismatch in the TBC sample lead to cracking of the bond-coat and eventual spallation of the TBC.

This study indicated that the eddy current testing has the potential to identify samples with extensive formation of mixed oxides. Thus, the technique is useful for inspecting service-exposed engine parts that are coated with TBC to screen out those containing severe oxidation of the bond-coat in order to reduce TBC failures during engine operation.

6.6.5 *INSPECTION OF NICKEL-ALUMINUM-BRONZE VALVES*

Nickel Aluminum Bronze (NAB) (9.5% aluminum, 5% nickel, 5% iron with balance copper) is used in piping and naval industries due to its corrosion resistance. However, when in contact with sea water, NAB parts suffer a specific type of subsurface corrosion called "de-alloying," which reduces the iron and nickel content and decreases the material strength and ductility. De-alloying corrosion occurs in

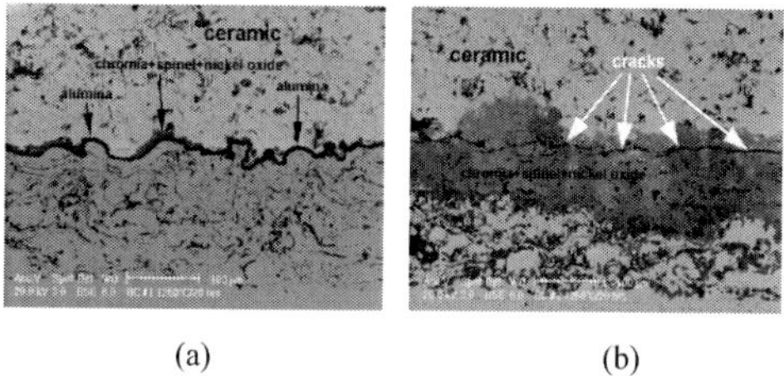

(a)   (b)

Figure 6.44 SEM micrograph of (a) sample "G" and (b) sample "P" after 20-hour exposure at 1250°C showing more oxides and cracking at the interfaces.

ship propellers, seawater pumps, and valves that are made of NAB alloys. This type of corrosion does not display the usual metal-loss characteristics, a fact that makes its detection very difficult.

In a feasibility study described in reference [33], electromagnetic testing was applied to the detection of de-alloying corrosion in a NAB valve, shown in Figure 6.45. Since de-alloying reduces the iron and nickel contents, it is expected to present changes in electrical conductivity and magnetic permeability, parameters that can be measured with electromagnetic techniques. In this study, a service-exposed NAB valve was cut in half and tested in nine locations marked in Figure 6.45. Five locations were chosen on the internal surface of the valve (nos. 2, 4, 5, 6, and 7), three locations on the cross-section surface (nos. 1, 3, and 8), and one location on the external surface (no. 9). Only the interior surface of the valve had been in contact with the sea water and was expected to be affected by the de-alloying process.

Three electromagnetic approaches were used for testing the valve at the chosen locations: electrical conductivity measurement, conventional eddy current testing using impedance plane diagrams, and pulsed eddy current inspection. For conductivity measurements, a commercial conductivity meter was used along with an absolute probe. The driving frequency was set to 240 kHz and, prior to each set of measurements, the instrument was calibrated on four conductivity test blocks (i.e., 0.97%, 3.51%, 9.32%, 29.56% IACS). In Figure 6.46, the electrical conductivity values measured at each of the nine locations are displayed. The locations chosen from the inside of the valve clearly showed an increase in conductivity by about 50% over the conductivity of the points selected on the cross-section or exterior of the part.

Conventional eddy current inspections at the same locations were performed using a commercial instrument with an absolute probe driven at 200 kHz frequency. Three materials were used as references for each individual test: a ferromagnetic 1018 steel block, a titanium block with a conductivity of 0.97% IACS, and Al-Si-bronze block with a conductivity of 9.32% IACS.

The phase and amplitude of the eddy current response was adjusted in such a way that the ferromagnetic response signal from steel moves upward and close to normal to the X axis, in other words, zero resistance (X axis being resistance and Y axis being inductance, see Figure 6.3 and Figure 6.41 for interpretation), while

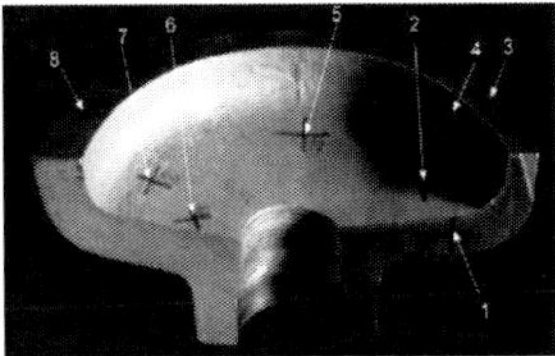

Figure 6.45 Pictures of the internal and external surfaces of a NAB valve. Crosses indicate locations of eddy current measurement.

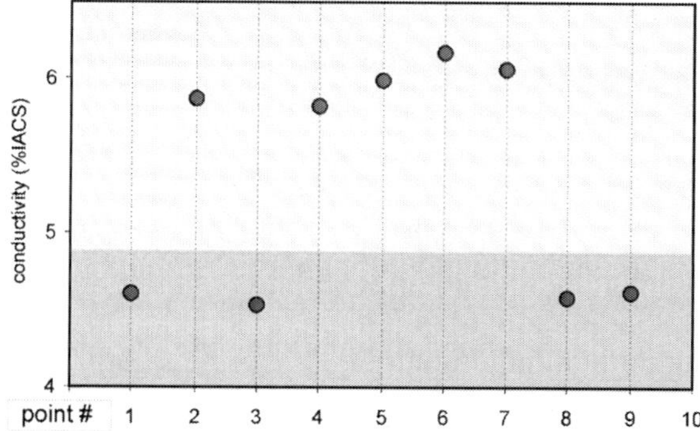

Figure 6.46 Electrical conductivity for the nine locations on the NAB quarter-valve.

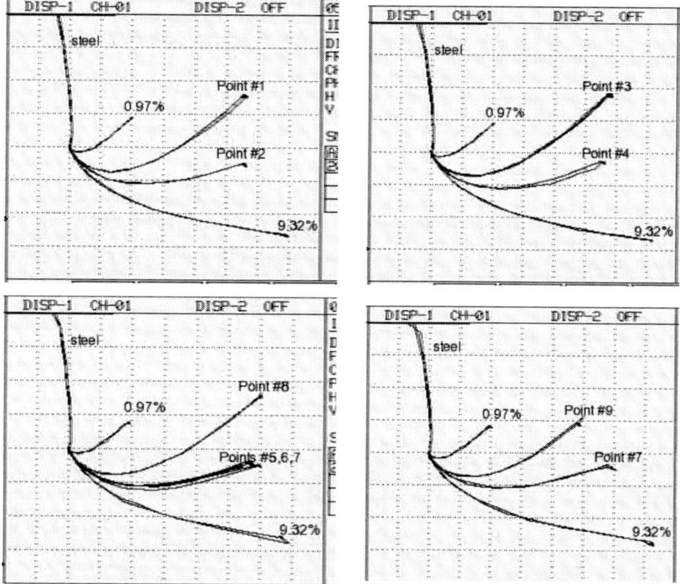

Figure 6.47 Impedance plane diagrams resulting from eddy current tests of various locations on the NAB valve along with the traces produced by the same test on conductivity reference blocks.

the traces of the two other reference blocks are at an angle as shown in Figure 6.47. The signals obtained from the different locations on the valve cross-section appear between the traces from the titanium and Al-Si-bronze references.

There is a clear distinction between the signal responses from locations 1, 3, 8, and 9 (i.e., exterior and cross-section) from those of the interior points 2, 4, 5, 6, and 7 on the impedance plane diagram. Here again, the locations on the cross-section or the external surface of the valve showed a lower conductivity than the points on the internal surface. In the above mentioned conductivity measurements, the ferromagnetic effect is not sensed. However, in impedance plane diagrams, this effect is displayed by the clockwise movement of the eddy current traces.

The pulsed eddy current approach was also employed for the evaluation of the de-alloying corrosion of the NAB valve. A commercial reflection probe, driven by a 1 ms period, half-cycle pulse, was used. Here, the response signal energy was plotted as shown in Figure 6.48. Again, there seems to be a clear separation between the internal locations 2, 5, 6, and 7 as compared to the others (1, 3, 8, and 9). This figure also indicates that the higher the PEC energy, the lower the likelihood of the location being de-alloyed.

The results of all three approaches confirmed that during the de-alloying process some of the less conductive elements (i.e., Fe, Ni, and Al) are released, resulting in an increase in the percentage of the more conductive copper, which causes an increase in conductivity. This is true for the internal locations 2, 4, 5, 6, and 7 that may have been affected by de-alloying and display higher conductivity than the unaffected external or cross-sectional locations 1, 3, 8, and 9. In addition, among the interior locations, no. 4, which corresponds to the side of the valve, shows the highest PEC energy of all the internal locations. This may indicate that the side of the valve is less affected than the rest of the valve's interior. Overall, the PEC method seems to be the most sensitive of the three and has the advantage of being capable of interrogating a larger volume of the material (due to a larger coil diameter and wider frequency band or deeper penetration) than the conventional eddy current or conductivity measurements.

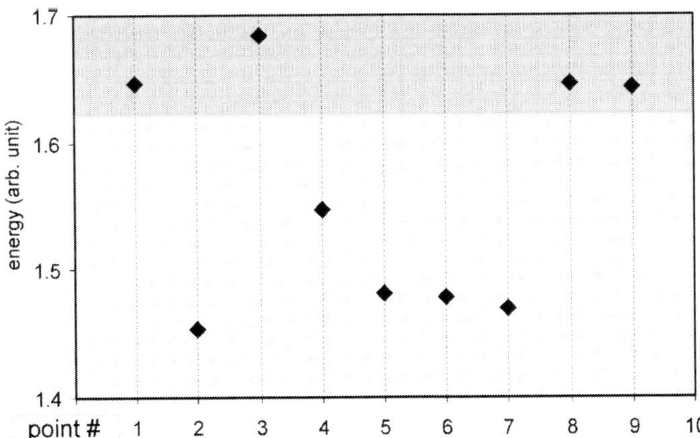

Figure 6.48 Pulsed eddy current energy levels measured at nine locations on the NAB valve.

6.6.6 CONDUCTIVITY MAPPING

Eddy current sensitivity to the electromagnetic properties of materials makes it a useful tool for identifying dissimilar conductive metals. This section presents two examples of mapping the electrical conductivity of dissimilar materials.

6.6.6.1 ALUMINUM-COPPER LINEAR FRICTION BONDS

Al-Cu assemblies are used as connectors in the electrical power generation industry. Such connectors are made by fusion of the two pieces using different processes, such as explosive welding or electron beam welding. However, in this type of connector, due to the difference in electrical conductivity of the constituents, a brittle inter-metallic phase of high resistivity forms at the Al-Cu interface, which reduces the efficiency and leads to eventual failure. Figure 6.49 shows an example of an Al-Cu joint manufactured through explosive welding before and after service. Linear friction welding (LFW) is a solid-state joining method that helps to overcome this problem.

The LFW process consists of oscillating one piece in a linear fashion while holding the second piece stationary and applying a constant axial force [34]. The friction between the two surfaces generates heat, which softens the material to the point of plasticity, though not fusion. Once sufficient plasticity occurs, the pre-set shortening distance is reached due to the axial force. At this point, the oscillation is halted, the two pieces aligned, and the forging power is applied to weld the two pieces together. The process is self-regulating in that heat is only generated as long as there is friction. Thus, if the material approaches the melting point, the viscosity is reduced causing a decrease in friction and the heat generation rate. The application of this emerging solid-state joining technology is particularly suitable for the assembly of dissimilar alloys. The rigorous control of the process parameters, which is possible with LFW, is especially attractive for minimizing the chemical reactions occurring at the interface that lead to the inter-metallic phase.

Two connector types manufactured using linear friction welding at high and low forging power inputs were provided for investigation along with one made by explosive welding. The micrographs of the cross-section of the two LFW joints

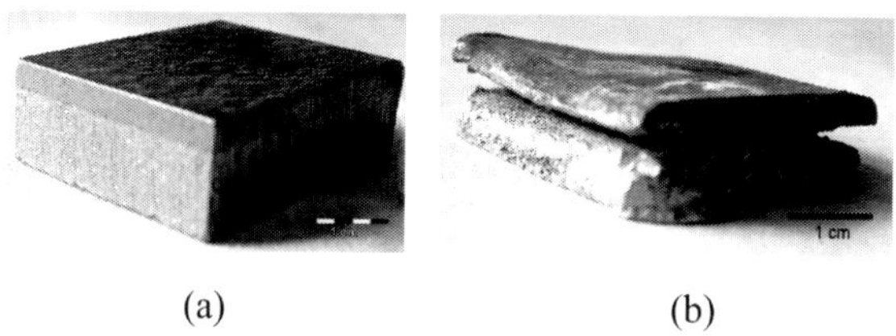

(a)　　　　　　　　　　(b)

Figure 6.49 Al-Cu connectors manufactured by explosive welding. (a): As-manufactured. (b): After 16 months of service [34].

are shown in Figure 6.50. Electrical conductivity measurements were carried out in order to evaluate the two axial force values used and the quality of the resulting LFW joints in the context of their suitability as electrical connectors. A commercial eddy current instrument equipped with an absolute coil-probe and the related conductivity reference blocks were used to measure the electrical properties of the welded joints. Since the eddy current signal phase is sensitive to conductivity changes, this characteristic of the signal was selected for joint evaluation.

Prior to the EC measurements, the conductivity of the probe was "nulled" in air and then placed on conductivity reference blocks. An area of 12 mm × 12 mm was scanned on each block, with an increment of 0.2 mm, resulting in 3600 individual measurement points. The phase angle for each reference block, which is related to its conductivity, was determined by averaging the phase values of the individual signals obtained while scanning the whole block. The averaging was done to minimize the effects of random errors.

Figure 6.51 presents the relationship between the measured average phase angle and the material conductivity for the seven reference blocks tested. This curve,

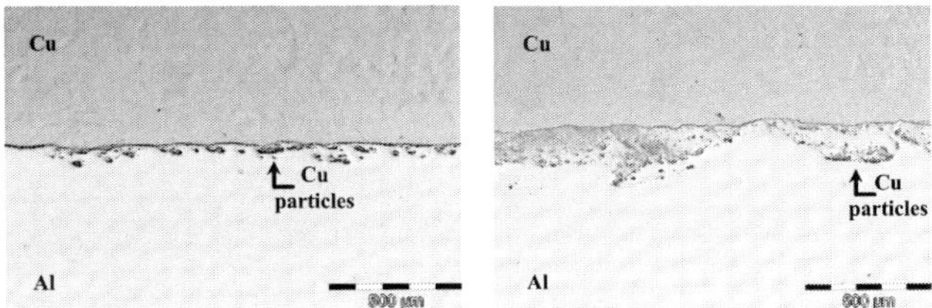

Figure 6.50 LFW Al-Cu specimens made using low power (left) and high power (right) [34].

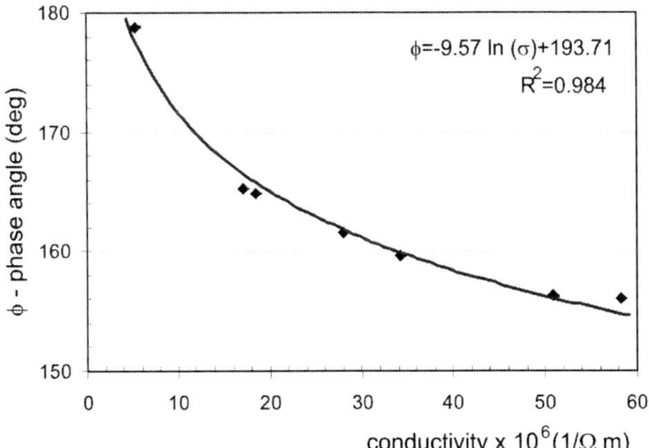

Figure 6.51 Calibration curve showing the eddy current phase angle as a function of electrical conductivity of the standard reference blocks.

which covers a wide range of conductivity values, was used as a calibration reference for determining the conductivity of the welded samples.

Eddy current area scans were performed to determine the conductivity of the three specimens made using linear friction welding at low power (LFW-LP), linear friction welding at high power (LFW-HP), and explosion welding (EW). The signal phase values measured across the weld were correlated with the absolute conductivity values obtained on reference blocks (Figure 6.51). Figure 6.52 shows the variation of conductivity across the interface for the three coupons referred to as conductivity profiles. The left and right vertical lines bordering the conductivity profiles are the conductivity values for aluminum alloy AA 6063 and OFE Cu (OFE stands for oxygen-free electronic copper that is 99.99% pure Cu with 0.0005% oxygen content). The conductivity values were normalized with respect to the conductivity of AA 6063 in order to allow comparison of the different welds.

It can be observed that there is very little difference between the conductivity profiles between LFW-LP and LFW-HP, but there is a small difference between the LFW and EW. In the LFW, the conductivity change is more gradual than that of the EW when approaching either Al or Cu conductivities. It is worth mentioning

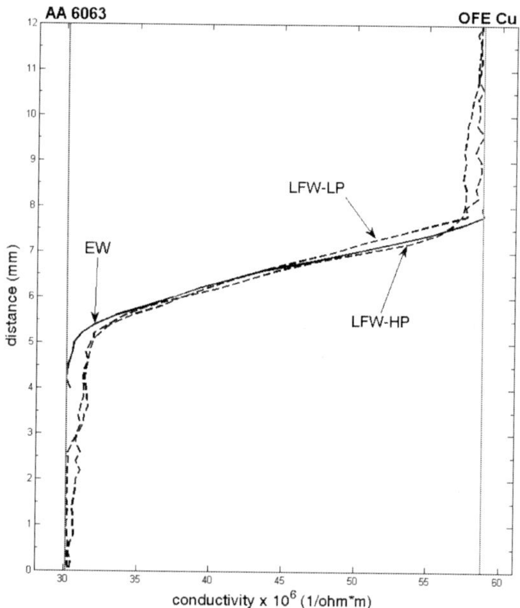

Figure 6.52 Conductivity profiling of the linear friction welded and explosive welded samples. The vertical borders represent conductivity values for aluminum alloy (left) and OFE Cu (right) [34].

that the width of the interface region, estimated on the basis of inflection points, is 2.4 mm ± 0.2 mm for both LFW welds and slightly larger (2.6–2.8 mm) for the EW weld. With a larger and more abrupt interface, the EW is expected to be less efficient and more prone to degradation than the LFW samples, and especially the LFW-HP sample.

6.6.6.2 Friction Stir Welding of Dissimilar Aluminum Alloys

In an attempt to represent a friction-welded structure of an aircraft (e.g., possible skin-stringer joint) that would be manufactured using different alloys, a lap-joint specimen was made of two aluminum alloys, as shown in Figure 6.53. The top Al layer was designated as UNS A97075 (UNS stands for Unified Numbering System), which had an electrical conductivity of 18.3 MS/m or 31.5% IACS and a thickness of 1.21 mm. The second Al layer designated as UNS A92024 (conductivity of 28.8 MS/m or 49.6% IACS) had a thickness of 2.57 mm. The two plates were friction stir welded to a depth of about half of the second layer with a weld surface width of about 6.2 mm as illustrated in Figure 6.53.

The pulsed eddy current technique was used with the goal of mapping the conductivity of the welded area that contained a mixture of the two Al alloys of different conductivities stirred at high temperatures. The PEC scan was performed from the top side of the weld (welding tool side) covering an area of 48 mm × 25 mm that corresponded to the beginning of a long weld line. As in the previous example, a calibration curve was established before the actual conductivity mapping using the lift-off point of intersection coordinates for seven conductivity standards having electrical conductivity values of: 5.4, 17.1, 18.5, 21.9, 28.1, 34.2, and 50.8 MS/m (or 9.32%, 29.52%, 31.82%, 37.78%, 48.44%, 59.01%, and 87.55% IACS).

The specimen was scanned twice, once with no lift-off and again with a 0.5 mm lift-off. As explained earlier, for each individual scan point, the LOI can be established as the intersection of the two resulting signals corresponding to 0 mm and 0.5 mm lift-off values. Either the LOI time or amplitude can be used for conductivity measurements. The LOI amplitude displayed a larger change; thus, it was chosen for conductivity mapping. An example of the resulting conductivity map is shown in Figure 6.54.

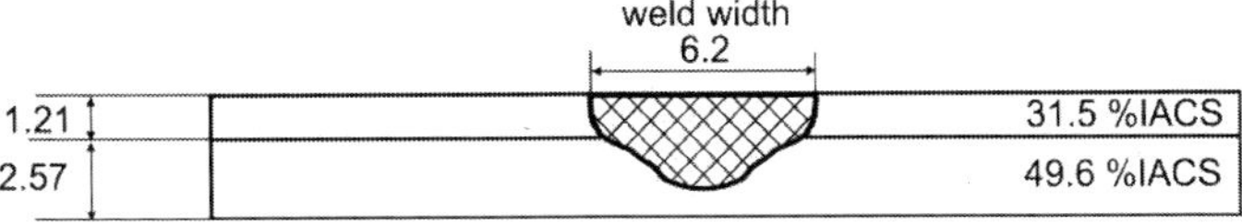

Figure 6.53 Schematic representation of a lap-joint consisting of two dissimilar aluminum plates that were friction stir welded [35].

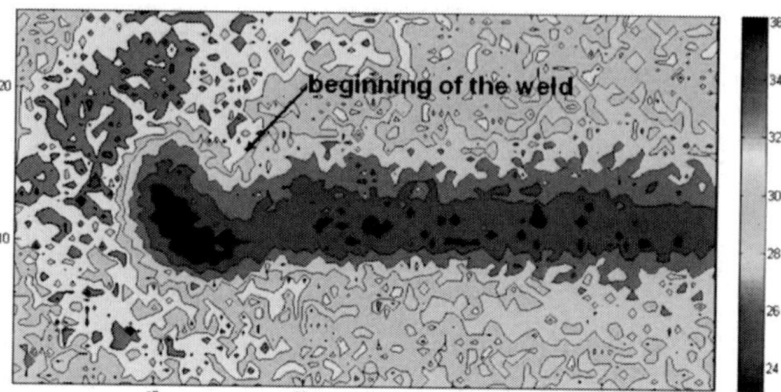

Figure 6.54 Pulsed eddy current surface map of conductivity of two dissimilar aluminum alloys friction stir welded together in the middle [35]. The units of both axes are mm.

This figure shows that in the region surrounding the beginning of the weld where the pin plunges into the material, the conductivity changes between 31% and 35% IACS (light and moderate gray area surrounding darker core). In this region, the conductivity is slightly higher than that of the top-layer but much lower than the second layer. On the weld itself (darker gray and black in center), the conductivity is 25–26% IACS, which is less than that of the top layer and much lower than for the second layer material. Away from the weld in the transverse direction, the conductivity is close to that of the bulk top material, indicating no significant influence from the weld.

The lower conductivity of the weld mixture as compared to both Al alloys may be due to the complex metallurgical changes that occur in this zone, such as the formation of elongated grains, which resist the flow of the free charges in the metal and lower conductivity. It must be noted that, although the conductivity was measured at the surface, the values represent the conductivity of the bulk material as the electromagnetic field produced and sensed by the pulsed eddy currents probe extends several millimetres into the material. The PEC-generated conductivity maps are not only influenced by the diffusion of the electromagnetic field, but also by the size of the coils used in the probe.

6.6.7 *Eddy Current Modeling*

Algorithms and software packages have been developed for modeling and simulating the electromagnetic techniques. ECSIM is an Eddy Current SIMulation software package that is based on boundary elements and solves quasi-static electromagnetic problems. Modeling using ECSIM requires input parameters, such as the dimensions, number of layers and windings of the coils, the geometry and electrical properties of the test piece, the probe or coil direction with respect to the test piece, the scan and index axes, etc. As output, one is able to simulate the coil response, the generated electromagnetic field in a conductive test piece, and the eddy current flow in the part. Simple examples are provided in Figure 6.55.

In the example presented in Figure 6.56, ECSIM is used to simulate eddy current signals when an absolute probe is used to detect a square-shaped notch in a flat conductive plate. The test plate is an aluminum alloy 2024-T3 with a conductivity of 31% IACS and relative permeability of 1. The notch is 1 mm × 1 mm and has an opening width of 0.1 mm. In this case, the coil has 100 turns, an inner radius of 0.5 mm, an outer radius of 1.5 mm, and a height of 2 mm.

The probe is excited with a 50 kHz AC current and is moved once along the notch length and then perpendicular to the notch length in such a way that the probe center follows the center line through and across the notch. Three different lift-off values (i.e., 0.20 mm, 0.25 mm, and 0.35 mm) are used. The simulated eddy current signals on impedance plane diagrams are presented in Figure 6.57. Note that no gain was applied in this simulation and the results are not normalized with respect to the probe impedance in air as is usually done in conventional eddy current measurement using impedance plane diagrams (e.g., Figure 6.47).

At 50 kHz frequency, which is typically used in eddy current inspections, the response signals for the three different lift-off values when scanned along the notch length are significantly higher in magnitude compared to those obtained when the probe is scanned perpendicular to the notch face. The sharp curvature of the notch tips creates a higher concentration and disturbance of the electromagnetic field

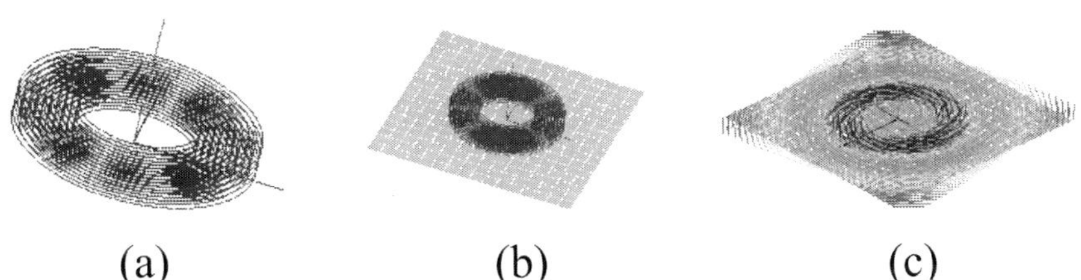

(a)   (b)   (c)

Figure 6.55 (a): An absolute eddy current probe as seen by ECSIM. (b): An absolute probe close to a conductive plate as seen by ECSIM. (c): Simulation of the induced electromagnetic field generated by an absolute probe in a conductive flat plate. Note that the field is strongest below the coil (dark gray ring), but the intensity gradually reduces both in the inward and outward directions.

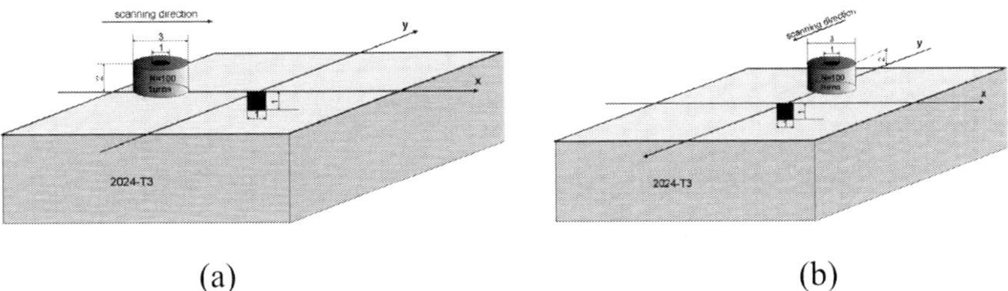

(a)   (b)

Figure 6.56 Simulation of a line scan eddy current inspection with an absolute probe. (a): Along an EDM notch direction. (b): Perpendicular to the notch plane.

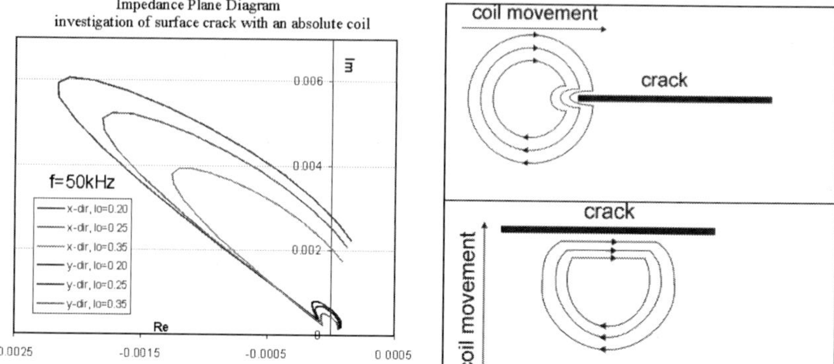

Figure 6.57 (Left): Simulation of eddy current signals on an impedance plane diagram for three lift-off values when the probe moves along the notch length (stronger signals) or across the notch length (weaker signals). (Right): Schematic of eddy current interaction with a notch (or crack) when the probe moves along or across the crack length.

than the notch face that is flat and thus a stronger signal. The simulation results indicate that the lift-off has a similar effect on the response signals regardless of the probe scan direction; that means the smaller the lift-off, the stronger the response signal.

In the second example, ECSIM is used to simulate eddy current inspection of a bolt-hole containing a semi-elliptical crack in its bore when a rotating probe is inserted into the hole. Since the ECSIM software package available at the time of this work was only capable of ECT simulation on flat pieces, the bolt-hole inspection using a cylindrically-shaped rotating probe was approximated to scanning a flat surface with a small flat probe (i.e., the bolt-hole was considered to have been cut along its length and opened into a flat plate). This approximation is graphically illustrated in Figure 6.58. In the simulation, the notch is considered to be in the axial direction of the hole that is the same as the probe insertion direction. Based on the previous results, in this direction the EC response would be stronger than in the circumferential or rotational direction.

For this simulation, the probe is considered to be a reflection-differential type that is designed for inspection of bolt-holes in the 4.75 mm to 6.35 mm diameter range with a comparatively small coil size. The reflection-differential probe consists of two coils: the outer cylindrical coil used as driver and the inner coil as pick-up. The diagram and dimensions are shown in Figure 6.59. The inner coil is split and wound in opposite directions on a ferrite core. This type of probe focuses the magnetic field and shows a higher sensitivity to cracks when the split line is parallel to the surface length or face of the crack. Such probes are used for inspection of bolt-holes in lap-joint structures as they are less sensitive to inter-layer gaps (coil split line must be parallel to the hole axis).

As in the previous example, the probe's characteristics and operating parameters, such as the coil dimensions, winding layers, and numbers, were entered as input parameters [36, 37, 38]. Also, the probe polarities, driving frequency, and core permeability, as well as the geometry and properties of the test piece, were entered into the software program and eddy current response signals pertaining to different conditions were simulated.

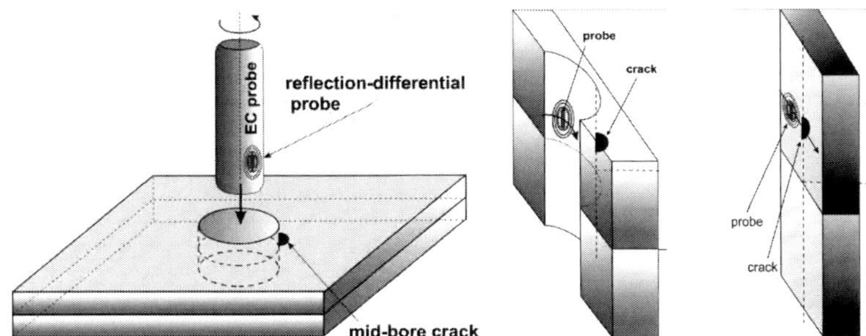

Figure 6.58 Graphical presentation of the transition of the bolt-hole inspection by a rotating coil-probe to scanning of a planar plate [37].

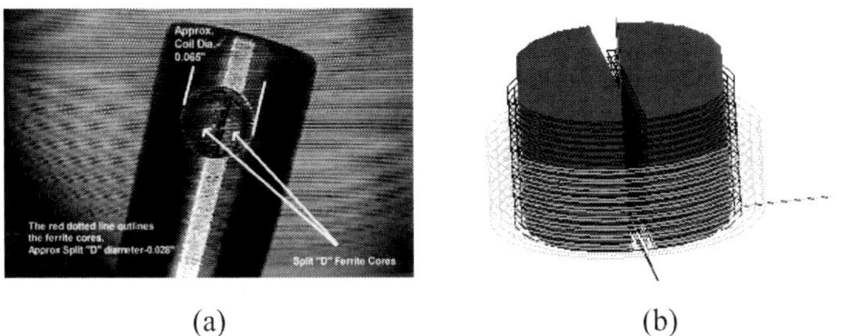

(a)   (b)

Figure 6.59 (a). A reflection-differential probe for bolt-hole eddy current inspection. (b): The reflection-differential split-D probe as seen by ECSIM.

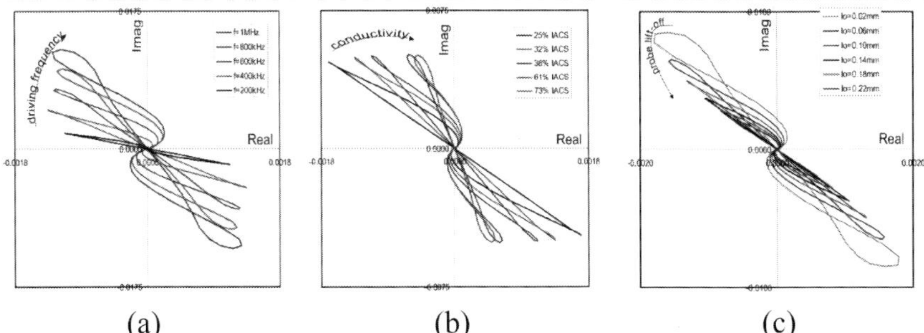

(a)   (b)   (c)

Figure 6.60 Simulation of eddy current response from a bolt-hole crack. (a): With change of driving frequency. (b): With change of material conductivity. (c): With change of probe lift-off.

Examples of simulated eddy current signals obtained at different drive frequencies, different material conductivities, or various probe lift-off values are shown in Figure 6.60. Note that in these examples, the crack face is parallel to the hole axis, and its size remains constant. Also, the ECT signal responses are not normalized.

As the driving frequency increases, the ability to penetrate into the material decreases, resulting in ECT concentration close to the surface; this causes larger response signals at the pick-up coil (i.e., higher amplitudes and larger phase angles with respect to the x axis). On the other hand, as the conductivity of the test piece increases, the "real" component of the signal that is related to resistivity decreases. This results in a clockwise phase change and lower amplitude, while the "imaginary" component remains almost unchanged. In contrast, when the probe lift-off increases, the signal phase remains almost the same while the signal amplitude reduces. Overall, modeling is a useful tool for understanding the influence of test parameters on the signal response and their optimization before the actual inspections [39].

6.7 CAPABILITIES AND LIMITATIONS

Capabilities:
- Non-contact and requires access to only one side of the test part.
- Simple or complex parts can be tested if electrically-conductive.
- Applicable to both aircraft engine and airframe components.
- Suitable for field, laboratory, or production inspections.
- Many types of probes, instruments, and approaches are available.
- Manual or automated inspections are possible.
- Can be used to detect and size surface or near-surface cracks or measure material thickness.
- Are capable of inspecting multi-layer structures with single-side access.
- Lap-joint corrosion in first and second layers resulting in thickness loss can be identified and measured.

Limitations:
- Only electrically-conductive materials can be tested.
- Very dependent on inspection settings (e.g., input signal, frequency).
- Probe lift-off, angle, and characteristics affect sensitivity.
- Depth of penetration is limited
- Discontinuities deep into the material cannot be detected, especially if they are small.
- Significant training of operators is required for proper inspection and interpretation of signals.
- Field applications are tedious and time-consuming.
- Three-dimensional presentation of results, as in tomography, is not possible.

6.8 MAGNETO-OPTICAL IMAGING (MOI)

6.8.1 Principles

The magneto-optical imaging principles are very similar to those of eddy currents, but in MOI, the current is induced by a thin, planar foil placed near and parallel to the surface of the specimen rather than a coil. Therefore, the induced eddy current is planar in nature, in othger words, sheet current [40]. In terms of detection, MOI uses a film-sensor and an optical polarization effect that occurs as a result of the magnetic field to detect and view the eddy currents. The technique can detect AC or DC magnetic fields generated by eddy currents, external DC magnetization, or simply the intrinsic material magnetic force.

Similar to the eddy current technique, the MOI method uses the principles of electromagnetic induction in an electrically conductive material. However, as schematically shown in Figure 6.61, the electromagnetic field in MOI is generated by a biasing coil and induced into the test piece through a bi-directional current foil, while a magnetic film-sensor is employed to detect the local flux density changes that are created by the material discontinuities. The current foil sheet can be excited in two orthogonal directions, but it is also able to produce a rotating eddy current by using two normal sinusoidal sources at 90° out of phase with each other, as shown in Figure 6.62. The film sensor is made up of a 3 μm thick bismuth-doped magnetic garnet, which, under the influence of the magnetic field, will be magnetized along the normal to the film's surface (referred to as the Faraday Effect). The film retains both negative and positive magnetic charges, and when a light is shined on the film, a preferential polarization of the light is achieved (i.e., the light beam is deflected according to the sign of the magnetization). The garnet film has an "easy" axis perpendicular to the film plane and a "hard" axis along the film's surface. This allows the sensor to create images of

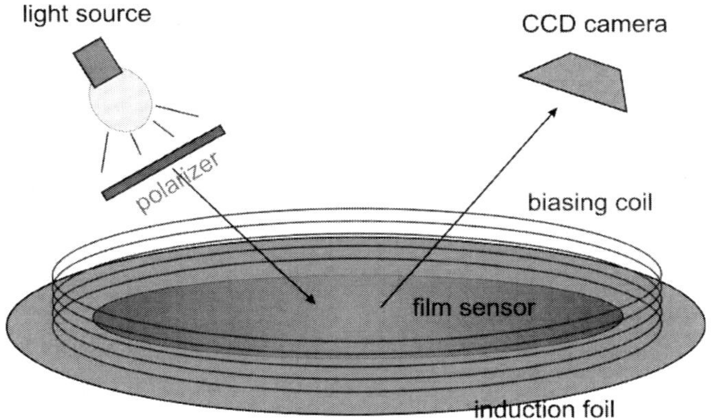

Figure 6.61 Schematic representation of the MOI method.

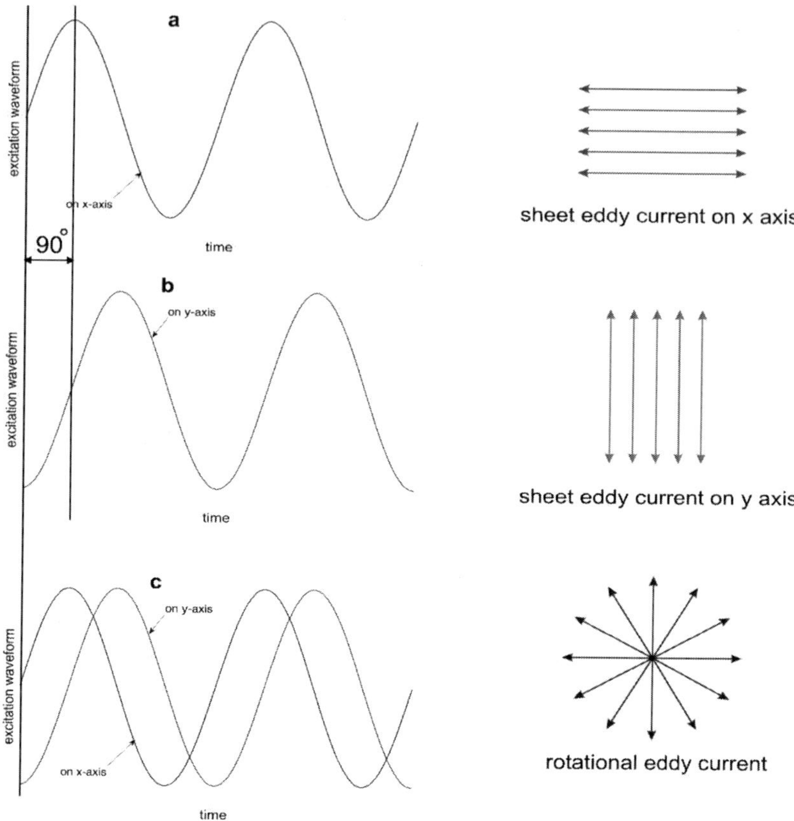

Figure 6.62 Possible in-plane eddy current excitation modes. (a) and (b): Linear excitation in two normal directions. (c): Rotational excitation as a result of simultaneous current injection in two orthogonal directions, out of phase by 90°.

the normal component of the magnetic field that is associated with eddy currents within the material.

The film-sensor has a low coercive force and a high Faraday Effect, which provides a high contrast visualization of the detected magnetic field. The image contrast is established by the AC biasing field, generated by the source coil, and induced through the induction foil (see Figure 6.61). This oscillating field keeps erasing old images at a rate of about 26 times per second. For crack detection using the two linear eddy current excitations, the current direction must be perpendicular to the crack face in order to be visible, as exemplified in Figure 6.63. The rotational eddy current is sensitive to cracks in different directions.

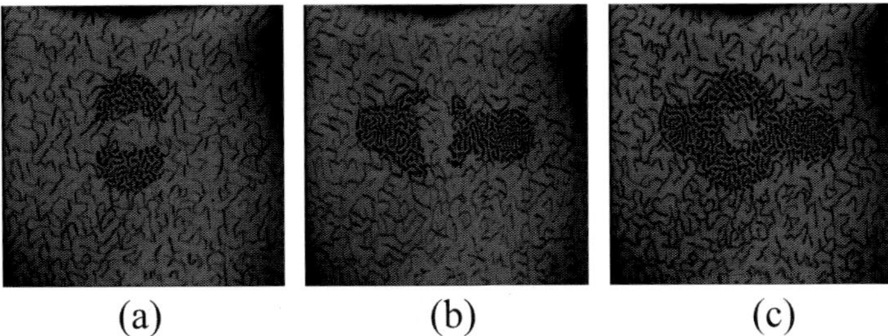

(a)   (b)   (c)

Figure 6.63 MOI images at different current excitation directions obtained for a crack extending from a rivet-hole. (a): Parallel to the crack face. (b): Perpendicular to the crack face. (c): Rotational.

6.8.2 Applications

Magneto-optical imaging (MOI) is a relatively new electromagnetic NDT method that may have applications where eddy current testing is applicable with the advantage of providing visual indications of discontinuities. MOI may be used as a fast and large area inspection tool for aircraft [41] to find cracks in airframe structures at rivet holes. The crack length is directly related to the extent of distortion of the MOI images while the depth can be determined by varying the frequency of excitation. Lowering the frequency allows inspections of deeper regions of the test piece and mapping the crack depth profile.

Some studies [41] have reported that the MOI method is capable of detecting 0.5 mm surface cracks. The approach may be used for corrosion detection too, and the same study has reported MOI sensitivity to corrosion in aluminum at 2.5 mm below the surface. While the MOI method is simple, intuitive, and fast, its major drawback is that only flat or nearly flat test pieces can be inspected.

6.8.3 Capabilities and Limitations

Capabilities:
- Uses three types of eddy current excitations: two linear and one rotational.
- Fast inspection approach suitable for large-area scans that reduces inspection time as compared to conventional eddy current tests.
- EC field visualization makes it simple to interpret the results.
- Applicable to non-magnetic, electrically-conductive materials, as well as ferrous components.
- Is capable of detecting cracks and corrosion in aircraft lap-joint structures.
- Estimates of the crack depth and length can be made.
- Detection of deep discontinuities may be possible by adjusting both the frequency and field strength.

Limitations:
* Only flat or nearly flat conductive parts can be inspected.
* Discontinuity sizing is not precise.
* Background noise may interfere with the visualization of discontinuities.
* A new approach that requires further qualification.

6.9 CHAPTER REFERENCES

[1] Status of Electromagnetic NDE Research at NRC Aerospace, C. Mandache, NRC-IAR Report: LTR-SMPL-2012-005, March 2012.

[2] The ABCs of NDT Development and Adoption for Aging Aircraft, D.S. Wilson and D.J. Hagemaier, Materials Evaluation, Vol. 57, No. 3, 1999, pp. 336–346.

[3] Electromagnetic Testing, NDT Handbook, American Society for Non-destructive Testing, Technical Editor; S.S. Upta and P.O. Moore, Vol. 5, 3rd Edition, 2004.

[4] Fundamentals of Physics, D. Halliday, R. Resnick, and J. Walker, John Willey & Sons, Inc., New York, 1997.

[5] Metals Handbook, 9th Edition, Vol. 17, NDE and Quality Control, ASM International, Metals Park, OH, 44073, 1989.

[6] NDT Resource Center of Iowa State University, http://www.ndt-ed.org/EducationResources/CommunityCollege/EddyCurrents/ProbesCoilDesign/diameter.htm.

[7] Magnetic Flux Leakage Investigation of Interacting Defects—Stress and Geometry Effects, C. Mandache, PhD thesis, Queen's University, 2004.

[8] Magnetoresistive Sensors for Non-destructive Evaluation, A. Jander, C. Smith, R. Schneiderb, Proceedings of the 10th SPIE International Symposium, Non-destructive Evaluation for Health Monitoring and Diagnostics, Conference 5770, 2005.

[9] Detection of Second and Third Layers Cracks under Fasteners using Pulsed Eddy Current and Magnetoresistive Sensors, C. Mandache, T. Byczkowski, M. Genest, and A. Fahr, 3rd International Symposium on Aerospace Materials and Manufacturing Processes, Montreal, Quebec, Canada, October 1–4, 2006.

[10] http://hyperphysics.phy-astr.gsu.edu/hbase/magnetic/hall.html.

[11] New Trends in Eddy Current Testing, C. Mandache, Canadian Institute for Non-Destructive Evaluation Journal, Feature Article, Vol. 29, No. 6, pp. 7–14, 2008.

[12] AC-driven AMR and GMR Magnetoresistors, P. Ripkaa, M. Tondra, J. Stokes, R. Beech, Sensors and Actuators A: Physical, Volume 76, Issues 1–3, pp. 225–230, 30 August 1999.

[13] Electromagnetic Enhancement of Pulsed Eddy Current Signals, C. Mandache, and J.H.V. Lefebvre, Proceedings of the Review of Progress in Quantitative NDE, Vol. 26A, Edited by: D.O. Thompson, and D.E. Chimenti, Vol. 894, pp. 318–324, 2006.

[14] Detection of Cracks Beneath Rivet Heads via Pulsed Eddy Current Techniques, J.S.R. Giguere, B.A. Lepine, and J.M.S. Dubois, Proceedings of the Review of Progress in Quantitative NDE, Vol. 21, Edited by: D.O. Thompson and D.E. Chimenti, pp. 1968–1975, 2002.

[15] Method and Apparatus for Eddy Current Detection of Material Discontinuities, J.H.V. Lefebvre, and C. Mandache, Unites States Patent #7,750,626, issued on July 6th, 2010.

[16] Transient and Harmonic Eddy Currents: Lift-off Point of Intersection, C. Mandache, and J.H.V. Lefebvre, NDT & E International, Vol. 39, No. 1, pp. 57–60, 2006.

[17] Generic Bolt Hole Eddy Current Testing Probability of Detection—Numerical Modelling Handbook, C. Mandache, M. Khan, M. Yanishevsky, A. Fahr, and D. Forsyth, NRC-IAR Report LTR-SMPL-2007-0121, 2007.

[18] New Method for Lift-off Independent Eddy Current Testing, C. Mandache, and J.H.V. Lefebvre, 14th ASNT Annual Research Symposium, Albuquerque, NM, March 14–18, 2005.

[19] Pulsed Eddy Current Measurement of Lift-off, J.H.V. Lefebvre, and C. Mandache, Proceedings of the Review of Progress in Quantitative NDE, Vol. 25, Edited by: D.O. Thompson, and D.E. Chimenti, pp. 669–676, 2005.

[20] Pulsed Eddy Current Thickness Measurement of Conductive Layers Over Ferromagnetic Substrates, J.H.V. Lefebvre, and C. Mandache, International Journal of Applied Electromagnetics and Mechanics, Vol. 27, No. 1-2, pp. 1–8, 2008.

[21] Non-destructive Inspection of T-56, 2nd Stage Turbine Wheel Lower Fir Tree Slots and Their Respective Blade Sets, M.R. Brothers, NRC-IAR Publication Number: LM-SMPL-2004-0113, 2004.

[22] Inspection of Aircraft Engine Components using Automated Eddy Current and Pattern Recognition Techniques, A. Fahr, C.E. Chapman, A. Pelletier and D.R. Roy, NRC-IAR Report: LTR-ST-1834, June 1991.

[23] GMR Sensor AAL002 from NVE Corporation, www.nve.com.

[24] An Investigation of Non-destructive Inspection Equipment: Detecting Hidden Corrosion on USAF Aircraft, J. Alcott, Materials Evaluation, pp. 64–73, January 1994.

[25] Subsurface Corrosion Detection in Aircraft Lap Splices Using a Dual Frequency Eddy Current Inspection Technique, J.G. Thompson, Materials Evaluation, pp. 1398–1401, December 1993.

[26] Eddy Current Sliding Probe Technology for the Detection of Corrosion in Multi-layer Aircraft Structures, paper no. 00296, Corrosion 2000, NACE International Annual Conference, Houston TX, March 2000.

[27] Friction Stir Welding Advances Joining Technology, M. Posada, J.P. Nguyen, D.R. Forest, J.J. DeLoach, R. DeNale, The AMPTIAC Quarterly, Vol. 7, No. 3, pp. 13–20, 2003.

[28] Friction Stir Welding of Aluminum Alloys, P.L. Threadgill, A.J. Leonard, H.R. Shercliff, P.J. Withers, International Materials Reviews, Published by Maney for the Institute of Materials, Minerals and Mining and the ASM International, Vol. 54, No. 2, pp. 49–93, 2009.

[29] Non-destructive Detection of Lack of Penetration Defects in Friction Stir Welds, C. Mandache, D. Levesque, L. Dubourg, and P. Gougeon, accepted for publication in Science and Technology of Welding and Joining, December 2011.

[30] Non-destructive Evaluation of Plasma Sprayed Thermal Barrier Coatings, A. Fahr, C. Mandache, and M. Genest, ICONIC 2007, St. Louis, MO, USA, June 27–29, 2007.

[31] Thermal Barrier Coatings, J. Thornton, X. Wu, S. Slater, M. Winstone, A. Fahr and D. Rowlands, Final Report for Operating Assignment TTCP-MAT-O27, TTCP Report TTCP-MAT-TP1-19-2001, September 2004.

[32] Detection of Thermally Grown Oxides in Thermal Barrier Coatings by NDE, A. Fahr, B. Roge and J. Thornton, Journal of Thermal Spray Technology, Vol. 15, No. 1, pp. 46–52, 2006.

[33] Overview of Potential Non-destructive Techniques for Inspection of Nickel-Aluminum-Bronze Valves, C. Mandache, and M. Brothers, NRC-IAR Report: LTR-SMPL-2007-0023, 2007.

[34] Linear Friction Welding of Al-Cu: Part 1—Process Evaluation, P. Wanjara, E. Dalgaard, G. Trigo, C. Mandache, J.J. Jonas, and G. Comeau, Canadian Metallurgical Quarterly, Vol. 50, No. 4, pp. 350–359, 2011.

[35] Pulsed Eddy Current Inspection of Friction Stir Welds, C. Mandache, L. Dubourg, A. Merati, and M. Jahazi, Materials Evaluation, Vol. 66, No. 4, pp. 382–386, 2008.

[36] http://www.ndetechnologies.com/

[37] Generic Bolt Hole Eddy Current Testing Probability of Detection—Numerical Modelling Handbook, C. Mandache, M. Khan, M. Yanishevsky, A. Fahr, and D. Forsyth, NRC-IAR Report LTR-SMPL-2007-0121, 2007.

[38] Bolt Hole Eddy Current Testing Probability of Detection, Part II: Numerical Modelling as a Cost- reduction Tool, C. Mandache, M. Khan, and A. Fahr, 12th International Conference on Fracture, Ottawa, July 12–17, 2009.

[39] New Trends in Eddy Current Testing, C. Mandache, Canadian Institute for Non-Destructive Evaluation Journal, Feature Article, Vol. 29, No. 6, pp. 7–14, 2008.

[40] Magneto-optic/ Eddy Current Imaging of Aging Aircraft: A New NDI Technique, G.L. Fitzpatrick, D.K. Thome, R.L. Skaugset, E.Y.C. Shih, and W.C.L. Shih, Materials Evaluation, Vol. 51, No. 12, pp.1402–1407, 1993.

[41] Characterization of Magneto-Optic Imaging Data for Aircraft Inspection, Y. Deng, X. Liu, Y. Fan, Z. Zeng, L. Udpa, and W. Shih, IEEE Magnetics Conference Publication, Intermag 2006, page 898, May 2006.

Ultrasonic Methods

7.1 PRINCIPLES

Ultrasonic testing (UT) is based on the transmission of high frequency (0.5–50 MHz) sonic waves through the test area and the measurement of the amplitude, frequency, or time of arrival of the returned echoes. This method is used to detect and size discontinuities, measure dimensions, or determine some properties of materials. A typical ultrasonic test set-up consists of several units, including a pulser/receiver, a transducer, and display devices. The pulser generates high voltage electrical pulses and drives the transducer. The transducer converts the electrical pulses into high frequency vibration or ultrasound. The ultrasonic energy is introduced into the test material and propagates through the materials in the form of ultrasonic waves. When discontinuities, such as cracks or voids, are encountered in the wave path, part of the ultrasonic energy is reflected back by the discontinuity. The reflected waves received by the transducer are converted back into electrical signals and after amplification at the receiver, are displayed on a screen as shown schematically in Figure 7.1 [1]. A basic screen display usually shows the signal amplitude in volts versus time. The signal amplitude, shape, and arrival time provide information about the size, shape, and location of the discontinuity.

7.1.1 WAVE TYPES

Ultrasonic waves are mechanical or stress waves, and their propagation in an elastic medium is by localized vibration or reversible displacement of the material from one point to the neighboring points. When the atomic or molecular segments

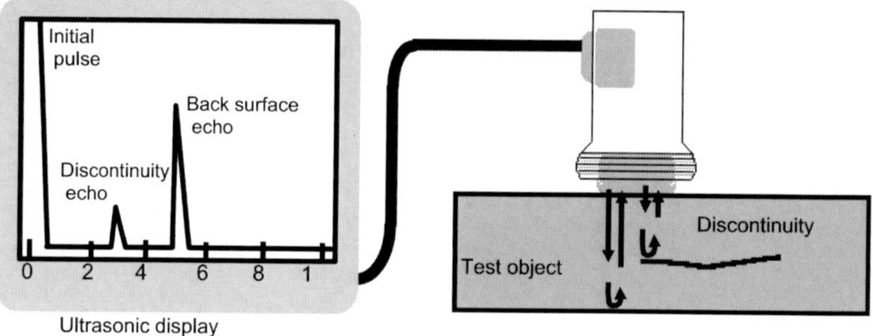

Figure 7.1 A simple ultrasonic test configuration showing a piezoelectric transducer in contact with the test part and the returning echoes [1].

or points of an elastic material are displaced from their equilibrium positions by any applied force, internal stresses act to restore the displaced points to their original positions. Because of the inter-atomic forces between adjacent points of the material, a displacement at one point introduces displacements at neighboring points, and so on; thus, propagating a wave [2]. Depending on how the material is made to oscillate, different wave types can be created. Compression or longitudinal waves are referred to as the wave types that propagate in the direction of the material movement. Longitudinal waves can be generated in solids, fluids, or gases and are widely used in NDT for discontinuity detection, dimension measurement, and property evaluation. Transverse or shear waves are the types that propagate perpendicular to the direction of the material movement. Unlike longitudinal waves that propagate by back and forth vibration of a material plane and collision or pushing of the neighboring planes, shear waves travel by up and down movement of a plane and pulling the adjacent planes. Thus, shear waves can only be generated in solids and are weaker and travel slower than longitudinal waves. Shear waves are usually generated from longitudinal. They are used when slower wave speeds or oblique angled beams are necessary to test the material. The particle movements responsible for the propagation of longitudinal and shear waves are illustrated below in Figure 7.2. This Figure also illustrates the wavelength (λ) that is the distance between repeating units of the wave patterns.

Sometimes the longitudinal or shear waves are made to travel along the surface of a solid generating a combined longitudinal and transverse motion that follows an elliptical orbit. The particle movements in this case concentrate near the surface of the solid covering a depth of approximately 1 to 1.5 wavelengths from the surface. These are referred to as surface or Rayleigh waves and are used to detect discontinuities located at or near the surface of the test article. When the waves are confined between two parallel surfaces of a plate that is a few wavelengths in thickness, they are called plate waves or Lamb waves. The vibration of particles in this case is a complex mixture of the different wave types and covers the whole

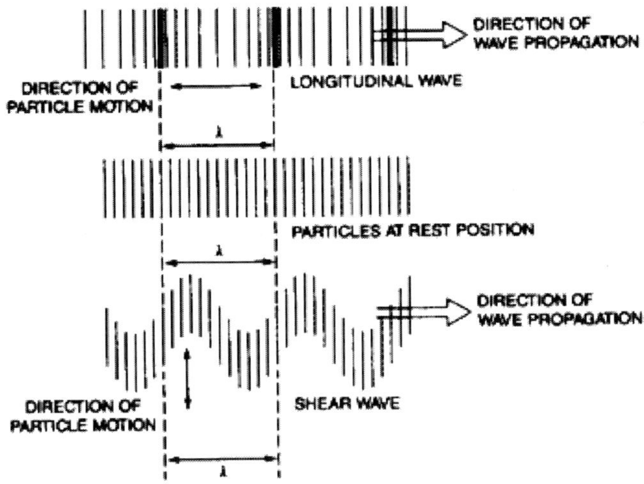

Figure 7.2 Particle movements in the longitudinal and shear waves [1].

thickness of the plate. Lamb waves are used when the entire plate thickness must be inspected over a large distance.

Different wave types travel at different speeds that vary depending on the material's elastic properties and density. The governing general relationship is:

$$V = \sqrt{C_{ij}/\rho} \qquad (7\text{-}1)$$

where V is the wave velocity, which is defined as the travel distance in a unit time, C_{ij} is the material's elastic constants, and ρ is the material's density.

The above equation may take a number of different forms depending on the vibration mode that creates the waves and the related elastic constants of the material. As sonic waves are stress waves, the same elastic constants used in mechanics are also applied in ultrasonic velocity calculations. Generally, elastic constants describe a material's behavior under mechanical stress or the stress-strain relationship, and include:

- *Young's Modulus* (*E*) is a measure of the material's elasticity (stiffness) or the tendency of an object to deform along an axis when opposing forces are applied along that axis. It is defined as the ratio of the tensile (or longitudinal) stress to the tensile strain (or elongation).
- *Shear Modulus* (*G*) is a measure of the material's rigidity or the tendency of an object to deform in shape when acted upon by opposing forces. It is defined as the ratio of shear stress to the shear strain.

- *Bulk Modulus* (*K*) is a measure of the volumetric elasticity or the material's ability to deform in all directions when uniformly loaded in all directions. It is defined as the ratio of volumetric stress to volumetric strain.
- *Poisson's Ratio* (*n*) is the ratio of transverse contraction strain to longitudinal extension strain or the material's contraction in the direction perpendicular to the applied stress over the material's extension in the same direction as the applied stress.

The subscript ij indicates the directional dependence of the elastic constants in relation to the wave types (e.g., longitudinal or shear) and wave propagation. In homogeneous and isotropic solids, the longitudinal and shear wave velocities are controlled by the material's Young's Modulus and Shear Modulus, respectively. In such materials, the elastic constants do not change with direction, and, thus, the longitudinal or shear wave velocities are identical in all directions within the material. However, in anisotropic materials they differ with the wave propagation direction, as the elastic constants are directionally dependent. For example, in rolled sheet-metals, the grains are elongated in the longitudinal direction and compressed in two transverse directions; thus, the elastic constants for the longitudinal direction are different from those of the transverse directions. Therefore, the ultrasonic wave velocity (both longitudinal and shear waves) traveling along the length of a rolled sheet-metal will be different from the same wave traveling across the metal plate or through the thickness. Attention to such differences is necessary for accurate interpretation of ultrasonic measurements.

For most isotropic materials, the shear wave velocity is approximately one half of its longitudinal wave velocity. Thus, if one is known, the other can be estimated. The ASTM Standard E 494 describes test procedures for measuring longitudinal and transverse wave velocities in isotropic materials [5]. The procedures are based on a comparison of the test material with a reference sample whose velocities are known. If a material density is known, the measurement of its ultrasonic longitudinal and transverse velocities enables determination of the elastic constants through the following equation [5]:

Young's Modulus: $E = [\rho\, V_s^2\, (3V_l^2 - 4V_s^2)] / (V_l^2 - V_s^2)$ (7-2)

Shear Modulus: $G = \rho\, V_s$ (7-3)

Bulk Modulus: $K = \rho\, (V_l^2 - 4/3\, V_s^2\,)$ (7-4)

Poisson's Ratio: $v = [1 - 2(V_s / V_l)^2] / 2[1 - (V_s / V_l)^2]$ (7-5)

where ρ is the material's density in kg/m³ (or lb/in.³), V_l is the longitudinal wave velocity in m/s (or in./s), and V_s is the shear wave velocity in m/s (or in./s). *E*, *G*, and *K* are in N/m²(or lb/in.²); and v is a unit-less value. To convert from metric to imperial units, the results must be divided by 386.4 in/s^2, which is the gravitational acceleration, g. The conversion factor is: 1 N/m^2 = 1.4504 × 10^{-4} lb/in^2.

7.1.2 ACOUSTIC IMPEDANCE

The product of the material's density (ρ) and its wave velocity (V) is a term called acoustic impedance (Z).

$$Z = V\rho \tag{7-6}$$

Acoustic impedance, measured in kg/m^{2}s (or lb/in.^{2}s), influences the amount of energy that is reflected or transmitted when the propagating ultrasound encounters a boundary between two media of different densities or elastic properties. This parameter directly affects the detectability of discontinuities. The larger the difference between acoustic impedances of the discontinuity and the host material, the greater the likelihood of detection becomes. Generally, the acoustic impedance of solids is higher than those of liquids, and those of gases or air are lower than those of fluids. Thus, voids, cavities, or cracks, that are essentially air pockets or gaps ultrasonically, are easier to identify than similar sized solid inclusions. Other factors also influence flaw detection as described in the ensuing sections.

7.1.3 REFLECTION AND TRANSMISSION AT BOUNDARIES

During a typical ultrasonic test, a beam of UT waves is incident to the test medium, either at a perpendicular angle or an oblique angle with respect to the surface. When an ultrasonic beam encounters a boundary of a second medium with different acoustic impedance, a portion of the acoustic energy is reflected back while the remaining energy is transmitted into the second medium. The amount of energy reflected or transmitted is dependent on the difference in acoustic impedances of the two media (i.e., impedance mismatch) as well as the beam incidence angle with respect to the surface. The reflection coefficient, R, is defined as the ratio of the intensity of the reflected wave to the intensity of the incident wave.

7.1.3.1 NORMAL ANGLE

When the ultrasonic beam is incident at a perpendicular angle to a boundary between two media with different acoustic impedances, R is given by:

$$R = [(Z_2 - Z_1) / (Z_2 + Z_1)] \tag{7-7}$$

where Z_1 and Z_2 are the acoustic impedances of the media 1 and 2, respectively. The transmission coefficient, T, at the normal angle is simply $1–R$, because all energy is either reflected or transmitted:

$$T = 1 - R = 1 - [(Z_2 - Z_1) / (Z_2 + Z_1)] \tag{7-8}$$

$$T = 4Z_2 Z_1 / (Z_2 + Z_1) \tag{7-9}$$

These relationships indicate that the lower the difference between the acoustic impedances of the two adjacent media, the more energy will be transferred from one material to the other and less reflection will occur at their boundary. If the impedances of two materials are the same, no reflections will take place. If the impedances differ significantly, there will be virtually complete reflection. For this reason, fluids are used between the UT probes and solid test pieces to fill the air gap and create coupling, as fluids have closer acoustic impedance to solids than either air or gases. On the other hand, in order to be able to detect a defect within a material by using reflected waves, the acoustic impedance of the defect must be greatly different from that of the host material. This implies that if the acoustic impedance of a defect is similar to that of the surrounding material, it will not be detected by using the ultrasonic technique in the reflection mode. An example is a foreign material inclusion that has a similar density and acoustic velocity to that of the host medium.

These characteristics are used to calculate the amount of energy reflected or transmitted at discontinuities. Reflection and transmission coefficients are often expressed in decibels (dB) to allow comparison of large changes in signal intensity ($1 dB = 10 \log R$). Signal intensity or amplitude values are usually measured in micro volts (μV) but presented in decibels (dB); $1 \mu V = 0 dB$, $10 \mu V = 20 dB$, $100 \mu V = 40 dB$, and so on.

7.1.3.2 OBLIQUE ANGLES

If the incidence angle of the UT beam is not perpendicular to the boundary of two media that have different acoustic impedances, the reflected and transmitted waves will undergo refraction (or change in wave propagation direction) and mode conversion (or change in wave type). Snell's Law provides the relationship between the reflected and transmitted wave angles and velocities of the two media as in the following expression and Figure 7.3 [1, 2]:

$$\frac{\sin \theta_1}{V_{L_1}} = \frac{\sin \theta_2}{V_{L_2}} = \frac{\sin \theta_3}{V_{S_1}} = \frac{\sin \theta_4}{V_{S_2}} \tag{7-10}$$

where V_{L_1} is the longitudinal wave velocity in medium 1, V_{L_2} is the longitudinal wave velocity in medium 2, V_{S_1} is the shear wave velocity in medium 1, and V_{S_2}

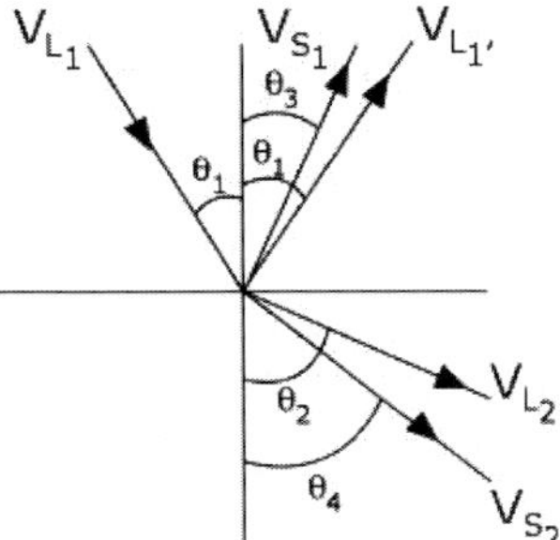

Figure 7.3 Reflection and mode conversion of an ultrasonic beam of longitudinal waves incident at an oblique angle onto a surface [1].

is the shear wave velocity in medium 2. θ_1 is the beam incidence angle, which is equal to the angle of reflected longitudinal waves, θ_2 is the refraction angle of longitudinal waves, θ_3 is the reflection angle of shear waves, and θ_4 is the refraction angle of shear waves.

Since the velocity of shear waves is less than that of longitudinal waves, both reflection and refraction angles for shear waves are smaller than those of longitudinal waves. The first critical angle is the angle of incidence at which the angle of refraction of longitudinal waves is 90 degrees. Beyond this critical angle, only shear waves propagate into the material. The second critical angle is the incidence angle that causes the shear waves to refract at a 90 degree angle. Beyond the second critical angle, the waves propagate as surface waves, which have similar velocity to shear waves. As in longitudinal waves, shear waves incident to a boundary of two media at an oblique angle will also undergo refraction and mode conversions. The relationships between the angles and velocities are similar to those mentioned above.

The reflection and transmission coefficients at oblique incidence angles are dependent on the wave types and their velocities as well as the beam angle, with respect to the boundary of two media. Algorithms to calculate these coefficients are well established and are available in published literature (e.g., [2, 3]). Mode conversion adds to the complexity of ultrasonic tests; thus, inspections at oblique angles are often carried out using a specific wave type. For example, angled-beam shear waves are used to reach areas of a component that cannot be examined using the normal longitudinal waves, or surface waves are employed to detect minute surface-breaking cracks that otherwise could not be identified using either longitudinal or shear waves. These approaches will be described in more detail under application examples.

7.1.4 *ATTENUATION*

As ultrasonic waves travel through a material, they lose their energy as the distance increases. Attenuation is defined as the loss of energy within the material due to factors, such as impedance differences at internal boundaries, diffraction at sharp edges, scattering by a material's microstructural features, absorption in the bulk material, and spreading of the ultrasonic beam.

Impedance differences at internal boundaries, such as free surfaces, change of materials, internal features, discontinuities or defects, etc., cause reflection, refraction, and mode conversion, as described above. In such cases, the loss of energy is dependent on the acoustic impedance mismatch of the two adjacent media as well as the size, type, shape, orientation, surface condition, and other physical characteristics of the media in the beam path. Attenuation due to impedance differences occur at discontinuities that are much larger than the UT wavelength (λ) and beam diameter, such as voids in metallic parts, delaminations in composites, or disbonding of structures. In these cases, measurement of the attenuation of the transmitted or reflected waves provides information about the discontinuity size, type, and location. Material discontinuities that are in the range of $0.1\ \lambda$ to $2\ \lambda$ or a beam size comparable to the discontinuity size scatter the waves in many directions. Scattering occurs at a material's microstructural features, such as grain boundaries, twin boundaries, porosity, and other small discontinuities. In such cases, attenuation measurement may provide information about the material, such as its porosity content, if other properties remain constant. When a wave front passes the edge of a reflector, it bends around the edge, creating diffraction. The effect depends on the roughness, size, and contour of the reflecting interface and is more pronounced when they are small relative to the wavelength. Absorption is the loss of energy due to the conversion of the elastic motion of the material into heat. While the above-mentioned factors are material dependent, beam spreading depends on the characteristic of the ultrasonic transducer.

Most commonly used ultrasonic transducers employ a piezoelectric element that vibrates under an applied voltage. However, near the face of the crystal, the sound intensity fluctuates rapidly due to the interference of waves generated by the edge of the crystal with those of the central portion of the element. The effect is a spatial pattern of intensity maxima and minima that is referred to as the near-field. The space between the intensity peaks becomes broader as the distance from the transducer face (d) increases. The length of the near-field N can be calculated from [2]:

$$N = (D^2 - \lambda^2) / 4\lambda \qquad (7\text{-}11)$$

where D is the diameter of the UT source and λ is the wavelength. At distances greater than N, referred to as the far-field, the intensity does not fluctuate but gradually decreases from its maximum at $d = N$ as the wave-front expands. This is known as beam spreading and is measured using [2]:

$$\gamma = 2 \sin^{-2}(0.5 \; \lambda/D) \tag{7-12}$$

where γ is the angle of divergence of the beam from the central axis.

The true or absolute attenuation is the summation of the energy losses due to all the above-mentioned processes and can be expressed by the following expression [2]:

$$I = I_0 \exp(-\alpha L) \tag{7-13}$$

where I_0 and I, respectively, are the ultrasonic intensity at the input and output sides of the test material, L is the distance between the input and output sides, and α is the attenuation coefficient. The absolute attenuation coefficient is difficult to measure, as it requires taking into account the material properties, specimen geometry, beam divergence, and instrumentation effects. However, apparent attenuation can be measured using the ASTM Standard E664-10 in flat parts with parallel surfaces. The approach described in this ASTM document is based on multiple reflections that occur between the two parallel surfaces of the test sample when the incidence beam is perpendicular to the surface (longitudinal waves) as illustrated in Figure 7.4. In such a case, the apparent attenuation in decibels (dB) per unit length is obtained using the following equation:

$$\text{Apparent attenuation} = 20 \log (A_m/A_n)/2T(n-m) \tag{7-14}$$

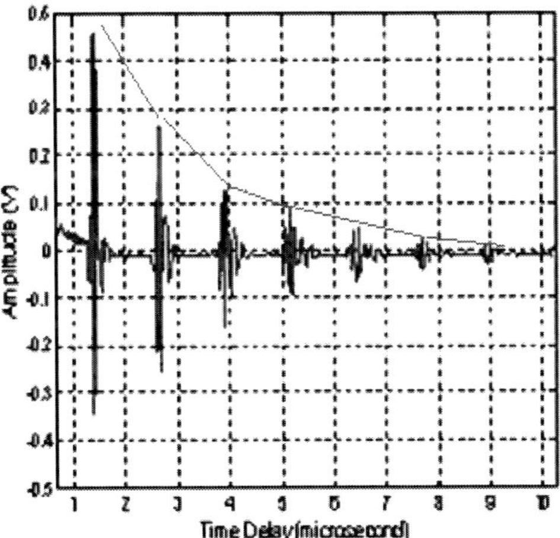

Figure 7.4 Multiple echoes in a sample with parallel surfaces. The drop in echo amplitude is related to ultrasonic attenuation in the material.

where A_m and A_n are the amplitude of the mth and n^{th} reflections returned ($n > m$) and T is the specimen thickness. Apparent attenuation may be used as a measure of changes in specimen's microstructure (e.g., average grain size, porosity content), elastic behavior (e.g., elastic constants), or internal flaws. Comparative assessment of the attenuation of two or more geometrically-identical samples may provide information about the above-mentioned material characteristics.

As mentioned earlier, ultrasonic attenuation is dependent on the wavelength, which, in turn, varies with the wave frequency (f) and the wave velocity in the material (V) through the following simple relationship:

$$\lambda = V / f \tag{7-15}$$

and generally increases proportionally to the square of frequency. At high frequencies, or when the $\lambda \leq$ the material's microstructural features (e.g., grain size), due to the high scattering-related attenuation, it may be difficult to penetrate the material to detect deeper discontinuities. On the other hand, at low frequencies, or when $\lambda \geq$ discontinuity size, it may be difficult to detect small discontinuities due to insufficient reflections. These factors affect the flaw detection sensitivity and resolution as described in the next section.

7.1.5   *SENSITIVITY AND RESOLUTION*

NDT sensitivity refers to the ability of a technique or an instrument to identify small changes in material characteristics and properties or to detect minute discontinuities. For defect detection, it is often defined in terms of the smallest flaw size detectable in a test component. The NDT indication obtained from a small discontinuity of known size that can be easily recognized from the background noise level with the instrument gain set at maximum provides a measure of the sensitivity. In NDT methods, such as ultrasonics, that use waves to inspect parts, a higher sensitivity is achieved by increasing wave frequency or reducing the wavelength.

NDT resolution is defined as the ability to identify and separate discontinuities that are close to each other or measure small property changes. For flaw detection, the resolution could be axial or lateral. Axial resolution is the minimum separation between two closely-located discontinuities in the direction parallel to the beam (or on top of each other) that can be identified as two discontinuities. Increased axial resolution can be achieved by using shorter UT pulses or increasing frequency. However, attenuation will increase with increasing frequency, as discussed above. Similarly, lateral resolution is the minimum separation that can be measured between two discontinuities located in the plane perpendicular to the beam direction. A high lateral resolution can be achieved by using high-frequency probes with a small diameter (D) and/or by focusing the beam. Lateral resolution

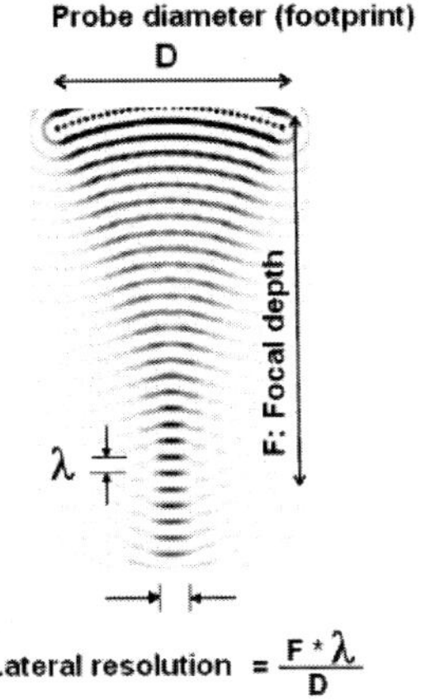

$$\text{Lateral resolution} = \frac{F * \lambda}{D}$$

Figure 7.5 Focusing of a UT beam to achieve increased lateral resolution. The relationship between lateral resolution and probe diameter, wavelength, and focal depth is also given [4].

is dependent on the focal depth (F), the wavelength (λ), and the probe diameter (D) as illustrated in Figure 7.5 and the expression below the diagram.

In ultrasonic testing, for effective flaw detection, it is generally necessary to use a wavelength smaller than half the flaw size (diameter, width, and length) in the direction facing the propagating waves. Usually, by using shorter wavelengths or higher frequencies, smaller flaws can be detected or smaller changes in properties/dimension can be measured. On the other hand, increase of the frequency results in higher attenuation and shorter range of wave propagation. Thus, it is important to choose a proper frequency in relation to the size of discontinuities that need to be detected and the range of inspection that must be covered for the specific material type that is being inspected.

7.1.6 Signal-to-noise Ratio

The signal-to-noise ratio (S/N) is a measure of the strength of the signal returned by a target (e.g., defect) as compared to the background noise (including noise due to material microstructure, test equipment, and inspection environment). S/N determines the detection capability of a flaw (detectability) or the ability to measure the difference in materials' properties.

For flaw detection, in practice, S/N must be greater than three in order to be able to identify defects with reasonable reliability and confidence. The NDT reliability and confidence, their measurement, and important factors affecting them are described in detail in a separate chapter. The strength of the UT signal from a defect in relation to the noise is dependent on many factors, including:

- Characteristics of the transmitting and receiving sensors (size, type, frequency, bandwidth, efficiency, focal depth, etc.).
- Inspection equipment and parameters (instrument characteristics and settings).
- Test method (e.g., direct echo reflection measurement, attenuation measurement).
- Test environment/conditions (e.g., laboratory vs. field, background noise level).
- Test material and microstructure (e.g., homogeneity, grain size).
- Component geometry (flat or curved surfaces, size, and complexity, etc.).
- Component surface conditions (e.g., smooth or rough, painted or not).
- Probe-to-target distance and coupling (coupling medium, thickness, type, etc.).
- Discontinuity (e.g., type, size, shape, orientation, location).

Careful attention must be paid to all the above-mentioned factors in order to achieve a good signal-to-noise ratio from a discontinuity and improve flaw detectability.

7.1.7   SCANNING AND PRESENTATION

Ultrasonic tests are carried out by introducing a beam of UT waves to the test object and measuring the received waveforms (reflected, transmitted, or scattered). The scanning of the probe and the measurement of the received waves may be carried out either manually or automatically. The received waveforms at the probe are converted to electrical signals and are analyzed after amplification. Signal characteristics, such as peak amplitude, frequency, and travel time, are used to obtain information about the test object or defects. The test object can be scanned in many different ways and the received signals can be analyzed and displayed in a number of formats. The most-commonly used scanning modes are A-, B-, and C-scan. These scan types were described earlier in Chapter 3 and are repeated briefly here, along with the other scan modes for comparisons.

7.1.7.1   A-SCAN

When the ultrasonic probe output corresponding to a single location of the test piece is presented in terms of amplitude (voltage) versus time (second), an A-scan signal is produced, as schematically illustrated in Figure 7.1. The signals may be generated by internal discontinuities or free surfaces of the object (e.g., front and rear surfaces). The flaw-signal amplitude is related to the size of the discontinuity (often directly proportional), and its location on the time axis is related to its distance from the sensor (always directly proportional). A-scan is the basis for all other ultrasonic scanning and measurement techniques. Size estimation is done by comparing the amplitude of the unknown discontinuity signal with that of the same flaw type of known size using reference samples.

The discontinuity location is often determined by measuring the position of the flaw-signal on the time axis with respect to either the front- or back-surface echoes, and by converting the time-scale measurements to distance using:

$$d = V(t_d - t_s) \qquad (7\text{-}16)$$

where V is the wave velocity, and t_d and t_s are the times taken for the waves to arrive from the discontinuity and the object surface (either front-surface or rear-surface), respectively. The quantity d is the distance between the reflecting-face of the discontinuity and the appropriate object surface.

7.1.7.2 B-SCAN

A B-scan refers to a linear scanning of the probe and display of the time-lapse between the flaw-echo and the object's front-surface echo as a function of the position of the probe along a chosen scan line, as shown schematically in Figure 7.6. The B-scan is basically a two-dimensional presentation of the cross-sectional position of internal discontinuities showing a time-slice through the specimen. A profile of the reflecting face of internal discontinuities or interfaces is displayed that indicates the distance, shape, orientation, and width of the discontinuity. B-scanning is sometimes used to show the cross-sectional distribution of flaws within a large structure prior to the more detailed and time-consuming C-scanning. In such cases, only selected segments of the part are examined using the C-scan.

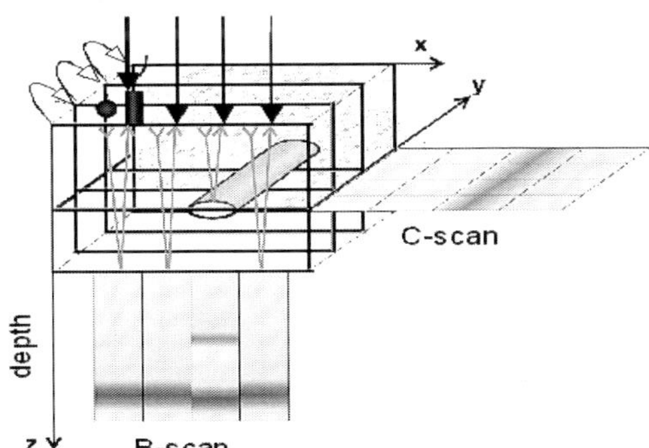

Figure 7.6 A schematic presentation of ultrasonic B-scan and C-scan. (http://www.ndt.net/article/ndtce03/papers/v023/v023.htm).

7.1.7.3 C-scan

C-scan refers to the two-dimensional scanning of the probe and display of either the amplitude or the time-location of the received signals as a function of the position of the probe. C-scanning is carried out by moving the probe linearly in one direction over the test object and repeating the process after the probe has been indexed by a small increment at the end of each scan (see Figure 7.6). In Figure 7.7 and Figure 7.8, the actual A-scan, B-scan, and C-scan presentations of two commercial ultrasonic instruments are provided. When signal amplitudes are plotted as a function of probe location, amplitude C-scans are generated. An amplitude C-scan shows a two-dimensional view of the interior of the test object identifying discontinuities perpendicular to the beam direction. An estimate of the cross-sectional size and planar location of discontinuities can be made from the amplitude C-scan images. The echo amplitude can be displayed in positive or negative, gray or color scales. Amplitude C-scans do not easily indicate how deeply flaws are located within the test object. However, such information may be obtained by time-gating, which essentially involves adjusting the instrument gate such that an appropriate depth range within the object is mapped. Alternatively, the time-lapse between the object's front-face and discontinuities is measured and displayed as a function of the probe position over the scan area. The resulting image is called a "time-of-flight C-scan." In this case, the depth location of the reflecting interfaces is shown in different colors or shades of gray. Figure 7.9 is

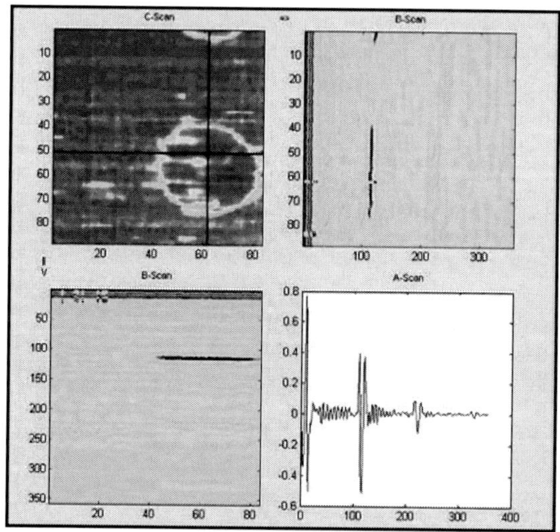

Figure 7.7 Actual screen displays of damage in a plate. (Top left): C-scan of the plate showing the damage in red, (top right): B-scan along the vertical line through the plate and damage, (bottom left). B-scan along the horizontal line through the plate and the damage, and (bottom right): A-scan at the point of the intersection of the vertical and horizontal lines.

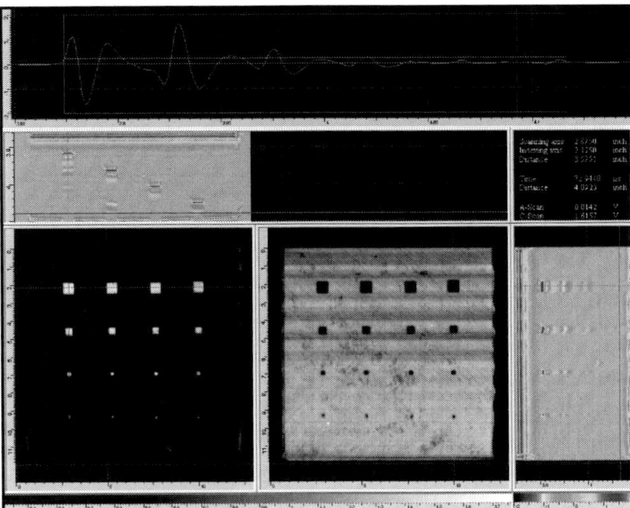

Figure 7.8 Actual screen displays of a composite panel with artificial defects of different sizes and depth locations obtained using a commercial ultrasonic instrument. (Top): A-scan at the point on the top-left corner defect, (middle-left): B-scan along the top-row defects showing their depth location, (bottom-left and middle): C-scans of the panel showing planar size and location of defects, and (bottom-right): B-scan along left-column defects showing their depth location.

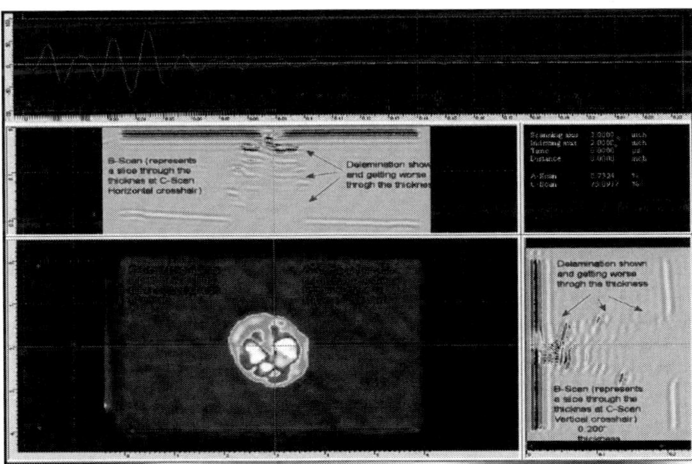

Figure 7.9 Additional actual screen displays of a composite solid laminate containing an impact damage. (Top): A-scan signal at a point on the damage, (middle): B-scan along the horizontal line, (bottom-right): time-of-flight C-scan of the damage and the panel, and (bottom-right): B-scan along the vertical line.

a screen display showing A-scan, B-scan, and time-of-flight C-scan of a composite panel with impact damage obtained using a commercial instrument. More description is provided in the Figure caption.

7.1.7.4 D-SCAN

D-scan is a term primarily used in weld inspection to display the cross-sectional view of the weld in its longitudinal direction (Figure 7.10). In this case, B-scan refers to the cross-sectional view of the weld transverse to its direction. A D-scan is generated by using an angled-beam (or wedge) transducer that is scanned along the weld and indexed in small increments in the direction normal to the weld-line after each scan. It presents the cross-section of the weld perpendicular to the scanning surface or perpendicular to the projection of the beam axis on the scanning surface [8]. In other words, the D-scan view is perpendicular to both the C-scan and the transverse B-scan of the weld and shows discontinuities in the whole length of the weld and their positions or depths from the scanning surface.

7.1.7.5 P-SCAN

P-scan is another term used in weld inspection and refers to a presentation that combines B-, C-, and D-scans in a single image, creating a complete view of the discontinuities, as shown in Figure 7.11 and Figure 7.12, ([9–12]). In P-scan, multiple angled-beam transducers operating independently are scanned along the length of the weld while each transducer covering a certain range of the weld or the part thickness. The P-scan view was originally devised for viewing welding flaws and was subsequently used for corrosion mapping as well. P-scan is a now a trade name used by FORCE Technology [10] and is assigned to its automated ultrasonic system designed primarily for pipe weld inspection.

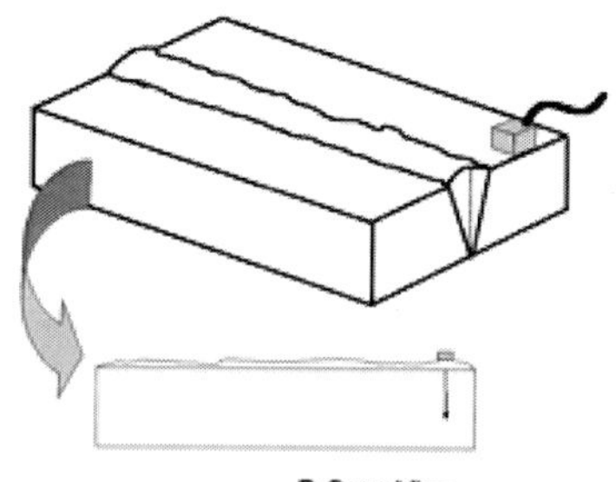

Figure 7.10 Schematics of a weld inspection and display using the D-scan [8].

Figure 7.11 Schematic presentation showing how a P-scan display is obtained using the top-view C-scan, the side-view D-scan, and the end-view B-scan [10].

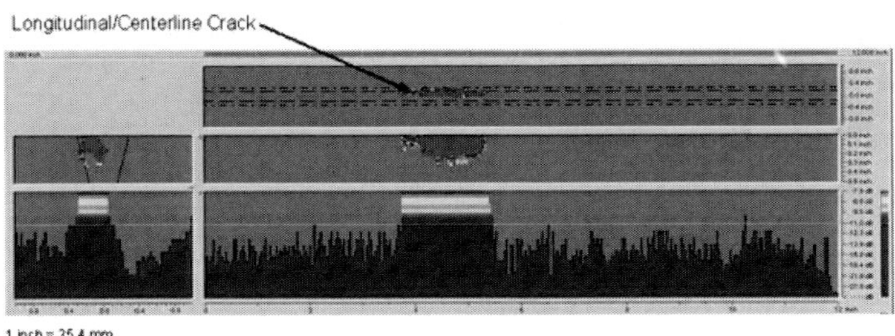

Figure 7.12 Presentation of weld inspection using; (top): C-scan, (middle-left): B-scan, (middle-right): D-scan, and (bottom-left and bottom-right): P-scans showing a longitudinal discontinuity in the weld [12].

7.1.7.6 POLAR SCAN

Polar scanning is used for inspecting disc-shaped components. It involves rotating the part around the disc's central axis on a turn-table while indexing the transducer by a small increment along the disc diameter after every revolution. A polar scan provides the same information as the C-scan.

7.1.7.7 T-SCAN

This term is sometimes used instead of time-of-flight C-scan. The T-scan presents layer-by-layer views of the cross-section of the test object. It is also used as a trade name for a series of commercial ultrasonic thickness gauges that provide such a capability [11].

7.1.8   ULTRASONIC COUPLING

Coupling of the probes to the test specimen can be achieved in different ways, including:

- Close contact using coupling fluids, jells, or soft materials.
- Immersion in water.
- Water-jets or water-columns.
- Non-contact approaches.

Contact, immersion, and water-jet approaches are described in the following paragraphs, but information on non-contact ultrasonic methods are provided at the end of this chapter.

7.1.8.1   CONTACT APPROACH

Inspection of small areas using piezoelectric probes is often done by putting the probe directly in contact with the test piece with a coupling fluid or jell in between and scanning the area of interest, either manually or using a small automated scanner (Figure 7.13). Normal or angled-beam probes or probe arrays may be used depending on the application. The manual contact approach is used for in-service inspection of small areas using simple amplitude versus time A-scan display for flaw detection or thickness measurement. The contact approach and A-scan display are often adequate for routine in-service inspections of small areas of aircraft skin or critical components of airframes, such as hinges and bolt-holes. Cracks and corrosion can be detected using the contact approach. Thickness loss due to corrosion can be measured in aircraft outer skins from the rear-surface echo location on the time axis of the A-scan signals. The results, however, are dependent on the operator skills in achieving good transducer-specimen coupling as well as interpreting signals accurately.

Automated scanning may be used for in-service inspection of small to medium size areas (~1 square meter) to speed up the test process. In this case, an automated scanner with at least two linear axes (x and y) equipped with a motion controller, a

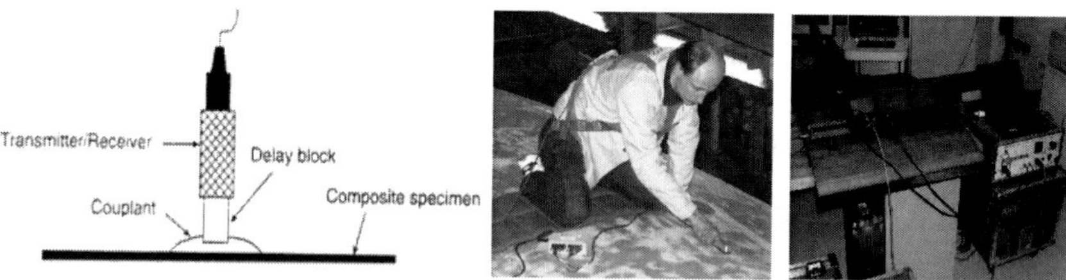

Figure 7.13 (Left): Schematic presentation of contact coupling approach, (center): manual inspection of aircraft, and (right): a typical automated x-y scanning equipment.

computer, and an appropriate software package is necessary to scan the ultrasonic probe over the test site. The inspection data are collected, processed, and displayed or recorded for interpretation by the operator. Two-dimensional images of the inspection results in A-, B-, and C-scan formats may be provided. Transducer coupling may be achieved by wetting the entire surface adequately with water or a coupling fluid before the inspection. It is also possible to provide continuous local coupling through the use of specialized attachments to the probe that supply water. Suction cups may be used to hold the scanner in place. Some systems use probes incorporated into a rubber-wheel device that provides both the transducer coupling as well as easy scanning. However, the signal-to-noise ratio of the rubber-wheel probes is not as good as the conventional fluid-coupling transducers due to the additional UT attenuation in the rubber wheel. Also, signals may vary due to the variation of the pressure applied to the probe or variation in the probe angle affecting the inspection results.

Capabilities and Limitations

Capabilities:
- Simple and inexpensive.
- Suitable for field applications.
- Normal or angled probes can be used.
- Provides instantaneous indication of flaws.
- Can be used on relatively complex parts or areas with limited access.

Limitations:
- Small coverage area.
- Surfaces must be clean and smooth.
- Coupling thickness variations could affect the results.
- Manual inspection is operator-dependent, time-consuming, and often does not provide a permanent record.
- Automated inspection may be difficult due to the transducer coupling problems.

7.1.8.2 Immersion Approach

In the immersion approach, both the component and the transducer are immersed in water for probe-to-component coupling. The probe is scanned automatically with respect to the test piece to perform the inspection (Figure 7.14). Automated scanning of the probe is carried out using a multi-axis system equipped with motion controllers, a computer, and the related software. Relatively large components or segments of aircraft that are flat and simple in shape can be inspected by scanning the probe at high speeds along one axis and indexing it by small increments after each scan in the perpendicular direction to the scan line until the inspection

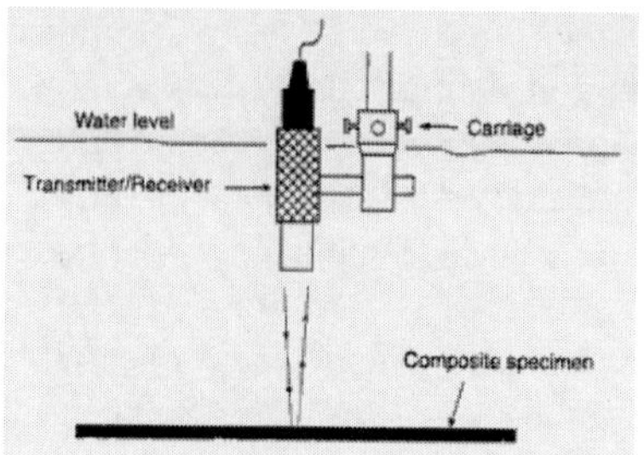

Figure 7.14 (Left): Schematic presentation of ultrasonic testing using the water immersion approach, and (right): the IAR automated immersion system being used during inspection of the Canadarm composite arm boom.

site is covered. Data collection, analysis, and display are also done automatically. A-scan, B-scan, and C-scan images of the inspection results are provided for interpretation by the operator. Most automated immersion systems consist of three linear and two rotational axes for probe manipulation. Some may also have a turn-table for inspection of circular or cylindrical parts and a bar-indexer for inspecting tubes. More information on the characteristics of automated inspection systems is provided later in this chapter under the heading "Equipment."

Capabilities and Limitations

Capabilities:
- Small or large area inspections are possible in the laboratory or manufacturing environments.
- Water coupling eliminates the variability associated with contact coupling.
- Suitable for automated inspection of relatively large but simple structures.
- Provides permanent images of the inspection sites.
- More efficient and reliable than the contact approach.

Limitations:
- Equipment is expensive.
- Cannot be used in the field or for in-service inspections.
- Inspection of complex parts is difficult and may be impossible.
- Some parts cannot be immersed in water.
- Presence of entrapped water bubbles could create false indications.

7.1.8.3　WATER-JET APPROACH

For inspection of large aircraft structures in the manufacturing environment, a water-jet or squirter system is often used. This approach is suitable for testing of honeycomb or foam-core sandwich structures that cannot be immersed in water. In this case, the transducer-to-component coupling is provided by columns of water squirted onto the part using specialized nozzles that surround the probe (Figure 7.15). Water columns act as wave guides. Often an assembly consisting of a combination of probe and nozzle is used, one to transmit and the other to receive the ultrasonic waves after passing through the part. Like the immersion approach, the probe/nozzle-pair assembly is scanned and indexed automatically with respect to the test piece to perform the inspection using a multi-axis computer-controlled system that provides A-, B-, and C-scan images of the part.

Capabilities and Limitations

Capabilities:
- Large area inspection is possible in the manufacturing environment.
- Suitable for parts that cannot be immersed in water.
- Provides permanent images of the inspection sites.
- Provides reliable results.
- Very efficient.

Limitations:
- Equipment is very expensive.
- Not suitable for laboratory or field use.
- Complex parts cannot be inspected.
- Water recycling is required.

7.2 TESTING PROCEDURES

Ultrasonic tests can be carried out using either normal incidence or oblique-angled beams depending on the object conditions, accessibility, test requirements, and other factors.

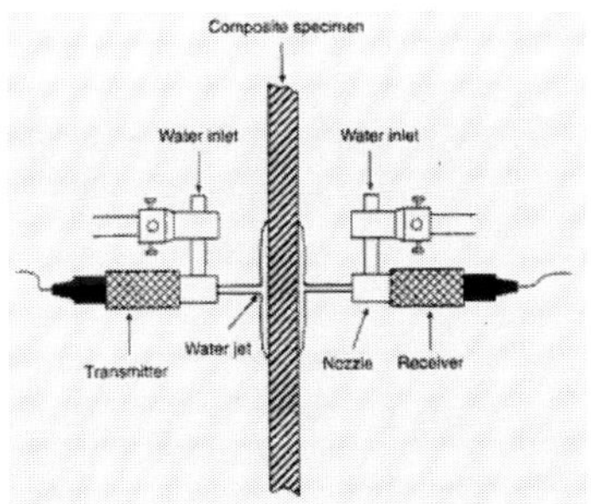

Figure 7.15 (Left): Schematic presentation of water jet or squirter coupling approach, and (right) a typical automated squirter system showing a composite part being inspected. Photograph is courtesy of Ultrasonic Sciences Ltd., http://www.ultrasonic-sciences.co.uk/index.htm.

7.2.1 Normal Beam

7.2.1.1 Pulse-echo Approach

In the pulse-echo method, a single transducer is used to send bursts of ultrasonic energy (pulses) into the test piece at regular intervals. The coupling of the transducer with the test piece is provided using one of the above-mentioned approaches. If the transmitted pulses encounter a reflecting surface, such as a discontinuity, a portion of the energy is returned and picked up by the same transducer (see Figure 7.1). The amount of energy that is returned to the probe is dependent on the acoustic impedance of the discontinuity, as described above, as well as its physical characteristics (size, shape, type, orientation, etc.) in relation to the UT waves (beam size, wave intensity, frequency, etc.). As mentioned earlier, the received waveforms or echoes are analyzed in terms of their amplitude and time of arrival to get an indication of the flaw size and location. Sometimes, the entire returned signals are digitized and analyzed to obtain additional information, such as frequency spectra, power content, phase changes, etc. Such analysis may provide information on flaw type, shape, or other characteristics, in addition to the size and location.

As mentioned earlier, the discontinuity size is estimated from the A-scan signal amplitude, and its location with respect to the object front (or rear) surface is determined from the A-scan signal location on the time axis. The depth-location is obtained by measuring the time it takes for the UT pulses, generated at the transducer and passed through the thickness of the material, to reach the discontinuity and, after reflection and passing through the material thickness once again, to return to the same transducer. In most applications, this time interval is a few microseconds or less. The two-way transit time measured is divided by two to account for the forward-and-return travel paths and multiplied by the ultrasonic velocity of the test material. The result is expressed in the following simple relationship:

$$d = Vt/2 \text{ or } V = 2d/t \qquad (7\text{-}17)$$

where d is the distance from the surface of the object where the transducer is located to the discontinuity surface facing the beam, V is the ultrasonic wave velocity in the material, and t is the measured round-trip transit time. Longitudinal, shear, or surface waves may be used depending on the inspection method employed. The ASTM Standard E494-05 provides guidelines for measuring ultrasonic velocity in materials using the pulse-echo approach by comparison with reference samples whose UT velocities are known [5]. This standard requires solid samples at least 5 mm in thickness with two parallel and smooth surfaces. Using this practice, one can measure both longitudinal and shear velocities. Once these velocities are measured, the Poisson's ratio and elastic constants of the material can be calculated using the equations provided above in Section 1.2.

The pulse-echo approach using longitudinal waves is used in most commercial thickness gauging instruments that operate at frequencies in the 500 kHz to 100 MHz range. Typically, lower frequencies are employed to achieve optimal penetration when measuring thick, highly attenuating, or highly scattering materials, such as cast metals, while higher frequencies are used to optimize resolution for thin, non-attenuating, non-scattering materials, such as metal cladding. Ultrasonic thickness gauging allows quick and reliable measurement of thickness without requiring access to both sides of the part. Accuracies as high as ±1 micron or ±0.0001 inch can be achieved in some applications.

Pulse-echo inspection for flaw detection can be performed using longitudinal, shear, surface, or Lamb waves. The beam can be perpendicular or at an angle with respect to the test piece depending on the flaw characteristics (type, size, location, orientation), as well as the part geometry and access for transducer placement. This approach is commonly used for field inspection of aircraft metallic parts for cracks and corrosion, as well as for testing composite structures for delamination or disbonding. In these applications, access to only one surface is necessary, but in the layered or sandwiched structures with interlayer air gaps, only the accessible surface can be tested.

The ASTM Standard E1901-08 provides guidelines for ultrasonic pulse-echo testing of materials using straight-beam longitudinal waves introduced by direct contact of the search unit with the test object [6]. This standard is for discontinuity detection and evaluation in volumetric materials and provides capability requirements for instrumentation, probes, couplant, and reference standards, as well as guidelines for standardization of apparatus, object surface examination and preparation, probe scanning, signal analysis, and interpretation of results.

The pulse-echo approach is also employed in the manufacturing environment, often using automated immersion systems. Figure 7.16 shows a simple pulse-echo immersion approach and full-waveforms reflected back by the parallel surfaces of a plate. Pulse-echo testing of flat parts immersed in a water tank is carried out using a beam perpendicular to the test part as described in the ASTM Standard E1001-06 [7]. This standard provides guidelines for the necessary equipment, references, examination, and evaluation procedures, as well as documentation and analysis of results. The approach can be used to inspect flat, slightly curved, or cylindrical parts made of metals or solid laminates. However, it cannot be used to inspect an entire sandwich structure; only the front face can be tested. Also, this method is not suitable for highly attenuating materials, such as porous cast metals or ceramics, as well as very thick structures, due the fact that UT waves travel twice through the test part losing their energy twice as much as in a single-path through-transmission approach, as described later.

It must be noted that in the pulse-echo configuration, the UT wavelength in relation to the object thickness should be considered during the tests. When the wavelength is much smaller than the thickness of the test piece, the front-surface and the back-surface signals can be easily resolved from each other as well as

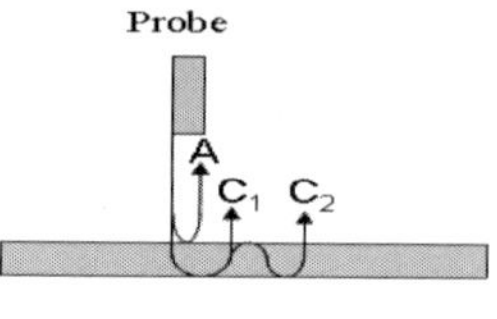

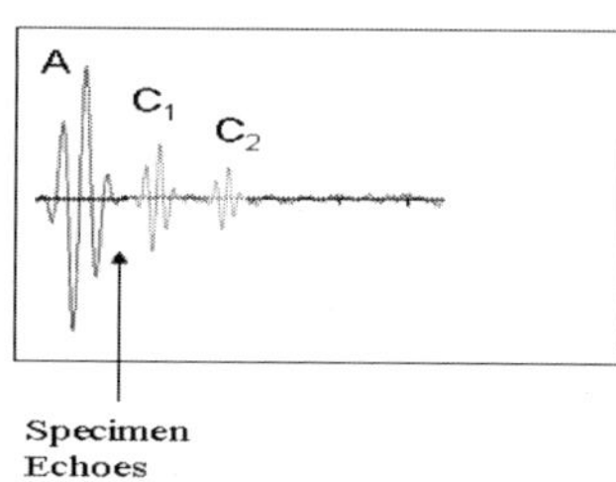

Figure 7.16 Schematic illustration of ultrasonic pulse-echo approach and multiple echoes from parallel surfaces.

from the internal discontinuity echoes that will appear between the two surface echoes. As indicated above, the time difference between the front-face echo and the flaw echo will give the location of the defect with respect to the front-surface, and the intensity of the flaw echo will provide information about the defect (e.g., size). However, when the wavelength is not sufficiently small (low frequency testing) or the object is very thin, echoes from surfaces and defects may overlap, making it difficult to apply this approach. Alternatively, methods such as the reflector plate pulse-echo approach that is described in the next section, can be used.

Capabilities and Limitations

Capabilities:
- Access to only one side is needed.
- Applicable to a variety of materials and geometries.
- Applicable in manufacturing environments or in the field.
- Manual or automated inspections are possible.
- Can be used to detect different flaw types (cracks, voids, corrosion, delamination, etc.), measure thickness, and determine some properties.
- Estimation of flaw size and location is possible.

Limitations:
- Coupling fluid is often needed.
- Object surfaces must be reasonably flat, smooth, and parallel.
- Difficult to inspect thin plates or detect near-surface flaws.
- In layered structures, such as lap joints, only the first layer can be inspected if an air-gap exist between the layers.
- Only the front-face of honeycomb or foam core structures can be inspected.

7.2.1.2 THROUGH-TRANSMISSION APPROACH

This method uses two transducers: one as a transmitter that emits repetitive pulses and the other as a receiver on the opposite side of the test piece that captures the UT waves transmitted through the test piece (Figure 7.17). The two transducers must be properly aligned. The through-transmission inspection of aircraft parts is often carried out in a water tank where both the part and transducers are immersed in water. Water-jets or squirter systems may be used to provide ultrasonic coupling instead of full immersion. In both cases, uniform coupling and accurate alignment are important.

In the through-transmission approach, UT waves travel only once through the material; thus, they suffer less loss of energy. This approach is most suitable for the inspection of materials with high attenuation, such as honeycomb or foam-core sandwich parts, and composite laminates. However, access to both sides of the test piece is required. It is widely used in aircraft manufacturing facilities for quality control inspection of composite structures, but cannot be used easily in the field. Also, the technique does not provide information about the through-the-depth location of the defects.

Capabilities and Limitations

Capabilities:
- Can be applied to thin or thick structures.
- Can be used to inspect a variety of materials.
- Honeycomb and foam-core structures can be inspected if edges are sealed.
- Estimation of the flaw size in the plane perpendicular to the beam direction is possible.

Limitations:
- Access to two opposite sides of the test piece is necessary.
- Water immersion or squirter systems are required.
- Not suitable for field use.
- Through-the-thickness flaw location is not provided.
- Not suitable for complex parts.

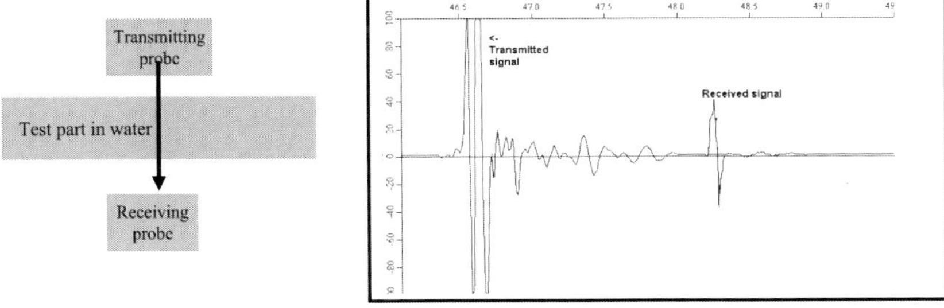

Figure 7.17 Through-transmission technique in immersion mode.

7.2.1.3 REFLECTOR PULSE-ECHO APPROACH

In case of thin tubular structures where the conventional pulse-echo or through-transmission approaches cannot be used, the reflector pulse-echo approach is employed. When the thickness of the test material is comparable to or smaller than the ultrasonic wavelength, if conventional pulse-echo is used, surface echoes may obscure echoes from internal flaws, making the pulse-echo inspections difficult and unreliable. Also, in geometries, such as long tubular structures, it is not easy to perform the through-transmission inspection since it requires placement of the receiving probe inside the part opposite to the transmitting transducer and scanning both probes simultaneously to cover the entire tube. In such cases, the reflector pulse-echo approach may provide a more practical solution. Thus, inspection of thin structures that have reasonably uniform, smooth, and parallel surfaces is often performed by using a flat and smooth reflector behind the test piece in a water tank and monitoring the return echoes from the reflector (Figure 7.18). Since the ultrasonic beam has to travel through the test material before being reflected by the reflector, the change of the reflector echo can be related to the UT attenuation within the material. However, the specimen to reflector distance and the reflection coefficient of the reflector must remain constant. In this case, the presence of discontinuities in the test material is identified by monitoring the reflector echo amplitude, which is directly related to the material's attenuation.

This approach is widely used for inspection of thin tubular structures made of composite laminates or sheet metals that cannot be tested using immersion pulse-echo or through transmission techniques. Like the conventional pulse-echo method, this method is not suitable for materials with high attenuation, such as very thick composite laminates, honeycomb, or foam-core structures, or porous metal castings, due to double-passing of the beam through the test part.

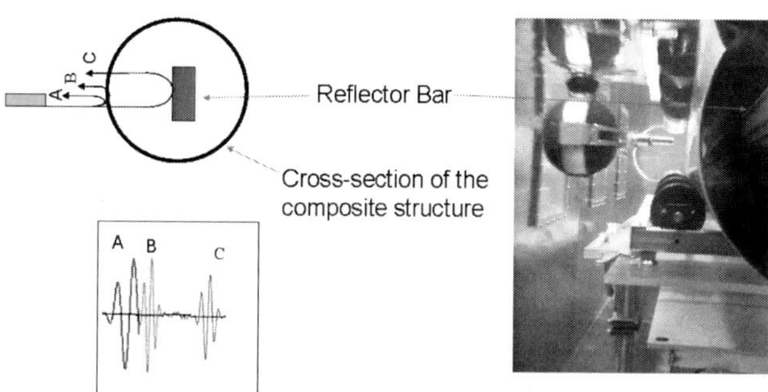

Figure 7.18 (Left): Schematic illustration of ultrasonic pulse-echo approach with a reflector bar inside the test piece in immersion mode and typical echoes expected, and (right) actual test arrangement showing the probe part and reflector.

Capabilities and Limitations

Capabilities:

- Can be applied to thin composite laminates or sheet metals.
- Can detect internal, surface, and near surface defects.
- Estimation of the flaw size and location is possible in the same plane as the test piece.

Limitations:

- Test part must have smooth, uniform, and parallel surfaces.
- Immersion of the test part in water is necessary.
- Not suitable for field use or manual inspections.
- Not suitable for porous metals, honeycomb, foam core, or very thick parts.
- Access to the back side of the test piece is necessary.
- Information on depth location of flaws is not provided.

7.2.2 Oblique-angle Beam

7.2.2.1 Pulse-echo Approach

When the incident beam enters the test piece at an oblique angle, unlike the normal incident case, the parts' front and back wall echoes are not usually seen, and only reflected or scattered waves from discontinuities or walls facing the beam are observed. The angled-beam approach using a single contact probe is used to detect internal discontinuities, such as voids, inclusions, and cracks, in metallic parts. Angled-beam transducers and wedges are typically used to introduce refracted longitudinal or shear waves into the test material, as seen in Figure 7.19 [1].

An angled sound path allows echoes from internal discontinuities to return before reflections from part walls, thereby identifying internal flaws facing the

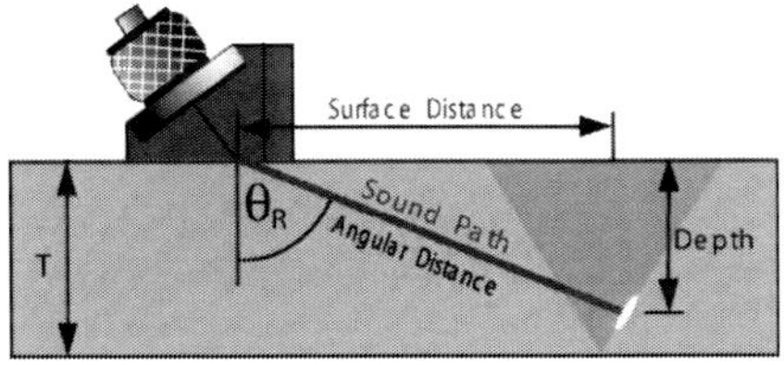

Figure 7.19 Detection of internal discontinuities using oblique-angle beam and contact probe [1].

beam. As indicated in this figure, knowing the wave velocity and the refracted angle of the material, it is possible to estimate the location of the discontinuity (surface distance from the probe and depth) by measuring the returned echo time. Also, flaw size can be estimated from the signal amplitude.

This approach is often employed when normal beam inspections cannot be carried out due to problems, such as unfavourable flaw orientation (e.g., cracks perpendicular to the part surface), hidden damage (e.g., cracks or corrosion under fasteners or bolts), or specific applications (e.g., weld inspections). In the latter application, the beam is first allowed to be reflected by the rear surface in order to cover the whole depth of the weld, as illustrated in Figure 7.20. In this case, the proper transducer distance from the weld or skip distance corresponding to the crown, sidewalls, and root of the weld is determined by using the refracted angle, beam index point, and material thickness.

In case of cracks, size estimates may be made more accurately using a technique known as tip diffraction. For example, the size of a crack originating from the backside of a flat plate (a) can be determined by linearly moving an angled-beam transducer over the test area and identifying returned echoes from the crack surface-opening and the end-tip (Figure 7.21). In this case, the principal echo comes from the scattering of the UT waves by the surface-opening of the crack, and this gives the crack position. A second, much weaker echo comes from the

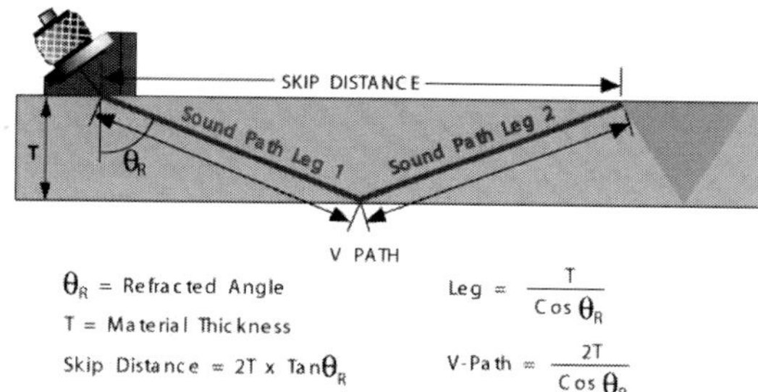

Figure 7.20 Schematic presentation of weld inspection using oblique angle-beam [1].

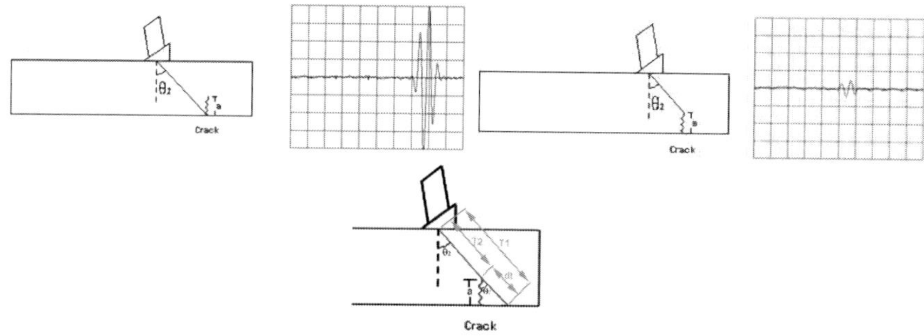

Figure 7.21 Crack detection and measurement using diffraction method [1].

end-tip of the crack, and since the distance traveled by the ultrasound is less, the second signal appears earlier in time. The time difference can be related to the size. Knowing the ultrasound velocity (v) in the material and the incident angle, θ, the crack length is obtained by measuring the difference in arrival times between the two echoes (*dt*) using the diagram and equation provided in Figure 7.21. For top-surface cracks, the beam is first allowed to reflect at the rear surface to reach the crack opening at the top-surface and then the crack end-tip once the transducer is moved away from the crack, similar to the approach illustrated in Figure 7.20.

The tip-diffraction approach is applicable when the geometry of the part is relatively simple and the crack orientation is known. This approach is also known as Time-of-Flight-Diffraction (TOFD) using the pulse-echo mode. TOFD can also be carried out using two transducers in the pitch-catch configuration as described in the next section.

$$a = \cos\theta \times \frac{(dt \times v)}{2} \qquad (7\text{-}18)$$

Capabilities and Limitations

Capabilities:
- Simple and inexpensive.
- Suitable for field inspection of metallic parts.
- Provides instantaneous indication of internal flaws.
- Flaw size can be estimated in simple geometries.

Limitations:
- Small coverage area.
- Surface must be clean and smooth.
- Coupling thickness variations could affect the results.
- Labor intensive and operator dependent.

7.2.2.2 PITCH-CATCH APPROACH

In the pitch-catch approach, two angled-probes are used: one to transmit and the other to receive sound waves. This approach can be used to detect and measure cracks that are nearly perpendicular to the surface, and is known as Time-of-Flight-Diffraction (TOFD) using pitch-catch. Figure 7.22 shows the basic configuration for the TOFD pitch-catch technique, and the generated signals as the probe-pair, held at a fixed separation, scans the test site.

Probes with a wide beam divergence angle are usually used since tip diffraction occurs regardless of orientation of the flaw. The first signal to arrive at the receiver is the lateral wave, which travels just beneath the upper surface of the specimen.

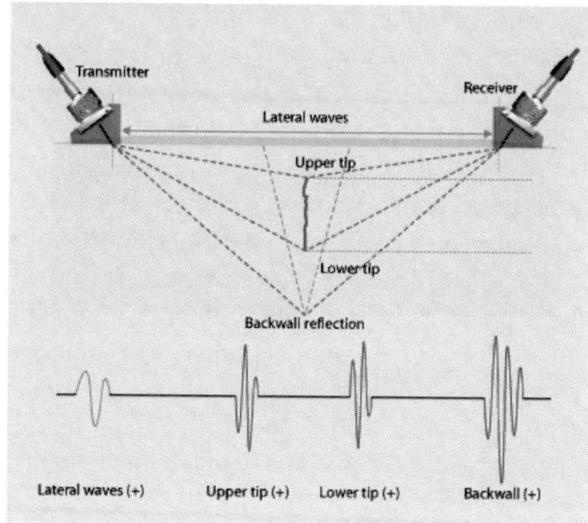

Figure 7.22 Basic arrangement for the time-of-flight-diffraction (TOFD) using pitch-catch approach and the resulting A-scan signals (picture courtesy of Olympus Inc. http://www.olympus-ims.com/en/panametrics-ndt-ultrasonic/tofd/).

In the absence of any cracks, the second signal to arrive is from the back wall surface. However, when a crack exists, diffracted energy from the upper and lower tips will be received before that of the back wall. Therefore, it is possible to determine, not only the through-wall height of the flaw, but also its location within the thickness of the specimen. Note that a phase reversal occurs between the lateral wave and back wall echo and between the upper tip and the lower tip of the flaw signals; this characteristic can help in the interpretation of TOFD data. The flaw size and position measurement accuracy is less if the defect is not perpendicular to the upper and lower surfaces of the test object.

Capabilities and Limitations

Capabilities:
- Simple and inexpensive.
- Suitable for field inspection of metallic parts (e.g., welds).
- Provides instantaneous indication of internal flaws.
- Flaw size and location can be estimated with good accuracy.

Limitations:
- Small coverage area.
- Surface must be clean and smooth.
- Thickness variation of coupling fluid could affect the results.
- Labor intensive and operator dependent.
- Flaw must have sharp tips to scatter the beam.
- Alignment of transducers is important.

7.2.2.3 GUIDED-WAVES APPROACH

Guided-waves or Lamb-waves are generated in parts with two parallel surfaces when the part thickness is less than the wavelength. Figure 7.23 shows the basic arrangement for the ultrasonic guided-wave approach. The guided-wave velocity varies as a function of the product of frequency and thickness. A large number of guided-wave modes exist in pipe- or plate-like geometries, and the acoustic properties of these wave-modes are a function of the test material, geometry, and the frequency. Predicting the properties of the wave-modes often relies on heavy mathematical modeling. The wave modes are typically presented in graphical plots called dispersion curves.

Guided-waves travel long distances in metals with little loss of energy, and, thus, are widely used in the inspection of long metal rods, pipes, rails, and plates. Guided-waves have also shown some potential for testing aircraft aluminum lap joints. Using an array of transducers, large areas can be covered. When launched across lap joints, at locations where there is a change of cross-section due to corrosion or a change in local stiffness due to the disbonding of the joints, changes may occur in the reflected and the transmitted waves. From the arrival time of the reflected echoes and the speed of the wave-mode at a particular frequency, the distance of a feature in relation to the position of the transducer array can be calculated [1]. GWT uses a system of distance amplitude curves (DAC) to correct for attenuation and amplitude drops when estimating the cross-section change from a reflection at a certain distance. The DACs are usually calibrated against a series of echoes from known features, such as fastener holes, with known signal amplitudes.

Capabilities and Limitations

Capabilities:
- Access to one side only.
- Suitable for long rods, tubes, rails, or plates.
- Large discontinuities perpendicular to the part surface can be identified.
- Easy to use in the field.

Limitations:
- A fluid or gel is necessary for transducer coupling.
- Probe alignment and pressure applied could affect discontinuity detection.

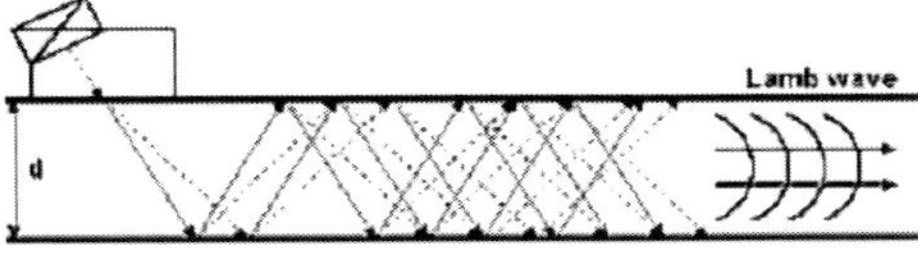

Figure 7.23 Guided or Lamb waves generation and propagation in a plate [15].

- Automated scanning is not easy.
- Interpretation of data is highly operator dependent.
- Small flaws, such as corrosion pits, cannot be easily found.
- Not very effective for inspecting complex parts or areas that are close to attachments.

7.2.2.4 SURFACE-WAVES APPROACH

If an ultrasonic beam is incident onto the surface of the object at the critical angle of shear waves, surface or Rayleigh waves are generated as illustrated in Figure 7.24. The governing relationship is known as Snell's law and is expressed as below:

$$\text{Sin} \, (\theta_c) = V_i / V_s \tag{7-19}$$

where θ_c is the critical angle of shear waves, V_i is the velocity of incident waves in medium 1 or the probe wedge material, and V_s is the velocity of shear waves in medium 2 (test material).

Surface waves cover a depth of about one wavelength below the material's surface, and if interrupted by surface breaking cracks or sharp edges perpendicular to their direction of propagation, they are reflected back. Thus, surface cracks close to normal to the beam direction can be detected by using contact or immersion probes (Figure 7.25).

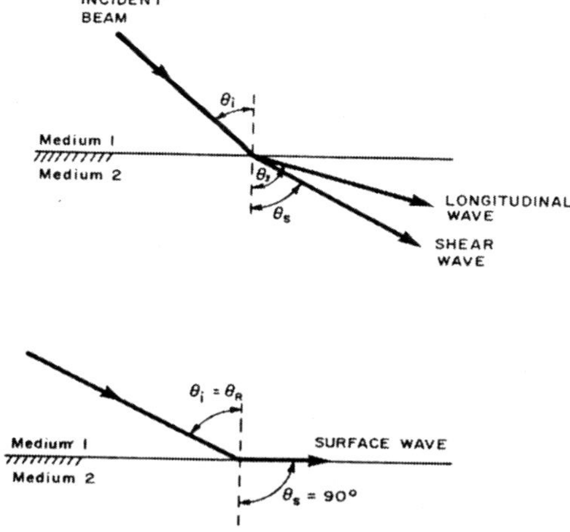

Figure 7.24 (Top): Mode conversion at oblique incident angle, and (bottom) generation of surface waves when the incident angle is equal to or larger than the critical angle of shear waves in the test material.

The immersion configuration is referred to as the "leaky surface wave approach" and will be described in more detail later. It is also possible to use two contact probes to generate and pick up surface waves in a pitch-catch configuration, as illustrated in Figure 7.26. This approach may be used to measure the depth of surface-breaking flaws.

Capabilities and Limitations

Capabilities:
- Needs access to one side only.
- Suitable for detection of surface-breaking defects.
- Hidden defects may be detectable.
- Flaw depth can be measured.

Limitations:
- Part surface must be clean and smooth.
- A fluid or gel is necessary for transducer coupling.
- Probe alignment and pressure applied on the probe could affect discontinuity detection.
- Automated scanning is not easy.
- Interpretation of data is highly operator dependent.
- Small flaws such as corrosion pits cannot be easily found.
- Not very effective for inspecting complex parts or areas that are close to edges.

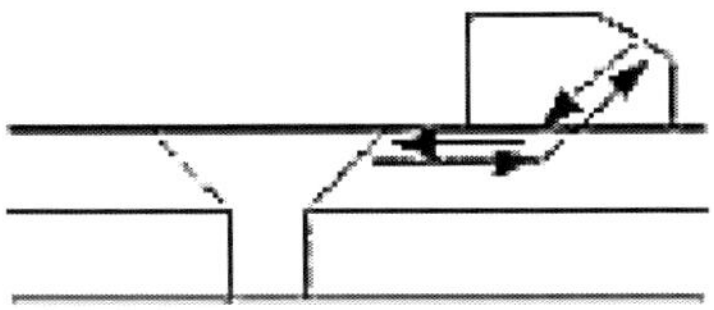

Figure 7.25 Surface wave approach using a contact transducer in pulse-echo mode.

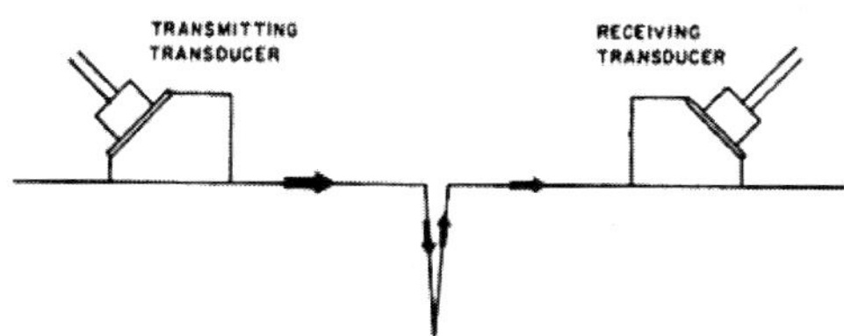

Figure 7.26 Surface wave approach using two contact probes in pitch-catch configuration.

7.2.2.5 LEAKY SURFACE-WAVES APPROACH

If both the part and the transducer are immersed in water and the transducer is placed at the critical incident angle with respect to the part, surface waves are generated. Such waves propagate very small distance at the solid-water interface, quickly losing their energy. When they encounter perpendicular surface-breaking flaws, they are reflected back and the reflected waves leak their energy back into the water, and this energy can be picked up by the same probe (Figure 7.27). This approach is referred to as the "leaky surface wave method" and is used to detect very small surface-breaking flaws. However, the part surface must be very smooth or polished.

The surface waves generated in this way lose their energy very rapidly, so only flaws that are located at the incident point of the UT beam are detected. This is an advantage when defects are close to each other or are near the edges of the part. Using this approach, C-scan images of surface discontinuities can be obtained; however, the images are usually distorted (elongated) due to the oblique incidence angle and the beam focal spot size. Examples will be provided later to illustrate such effects. It is also possible to generate and pick up leaky surface waves by using a large aperture focused probe that is placed perpendicular to the part such that the focal point is below the part surface (or defocused) as shown in Figure 7.28. In this configuration, in addition to the direct reflection of the normal incident beam (or specular waves) and the waves radiated from the transducer edges, leaky Rayleigh waves are generated from the beams striking the material at or above the critical angle. Since the Rayleigh waves are generated in a full circle, surface discontinuities in all directions can be detected. Using this configuration, C-scans of surface or near surface flaws can be obtained without distortions.

Capabilities and Limitations

Capabilities:
- Access to only one side is needed.
- Sensitive to surface flaws.
- Flaw depth measurement is possible.

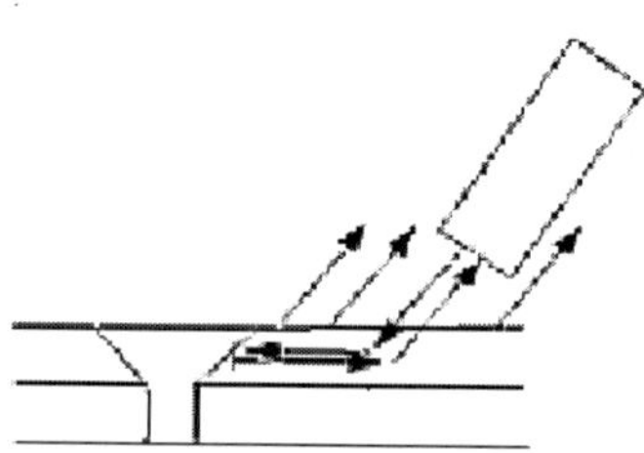

Figure 7.27 Generation and detection of surface waves in immersion mode.

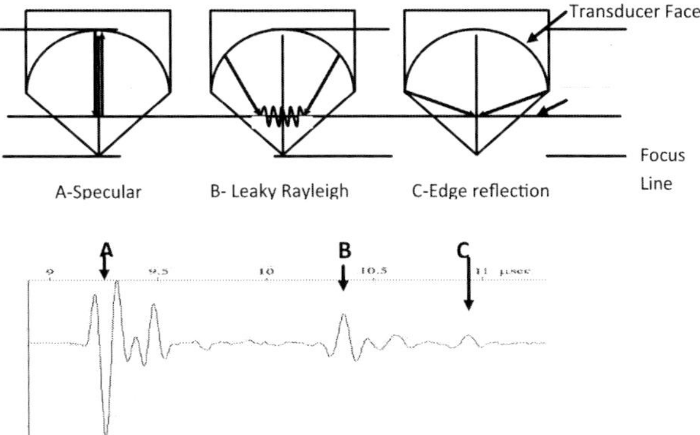

Figure 7.28 Ultrasonic leaky surface waves generated and picked up by a large-aperture focused probe placed such that the focal point is below the surface (14).

Limitations:

- Deep flaws cannot be detected.
- Flaw orientation is important; cracks parallel to the beam cannot be detected.
- Short inspection range, especially in the immersion mode.
- Applicable to smooth surfaces only.

7.2.2.6 BACK-SCATTERING APPROACH

If the probe is placed at an oblique angle with respect to the test piece in immersion mode, the back-scattered energy from material discontinuities or other features that are close to the normal angle to the incident beam may be detectable. This approach can be used to detect distributed internal or surface discontinuities, such as porosity and micro-cracks. This approach is also used to establish fiber orientation in composites. This is achieved by measuring the backscattered energy while the part is rotated on a turntable as described later under application examples. The return signal amplitude peaks every time the beam is perpendicular to the fiber direction. This approach is referred to as polar backscattering.

Capabilities and Limitations

Capabilities:

- Access to only one side is needed.
- Sensitive to distributed materials discontinuities.
- Sensitive to fiber orientation in composites.

Limitations:
* Automated immersion C-scan is required.
* Flaw quantification is difficult.
* Useful as a laboratory technique only.

7.3   EQUIPMENT

Ultrasonic test equipment generally consists of a probe or transducer, a pulser, a receiver-amplifier, a display oscilloscope, an electronic clock, and the associated power supply units. Most modern instruments are computer-based and have signal and image processing capabilities. In automated equipment, the UT instruments are integrated with a scanning system that enables the movement of the probe in a raster mode over the test part to generate two dimensional images of the ultrasonic results. Instruments made by different manufacturers vary in capabilities and functions as well as their ability to inspect different materials or geometries. Thus, the choice of the equipment should be based on the applications. Portable instruments and hand-held probes are suitable for in-situ inspections of small areas or complex geometries. However, for large area inspections in the laboratory or manufacturing environments, due to the limitations of hand-held sensors in terms of scanning speed and data recording, automated scanning systems are more appropriate.

Often ultrasonic instruments are designed with certain applications in mind, and their characteristics, such as the UT wave frequency range, the probe type, and coupling, as well as the size and configuration of the scanning system, are chosen accordingly. Several instruments and scanning configurations with different capabilities may be necessary to cover a range of applications. Figure 7.29 to Figure 7.31 show examples of ultrasonic equipment and scanning systems. The equipment includes a portable instrument with a transportable x-y linear scanner for in-situ inspections of small areas and two stationary systems designed for automated inspection of large samples or components in the laboratory. The system shown in Figure 7.31 is applicable to both large airframe metallic and

Figure 7.29 A portable ultrasonic instrument with an X-Y linear scanner

Figure 7.30 An automated X-Y linear scanner for ultrasonic scanning of simple parts in the contact mode.

Figure 7.31 An automated ultrasonic system for inspection of simple flat or contoured parts in immersion mode.

composite structures as well as small engine parts. Plates, tubes, and more complex geometries can be tested for flaw or damage detection, measurement of dimensions, and evaluation of some material properties, such as Young's modulus and density. For testing of curved parts, the transducer angle with respect to the part as well as the scan profile must be adjusted such that the UT beam is always perpendicular to the part surface and the transducer-part distance remains constant at all times (Figure 7.32). In automated systems, these are achieved through a software program.

This immersion system is equipped with three linear axes for scanning and indexing, and two rotational transducer manipulation axes for maneuvering the probe head. These are used to inspect a variety of materials and component geometries in pulse-echo, through-transmission, or angled-beam configurations. In addition, a separate turntable and a bar-indexer are available for the inspection of

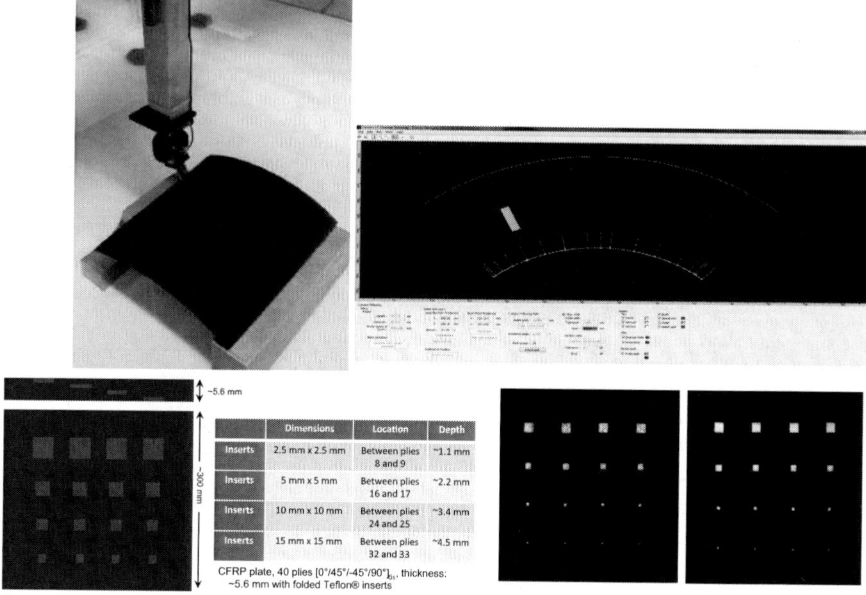

	Dimensions	Location	Depth
Inserts	2.5 mm x 2.5 mm	Between plies 8 and 9	~1.1 mm
Inserts	5 mm x 5 mm	Between plies 16 and 17	~2.2 mm
Inserts	10 mm x 10 mm	Between plies 24 and 25	~3.4 mm
Inserts	15 mm x 15 mm	Between plies 32 and 33	~4.5 mm

Figure 7.32 Automatic adjustment of (top left): transducer angle and (top right): scan profile for ultrasonic testing of curved parts. Ultrasonic C-scan images of two reference panels containing artificial defects of different sizes. (Bottom left): flat panel and (bottom right): curved panel.

circular parts (e.g., engine discs) or cylindrical structures (e.g., composite tubes). The system is equipped with a dedicated ultrasonic signal/image processing software package that enables full signal capturing and analysis that provides A-, B-, and C-scan results. Both amplitude and time-of-flight C-scan imaging is possible. The full waveform capturing allows post-processing of the archived data in time or frequency domains for flaw or material characterization.

7.3.1 PIEZOELECTRIC TRANSDUCERS

The transducer is the essential part of conventional ultrasonic systems. It converts electrical pulses into mechanical vibrations in the ultrasonic frequency range and vice versa. There are a variety of transducer types, the most commonly used are based on piezoelectric ceramic elements and are used for both contact and immersion inspections. The contact transducers are made to generate straight-beam or angled-beam longitudinal, shear, or surface waves, depending on how the transducer ceramic element is cut and how it is positioned in the probe unit

Some transducers contain two-elements: one to send and the other to receive sound waves, and both are housed in one unit. The piezoelectric element is covered on the side of the transducer face by a plastic film or sheet to provide protection and also for better acoustic impedance matching with the coupling fluid. A thick plastic face can also act as a delay line to provide an adequate time gap between the initial vibration signal of the ceramic element and echoes returned from the test piece. The delay lines are necessary for thickness gauging, which is based on accurate measurement of the time gap between the specimen front and back face echoes. The plastic face-cover can be wedge-shaped to enable the UT beam to enter the test material at an angle. It can also be shaped as a concave lens to provide a focused beam. On the opposite side of the transducer face, the piezoelectric element is backed by a high acoustic damping compound to absorb vibrations towards the interior of the transducer and to allow wave propagation in the transducer face direction only. Figure 7.33 shows the basic construction of conventional UT transducers.

At IAR, a wide variety of UT transducers is available that includes contact probes (normal and angled-beam) as well as immersion transducers (straight or focused-beam). The contact probes are in the 1–5 MHz frequency range while the immersion transducers range in frequency from 1 to 50 MHz.

7.3.2 TRANSDUCER ARRAYS

Conventional UT transducers often have only one element and occasionally two elements. However, recently, transducer arrays have been manufactured by placing a large number of elements in certain patterns in one single unit. The elements are electronically triggered one after another with a pre-determined phase difference to provide slightly different beam angle so that a larger section of the test part can be interrogated in a single inspection. This UT beam steering is usually referred to as S-scan and is similar to the imaging approach used in medical ultrasound. The elements can be arranged in linear, circular, or more complex array forms. Figure 7.34 shows the basic principle of a linear transducer array.

The electronic scanning in linear arrays (E-scan) allows a faster coverage of the test area than a mechanical system using a single probe. Thus, these multi-element

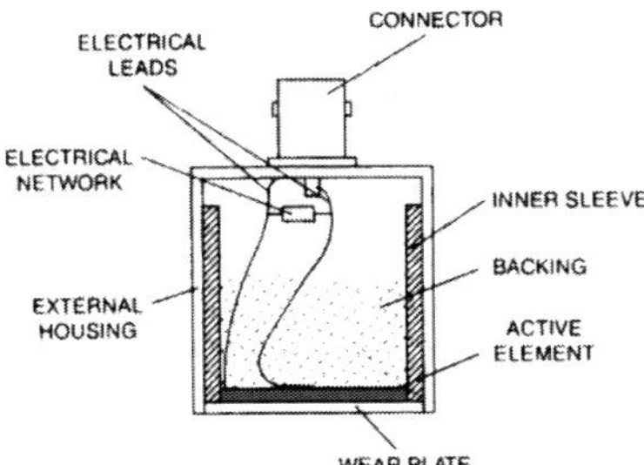

Figure 7.33 Basic construction of an ultrasonic transducer [1].

transducers are used to increase scanning efficiency for large area inspections where the primary concern is defect detection rather than characterization. Linear arrays may be used for identification of defects, such as corrosion in aluminum lap-joint structures or impact damage in composite parts. For crack detection, circular or steering arrays are often employed. Commercial instruments are available to identify bolt-hole cracks in aircraft lap-joint structures (e.g., Figure 7.35). In this case, the beam steering capability allows the generation of ultrasonic waves at different refraction angles but all in the same vertical plane. The approach can detect cracks of different orientations, even those that are not perpendicular to the beam direction.

Large-scale systems based on transducer arrays have been developed at Boeing for inspection of the entire aircraft (Figure 7.36). In Canada, the RD Tech Division of Olympus Inc. manufactures such transducers and the associated instruments commercially. The disadvantage of the piezoelectric-based UT arrays is the need for a coupling fluid between the probe and the test material that may reduce the inspection speed or create false indications if the coupling is lost.

Figure 7.34 The basic principle of the linear transducer array. Elements are along the top and the active beam is identified by the dark gray (courtesy of RD Tech).

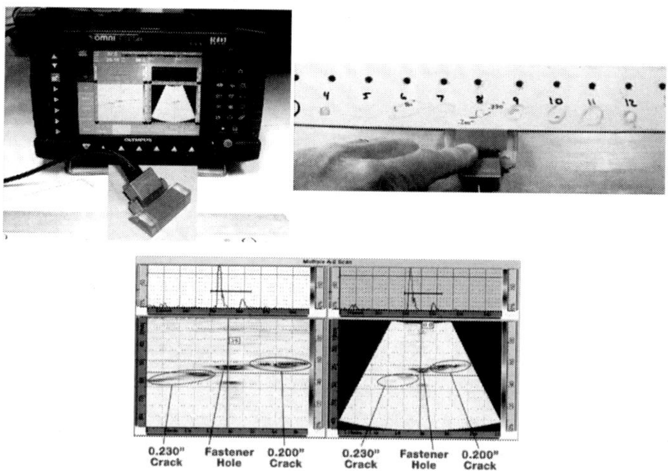

Figure 7.35 A commercial ultrasonic instrument (Olympus Omniscan) equipped with a transducer array used for crack detection in aircraft aluminum lap joints (www.olympus-ims.com).

Figure 7.36 Aircraft inspection using an ultrasonic phased array system at Boeing.

7.4 APPLICATION EXAMPLES

Ultrasonic testing is widely used in aeronautical applications to inspect and evaluate various types of materials and components, both during manufacture and in service. The choice of the specific procedure or test parameters is dependent on many factors, including the part's material, geometry, surface condition, and accessibility to the inspection site, as well as the test requirements (e.g., flaw detection, property measurement, or dimension gauging) and the target characteristics (e.g., flaw type, size, location, and orientation). In the following paragraphs, a few examples of UT applications for flaw detection/sizing and property measurements are provided with emphasis on important factors that may affect the outcome of the test. The examples are mostly selected from the extensive work carried out by the author and colleagues, and cover a wide range of materials including aerospace composites, metals, and coatings as well as part geometries and target characteristics.

7.4.1 PROCEDURES AND PARAMETERS

The ultrasonic results may change depending on the employed procedures and parameters as demonstrated in the following two examples.

7.4.1.1 FREQUENCY EFFECTS

For the majority of aerospace materials, nominal frequencies in the range of 0.5 to 20 MHz are usually used for detecting discontinuities or measuring properties and dimensions. Frequencies < 5 MHz are generally used for inspection of inhomogeneous materials, such as composites or thick structures, for optimal penetration depth, while frequencies in the range 5–20 MHz are often applied to homogeneous materials for detecting smaller defects. Occasionally, frequencies higher than 20 MHz are needed to identify very small defects that may be considered critical or for thickness gauging of thin coatings. For example, in some high temperature structural ceramics, such as silicon nitride or silicon carbide, discontinuities larger than 0.3 mm may be considered critical, and, thus, frequencies as high as 100 MHz may be necessary. In contrast, frequencies in the kHz range are adequate for inspection of materials that have high attenuation rates, such as honeycomb or foam-core sandwich structures.

Although the use of high frequencies generally results in greater sensitivity and resolution that enables the detection of smaller discontinuities, this is not always desirable or achievable. For example, in Figure 7.37, the ultrasonic C-scan test results on a CFRP composite reference piece containing embedded Teflon tapes of different sizes and depth locations for two different frequencies are presented. At 1 MHz, minor changes in the material or geometry have been identified by the different background colors representing variations of the returned signal amplitude. Nonetheless, all the Teflon inserts are detected. At the higher frequency of 5 MHz, the minute material or geometry variations affect the returned echoes to a larger degree, resulting in more pronounced background color changes, which

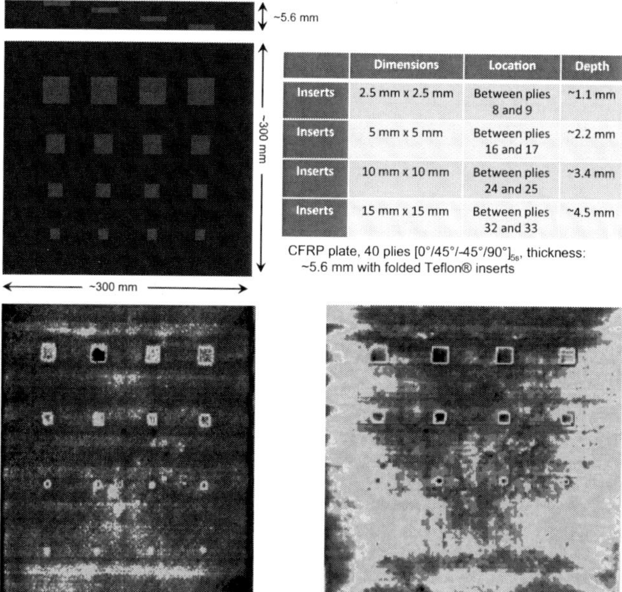

	Dimensions	Location	Depth
Inserts	2.5 mm x 2.5 mm	Between plies 8 and 9	~1.1 mm
Inserts	5 mm x 5 mm	Between plies 16 and 17	~2.2 mm
Inserts	10 mm x 10 mm	Between plies 24 and 25	~3.4 mm
Inserts	15 mm x 15 mm	Between plies 32 and 33	~4.5 mm

CFRP plate, 40 plies [0°/45°/-45°/90°]$_{5s}$, thickness: ~5.6 mm with folded Teflon® inserts

Figure 7.37 (Top): A 40-layer CFRP solid laminate reference panel with embedded discontinuities simulating delamination (folded Teflon inserts) of various sizes and depths, and (bottom): Corresponding ultrasonic C-scan images of the above panel obtained using (bottom left): 1 MHz and (bottom right): 5 MHz transducers. This panel is curved and the parallel traces in the background are due to the change of transducer angle to maintain beam normality.

obscure indications of the smaller inserts. In this case, higher frequency or higher sensitivity has adverse effect on flaw detectability.

Furthermore, the depth of penetration of the ultrasonic waves into the test material reduces with increasing frequency. In the pulse-echo configuration, this may cause defects to be missed that may exist at deeper sections. Therefore, the proper selection of the transducer/equipment frequency is important and should be based on the required sensitivity, resolution, and penetration depth (or inspection range), and should be verified using appropriate reference standards.

7.4.1.2 PROCEDURE EFFECTS

A number of ultrasonic approaches are available for testing aerospace materials and components after manufacturing or during service. For example, composite parts are tested to identify manufacturing defects or service damage using ultrasonic pulse-echo, through-transmission or reflector-assisted pulse-echo approaches. Each approach may provide different results as described in the following examples.

Figure 7.38 shows ultrasonic C-scan images of a CFRP solid laminate reference panel with embedded Teflon inserts of various sizes and depth locations. The inspections were carried out using the through-transmission (TT) and pulse-echo (PE) techniques. These images are obtained by mapping the amplitude of the received signals within a pre-selected gate. For the TT approach, the gate is placed on the overall signal received at the receiving probe (see Figure 7.17), while in the PE case, the gate is adjusted to cover the material volume between the front and rear surfaces excluding surface echoes. In these two-dimensional images, the gray shades are related to the intensity of the gated signal. Both methods easily identify the larger inserts; however, the pulse echo method provides a better resolution as, in this case, the effects of surface roughness are not seen. In contrast, the TT image that is based on the total transmitted signal and represents

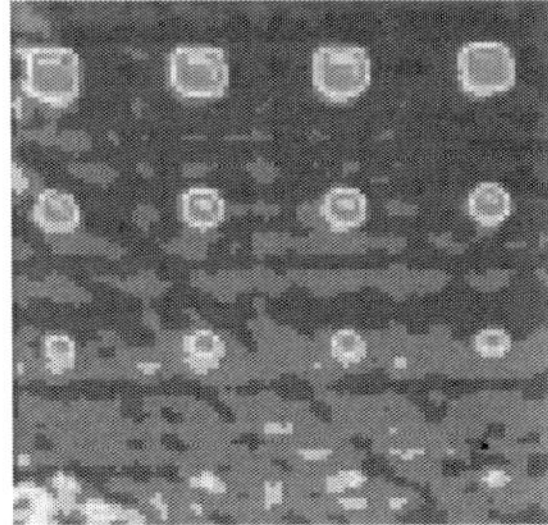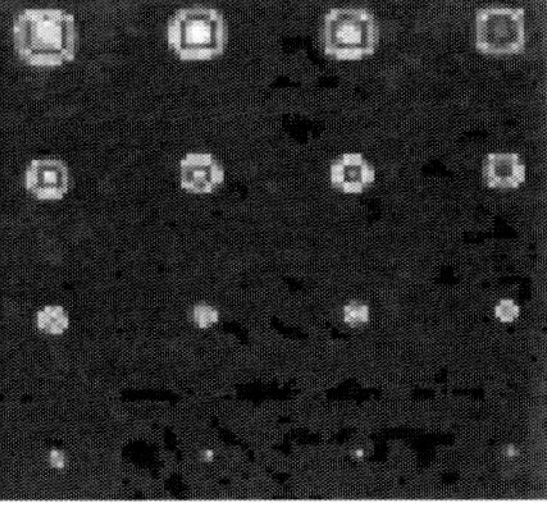

Figure 7.38 Ultrasonic C-scan inspection of a flat solid laminate reference panel containing Teflon inserts of different sizes at different depths using (left) through-transmission and (right) pulse-echo approaches. Note that the gate in the latter case is in-between the front and rear-surface echoes.

the entire material (including surfaces) displays a higher background noise due to surface features obscuring traces of small inserts. Thus, the proper selection of the inspection approach and the related signals is important in obtaining the required information. This will be elaborated further in the next example. Both the TT and PE amplitude images provide planar indications of the flaw size but not the depth locations.

Figure 7.39 shows the cross-sectional view of a composite structure which is a near-tubular part made of CFRP solid laminates reinforced by three full-circle and one half-circle internal stiffeners. This component is inspected using the conventional pulse-echo and reflector-assisted pulse-echo approaches, and the corresponding C-scan results are also presented in this figure. These C-scans are obtained after subjecting the part to impact tests at different energy levels and on several sites. Inspections are carried out by scanning the transducer along the

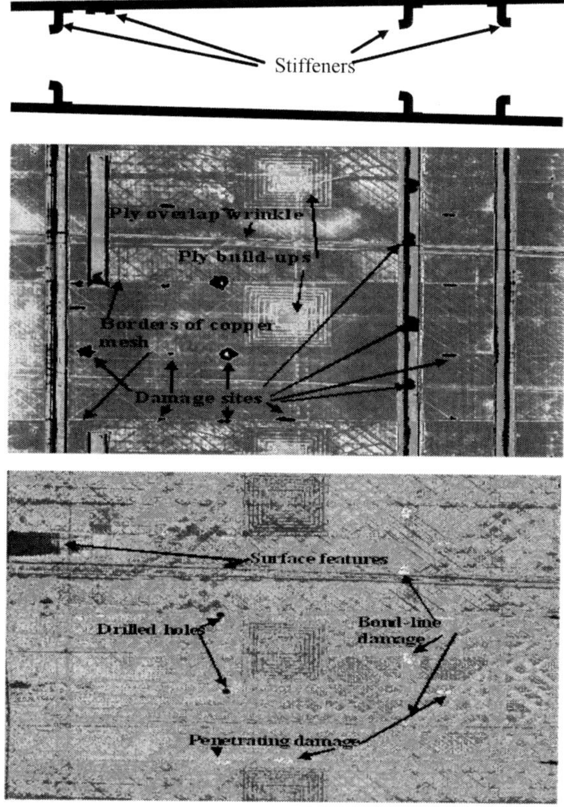

Figure 7.39 (Top): A schematic diagram showing a near-tubular composite structure with internal stiffeners, (middle): ultrasonic C-scan image obtained using reflector-assisted pulse-echo approach, and (bottom): ultrasonic C-scan image obtained using conventional pulse-echo approach.

length of the tube while the tube is rotated on a bar-indexer by about 0.5° after each scan until the entire tube is covered. Reference samples (such as the one described in the previous section) are first used to calibrate the inspection instruments and methods.

The C-scan image seen in the middle of Figure 7.39 is obtained by monitoring echoes returned from a stainless steel reflector bar placed inside the tube (refer to Figure 7.18). In this case, the UT beam travels through the part to get to the reflector and back to the transducer; thus, the resulting image provides details of the entire structure including the location of the internal stiffeners, the ply-overlaps, ply-built-up areas, material changes, such as copper mesh, installed for lightening protection, and drilled holes, as well as the location, shape, and extent of damage caused by impact. However, this image does not indicate how deeply the damage has penetrated into the material and whether or not the bond-lines between the tube and stiffeners have been affected. To obtain the additional information, the conventional pulse-echo approach is used. In this case, the image is generated by choosing a narrow gate, such that only the inner surface echoes are monitored. Thus, the resulting C-scan shows features, such as the ply build-up areas of the inner surface, holes that have penetrated the inner surface, as well as damage to or close to the inner surface (including the bond lines), but not the internal ribs or damage close to the outer surface seen in the previous overall C-scan image. However, in this case, unlike the previous example, the gate covers the inner surface, and features, such as surface wrinkles, roughness, and attached tapes, can be seen in the background. As a result, damage traces are not as visible as in the previous image and could be easily missed. Thus, sometimes it is necessary to perform multiple inspections using different test procedures, parameters, or set ups.

Another example is provided in Figure 7.40 where an actual aircraft rudder is tested using ultrasonic through-transmission and ultrasonic pulse-echo approaches. The rudder is made of a composite honeycomb material (aluminum honeycomb and CFRP skin laminates). The rudder is susceptible to water ingress through hinges that are bolted to the structure in three locations. Penetration of water into the structure may cause the corrosion of aluminum honeycomb, which affects the structural integrity and must be detected. Ultrasonic C-scan inspection of the rudder using the through-transmission approach enables the detection of water intrusion into the honeycomb core, as illustrated in Figure 7.40 (middle). In this mode, the UT waves travel through the solid laminate skin and the Al cell walls. If water is present in some cells, a portion of the UT energy is dissipated in the fluid, thereby affecting the signals at the receiving sensor, as seen here. However, this approach cannot be easily applied in the field, and the rudder must be removed from the aircraft for inspection in a hanger or laboratory environment.

In contrast, the pulse-echo method is easier to use in field applications, but only echoes from the skin facing the probe can be captured and monitored, resulting in a lack of sensitivity to entrapped fluid within the honeycomb core as illustrated in the same Figure (bottom image). However, the pulse-echo approach is more

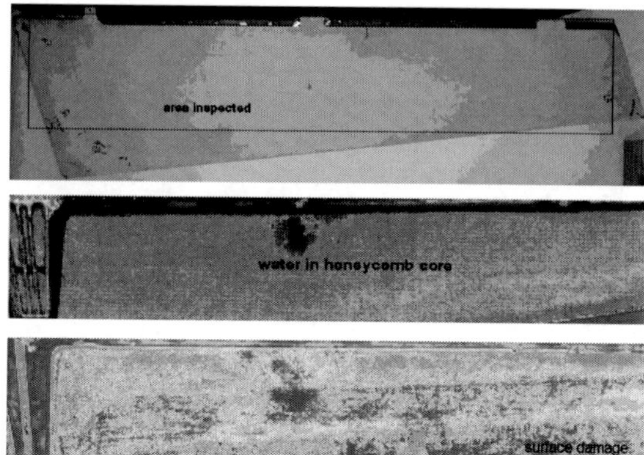

Figure 7.40 (Top): Photograph of an aircraft rudder made of Al honeycomb core and CFRP skins, (middle): ultrasonic through-transmission inspection showing water ingress, and (bottom): ultrasonic pulse-echo inspection of the same part showing features within the upper skin layer.

sensitive to the skin damage. Therefore, both approaches are needed to assess the entire structure.

7.4.2   FLAW DETECTION AND SIZING

As mentioned before, there are many different ultrasonic approaches for flaw detection and sizing; the choice of the proper method is dependent on the material, geometry, flaw type, size, and location. Three examples are provided here that use ultrasonic surface waves generated using the direct contact approach, immersion method, and a novel technique based on a defocused beam.

7.4.2.1   SURFACE-BREAKING CRACKS IN METALS

Reference [16] describes an approach developed for detection and sizing of surface-breaking cracks in large metal plates. The approach uses two contact wedge-probes: one to generate and the other to pick up surface waves in pitch-catch configuration, as seen in Figure 7.41. The part inspection involves simultaneously scanning the area of interest with a pair of wedge-probes (with appropriate critical angle) located face-to-face at a fixed distance from each other and monitoring the received signal. When the probe-pair is moved from the flaw-free area to the flaw site (flaw site must be between the two probes), the change of the signal amplitude indicates the presence of the surface-breaking defects, and the time-shift provides an indication of the depth of the defect. This time-shift is twice the surface wave travel time from the flaw surface opening to the end-tip. By knowing the surface wave velocity of the test piece, the flaw depth can be calculated with good accuracy.

7.4.2.2 CRACKS UNDER FASTENER HEADS

References [18] and [19] describe an approach based on ultrasonic surface waves to detect cracks under fastener heads in aluminum lap joints. Samples simulating aircraft fuselage skin are made of two aluminum plates fastened together with two rows of fasteners (e.g., Figure 7.42). EDM notches and fatigue cracks of various sizes (0.25 mm to 1.5 mm) were present in the upper plate under the fastener heads. A row of fastener holes, also containing cracks or EDM notches, is seen without the fasteners for comparison.

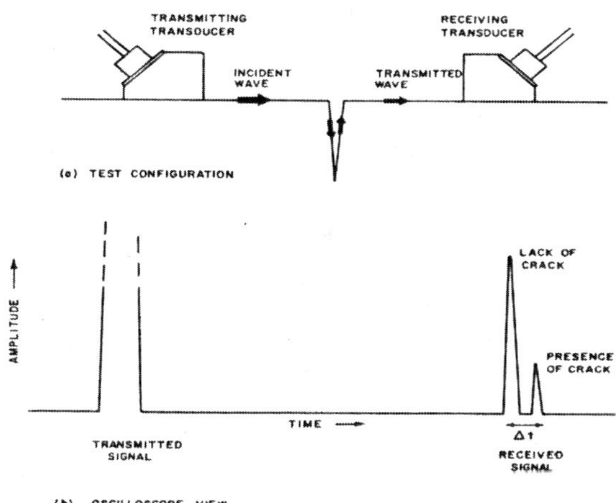

Figure 7.41 Detection and measurement of surface discontinuities (e.g., cracks) in metal plates using ultrasonic surface waves.

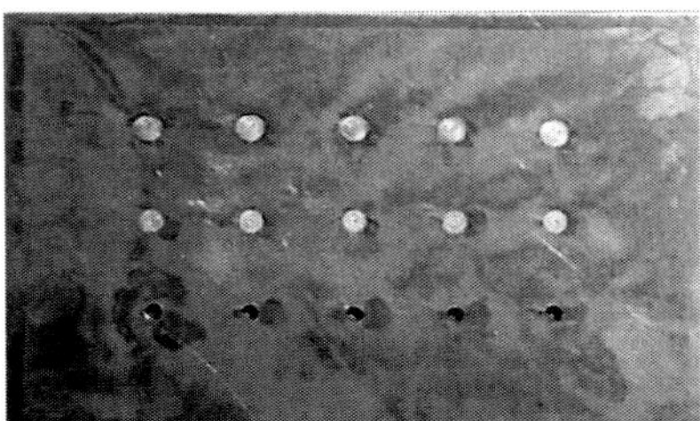

EDM length 0 mm 25 mm 0.75 mm 1.25 mm 1.75 mm

Figure 7.42 A photograph of an aluminum sample with two rows of fasteners (different sizes) simulating aircraft lap joints. Except for the first column on left, all holes of the top plate contain EDM notches on both sides.

Leaky Rayleigh waves are generated by placing a focused ultrasonic transducer at an oblique angle with respect to the test sample (see Figure 7.43), as described in Section 7.2.2.4, using the immersion approach. Inspection is carried out by moving the transducer along a single scan-line parallel to a row of fasteners and monitoring the returned signals. By gating the appropriate signals, it is possible to determine the center position of the fastener hole and discontinuities, and to produce C-scan images of the bolt-holes and their flaws. When the transducer is scanned from left to right, the crack signals will appear before the signal from the edge of the bolt-hole if the crack is on the left-side of the hole, or after the hole-edge signals if the crack is on the right side of the hole (Figure 7.44). Thus, the position of cracks can be identified. Unlike the normal C-scan images that resemble the target shape, in this case, images will be elongated (see Figure 7.45) due to the fact that the UT beam strikes the target at an angle and has an elliptical shape at the impingement point rather than circular. Butterfly-shaped indications are produced when the beam passes over the bolt hole, the butterfly body being a hole (or a fastener) and the wings being the cracks or EDM notches. With this approach, EDM notches or fatigue cracks as small as 0.75 mm can be detected with fasteners still in place. The sensitivity increases to 0.25 mm when fasteners are removed. However, the size and shape of the fasteners and the crack position as well as their orientation influence detectability.

7.4.2.3 GUIDED WAVES FOR LAP-JOINTS

Economical inspections of aircraft fuselages or wings require methods that can cover large areas. Ultrasonic techniques using guided waves have the potential for large area inspection. As mentioned earlier, guided waves or Lamb waves are created in thin plates when oblique incident angles are used. Defects between two

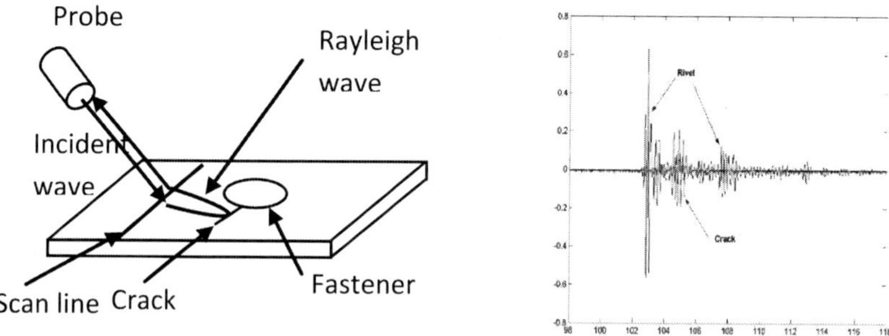

Figure 7.43 Schematic illustration of immersion Rayleigh wave inspection of cracks emanating from fastener holes. The Rayleigh waves must be close to perpendicular to the crack face.

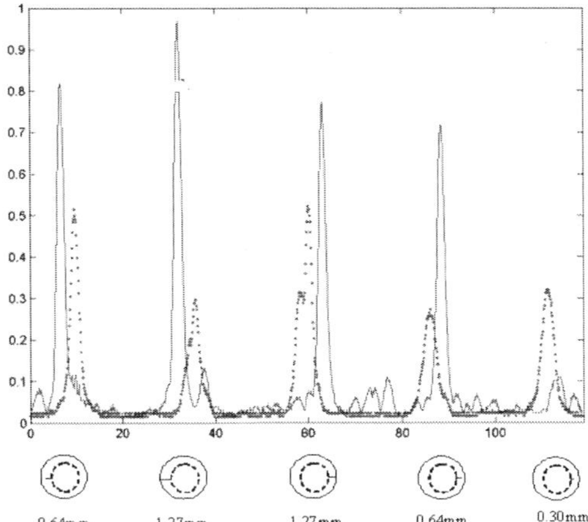

Figure 7.44 Return signals from hole edges in red and cracks in blue. Here, when the crack is on the left side of the hole, its signal appears before the hole-edge signal, and when the crack is on the right of the hole, its signal appears after the hole-edge signal.

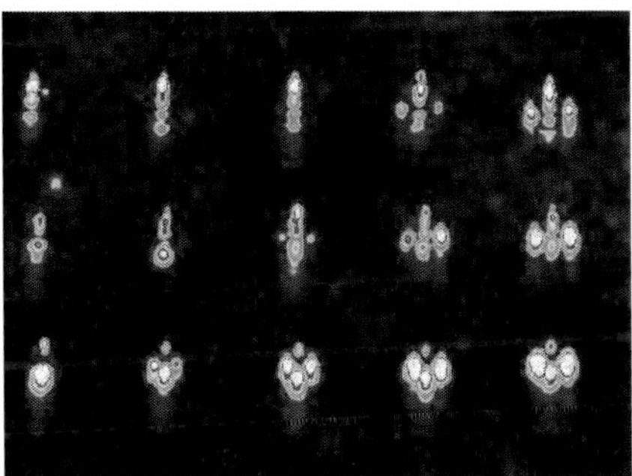

Figure 7.45 Ultrasonic leaky Rayleigh wave C-scan image of the above specimen showing holes and EDM notches. Note that the hole and notch images are elongated due to the oblique incident angle.

layers, such as corrosion or disbonds, are seen as thickness changes and, therefore, are detected.

Practical application of this approach for large area aircraft inspections requires probes that will provide a good signal to noise ratio and a stable scanning mechanism that will ensure a constant coupling of the probes. Several probe types and scanning configurations were examined on simulated and real aircraft lap joint specimens, and the pitch-catch configuration (Figure 7.46) was found to be the most practical. This approach involves placing transducers on either side of the joint and scanning the transducer pair along the length of the joint. This is significantly faster than raster scanning methods. When there is no disbond or corrosion, a major portion of the UT energy is transferred through the adhesive bond into the second layer resulting in a strong RF signal. However, when a disbond or corrosion exists between the two bonded layers, the transfer of energy is reduced and the signal intensity drops.

It must be noted that with conventional piezoelectric wedge probes, the overall signal amplitude may vary greatly due to the variation of the amount of the fluid used for transducer coupling, as well as probe alignment or the pressure applied. These problems may be reduced by using rubber-wheel probes or non-contact electro-magnetic acoustic transducers (EMATs). The EMAT approach will be described later in this chapter.

7.4.2.4 SMALL DEFECTS

Reference [21] describes a novel UT approach for detecting small surface and near-surface defects. The test piece is a 9 mm thick steel plate spray-coated with a 400-μm thick ceramic-metal (cermet) coating consisting of 86% tungsten carbide, 10% cobalt, and 4% chrome. Two artificial flaws were introduced in the specimen, as shown in Figure 7.47. The flaws consisted of a 5-mm-diameter side-drilled

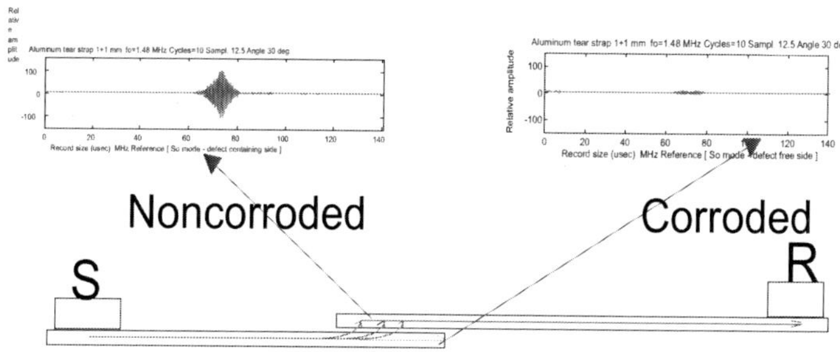

Figure 7.46 Ultrasonic guided wave inspection of a lap joint in pitch-catch configuration and corresponding signals for corroded and non-corroded areas.

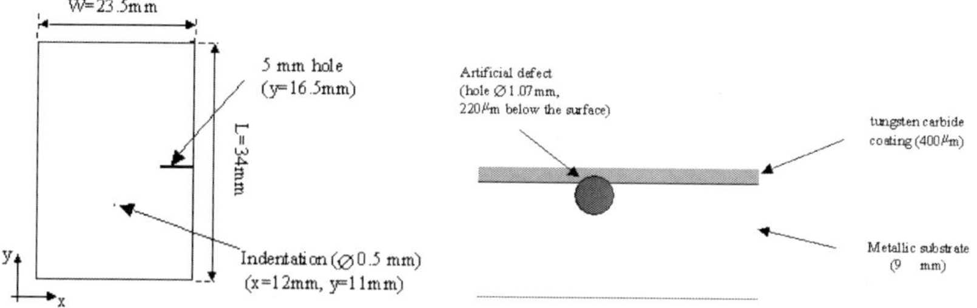

Figure 7.47 Schematic of the cermet test specimen. (Left): dimensions and flaw locations, and (right): side view showing the cermet coating thickness, the substrate thickness and the electro-discharged machined (EDM) hole at the substrate-coating interface.

hole that simulated cavities and a 0.5 mm diameter indentation on the surface that simulated an impact dent.

The ultrasonic pulse-echo technique was first used to provide some indication of the presence of defects (Figure 7.48); however, the indications were not quite obvious. Thus, further tests were performed using ultrasonic leaky surface waves (or Rayleigh waves) as described earlier. For these tests, a large aperture (45°) probe with a focal length (F) of 12.7 mm and a center frequency of 10 MHz was used. During the test, the probe was moved towards the sample surface in order to defocus the beam. This enabled separation of the leaky Rayleigh wave signals from the unwanted direct reflections (specular waves) and edge waves. A windowing procedure was applied to isolate the signal corresponding to the leaky surface waves.

Figure 7.49 presents the ultrasonic inspection results corresponding to different probe positions between –F/2 and –F/2+3.0 mm. This method clearly identifies the defects; however, the image contrast varies with the probe position. In addition to the artificial defects, the surface wave inspections revealed a scratch on the substrate surface located along the length of the specimen at about 7 mm location on the x- axis.

7.4.3 MATERIAL CHARACTERIZATION

As mentioned earlier, ultrasonic velocity in a material is related to its density and elastic constants, and the UT wave attenuation is dependent on the material's microstructure. Thus, UT velocity or attenuation measurements may be used to characterize materials. A few examples of such measurements are provided in the following sections.

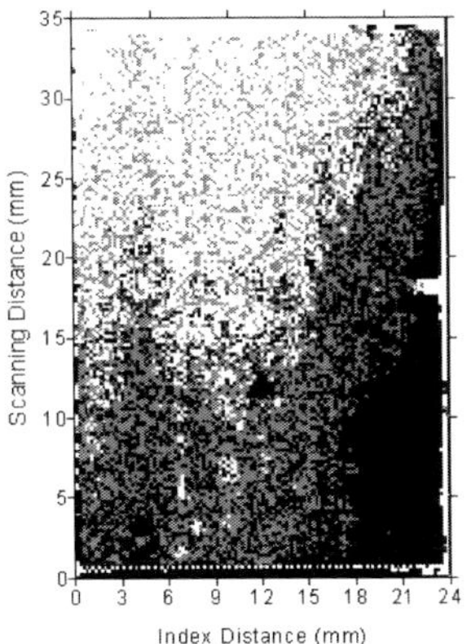

Figure 7.48 Ultrasonic pulse-echo C-scan.

7.4.3.1 Density Measurement in Coatings

An ultrasonic technique was used to obtain an estimate of the coating density using the approach suggested by Kinra and Zhu [22]. This approach is based on the comparison of the theoretical and the experimental transfer functions of the ultrasonic signals and calculating the error between the two until it is a minimum. Figure 7.50 illustrates ultrasonic immersion test of a coated sample at normal incident angle where $u(t)$ represents the incident waves, $f(t)$ represents multiple echoes produced as a result of reflections by the coating, and $g(t)$ represents reflections off the back-wall of the substrate. The fast Fourier transform (FFT) of $f(t)$ and $g(t)$ are $F(\omega)$ and $G(\omega)$, respectively. The transfer function is defined by:

$$H(\omega) = \frac{G(\omega)}{F(\omega)} \tag{7-20}$$

As described before, the reflection and transmission coefficients at a boundary between two media when the waves travel from medium i to medium j and

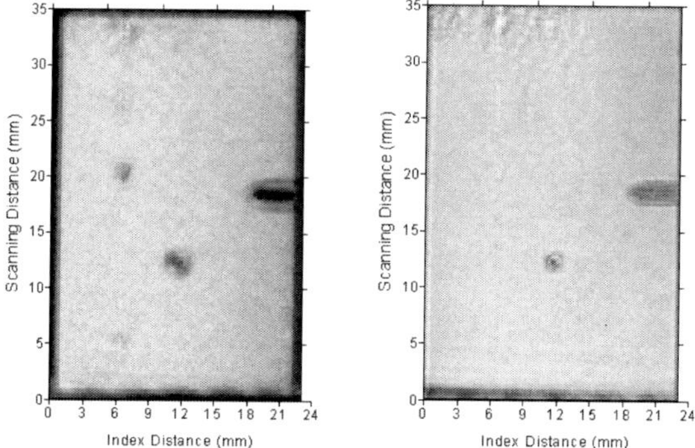

Figure 7.49 Surface wave C-scan results for two different probe positions; (left): −F/2+2.0 and (right): −F/2+3.0 mm.

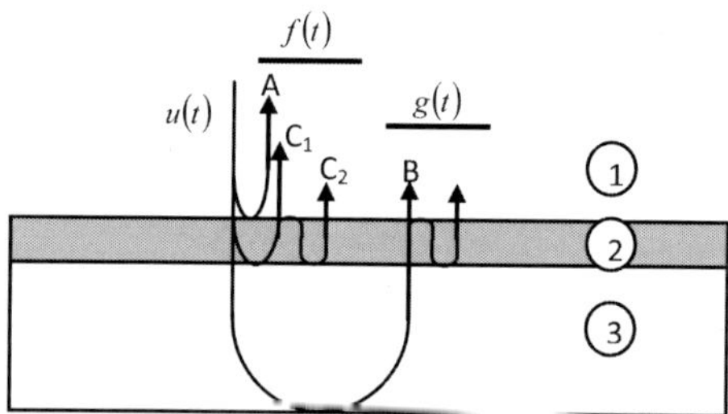

Figure 7.50 Ultrasonic multiple reflections at normal incident angle in a coated sample immersed in water. 1, 2, and 3 indicate water, coating, and substrate, respectively.

the beam is perpendicular to the boundary surface, are functions of the material's acoustic impedance expressed by:

$$R_{ij} = \frac{Z_j - Z_i}{Z_j + Z_i} \text{ and } T_{ij} = \frac{2Z_i}{Z_j + Z_i} \tag{7-21}$$

Taking into account the reflection and transmission coefficients at all three boundaries, the theoretical transfer function from the coating side can be expressed as:

$$H(\omega) = \frac{R_{31}T_{12}T_{21}T_{23}\exp(-2\alpha h)\exp-2i(k_2 h + k_3 L)}{\left(1 + R_{12}R_{23}\exp(-2\alpha h)\exp(-2ik_2 h)\right)\left(R_{12} + R_{23}\exp(-2\alpha h)\exp(-2ik_2 h)\right)} \tag{7-22}$$

where indices 1, 2, and 3 correspond to the water, coating, and substrate, respectively. R_{ij} and T_{ij} are the reflection coefficient and transmission coefficient between media "i" and "j", respectively. α is the attenuation coefficient in the coating, and h and L are the thickness of the coating and the substrate, respectively. k_2 and k_3 are the wave numbers ($k = V\omega$) of the coating and the substrate, respectively. V is the ultrasonic velocity in the respective medium and $\omega = 2\pi f$ where f is the ultrasonic frequency.

Using the above-mentioned cermet coating as a sample, ultrasonic measurements were carried out at normal incident angle while the specimen was immersed in water. A 10 MHz frequency flat transducer was employed during the measurements. Echoes created by the coating, $f(t)$, and the substrate back surface, $g(t)$, were measured and converted to the frequency domain to obtain the experimental transfer function. Using the above equation, the theoretical transfer function was calculated where medium 1 was water with ultrasonic velocity of 1480 m/s, and density of 1000 kg/m³, and medium 3 was a steel substrate with a velocity of 5890m/s, a density of 7900kg/m³, and a thickness of 9 mm. The medium 2 or cermet coating density and ultrasonic velocity values were calculated by iteration and are presented in Figure 7.51. The values of the root-mean-square error between the calculated and measured transfer functions provide the density and the UT velocity of the coating. The minimum error value in this figure, shown by a dot, corresponds to the density of 14540kg/m³ and ultrasonic velocity of 4820m/s. The cermet UT velocity was measured to be 4880m/s and its literature-reported density is 14670kg/m3. The difference is less than 1.5% in both cases.

7.4.3.2 Density / Porosity Mapping

The report [30] describes an approach that may be used to map the local variations in density or porosity content. This approach is based on the method proposed by

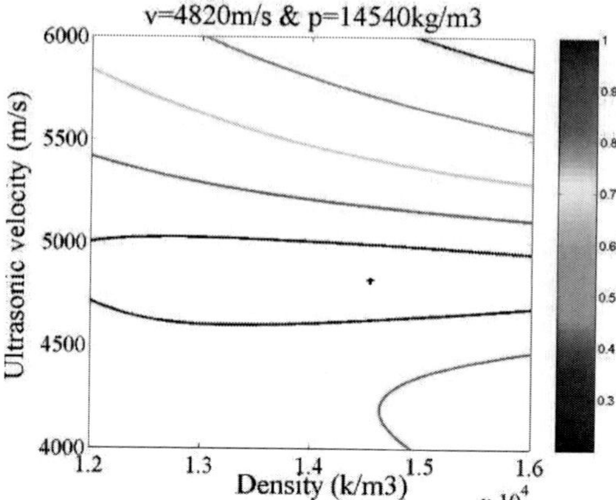

Figure 7.51 Values of the root-mean-square error between the experimental and theoretical transfer functions.

Hsu and Hughes [31] for simultaneous measurement of the ultrasonic velocity and thickness of a sample using the through-transmission method, as illustrated in Figure 7.52. In this approach, the transmitting and receiving transducers are capable of functioning in both pulse-echo and through-transmission modes. The method is applied to a simple, flat test piece and the time-of-flight of longitudinal ultrasonic waves with and without the sample is measured.

The definitions of the experimental notations are as below:

- x_1 is the distance between the transmitter and the left face of the test piece.
- x_2 is the distance between the receiver and the right face of the test piece.
- L is the distance between the transmitter and receiver.
- d is the thickness of the test piece.
- t_1 is the time-of-flight of the UT waves from the transmitter to the test piece and back to the transmitter.
- t_2 is the time-of-flight of the UT waves from the receiver to the test piece and back to the receiver.
- t_m is the time-of-flight of the UT waves between the transmitter and receiver when the test piece is in place.
- t_w is the time-of-flight of the UT waves between the transmitter and receiver in water (when the test piece is absent).
- v_w is the UT wave velocity in water.
- v is the longitudinal UT wave velocity in the test piece that must be measured.

In through-transmission mode with the presence of the test piece, the UT waves travel from the transmitter through water-specimen-water to the receiver and t_m is:

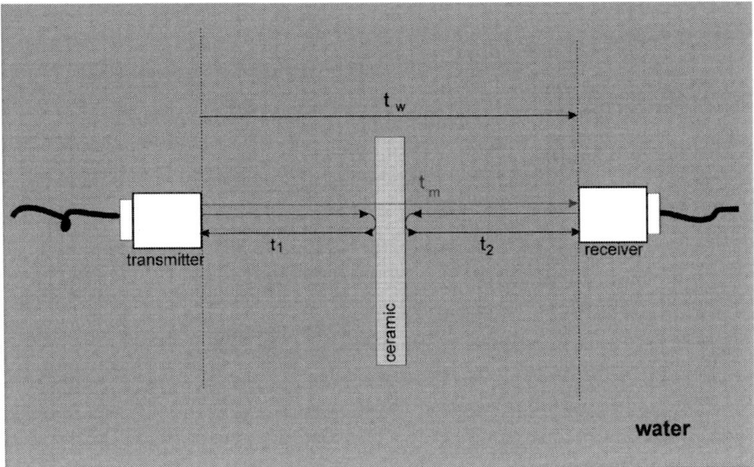

Figure 7.52 Experimental set-up for simultaneous measurement of UT velocity and beam path.

$$t_m = \frac{x_1}{v_w} + \frac{d}{v} + \frac{x_2}{v_w} \qquad (7\text{-}23)$$

In the through-transmission mode with the absence of the specimen, the UT waves travel from the transmitter through water to the receiver and t_w is:

$$t_w = \frac{L}{v_w} = \frac{x_1 + d + x_2}{v_w} \qquad (7\text{-}24)$$

In the pulse-echo mode with the presence of the test piece, when the UT waves are emitted by the transmitter, reflected back by the specimen, and picked up by the transmitting transducer:

$$t_1 = \frac{2 \cdot x_1}{v_w} \qquad (7\text{-}25)$$

In the pulse-echo mode with the presence of the test piece, when the UT waves are emitted by the receiving probe, reflected back by the specimen, and picked up by the receiver:

$$t_2 = \frac{2 \cdot x_2}{v_w} \qquad (7\text{-}26)$$

All the time-of-flight values can be measured experimentally from which the local UT velocity and thickness of the test piece can be obtained using:

$$v = v_w \cdot \frac{\left(t_w - t_1\right) + \left(t_w - t_2\right)}{\left(t_m - t_1\right) + \left(t_m - t_2\right)} \tag{7-27}$$

$$d = \frac{v_w}{2} \cdot \left[\left(t_w - t_1\right) + \left(t_w - t_2\right)\right] \tag{7-28}$$

For this purpose three different scans are needed:

- First is a scan in the through-transmission mode with the transmitter generating the UT waves and the receiver detecting the waves.
- Second is a scan in the pulse-echo mode with the transmitting probe generating the UT waves and also detecting them after being reflected by the left surface of the part.
- Third is a scan that is the same as the second one, but the receiving probe is generating the UT waves and is also detecting them after being reflected by the right face of the part.

The time-of-flight through water and the time-of-flight through the specimen are obtained in one test by having a scan area larger than the part. This approach was used on a flat piece of ceramic (alumina), 5.7 mm × 5.7 mm × 6.8 mm, immersed vertically in water, as shown in Figure 7.52. In this experiment, 12.7 mm diameter flat probes with a nominal centre frequency of 2.25 MHz were scanned at 0.25 mm increments in such a way that the scan area was larger than the surface of the specimen.

Figure 7.53 shows how t_m and t_w are determined from the two UT signals generated in the through-transmission mode when the UT beam passes through the water-specimen-water or water alone, respectively. Similarly, t_1 and t_2 are found

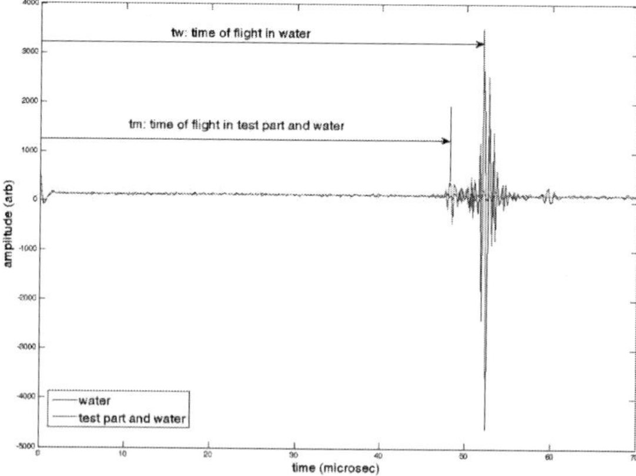

Figure 7.53 Ultrasonic echoes generated when the beam passes through the water and the test piece and when the beam passes through the water alone.

from the pulse-echo measurements when the transmitter or receiver is used for both transmitting and receiving waves.

After measuring t_1, t_2, t_w and t_m, v, and d values were determined using equations (27) and (28) and displayed in C-scan format. The UT velocity in water, v_w, is assumed to be constant and equal to 1485 m/s at a temperature of 21°C [31].

Although the above-mentioned equations provide simultaneous measure of the specimen local velocity and thickness, since the specimen was flat with two parallel faces and constant thickness, the "d" measured here is, in fact, the effective beam path length. Assuming that the test piece is isotropic, homogeneous, and dispersion-free with a constant thickness, the changes in v or d can be related to local variations in the material's density or the percentage of porosity content. Based on reference [31], a region of high porosity content is associated with an effective beam path length larger than the actual specimen thickness, as seen in Figure 7.54 (left). Similarly, it is known that velocity decreases when the porosity increases, as it is apparent from the results presented in Figure 7.54 (right). In fact, a linear inverse proportionality relationship exists between the ultrasonic longitudinal wave velocity and the porosity content of alumina.

In summary, variations in the UT longitudinal wave velocity or the effective beam path length can be indicative of local changes in density or porosity content, and maps like Figure 7.54 may be used to check the uniformity of the material. This approach does not require a calibration reference sample.

7.4.3.3 HEAT-DAMAGE IN 7050 ALUMINUM ALLOY

7050-T7451 aluminum alloy is widely used in airframe structures. The exposure of this alloy to extreme temperatures, such as accidental engine exhaust leaks or fires, may adversely affect the mechanical properties that have been optimized by the manufacturer's heat treatments. Thus, the exposure of 7050 aluminum alloy to heat damage was studied using the relationships of ultrasonic attenuation with static material properties and material microstructure [23]. Samples were cut from a cold-rolled 7050-T7451 aluminum plate and then heated in a furnace at a given

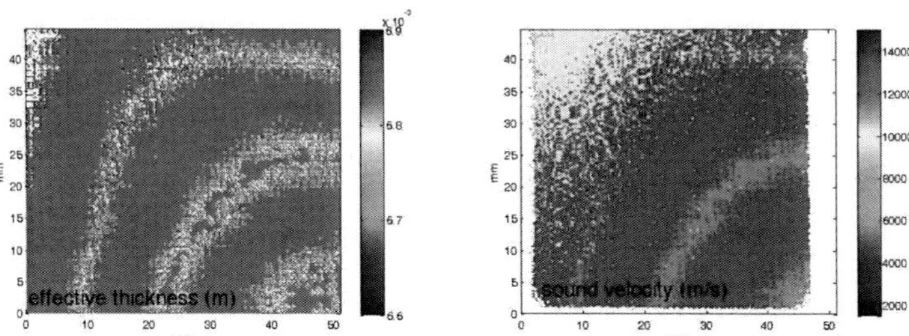

Figure 7.54 Maps of variations the UT effective beam path length and velocity in an alumina test piece.

temperature for a specific duration to simulate different degrees of heat damage or achieve a particular heat-treatment, as provided in Figure 7.55 (left). It was shown that attenuation measurements at high frequencies (> 20 MHz) can be related to the material yield strength through relationships, as shown in Figure 7.55 (right). Letters A through G indicate different heat-treatment/heat-damage conditions, as described in the figure caption. This figure indicates that microstructural changes taking place at extreme temperatures affect the material strength as well as the UT wave attenuation. Thus, by measuring the UT attenuation of a heat-affected zone of an aircraft as compared to the similar intact section of the same or a different aircraft, one may be able to estimate the extent and severity of damage.

7.4.3.4 FIBER ORIENTATION IN COMPOSITES

Composite laminates are laid up in different orientations to achieve certain elastic properties and mis-orientation of one or a few plies may affect the overall properties. Ultrasonic back-scattering at oblique angles can be used to establish ply orientation in composites. This is achieved by measuring the backscattered energy while the part is rotated, as shown schematically in Figure 7.56. The return signal amplitude peaks every time the beam is perpendicular to the fiber direction; thus,

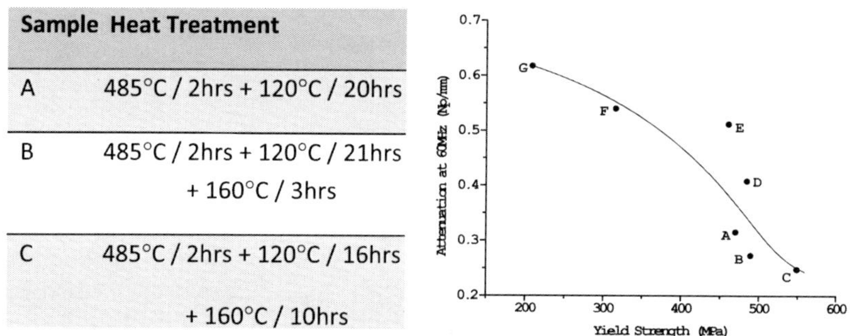

Sample	Heat Treatment
A	485°C / 2hrs + 120°C / 20hrs
B	485°C / 2hrs + 120°C / 21hrs + 160°C / 3hrs
C	485°C / 2hrs + 120°C / 16hrs + 160°C / 10hrs

Figure 7.55 Ultrasonic attenuation at 60 MHz frequency vs. yield strength for 7050 aluminum alloy. A to E are standard heat-treatments while F and G are over-aging treatments.

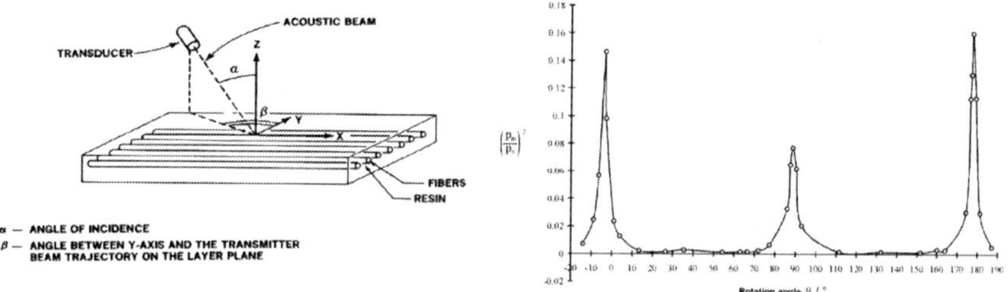

Figure 7.56 Ultrasonic backscattering at an oblique incident angle on a cross-ply composite laminate.

by mapping the backscattered signals while the part is rotated on a turn-table, polar C-scan images of fiber orientation can be obtained. Figure 7.57 shows a polar C-scan of a composite laminate with 0 degree, ±45 degree and 90 degree fiber orientations.

7.5 NON-CONTACT ULTRASONIC TESTING

Using a fluid to create coupling between the transducer and the test part may be problematic for some aircraft structures made of honeycomb or foam core materials. If the edges are not sealed completely, water could penetrate into the core during immersion or water-jet inspections. In addition, immersion or water-jet C-scanning cannot be easily performed on complex or large structures. Non-contact ultrasonic techniques do not require a fluid between the probe and the part. Several non-contact UT techniques are available [24], including air-coupled ultrasonic, laser ultrasonic, and EMAT (electro-magnetic acoustic transducer) ultrasonic. These three approaches will be described briefly in the following sections.

7.5.1 *AIR-COUPLED ULTRASONIC TESTING*

In the air-coupled ultrasonic method, the ultrasound is coupled to the test piece via the ambient air (or other gases), eliminating the need for a liquid coupling. Some air-coupled UT transducers are based on micro-machined capacitance devices, while others use an efficient air-to-solid impedance matching approach along with low frequencies to generate ultrasound in air. The micro-machined devices use a polished silicon element containing an array of small pits (see Figure 7.58) [25]. The pits are created by micro-machining techniques in order to introduce small resonant air pockets for improved sensitivity. A thin metallic coating is deposited on the surface of the element on one side between the pits to form one of the electrodes of the capacitor. A thin polymer membrane is then placed on the opposite side of the element that is metal-coated on its outer surface only. Application of

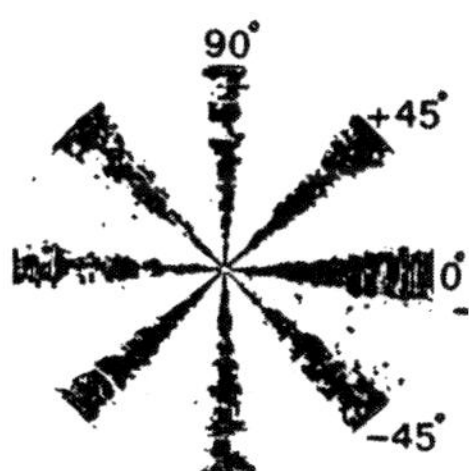

Figure 7.57 A polar C-scan image of a composite laminate using the back-scattering method. Speckling between 0° and –45° traces indicates the presence of a misoriented ply.

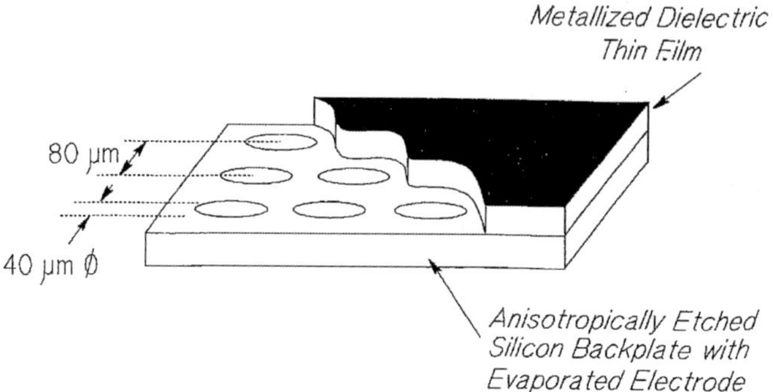

Figure 7.58 A small section of a micro-machined air-coupled ultrasonic device. Note that the thin polymer has been removed on the left side for clarity.

a DC bias voltage to the device attracts the membrane to the back-plate, forming the capacitance structure, which can now act as either a source or receiver. The frequency band and sensitivity are dependent on the design of the transducer, among other things.

Generally, the air-coupled ultrasonic method is more effective on materials of low acoustic impedance (e.g., plastics); however, very thin plates of composites and metallic materials can also be inspected [26]. The technique has been applied for flaw detection, property evaluation, topography, and thickness gauging in several aerospace materials, and a few examples of results are provided in reference [25]. The signal-to-noise ratios are very small and significant filtering of background noise and signal processing are needed to extract information. The approach has very limited application in aerospace due to the enormous energy loss that occurs at air-solid interfaces as well as high attenuation of ultrasound in air at frequencies higher than 0.5 MHz that are used in NDE.

Capabilities and Limitations

Capabilities:
- Non-contact inspections at close proximity to the test piece.
- No fluid coupling is necessary.

Limitations:
- Very small signal-to-noise ratio.
- Only very thin plates can be tested.
- Very limited use in aeronautical applications.

7.5.2 *Electro-Magnetic Acoustic Transducers (EMAT)*

The difference between piezoelectric and EMAT transducers is schematically illustrated in Figure 7.59 [27]. As mentioned earlier, a piezoelectric transducer uses a PZT material that vibrates when repeated electrical pulses are applied. An EMAT on the other hand uses a permanent magnet to produce a steady magnetic field while a coil of wire carrying radio frequency (RF) currents introduces eddy currents into the surface of a conductive specimen. The interaction of the eddy currents with the magnetic field produces Lorentz forces that cause the specimen to vibrate in phase with the applied RF signals, creating ultrasonic waves in the specimen. Here, there is no external vibrating transducer or coupling fluid; the UT waves are created by the specimen vibration due to the applied electro-magnetic forces.

Similar to PZT probes, EMATs can be used as both transmitter and receiver. In the receiving mode, the vibrating specimen can be regarded as a moving magnetic field that generates currents in the coil. EMATs are generally much better detectors than generators. One major problem with EMAT is that the efficiency rapidly decreases with the coil-specimen distance (lift-off), as this distance affects the strength of the magnetic field. EMAT probes are particularly suitable for non-contact inspection of moving parts, rough surfaces, or hot specimens. The ultrasonic signals produced by EMAT probes are significantly weaker than the conventional PZT transducers, and higher input powers are needed. EMATs are not widely used in aerospace.

Capabilities and Limitations

Capabilities:
- Non-contact inspection from close proximity to the test piece.
- Applicable to rough surfaces and moving parts.
- Access is needed to only one surface.

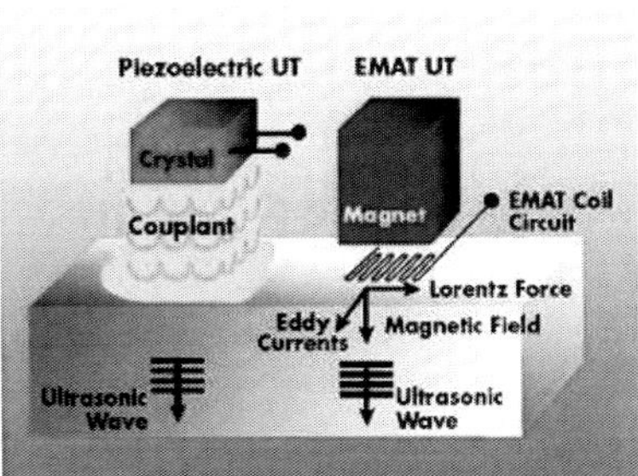

Figure 7.59 Schematic presentation of ultrasonic wave generation using piezoelectric and EMAT transducers [27].

Limitations:
*   Only electrically conductive materials can be tested.
*   Sensitivity is lower than for conventional piezoelectric probes.
*   Results are affected by small changes in probe lift-off.
*   Not suitable for field applications.

7.5.3 *Laser Ultrasonic Testing*

In the laser ultrasonic technique, a pulsed laser is used to heat a spot on the surface of the test part and the rapid thermal expansion of the material generates stress waves perpendicular to the surface. The frequency of the stress waves is related to the laser pulse rise-time, and with the use of proper pulse duration (typically 5–10 ns), ultrasonic waves are generated. A Q-switched Nd YAG laser is used to generate ultrasound up to ~15 MHz. The generated ultrasonic waves propagate perpendicular to the surface of the test piece regardless of the direction of the incident laser beam. The UT waves are reflected back by internal discontinuities in the same manner as in conventional UT. The arriving waves at a surface cause a small displacement, which is detected optically using a continuous wave laser interferometer. Figure 7.60 shows the basic diagram of laser ultrasonic generation and detection.

Since in laser ultrasonic approach the source of ultrasound is the surface of the material itself and the detection of the returned ultrasonic waves is also performed off the same surface, the normal incidence requirement of conventional piezo-electric-based ultrasonic methods is always met for simple or complex parts. This allows easy optical scanning of both flat and contoured surfaces in one inspection using a revolving mirror (Figure 7.61) [28]. Thus, complex parts with sharp corners, that are not easily inspected using conventional ultrasonic C-scanning, can be tested. Figure 7.62 illustrates a laser UT C-scan inspection result of a U-shaped composite part with embedded flaws at different depths in the flat and curved

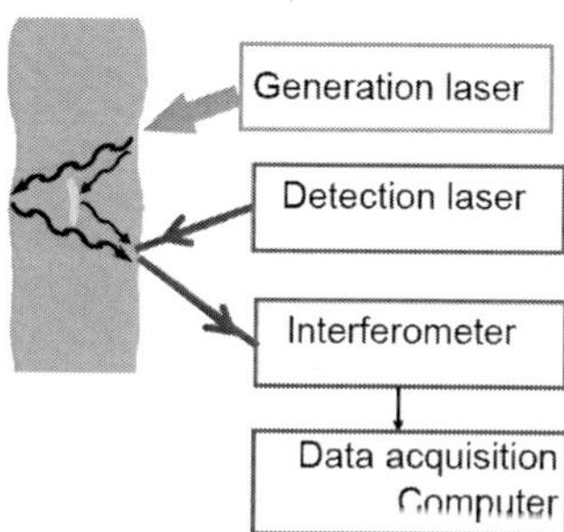

Figure 7.60 Schematic of a laser generated ultrasound pulse echo set up.

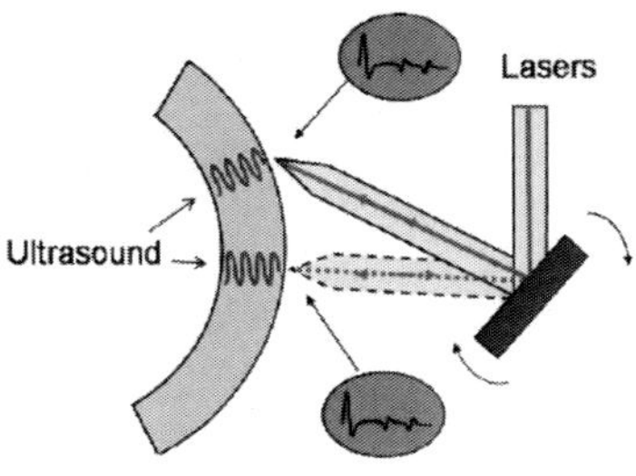

Figure 7.61 Laser-ultrasonic inspection of a curved part. The two arrows indicate the generation and detection laser beams, which are shown at two locations, determined by the optical scanner. The inserts indicate schematically the ultrasonic echoes, which can, for example, be observed on an oscilloscope [28].

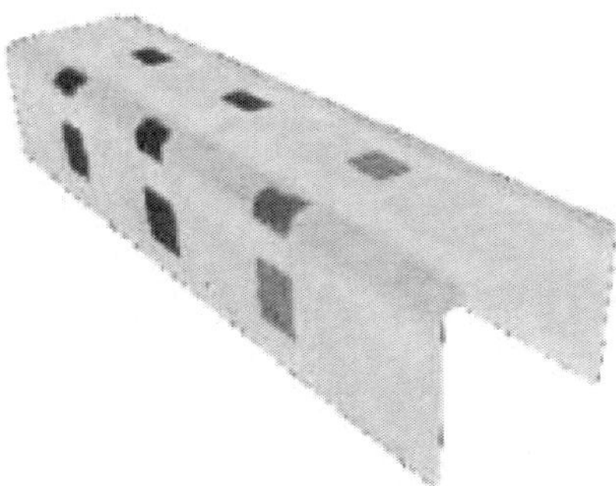

Figure 7.62 Laser ultrasonic C-scan image of a composite part with embedded flaws in flat and curved areas [28].

sections. The image also shows a color change at the corners that is associated with the change of the thickness.

Similar to the conventional UT, in addition to C-scan images, B-scan and time-of-flight C-scan maps can be generated as illustrated in Figure 7.63. These images show typical delamination damage caused by an impact on a flat composite specimen. Also, laser ultrasonic generation and sensing do not require direct contact with the test part or a coupling medium and allow non-contact measurements from considerable distance at high speeds. The technique is very attractive for process control and inspection of complex parts.

The laser ultrasonic approach has applied to many industrial applications including inspection of aircraft components. Examples are provided in Figure 7.62 and Figure 7.63 [28]. The technique can be used on various composite materials (graphite epoxy, Kevlar epoxy, glass epoxy, honeycomb panels), with a bare surface, a painted surface or different geometries (U-shape, T-shape, and sine wave cross sections). However, depths of only a few millimeters below the surface can be tested; thus, thick parts must be inspected from both sides for full thickness coverage. Defects, such as porosity, disbonds, delamination, and impact damage, can be detected.

The technique has also been demonstrated on actual aircraft parts from a fuselage, wing, and stabilizer in a laboratory environment (Figure 7.64). The technology is potentially applicable to full-size composite aircraft structures, such as fuselage segments that cannot be easily inspected using the conventional ultrasonic C-scan facilities. In such applications, the inspection head of the system could be extended by several meters allowing inspections from inside or outside of tubular structures, as may be seen in Figure 7.65. While the technique is successful in finding delamination in composite solid laminates, difficulties may exist in detecting disbonds in honeycomb and foam core structures, particularly when they occur in the back skin. Recently, a new approach has been developed that uses a low frequency pulsed-laser to create bulging and vibration of the disbonded area due to thermoplastic effects and a two-wave mixing photo-refractive laser interferometer to detect the resulting surface displacements [29]. This approach,

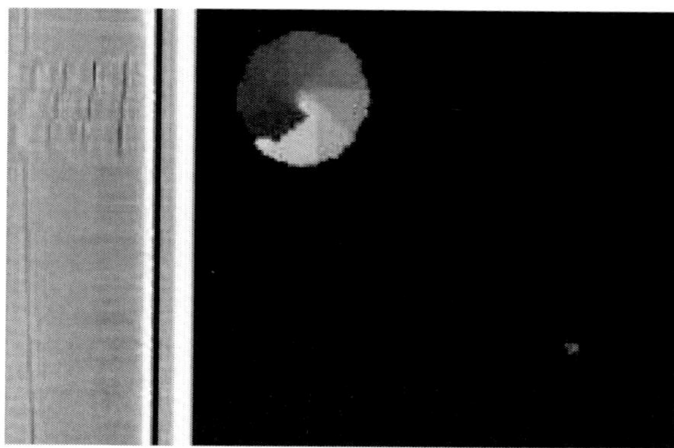

Figure 7.63 Laser ultrasonic B-scan and Time-of-flight C-scan of an impact damage in a composite plate [28].

Figure 7.64 A laser ultrasonic system designed to inspect large-size composite structures that is being used during inspection of a wing panel (left) and a composite prototype fuselage (right).

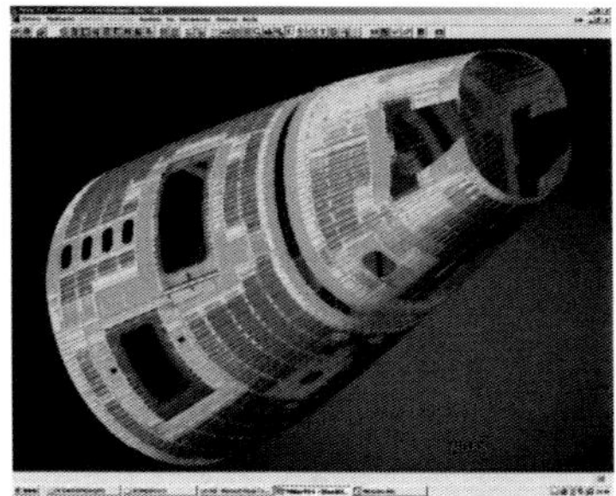

Figure 7.65 A laser ultrasonic inspection results of a fuselage section (from http://www.qmtmag.com/display_eds.cfm?edno=1066832)

which is called "laser tapping," uses the same hardware as laser ultrasonic and could be applied concurrently to detect skin to honeycomb debonding.

Capabilities and Limitations

Capabilities:
- Non-contact inspection from a distance.
- Applicable to curved surfaces or complex parts.
- Access to one surface is needed.
- Can be used on composite or metallic structures.
- Surface preparation and paint removal are not necessary.
- Can be used in ambient or high temperatures.

Limitations:
- Equipment is expensive and relatively complex.
- Depth of coverage is limited, especially for highly attenuating materials.
- Sensitive to environmental vibrations.
- Only flaws close to surface can be detected.
- Not suitable for field use due to limited portability.

7.6 CHAPTER REFERENCES

[1] Introduction to Nondestructive Testing, NDT Resource Centre, Center for Non-destructive Evaluation, Iowa State University, Ames, Iowa 50011, USA.

[2] Metals Handbook, Non-destructive Evaluation and Quality Control, Ninth Edition, Volume 17, 1989.

[3] American Society for Non-destructive Testing. Non-destructive Testing Handbook, 2nd Edition, Volume 7, Ultrasonic Testing, 1991.

[4]   Basic Ultrasound, Echocardiography and Doppler for Clinicians, Asbjørn Støylen, http://folk.ntnu.no/stoylen/strainrate/Ultrasound/#Attenuation.

[5]   ASTM Standard Practice for Measuring Ultrasonic Velocity in Materials, ASTM E494-05, Volume 03.03, October 2010, ASTM International, 100Barr Harbor Drive, PO Box C700, West Conshohocken, PA 19428-2959, USA.

[6]   ASTM Standard Guide for Detection and Evaluation of Discontinuities by Contact Pulse-echo Straight-beam Ultrasonic Method, ASTM E1901-08, Volume 03.03, October 2010, ASTM International, 100Barr Harbor Drive, PO Box C700, West Conshohocken, PA 19428-2959, USA.

[7]   ASTM Standard Guide for Detection and Evaluation of Discontinuities by the Immersed Pulse-echo Ultrasonic Method Using Longitudinal Waves, ASTM E1001-06, Volume 03.03, October 2010, ASTM International, 100 Barr Harbor Drive, P.O. Box C700, West Conshohocken, PA 19428-2959, USA.

[8]   Sizing of Surface-breaking Cracks in Complex Geometry Components by Ultrasonic Time-of-Flight Diffraction (TOFD) Technique, S.K. Nath et al., Insight, Vol. 49 No 4, pp. 200–206, April 2007.

[9]   http://www.ndtnet.org

[10] http://www0.force.dk/p-scan/

[11] http://www.silverwinguk.com/

[12] Internet site of the US Department of Transportation, Federal Highway Administration, Office of Research, Development, and Technology, http://www.fhwa.dot.gov/publications/research/infrastructure/structures/04124/06a.cfm.

[13] Ultrasonic Sciences Ltd. Unit 4, Springlakes Industrial Estate, Deadbrook Lane, Aldershot, Hants, GU12 4UH, England, http://www.ultrasonic-sciences.co.uk/index.htm.

[14] Lamb Wave Propagation Across a Lap-joint, Z. Chang, D. Guo, and A.K. Mal, Review of Progress in Quantitative Nondestructive Evaluation, Vol. 15, D.O. Thompson and D.E. Chimenti, eds. Plenum Press, New York, 1996, pp. 185–192.

[15] A Study on the Guided Wave Mode Conversion Using Self-calibrating Technique, C, Younho, P, Jung-Chul, J.L. Rose, and D.D. Honerholt, www.ndtnet.

[16] NDE of Hidden Flaws in Aging Aircraft Structures Using Obliquely Backscattered Ultrasonic Signals, Y. Bar-cohen, A.K. Mal, and M. Lasser, NDT Net, Vol. 4, No. 1, Jan. 1999.

[17] Ultrasonic Surface Wave Measurement of Surface Breaking Cracks in a Railway Wheel, A. Fahr, A.R. Kotanen and A.M. Charlesworth, Journal of the Canadian Society for Nondestructive Testing, Vol. 7, No.1, 1986.

[18] NDI of Cracks Under Installed Fasteners, B.A. Lepine and A. Fahr, Institute for Aerospace Research, National Research Council Canada, RP-SMPL-2001-0136.

[19] Review of IAR NDI Research in Support of Ageing Aircraft, A. Fahr, D.S. Forsyth, J.P. Komorowski, and C.E. Chapman, Institute for Aerospace Research, National Research Council Canada, CP-SMPL-1998-0018, 2/9/1998.

[20] Nondestructive Evaluation of Cermet Coatings using Eddy Current and Ultrasonic Techniques, B. Roge, A. Fahr, J.S.R. Giguere, and K. I. McRae, Institute for Aerospace Research, National Research Council Canada : CPR-SMPL-2001-0095, 2001.

[21] Nondestructive Evaluation of Cermet Composite Coatings, J.S.R. Giguere, B. Roge, K.I. McRae, and A. Fahr, Institute for Aerospace Research, National Research Council Canada CPR-SMPL-2001-0060, 2001.

[22] Ultrasonic Nondestructive Evaluation of Thin (sub-wavelength) Coatings, V.K. Kinra and C. Zhu, Journal of Acoustical Society of America, 93 (5), pp. 2454–2467, May 1993.

[23] Ultrasonic Assessment of Heat-Damage in 7050 Aluminum Alloy, R.D. LeBlanc, Thesis for Masters Degree, School of Engineering, Moncton University, 1996.

[24] Non-contact Ultrasonic NDE Systems for Aging Aircraft, R. E. Green and B.B. Djordjevic, Proc. 2nd Joint NASA/FAA/DoD Conference on Aging Aircraft, Williamsburg, VA, 31Aug - 03 Sept 1998, NASA/CP-1999-208982/PART1, pp. 244 – 249, 1999.

[25] Development of Air-coupled Ultrasonic Techniques for Inspection of Aircraft Structures, A. Fahr, and D.W. Schindel, Institute for Aerospace Research, National Research Council Canada, LTR-SMPL-1999-0014, 1/19/1999.

[26] Air-Coupled Ultrasonic NDE of Bonded Aluminum Lap Joints, D.W. Schindel, D.S. Forsyth, A. Fahr. and D.A. Hutchins, Institute for Aerospace Research, National Research Council Canada, JA-SMPL-1996-0055, 6/25/1996.

[27] Internet site http://www.citizendia.org/Electro_Magnetic_Acoustic_Transducer.

[28] Laser Ultrasonic Inspection and Characterization of Aeronautical Materials, J-P. Monchalin, C. Neron, M. Choquet, D. Drolet and M. Viens, NDT.Net-Vol. 3, No. 11, November 1998.

[29] Application of Laser Tapping and Laser Ultrasonics to Aerospace Composite Structures, A. Blouin, C. Neron, B. Champagne and J-P Monchalin, Insight, Vol. 52, No. 3, pp. 130–133, March 2010.

[30] Concomitant Measurement of Sound Velocity and Path Length in a Ceramic Test Piece by Ultrasonic Waves, C. Mandache, M. Brothers, and A. Merati, NRC Report: LM-SMPL-2012-0071, 2012.

[31] Simultaneous Ultrasonic Velocity and Sample Thickness Measurement and Application in Composites, D.K. Hsu, and M.S. Hughes, Journal of Acoustical Society of America, Vol. 92, No. 2, pp. 669–675, 1992.

[32] Ballistic Impact into Fabric and Compliant Composite Laminates, B.A. Cheesman, and T.A. Bogetti, Composite Structures, Vol. 61, pp. 161–173, 2003.

[33] Characterization of Alumina Ceramics by Ultrasonic Testing, L.S. Chang, T.H. Chuang, and W.J. Wei, Materials Characterization, Vol. 45, pp. 221–226, 2000.

CHAPTER 8

Acoustic Techniques

In this publication, the term "acoustic techniques" refers to NDT methods that use acoustic waves below 500 kHz. The waves may be produced by an external mechanical device, a piezoelectric transducer, or by the material itself in the form of acoustic emissions. Methods described in this chapter include sonic tests (e.g., tap testing, mechanical or acoustic impedance analysis), acoustic emission, and acousto-ultrasonic techniques.

Sonic tests refer to approaches that use repeated mechanical impacts, acoustic impulses, or RF waveforms to generate low-frequency acoustic waves in order to detect relatively large discontinuities, such as disbonds, in aircraft sandwiched structures. These tests are generally based on the principle that disbonds in a bonded part cause local change in the stiffness, which, in turn, affects the vibrational characteristics of the component. This principle is the basis of several commercial bond testing instruments that are primarily designed to inspect honeycomb or foam-core sandwich structures for the disbonding of the face sheets from the core. Such instruments often use different modes and frequencies to enable detection of various sizes and locations of defects. The most widely used sonic tests are tap-tests, mechanical impedance analyses, and acoustic resonance approaches. Although acoustic emission and acousto-ultrasonic methods also operate in a similar frequency range, their principles and applications are different, as described in the following sections.

8.1 TAP TESTING

8.1.1 PRINCIPLES

In tap testing, the inspector simply listens to the sound made when the surface is tapped by a solid object like a small hammer or a coin (Figure 8.1). In this approach, essentially, the damping of the sound waves is recognized by the inspector when a damaged area is impacted. A duller sound is produced at damage sites when compared with good areas due to the absorption of the high frequency components of the sonic waves.

8.1.2 APPLICATIONS

Tap testing is a simple technique that is often used to check for large planar discontinuities located close to the surface. Experienced inspectors can identify such damage, but the method is highly operator-dependent. Commercial computer-aided tap testing instruments are available for improved damage detection. Being a simple and inexpensive approach, tap testing is used widely by the aircraft operators to detect delamination and disbonding of composite skin in honeycomb parts. More controlled versions of tap testing are described by terms, such as acoustic impedance, membrane resonance, or mechanical impedance analysis.

The reliability of simple tap testing may be improved by recording and analyzing the acoustic response with microphones or air-operating acoustic transducers. Also, using automated or instrumented impact machines is expected to provide more reliable results. Acoustic impact C-scan images can be produced in a similar manner to those produced by low frequency ultrasound.

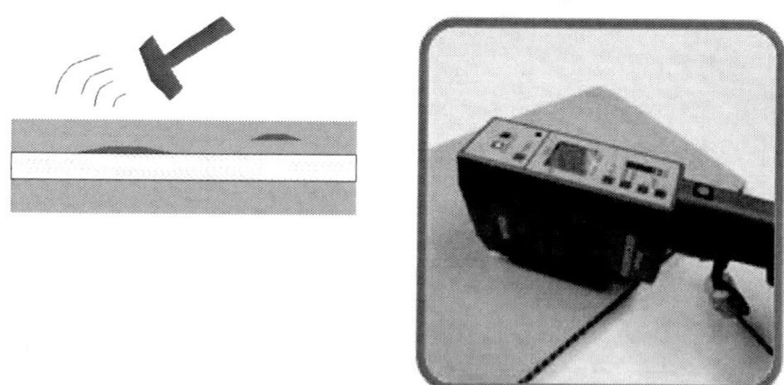

Figure 8.1 (Left): Principles of tap testing method (www.compositenet.com). (Right): A commercial tap testing device (www.jrtech.co.uk).

8.1.3 *CAPABILITIES AND LIMITATIONS*

Capabilities:
* Simple and inexpensive.
* No coupling fluid is needed.
* Can be applied in the laboratory or in the field.
* Limited training is needed.
* Good first-line inspection tool for sandwich parts.

Limitations:
* Difficult to identify damage when the structure is constrained.
* Operators' good hearing and experience are important.
* Highly operator-dependent.
* Damage size cannot be determined.
* Possibility of damage growth if impacted hard.
* Lacks reliability.
* Requires verification by other NDT methods.

8.2 MECHANICAL IMPEDANCE ANALYSIS (MIA)

8.2.1 *PRINCIPLES*

The mechanical impedance approach uses a single dry-coupled contact probe that measures the local change of surface stiffness. The local mechanical impedance (Z) of a part is defined by [22]:

$$Z = F/V \tag{8-1}$$

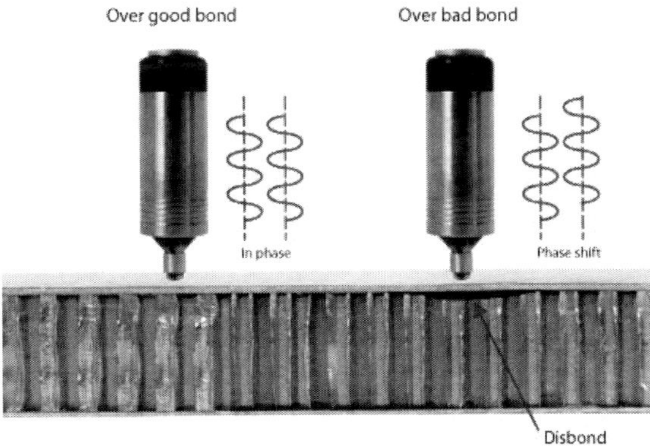

Figure 8.2 Typical MIA probes for disbond detection in sandwich parts [1].

where F is the harmonic input force and V is the resultant local velocity (displacement in unit time) at the force application point. Measurements are carried out at a single frequency, typically between 1 and 10 kHz, using specially-designed displacement- sensing probes. The MIA probe uses a spring-loading mechanism to ensure constant pressure on the sensor during the test. Since the MIA measures the change of local stiffness, it is well-suited for testing sandwich parts that are generally designed to be stiff and lose their stiffness as a result of damage.

The pitch-catch MIA method employs a dual-element, point-contact, dry-coupled probe instead of the single element probe used in the conventional approach. In this case, one element transmits a burst of acoustic energy into the test part, and a separate element receives the sonic waves propagated across the test piece between the probe tips, as illustrated in Figure 8.3 [1]. The bond condition beneath the two contact points affects the characteristics of the transmitted acoustic energy that, in turn, change the amplitude and phase of the received signals. The changes of the received signal, displayed in the form of the signal amplitude, vector, or swept frequency, may indicate disbonds. A well-bonded condition causes part of the acoustic energy to dissipate into the core, resulting in low signal amplitude. In a disbond condition, the acoustic waves travel between the two elements with little loss of energy into the core, resulting in higher signal amplitude. This approach generally uses a higher frequency range than the single probe MIA (e.g., 5 kHz to 100 kHz).

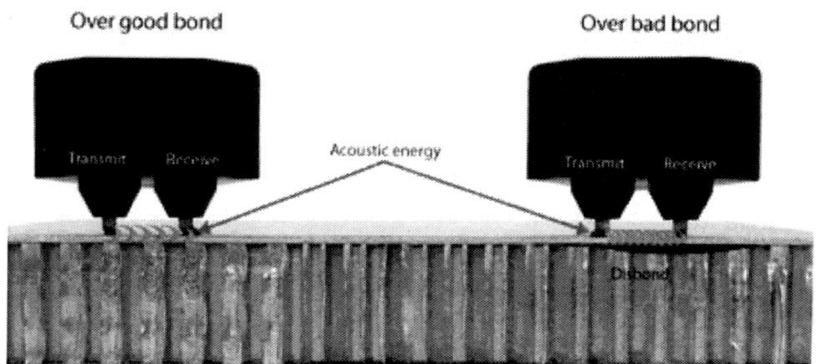

Figure 8.3 Typical pitch-catch MIA probes for disbond detection in sandwich parts [1].

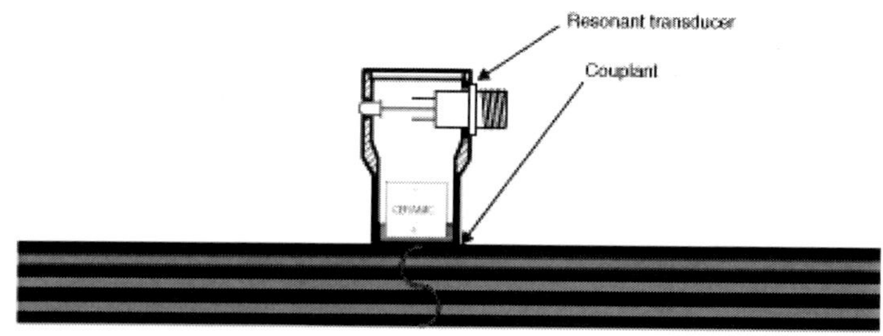

Figure 8.4 Principles of the acoustic resonance approach [1].

8.2.2 APPLICATIONS

Typical applications of MIA include inspection of honeycomb or foam-core structures for disbonding of the face-sheets from the core. The local impedance measurement sensitivity varies with the mechanical impact wavelength and the choice of a proper test frequency is critical for obtaining good results. During instrument setup, the drive frequency is often swept through the full frequency range to determine the optimal value for the test. The operator selects the frequency that provides the greatest difference in the response signal phase or amplitude between the well-bonded and poorly-bonded segments of the representative reference sample. Testing of the real test object is then performed at this fixed frequency.

8.2.2.1 APPLICATION EXAMPLES

In the work described in reference [22], MIA measurements were carried out on foam-core aluminum panels using a commercial system equipped with a hand-held contact probe. The probe was moved over the test part, and areas with different mechanical impedance were indicated by the position of a flying-dot on the instrument CRT screen (e.g., Figure 8.5). The samples were made of a 38 mm-thick PVC foam-core adhesively-bonded between two layers of 1 mm-thick aluminum. Discontinuities included Teflon tapes of different sizes at the Al-core interface, as well as unbonded areas created by removing a small portion of the foam before fabrication. Also, real fatigue- and impact-induced disbonds were present. Except for the unbonded areas and small Teflon inserts, the remaining discontinuities were detected by the MIA using either manual or automated inspections. Generally, if a discontinuity is large enough to change the local stiffness, it is expected to be detected by the mechanical impedance analysis approach.

8.2.3 CAPABILITIES AND LIMITATIONS

Capabilities:
- Measures the local stiffness of the material.
- Suitable for detecting disbond in honeycomb or foam-core structures.
- No coupling fluid is needed.
- Equipment is relatively inexpensive.
- Can be applied in the laboratory or in the field.

Limitations:
- Only applicable on sandwich structures.
- Manual contact inspection requires skilled operators.
- Is not quantitative.
- Disbonds must be large enough to change the local stiffness to be detected.

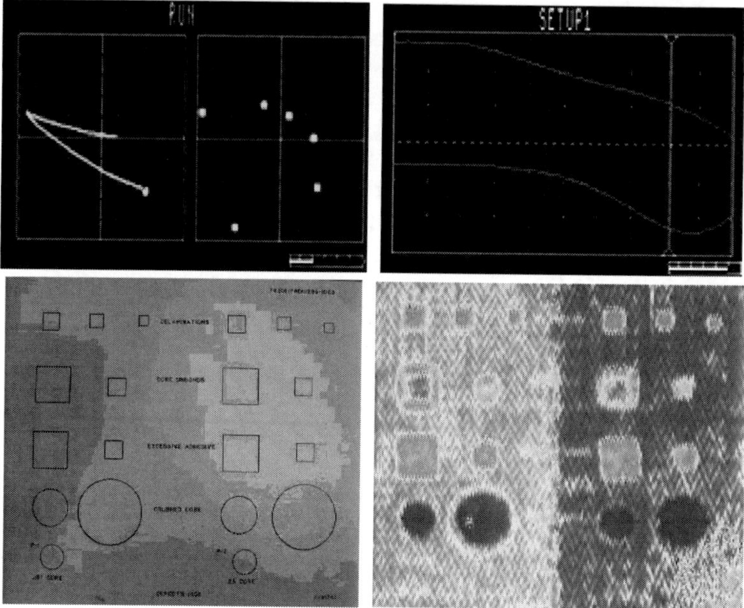

Figure 8.5 (Top): Screen displays of a commercial bond testing instrument using MIA (left) and acoustic resonance (right) approaches. The signal shift from the calibration position indicates the presence of a disbond, and the angle or amplitude of the change is related to the depth of the disbond. (Bottom-left): A reference sample representing the CF-18 rudder materials with simulated defects. (Bottom-right): BondMaster inspection results of the above sample using the resonance mode [1].

8.3 ACOUSTIC RESONANCE

8.3.1 *PRINCIPLES*

The resonance method uses a narrow-band piezoelectric contact probe that requires a coupling fluid, as indicated in Figure 8.4. The frequency of continuous acoustic waves, which are injected locally, is changed until a resonance condition is produced. At resonance, the following relation exists between the part thickness, T, and the wavelength, λ;

$$T = (n+1)\,\lambda \qquad (8\text{-}2)$$

λ is related to the frequency through the following equation;

$$\lambda = V / f \qquad (8\text{-}3)$$

where V is the acoustic velocity of the material, and f is the resonance frequency at different harmonics corresponding to $n = 1, 2, 3, 4...$ This approach is mainly applied to bonded laminates where the presence of a delamination or disbond is seen as a change of thickness. Delamination or disbonds closer to the top surface are identified by their higher resonance frequency.

Some commercial bond testing instruments (e.g., Olympus BondMaster™ 1000e+) operate in all the three modes (MIA, pitch-catch MIA, and acoustic resonance). They allow the user to select the optimal method for the particular application at hand. In Figure 8.5, a typical screen display of a commercial resonance instrument is provided. In this case, the signal shift can be related to the presence of a disbond. Disbond and delamination in composite laminates or the face sheet of sandwich parts can be identified.

8.3.2 APPLICATIONS

While the MIA is only applicable to sandwich parts, the acoustic resonance approach may be used on composite laminates or bonded structures to identify planner discontinuities, such as delamination or disbonds. Examples provided below are obtained with the above-mentioned commercial bond testing device (BondMaster) using the resonance method [1].

Figure 8.5 shows a honeycomb sandwich reference sample representing materials used in the CF-18 aircraft rudder. The sample contains various types and sizes of defects, including skin delamination, core unbonded areas, and crushed core, as identified in the photograph. The sample is made of two different honeycomb sizes. The BondMaster results are also shown in this figure, identifying all the larger defects and some of the smaller ones. This example provides some indication of the sensitivity of this approach.

In Figure 8.6, an actual CF-18 rudder is being inspected using an automated scanning system equipped with a BondMaster. The corresponding inspection image of the part obtained by using the resonance mode of the BondMaster is also shown in this figure. No major defects are evident in the image shown here, and color changes seen are due to the local differences in the material and construction of the component [20].

8.3.3 CAPABILITIES AND LIMITATIONS

Capabilities:
- Measures the local changes in resonance frequency.
- Suitable for detecting planner discontinuities or defects, including delamination in composite laminates and disbonds in honeycomb structures.
- Equipment is relatively inexpensive and easy to use.
- Multi-mode commercial instruments are avalable.
- Can be applied in the laboratory or in the field.
- C-scan images of the damage can be obtained for documentation.

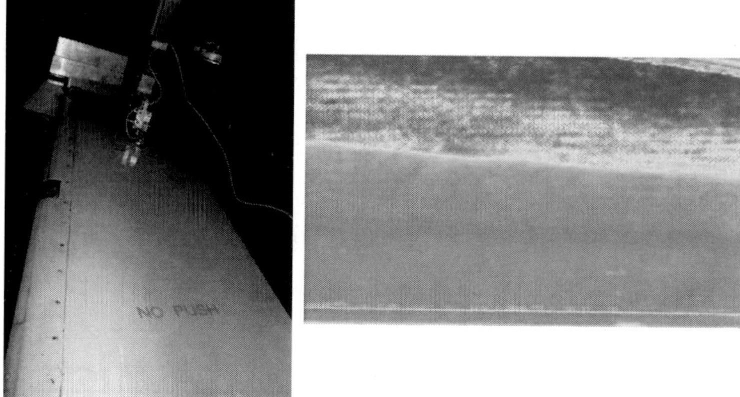

Figure 8.6 Automated inspection of a CF-18 rudder with a BondMaster and the corresponding results using the resonance mode. No defects are present in the segment shown here but changes in the material and construction are evident from the differences in colors [1].

Limitations:
- Coupling fluid is needed.
- Is not quantitative.
- Damage must be close to the surface to be detected.
- Small defects may not be detectable.

8.4 ACOUSTIC EMISSION TESTING

8.4.1 *PRINCIPLES*

Acoustic Emission (AE) is a result of stress waves generated by an object when it is subjected to an external load, pressure, or temperature change. The waves propagate within the material in all directions and the portion that reaches the surface is picked up by AE sensors and analyzed. In metals, AE is produced during processes like melting, phase transformation, twinning, dislocation glide, deformation and crack nucleation, and propagation. In composites, processes like matrix cracking, fiber splitting and breakage, delamination, and debonding result in acoustic emissions. Figure 8.7 schematically illustrates the principles of the acoustic emission method [1].

The detection and analysis of the AE signals may provide information about the onset of the above-mentioned processes and, in some cases, may identify their origin or extent. If the source of AE is damage-related, the detection and monitoring of emissions may be used to predict material failure. However, quantitative information often cannot be obtained easily. As a research tool, AE has been used extensively in aerospace mechanical testing of metallic and composite materials,

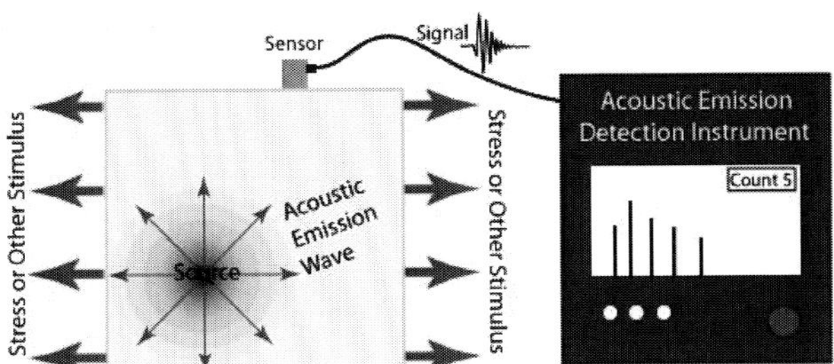

Figure 8.7 Principles of the acoustic emission method [1].

as well as during full-scale fatigue testing and in-flight monitoring of aircraft structures. It has also been used for structural integrity assessments, detecting flaws, testing for leaks, and monitoring the quality of engineering materials or processes. Acoustic emission differs from other acoustic or ultrasonic NDT methods in two ways. First, unlike acoustic and ultrasonic methods that are based on transmitting sonic or ultrasonic energy into the material and measuring the reflected or transmitted waves to obtain information about the material, the AE waves are generated by the material itself. Second, AE provides only information about dynamic processes, such as crack growth, rather than stable pre-existing discontinuities or conditions that are measured by the other NDT methods. Therefore, to detect material flaws, it is necessary to supply adequate energy (e.g., load) to allow the flaw to grow. For this reason, AE tests are often performed on specimens or components during a controlled stimulation (e.g., mechanical testing) or during operation of a structure (e.g., in-flight monitoring).

8.4.2 EQUIPMENT

An AE system typically consists of a sensor, a preamplifier, filters, an amplifier, and the related measurement, display, and storage equipment [3]. AE sensors are similar in construction to ultrasonic piezoelectric probes; however, they operate at much lower frequencies. AE sensors operate in 100–500 kHz frequency range and are more sensitive to dynamic processes that are the sources of the AE waveforms. Sensors and preamplifiers are designed to filter out the unwanted signals outside this range. The preamplifier is either integrated with the transducer or placed close to it, and it provides signal pre-amplification. The signal is relayed through a band-pass filter for elimination of low frequencies (common to background mechanical noise) and very high frequencies (common to background electromagnetic noise). Further amplification, filtering, signal processing, and analysis are carried out by the AE measurement system and the associated computer. The results are displayed in real-time and saved or recorded. Figure 8.8 illustrates the basic instruments of a single channel AE measurement system.

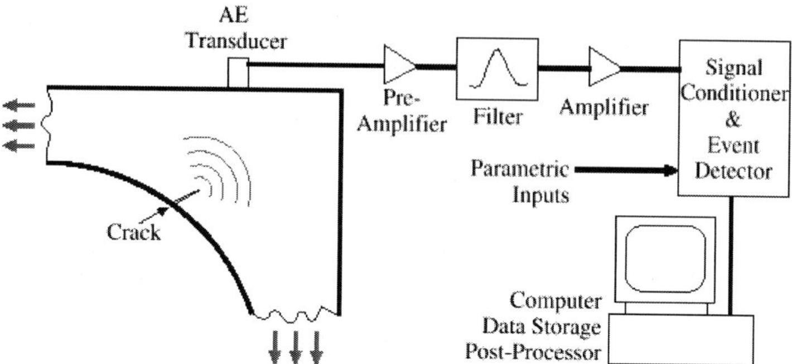

Figure 8.8 Schematic diagram of a basic AE measurement system [4].

Multi-channel systems involving several transducers and the associated devices are also available for testing large structures or for locating a damage source by AE measurements.

8.4.3 TEST PROCEDURES

For AE tests, the sensor is coupled to the test surface using a coupling fluid and is held in place with tapes or a clamp. Adhesives are also used to attach sensors and provide coupling at the same time. The threshold level, the gain, and necessary filter bands are selected based on the background noise level, and appropriate calibrations are carried out. Then, the load is applied on the test component and the AE signals are monitored and recorded. Most AE systems use hit-driven architecture and provide signal characteristics, such as maximum amplitude, rise-time, duration, area under the signal, and related statistical information on the measured parameters. Many types of commercial equipment also provide the time and load readings. Some systems may even provide methods or algorithms for identifying the different sources of AE signals using specialized software packages, such as pattern recognition analysis.

8.4.3.1 SIGNAL PROCESSING

Acoustic emissions are generated either continuously or in bursts and are detected by AE sensors. A single AE signal at the sensor out-put point is a series of oscillations, as shown schematically in Figure 8.9. Each signal is called an event that has a number of oscillations or counts, a peak-amplitude (in volts), a rise-time (in seconds), a time-duration, and an envelope surrounding the whole event. To capture and measure AE events, a threshold voltage is first selected and set just above the background noise level. Then, each time the threshold value is exceeded, a digital pulse is generated. The first pulse is used to signify the beginning of an AE event

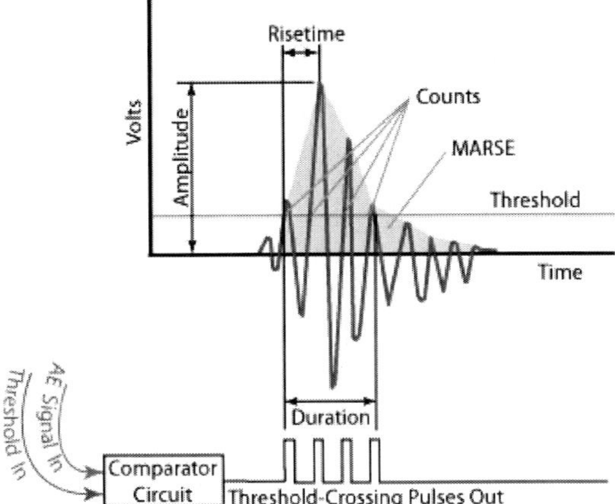

Figure 8.9 Principles of AE signal analysis showing different signal parameters used in AE measurements.

(hit). Pulses will continue to be generated when individual oscillations exceed the threshold voltage. Once this process has stopped for a time duration that is predetermined, the hit is finished.

The data from the hit is then read into a microcomputer, and the measurement circuit is reset. During periods of inactivity, the system lies dormant, and once a new signal is detected, the system records the hit or hits, and the data is logged for real-time and/or future display. AE signal parameters that are usually measured are described below:

- **Number of counts**, N, refers to the number of times the threshold is exceeded in an event or the number of digital pulses generated in each event. The number of counts depends on the magnitude and duration of the AE activity, which, in turn, are dependent on the test material, as well as the reverberation of the sensor that is a characteristic of the sensor.
- **Peak-amplitude**, A, is the maximum measured voltage of the signal in decibels (dB) and is directly related to the magnitude of the source event. Signals with amplitudes below the threshold will not be recorded. AE amplitude values are measured in micro-volts $1\mu(V)$ and presented in decibels (dB): $1\mu V = 0dB$, $10\mu V = 20dB$, $100\mu V = 40dB$, and so on.
- **Rise-time**, R, is the time interval between the first threshold crossing and the signal peak. Rise-time is related to the way AE waves propagate from the source to the sensor within the material (e.g., material attenuation). Being affected by the wave path within the material, this parameter can be used as a criterion to identify emissions from the objects and to filter out the unrelated noise.

- **Duration**, D, is the time difference between the first and the last threshold crossings. Like counts (N), this parameter depends on the source magnitude, which, in turn, is dependent on the acoustic properties of the test material and the sensors' reverberation. It can be useful for recognizing long-duration failure processes, such as plastic deformation. Brittle failures often generate short events of high amplitude.
- **MARSE**, E, refers to the area under the signal envelope and is related to the energy of the emission. MARSE will vary with the signal duration and amplitude, but is independent of the number of counts and the threshold.

8.4.3.2 DATA DISPLAY

In a software-based, hit-driven AE system, the measured parameters can be displayed in different ways, as shown in Figure 8.10. This Figure illustrates a typical screen presentation of a commercial AE system. Information, such as the following, is generally provided by AE equipment:

- **Transient signals** show the actual AE events from individual sensors in the time-domain (amplitude vs. time, top-left display) and/or in the frequency-domain (amplitude vs. frequency, bottom-left display).
- **History plots** show the AE activities from the start to finish of the test in terms of the number of events (or counts) as a function of time (bottom-right display). Either cumulative or individual data can be displayed for all sensors or for a specific sensor. Also, peak-amplitude (or energy) of individual events can be displayed as a function of time.

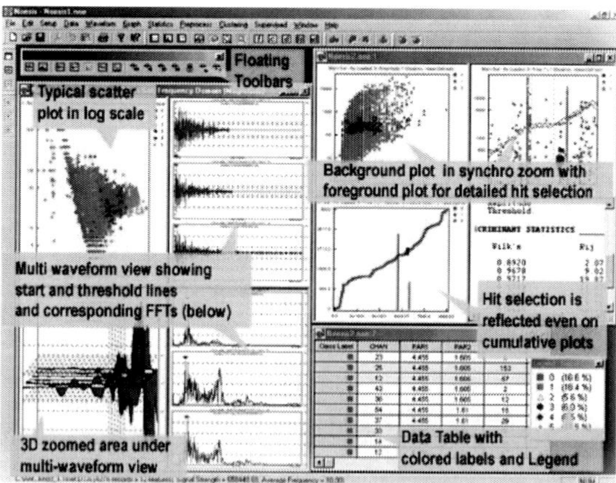

Figure 8.10 Examples of AE data displays showing multi-waveform view of transient AE in time-and frequency-domains, as well as cumulative activity and scatter plots [5].

- **Histograms** show the distribution of the number of emissions as a function of their amplitude (middle-right display) for several or a single transducer. Similar histograms can be displayed for duration, rise time, or other parameters of the AE waveforms.
- **Point plots** show the correlation between the different AE parameters for all sensors (top-right display) or an individual sensor.
- **Location displays** show the position of the AE source with respect to the sensors.

8.4.3.3 AE SOURCE LOCATION

Locating the source of acoustic emissions is often the main goal of an inspection. AE source location requires multiple sensors and a multi-channel system. As hits are recorded by each sensor/channel, the source can be located by knowing the velocity of the AE waveforms in the material and the difference in the arrival times of emissions at the different sensors, as measured by the hardware circuitry or the computer software. Source location techniques assume that AE waves travel at a constant velocity in every direction in a material and do not usually consider reflections or mode conversions at boundaries. However, various effects may alter the expected velocity of the AE waves (e.g., local changes of the material density, internal features, or boundaries) and can affect the accuracy of the measurements. Therefore, the material and geometric effects of the structure being tested must be considered in choosing the number and location of the transducers. Linear location technique, shown in Figure 8.11, requires only a two-channel system and is used for beam-like structures (shafts, struts, bars, trusses). The location of the

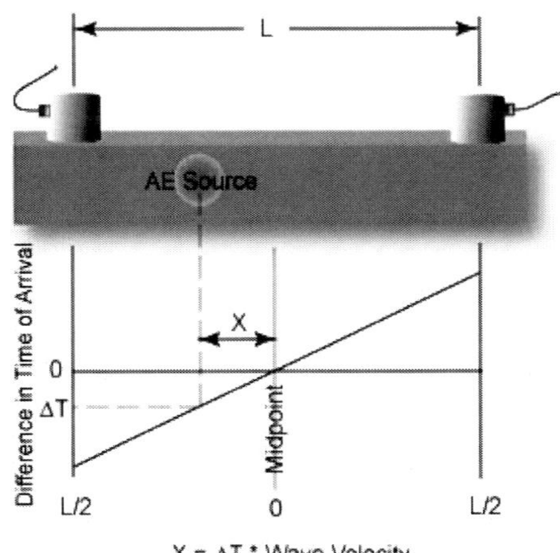

Figure 8.11 AE source location in a simple beam using two transducers [1].

source is determined by establishing which sensor first records the emissions and the time difference between the events, as recorded by the two channels.

Zone location technique, as the name implies, aims at tracing the emissions to a specific zone or region around the sensor that first detects emissions among an array of multiple sensors. This method is used in anisotropic materials when high material attenuation affects the quality of signals or in large structures where sensors can be placed far apart. Zones can be lengths, areas, or volumes, depending on the dimensions of the sensor array. Figure 8.12 illustrates a planar sensor array with seven sensors covering the entire part. In this case, sensor 4 first detects the emissions, and the source can be assumed to be within the region around this sensor and less than halfway to other sensors. By adding more sensors and measuring the arrival times and amplitudes, one can determine the source zone more accurately.

Point location technique requires that AE signals be detected by at least two sensors for linear, three sensors for planar, and four sensors for volumetric components. The arrival times at each sensor are measured in order to pinpoint the source of the emissions. Arrival times are determined by using peak amplitude or the first threshold crossing. Knowing the wave velocity and exact positions of the sensors, the source location can be determined accurately using the following simple diagrams (Figure 8.13).

8.4.3.4 CALIBRATION

Calibration of the AE sensors and electronic equipment is necessary, and verification of the performance of the entire test system is very important. The ASTM preferred approach for conducting performance verifications is a pencil lead break (PLB) test. The ASTM E-975 specifies breaking of a pencil lead (2H, 0.3 mm in diameter and 2–3 mm in length) on the test object at 10 to 15 cm distance from each sensor [6]. Typically, the peak amplitude of the signals from each sensor is recorded for three identical PLB tests, and the average of each channel in relation

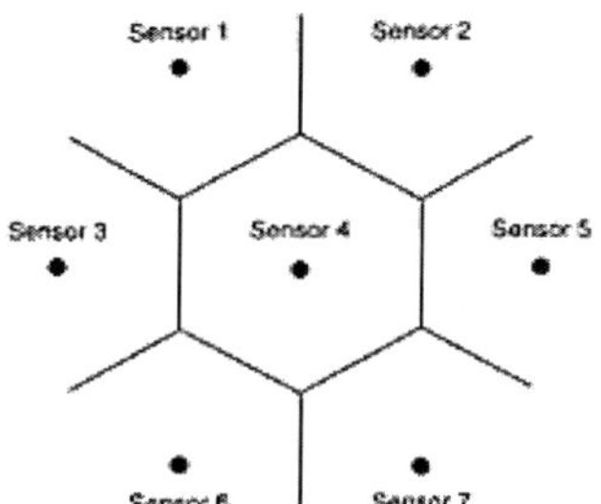

Figure 8.12 Zone AE source location method involving several sensors placed on a test part [1].

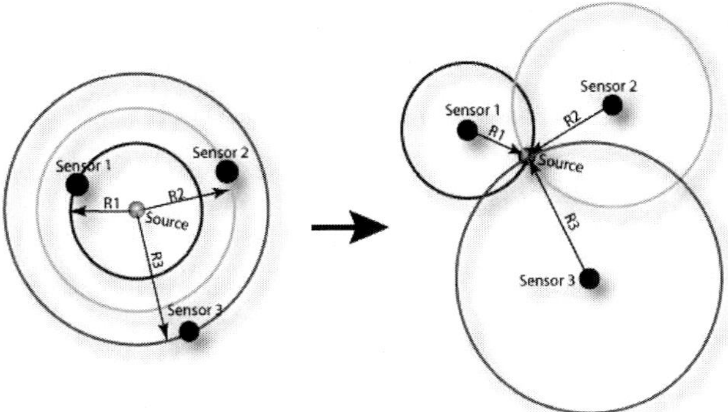

Figure 8.13 AE source location in a simple sheet using three transducers [1].

to the average of all the channels is calculated. This value must be within a certain range to be considered as an acceptable test.

8.4.4 Applications

Acoustic emission is a useful tool for materials study in a laboratory environment, especially for monitoring materials deformation and failure during mechanical testing. It provides an immediate indication of materials' behavior under stress that can be related to the failure process (e.g., dislocation glide, crack nucleation, and propagation). It is also used in practical applications, such as proof-testing of components (e.g., gas bottles or pressure vessels), as well as for process monitoring (e.g., friction stir welding) and during structural tests (e.g., full-scale testing or in-flight monitoring of aircraft). There is a number of ASTM standards and guidelines related to AE testing and applications. Examples include:

- ASTM E 1932 for the AE examination of small parts.
- ASTM E1419-00 for the method of examining seamless, gas-filled pressure vessels.
- ASTM E569-07 for monitoring of structures during controlled stimulation.
- ASTM E 2661/E 2661M-10 for aerospace composite panels and plate-like elements that are made of fiber/polymer materials.

The ASTM Standard E 569-07 [6] provides guidelines for acoustic emission monitoring of structures during controlled stimulation by mechanical or thermal stresses. Such stimulations produce changes in the stress levels within the structure that may create emissions at stress concentration sites, such as discontinuities. The basic functions of the AE monitoring are to detect, locate, and classify emission sources. AE may originate from materials' defects or other areas of stress concentration within the test component, as well as from sources of the background noise in the test site. The major sources of background noise are movements and friction of loose joints and attachments in the object, contact points of the loading jigs with the object, acoustic noise generated by hydraulic systems, fans, and other components of the loading equipment, and electromagnetic noise from electronic equipment. It is important to eliminate or reduce the background noise level as much as possible.

Acoustic noise can be reduced or eliminated by applying acoustic impedance-mismatch barriers or damping materials at strategic locations between the source and the test part. Electronic noise can be decreased by proper grounding and shielding practices or by using differential sensors with built-in preamplifiers. Movement and friction noise can be reduced by minimizing the number of joints, attachments, and contact points, or through modifications to the loading hardware. In addition to the above, adjustments of the frequency band-pass, sensitivity, or threshold, as well as selective signal accept/reject criteria based on load, time, and origin may be used. Since background noise often has different characteristics from the true AE, it may be possible to apply advanced signal processing and pattern recognition techniques for their differentiation.

Overall, the AE technique is limited by the amount of undesirable background noise present during testing. It is important to identify the background noise sources and eliminate or reduce the noise through appropriate actions, such as placing AE sensors in locations away from the noise source, protecting the sensors and the AE system with a grounded-mesh, using electronic filters, choosing an appropriate frequency gate, eliminating emissions outside the range of the correct arrival times, etc. The ASTM E 2661/E 2661M-10 for aerospace composite panels and plate-like elements is aimed at detecting and locating emission sources within a composite structure. This standard may be used for the purpose of inspection, proof testing, quality control, or termination of the structural-test prior to failure.

8.4.4.1 LABORATORY USE

Acoustic emission testing may be carried out as part of mechanical testing of samples, proof testing of components, or full-scale testing of structures. Acoustic emission response of different materials and components under various loading conditions (e.g., tension, compression, cyclic, etc.) vary significantly. Factors, such as material composition, ductility, homogeneity, microstructure, and discontinuities, as well as loading configuration and failure process, affect the AE response. A few examples of simple specimen tests in the laboratory environment are provided below.

During tensile testing of ductile materials, as illustrated in Figure 8.14, acoustic emissions generally begin soon after the load application and continue at a steadily increasing rate until yielding occurs. After the material yields, the AE activity continues at a decreased rate until the material fails [4]. The early AE

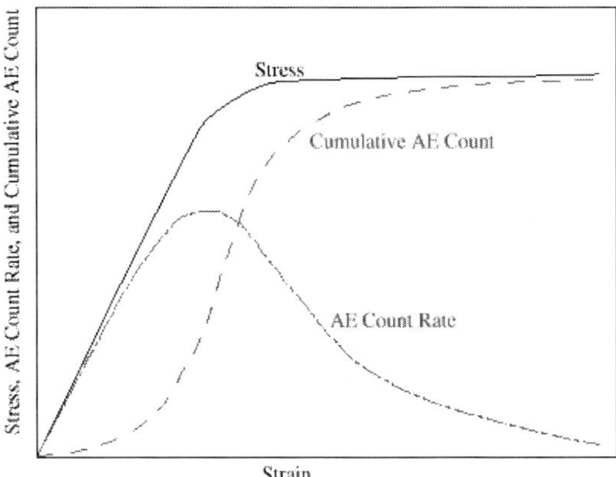

Figure 8.14 Typical acoustic emission response of a ductile material during tensile testing [5].

activity during the loading stage is associated with microstructure changes, such as dislocation glide and plastic deformation at stress concentration sites. This is followed by crack nucleation and propagation that create sudden bursts of AE activity and a higher emission rate representing a release of elastic energy until the final failure.

If repeated load is applied at the early stages of loading, AE activities generally begin when the previously exerted load is reached. In other words, microstructure changes in the material do not increase or move until the former stress is exceeded [3]. This phenomenon is known as the Kaiser Effect and was used in the early AE tests to eliminate initial emissions due to the settling of the test specimen and the loading system. Later experiments showed that at high loads, discontinuities that had reached a significant size would continue to generate emissions below the previous maximum load or when the load is held at a constant level. This phenomenon is known as the Felicity Effect.

The Kaiser and Felicity Effects are schematically illustrated in Figure 8.15, where load is plotted versus the cumulative AE activity. This plot shows emissions at the initial loading stage (AB) and no emissions upon unloading (BC). When the specimen is re-loaded, there is no AE activity until the previous maximum load is reached (point B) that is in accordance with the Kaiser Effect. As load is increased beyond point B, damage is accumulated at an increasing rate, as indicated by the higher emission rate until the next unloading (point D), and no AE activity occurs during unloading to point E. Upon re-loading from point E, emissions start at point F that is below the previous maximum load and continue during loading (FG), as well as when load is held at a constant level (GH). The FH portion of the AE activity shows the Felicity Effect, occurring as a result of discontinuities that have grown to a significant or critical size.

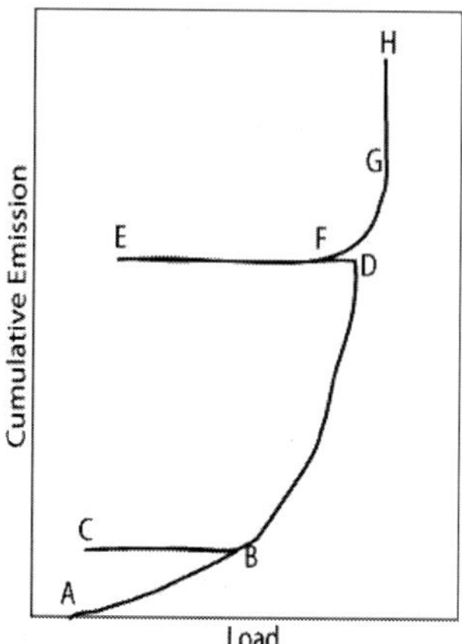

Figure 8.15 Cumulative acoustic emission vs. load showing AE activity during initial loading (AB), the Kaiser effect (BCB), AE activity during the secondary loading (BD), Felicity effect (DEF), and AE activity during load hold (GH) [3].

These types of loading, unloading, and reloading cycles may be used along with acoustic emission monitoring to determine if a structure contains structurally-significant defects or damage. The Kaiser and Felicity Effects do not apply when the damage is a result of a time-dependent process, such as corrosion or hydrogen embrittlement [3]. However, in the latter case that involves several sequential processes, including hydrogen absorption, embrittlement, and eventual cracking, AE rate may be used to determine the time for the onset of rapid fracture [7].

Acoustic emission monitoring during delayed failure of hydrogen-charged, low-alloy, high-strength steel at room temperature has shown slowly increasing low-amplitude (35–55 dB) discrete AE counts with increasing time at the initial stage of the embrittlement process and a rapid increase of high amplitude (60–100 dB) AE signals in the later stage prior to fracture [8]. Also, some materials, such as ceramics, composites, or hybrid materials, may not exhibit such behavior.

More complex loading modes, such as those experienced by materials during cyclic fatigue tests, often result in a greater number of emissions that include undesirable noise from the loading system. In such cases, machine noise may be identified using guard sensors around the gauge sensor and screen-out based on the start times of the AE activity at the different sensors. Noise from the test machine is expected to arrive at the guard sensors before reaching the gauge sensor.

An example of the laboratory application of AE is provided in reference [10] where the technique was employed on simple adhesively-bonded single-lap joint specimens during shear strength testing. The objective of the study was to determine if information about the bond quality could be obtained from the AE response. The specimens were made of aluminum adherents and two commercial adhesives, namely, the room-temperature curing HYSOL 9320 and the high-temperature curing FM300 [10]. Different surface treatments were used to achieve variations in bond strength in order to determine if AE could identify the differences.

The AE monitoring of the samples was carried out in the laboratory environment under controlled conditions and the signals were analyzed in terms of the cumulative number of events and the peak amplitude distribution. Figure 8.16 and Figure 8.17 illustrate the results obtained for the two adhesive types and the

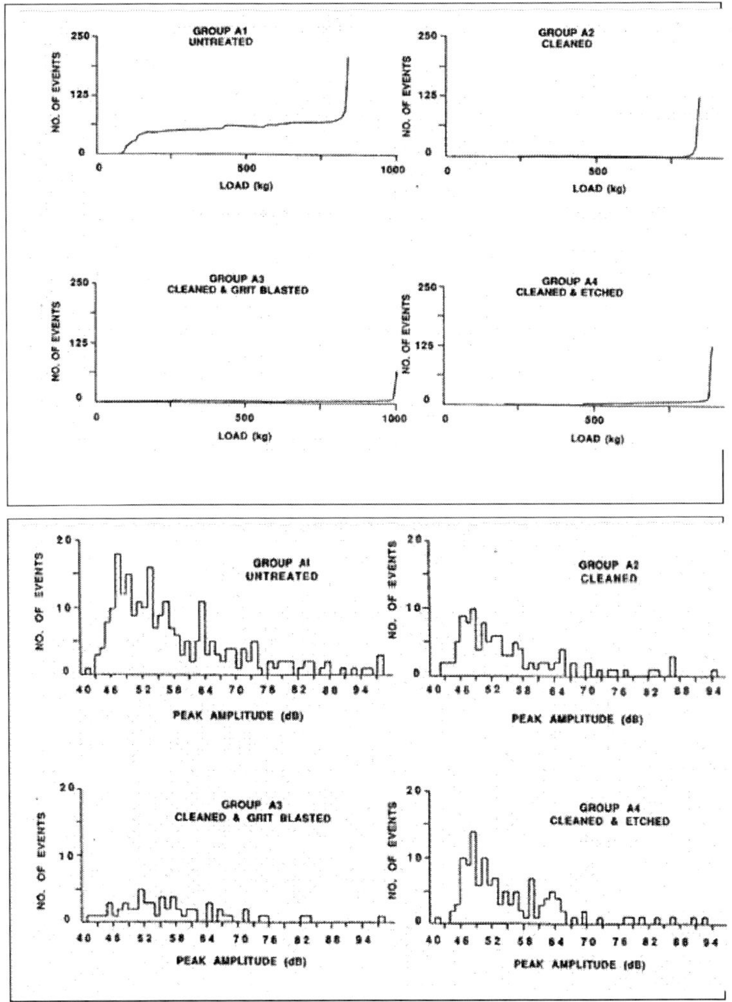

Figure 8.16 Cumulative number of AE events vs. load (top) and amplitude distributions (bottom) for single lap-joint specimens made of Al adherents and HYSOL 9320 adhesive using different surface treatments [10].

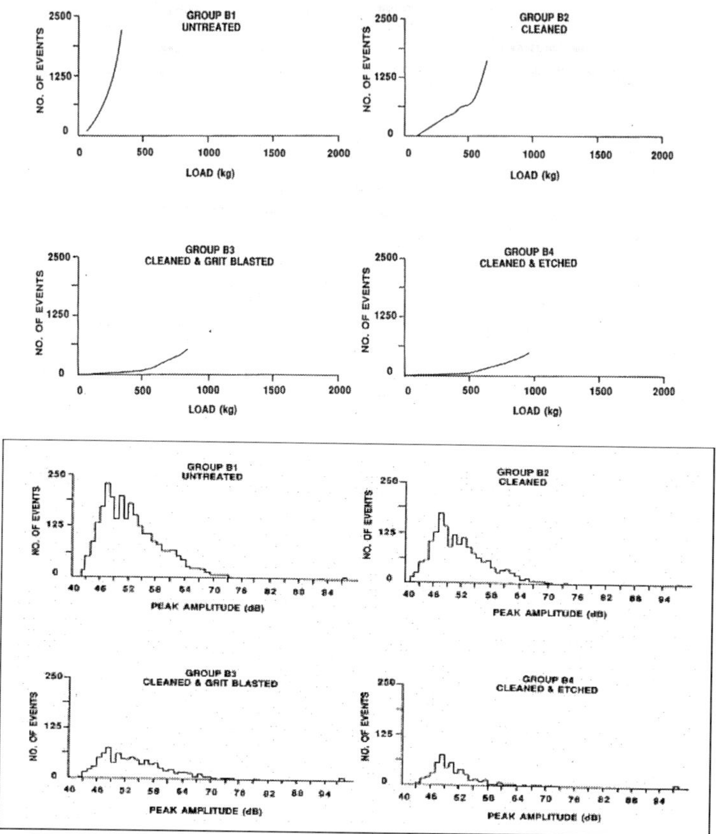

Figure 8.17 Cumulative number of AE events vs. load (top) and amplitude distributions (bottom) for single lap joint specimens made of Al adherents and FM 300 adhesive using different surface treatments [10].

different surface treatments. As seen in these figures, the results indicate that the AE behavior is dependent on the type of the adhesive as well as the bonding process.

The specimens made of HYSOL 9320 adhesive showed a low emission rate until just before failure when the rate increased dramatically. In the specimens prepared using the FM 300, the emission rate remained almost constant throughout the test, and the total number of events was significantly higher than for the HYSOL 9320, especially for untreated or poorly-treated samples.

The study showed that, for a given adhesive system, the number of emissions prior to failure correlates well with the joint surface treatment and the ultimate strength. Generally, weaker joints produced a larger number of emissions. However, variation in AE response from sample-to-sample and test-to-test make it difficult to correlate the results with the actual failure process. Nevertheless, the early AE activities prior to failure in the untreated samples were associated with the interfacial failure that starts at lower loads in these specimens and continues until complete failure.

8.4.4.2 HEALTH MONITORING

AE has also found applications in health monitoring of aerospace structures since sensors can be attached to damage-prone sites that are not easily inspectable by other NDT methods. However, AE monitoring of aircraft structures is not an easy task. This is due to the fact that most aircraft structures consist of complex assemblies of numerous components that are joined using rivets, fasteners, bolts, or adhesives. They are designed in this manner in order to carry significant loads while being as light as possible. The numerous mechanically-fastened structures make AE monitoring of aircraft and damage source-location very difficult, due to the associated frictional noises. In-flight applications are even more difficult due to the additional noise created by the hydraulic systems, electrical devices, airframe vibrations, air flow, and engine sound. Despite the difficulties, numerous attempts have been made to use AE during laboratory structural testing of aircraft components as well as in-flight monitoring of airframe structures with limited success.

In reference [11], the application of acoustic emission during fatigue testing of F-15 aircraft has been described. The main areas of interest in this study were the connecting lugs between the wings and the fuselage that bear extremely high loads during operations and must be inspected regularly in the field. Fatigue cracks are often difficult to locate in these places; thus, AE could potentially help inspectors to determine when and where to look for cracks. The wing-to-fuselage attachment lugs are integral to the wing spars and fuselage bulkheads and are machined from 2124 aluminum alloy, 7075 aluminum alloy, and 6Al-4V titanium forgings. Figure 8.18 shows the aircraft under test, as well as the sensors and preamplifiers located on a section of the aircraft. Figure 8.19 shows the sites of AE activities

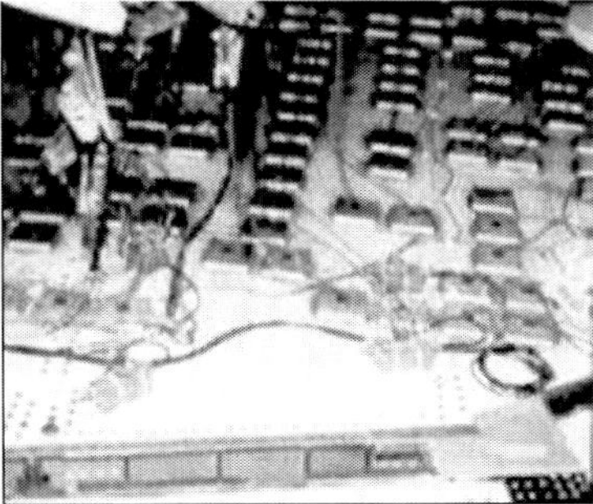

Figure 8.18 An F-15 aircraft being set-up for full-scale fatigue testing (left) and a view of a section of the same aircraft with acoustic emission sensors/ preamplifiers attached to the structure (right) [11].

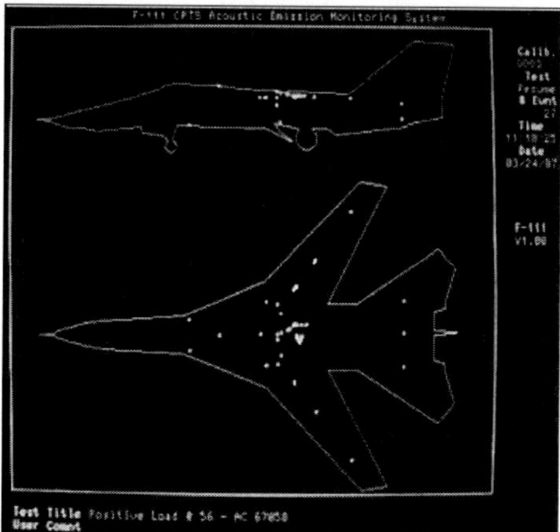

Figure 8.19 Areas of high AE activity obtained during full-scale fatigue testing of the above-mentioned F-15 aircraft [11].

on the aircraft as obtained during full-scale fatigue tests. While emissions were recorded in many locations, some areas produced more emissions than others. The regions of highest AE activity were believed to be the areas of most concern and primary sites for further inspection using other NDT methods.

Overall, acoustic emission monitoring is a valuable technique for laboratory studies of materials during mechanical testing of simple samples; however, special attention must be paid to reduce the interference of the environmental or mechanical noise with the actual material's AE. Also, a good understanding of the physics of the AE generation and propagation within the materials is essential for accurate interpretation of results. Similarly, AE monitoring of components during proof-testing or full-scale structural tests can provide useful information; however, proper sensor locations and noise filtering are important to improve repeatability and reliability of results. Verification by conventional NDT methods or by optical microscopy is often necessary to establish confidence in the AE data.

8.4.5   CAPABILITIES AND LIMITATIONS

Capabilities:

- Dynamic processes, such as materials deformation, crack growth, and failure, can be detected in real-time.
- May provide warning of incipient failure under controlled test conditions.
- Useful for monitoring of test specimens during mechanical loading in the laboratory environment.
- Useful for proof-testing of certain components, such as gas bottles and pressure vessels.
- May be useful for real-time health monitoring of some components or structures.

Limitations:

- Limited by the background noise that could result in missing small-amplitude emissions or producing false indications.
- Requires loading of the test object (mechanical or otherwise).
- Cannot detect stable or stagnant processes or discontinuities.
- Interpretation of results is difficult and requires skilled personnel.
- Not very reliable and often requires verification by other NDE methods.

8.5 ACOUSTO-ULTRASONIC TECHNIQUE

8.5.1 PRINCIPLES

In the acousto-ultrasonic (AU) technique, a broadband piezoelectric transducer is used in contact with the test piece to transmit a repetitive series of ultrasonic pulses and a sensitive acoustic-emission (AE) sensor is employed to capture the propagating stress waves (Figure 8.20) [12]. Unlike the acoustic emission testing that relies on dynamic processes to generate AE waves, in the AU method, stress waves are transmitted into the material, as in ultrasonic testing (UT). The difference between the AU and the UT methods is only in the way the ultrasonic waves are transmitted and received, as well as in the manner the signals are analyzed.

In the ultrasonic approach, the UT waves are transmitted in a particular direction and either the reflected or transmitted waves are picked-up by the same or another UT probe. Here, specific echoes are analyzed and related to specific defects or material properties. In the acousto-ultrasonic approach, UT waves are injected into the material such that they propagate in all directions and a portion of the energy is picked-up by an AE sensor, which is more sensitive than UT probes. In this case, the whole received signal is analyzed in a similar manner as the AE signals and related to the material microstructure, properties, and diffused discontinuities that lie in the path of the propagating waves. Thus, the AU method combines aspects of ultrasonic material characterization and discontinuity detection capabilities with acoustic emission signal analysis approach to provide an indication of the overall condition of the test object.

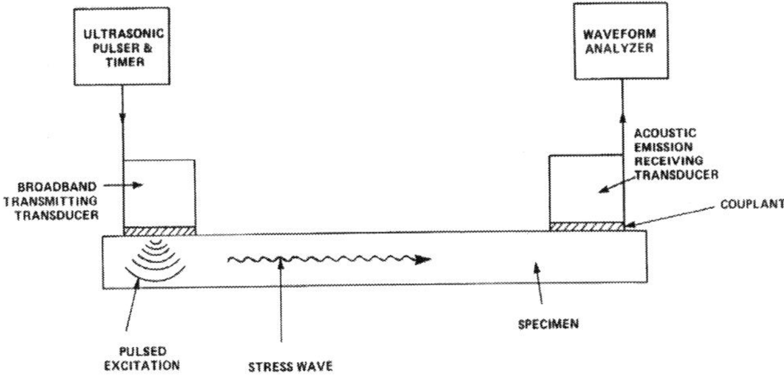

Figure 8.20 Principles of the acousto-ultrasonic testing approach.

Like ultrasonic through-transmission and guided-wave methods, the AU approach uses the attenuation of stress waves to obtain information about the materials' subtle defects or property changes that are distributed or diffused and together affect the wave propagation characteristics. For signal analysis, since AE sensors and instruments are used to capture the AU stress waves, the same AE signal processing approaches and definitions also apply to the acousto-ultrasonic signals. The basic AE signal parameters include the amplitude, ring-down counts, and duration as defined in the previous chapter. These parameters are used either individually or in combination to calculate a parameter referred to as the "Stress Wave Factor (SWF)." The SWF is basically a measure of the propagation efficiency or attenuation of the stress waves and can be calculated in many different ways.

One way of calculating the SWF is to take into account both the ring down counts and the amplitudes of the total received signals. This is referred to as the "Acousto-Ultrasonic Parameter (AUP)" [12] and is basically a weighted ring-down count computed in the following manner. First, the noise level is established and a threshold is set just above the noise (V_0). Then the total number of counts above V_0 is determined. The threshold is increased slightly and the number of counts above the new threshold is found. The difference of the two counts is obtained and multiplied by the amplitude at that level. This process is repeated until the threshold is equal or greater than the peak amplitude. This procedure is illustrated in Figure 8.21 and can be expressed by:

$$AUP = \sum_i V_i (C_i - C_{i+1})$$

(8-4)

Where V_i and C_i are the threshold and number of counts at the i-th level, respectively.

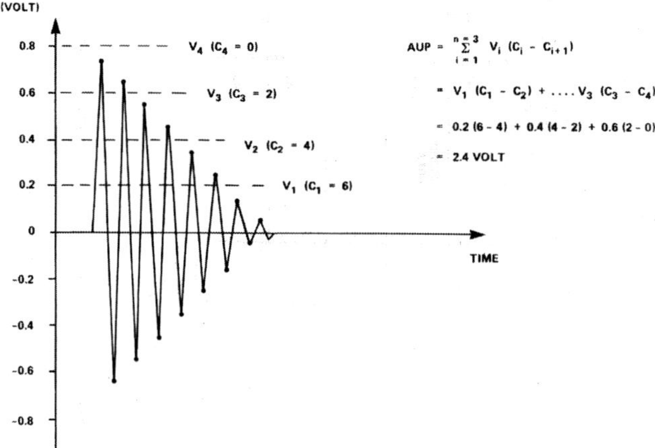

Figure 8.21 Illustration of the AUP calculation.

8.5.2 TEST PROCEDURE

The ASTM E 1495-02 provides a standard guide for acousto-ultrasonic assessment of composite laminates and bonded joints [13]. A typical AU test arrangement is shown in Figure 8.22. Both transmitting and receiving transducers are attached to the surface of the test piece using a spring-loaded device to apply constant pressure onto the transducers and a fluid to achieve acoustic coupling. The transmitting probe is optimized for wave generation while the receiving transducer is optimized for sensing. Both probes are often placed on the same side of the part at the normal angle with respect to the surface. In this approach, reproducibility of the AU data is very important as many factors could affect the results. The main factors include:

- The type, amount, and thickness of the coupling fluid and the presence of bubbles in the couplant.
- Specimen surface roughness and texture.
- The pressure applied to the transducers.
- The location, alignment, and spacing of the transducers.
- The equipment type and settings (e.g., frequency range, gain, threshold, etc.)
- Environmental parameters, such as temperature, vibration, noise, etc.
- Test part material, geometry, microstructure, homogeneity, etc.

Proper reference standards are required for the calibration of the AU measurement, and such references must exhibit the full range of expected material conditions or discontinuity states that are being investigated. Using the calibration references, a base-line SWF or AUP value representing the optimal condition of the material is established. When nominally identical samples are being examined, the AU measurements are usually normalized with respect to the base-line value, and differences between the samples are observed by comparing the normalized AU results. For improved repeatability and reliability, the AU measurements must be carried out under well-controlled conditions, such that all the factors related to the transducer-specimen coupling, test conditions, settings, equipment, environment, and geometry are maintained constant. In this way, subtle changes in the test object may be identified through the measurements of the AU signals. A few examples are presented below.

8.5.3 APPLICATIONS

Acousto-ultrasonic approach is often used when the conventional ultrasonic or acoustic methods are found to be inadequate. Two examples are provided below.

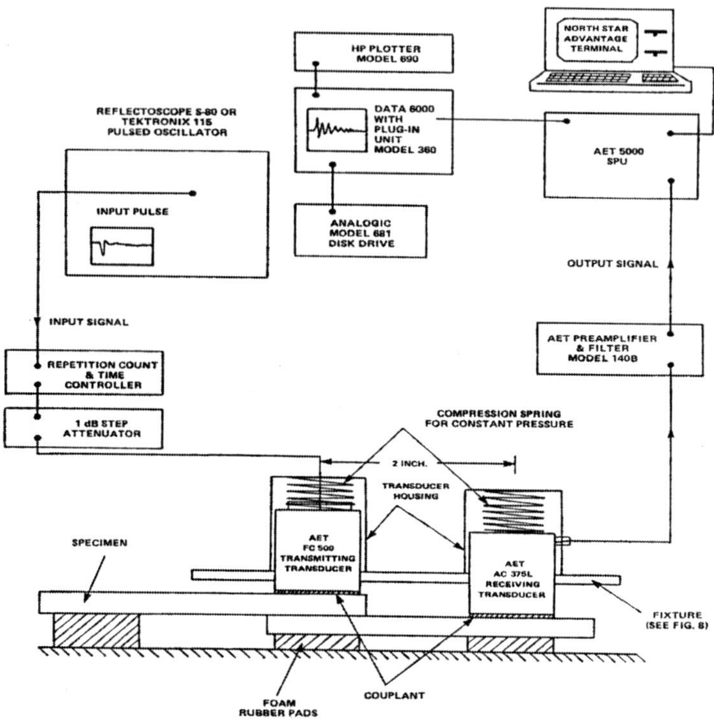

Figure 8.22 Typical experimental setup for acousto-ultrasonic measurements.

8.5.3.1 ADHESIVE BOND CHARACTERIZATION

Characterization of adhesively-bonded joints in terms of bond quality and strength is a continuing challenge for NDT. Reference [14] describes the use of the acousto-ultrasonic method on simple single-lap joint specimens made from steel adherents and the Cytec FM 300 film-adhesive. This is a structural adhesive that is widely used in aerospace and was used in the original manufacture of most bonded structures of the CF-18 aircraft, including the rudders and horizontal stabilizers [15]. The adhesive is typically cured at 177°C and is expected to hold its strength at temperatures below 150°C. The aim of the study described in Reference [14] was to determine if the bond strength remains unchanged or deteriorates when the temperature approaches the maximum limit, and if the AU technique is capable of detecting the changes.

For this purpose, a number of single-lap joint specimens were heated to temperatures ranging from 30 to 150°C and the AU measurements were carried out in intervals of 10°C. After AU tests, the specimens were loaded in tension at the same elevated temperatures used for the AU measurements until their failure. Both the

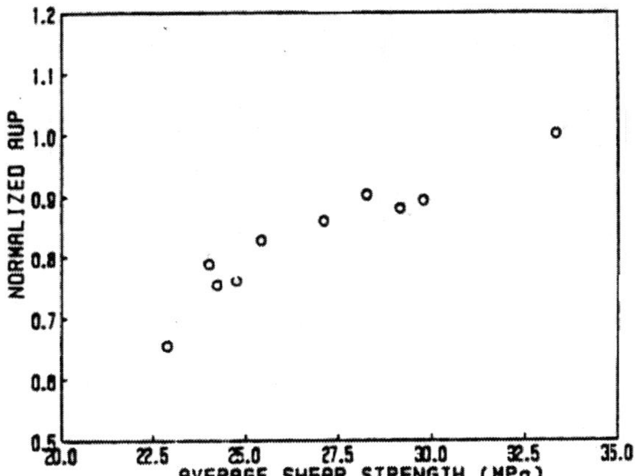

Figure 8.23 Correlation between the acousto-ultrasonic parameter (AUP) and bond strength for single lap joint specimens subjected to high temperatures.

AUP and the residual shear strength of the single lap-joint specimens showed a decline with increasing temperature beyond about 100°C, which was associated with the softening of the adhesive that affects both the strength and the AU wave propagation. The average shear strength at 100°C was 76% and at 150°C was 37% of the room-temperature value. The correlation between the normalized AUP and the average shear strength was close to linear, as seen in Figure 8.23.

Having found a reasonably good correlation between the bond strength and the AUP, the next experiments aimed at assessing the effect of repeated thermal exposure on the bond quality and if the AU approach could identify a possible change of bond strength. This was of interest for situations when a bonded aircraft structure is repeatedly exposed to higher than normal temperatures (e.g., engine exhaust, extreme desert temperatures, fire), but the damage is not detectable using conventional NDT methods.

For this purpose, similar specimens were exposed repeatedly to150°C for one hour and then cooled down to room temperature for AU and mechanical tests. The results were compared with those of the as-fabricated control specimens. Inspection by conventional ultrasonic C-scan did not reveal any detectable damage. However, the acousto-ultrasonic measurements showed that the AUP of specimens subjected to cyclic thermal exposure was 12% lower than the controls, while their average shear strength was 20% less than the as-fabricated ones [14]. Examination of fracture surfaces by optical microscopy indicated more evidence of interfacial failure in the thermally-cycled specimens than the controls.

Further work was carried out to investigate the effect of adhesive aging on strength and the AUP. Aircraft adhesives generally must be kept in a freezer; however, during the fabrication of composite parts, which is usually a very time-consuming manual process, they are moved in and out of the freezer repeatedly and left at room temperature for some time. The cumulative effect of adhesive aging at room temperature was of interest. Thus, FM 300 adhesive-films were aged at room temperature for periods of 3 to 8 weeks. Single lap-joint specimens were then made using CFRP composite adherents and the aged adhesives [12]. After cure, the bonded specimens were inspected for any flaws using ultrasonic C-scan.

The C-scan tests showed no difference among the specimens, including the control ones. However, the AU measurements showed that the specimens with aged adhesives exhibited a linear drop to about 45% of the room-temperature value after 8 weeks of adhesive aging. The joint average shear strength showed a reduction of about 15% for the same aging period. This relatively small reduction in strength with aging time was associated with the fact that slow curing of the adhesive takes place at room-temperature, which is much less than the cure temperature of this adhesive. Despite the small strength degradation, the AU stress wave propagation was affected to a higher and more measurable degree. The change in the AU signals was associated with minute changes in the overall adhesive and interfacial prosperities. Fractographic examination of the fracture surfaces showed that specimens with longer periods of adhesive aging failed predominantly at the interface.

Acousto-ultrasonic measurements were also carried out in conjunction with pattern recognition analysis on specimens with the adhesive bond lines degraded in different ways. The aim was to find out if the defective samples could be differentiated from the well-bonded ones using their AUP signatures for quick manufacturing quality control purposes. For these tests, single lap-joint specimens were made of aluminum adherents and two commercial adhesives (FM 300 and Hysol 9320) [15].

Degradation of the bond line was achieved by deliberate introduction of voids in the joint, contamination of the adhered surfaces, modification of the standard surface preparation prior to bonding, or variation of the adhesive layer thickness [17, 18]. AU signals from several measurements on each set of specimens were digitized and analyzed in both time and frequency domains. Several features of the AU signals were selected for classification into groups using a commercial pattern recognition software package (Statistical Analysis System or SAS). Finally, specimens were loaded to failure and classification of the AU data based on the failure loads was carried out.

As shown in Figure 8.24, by using pattern recognition on the AU signals, it was possible to classify the samples into separate groups based on their AU response signals with reasonable recognition rates. For example, samples having a different joint thickness from the reference group were classified into a different group, or specimens that failed at lower loads formed separate groups from those that had higher failure loads on the basis of their acousto-ultrasonic signals.

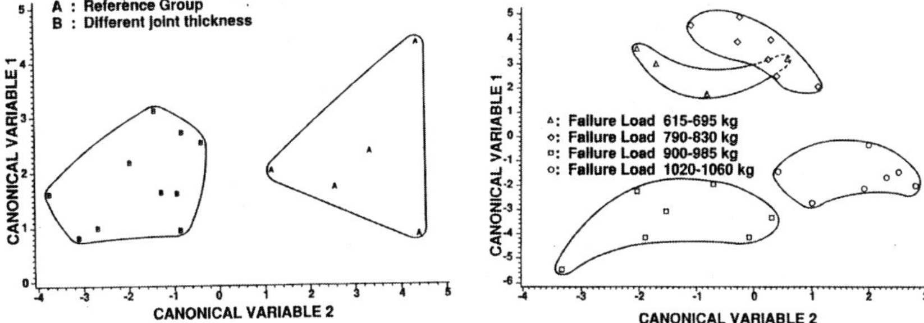

Figure 8.24 Classification of AU signals by the sample group (left) and failure load (right).

The above-mentioned AUP-strength correlations and the AU signal classifications indicate that, in a simple single lap-joint configuration, the AU stress wave propagation efficiency, which is reflected in the AU signal strength and is measured by the AUP, is affected by the bond features, such as discontinuities, thickness variations, or degradation. The effect on stress AU wave propagation efficiency is in the same manner as these features affect the bond shear strength or the load transfer efficiency. Thus, acousto-ultrasonic tests provide a potential NDT approach for adhesive bond evaluation. The above-mentioned work, along with similar AU studies elsewhere on different materials and configurations (e.g., [19, 20, 21]), proved the viability of the acousto-ultrasonic approach and lead to the development of the ASTM Standard E 1495-02 on this technique.

8.5.3.2 Inspection of Foam Core Aluminum Panels

Sandwich structures made of thin metallic or composite skins and a PVC foam core are used in small light-weight aircraft, since they are less expensive than the more conventional honeycomb sandwich materials. In addition, the foam core provides better thermal insulation and noise absorption. However, such materials are susceptible to impact and fatigue that create debonding of the face sheet from the core. In the work described in reference [22], the acousto-ultrasonic technique is compared with the conventional ultrasonic pulse-echo and a commercial bond testing instrument (Mechanical Impedance Analyzer or MIA). The aim is to establish the applicability of the AU method for the detection of disbond in a foam-core sandwich part made of very thin aluminum skins and a PVC foam core. Descriptions of the ultrasonic pulse-echo technique and the mechanical impedance analysis are provided in separate sections in this publication.

The samples were made of 38 mm-thick PVC foam core adhesively-bonded between two sheets of aluminum, 1 mm in thickness. Both simulated and real disbonds were examined. The simulated disbonds were introduced by embedding Teflon tapes of different sizes at the Al-core interface as well as by removing a small portion of the foam before fabrication that creates cavity-like unbonded areas. Real disbonds were introduced by either cyclic loading of the panels in three-point bending or subjecting them to impact damage under controlled conditions.

The ultrasonic pulse-echo method was capable of detecting Teflon inserts and the cavity resulting from the foam removal, but it was ineffective in detecting disbonds created by the impact or fatigue. The inability of the pulse-echo UT to detect real disbonds is due to the fact that disbond occurs in the foam just below the adhesive, creating surfaces that are rough, porous, and non-reflective. While the MIA was able to detect the real disbonds due to the drop in the local stiffness, the unbonded areas and the Teflon inserts were not detected, as these were not large enough to affect the local stiffness. The acousto-ultrasonic method was capable of detecting both the artificial as well as the real fatigue and impact disbonds. As mentioned earlier, the AU is based on the measurement of the stress wave transmission efficiency that is affected by all discontinuity types present in the wave path. For the AU tests, it was possible to use a pair of rubber-wheel probes that were dry-coupled and rolled over the smooth Al face-sheet during the inspections. With this kind of probe, the AU inspection of such panels could be carried out in a fast and effective manner. Constant pressure on the probes must be maintained for reliable results.

8.5.4 *Capabilities and Limitations*

Capabilities:
- Requires access to one side only.
- Provides information about the overall condition of the materials.
- Useful for detection of subtle and distributed discontinuities.
- Has potential for assessment of bond quality.
- Probe alignment is not important.
- Dry-coupling and rubber wheel-probes may be used for ease of application.

Limitations:
- Many factors including transducer coupling and pressure can affect the results.
- Measurements must be made under well-controlled conditions.
- Individual discontinuities cannot be located.
- Automated scanning and imaging is not easy.
- Signal interpretation is difficult.
- Requires skilled personnel to analyze and interpret the results.
- Verifications by other NDT methods may be needed.

8.6 CHAPTER REFERENCES

[1] Multi-mode Adhesive Bond Testing, Olympus Corporation Website, http://www.olympus-ims.com/en/bondmaster1000eplus/.

[2] Introduction to Non-destructive Testing, NDT Resource Centre, Center for Non-destructive Evaluation, Iowa State University, Ames, Iowa 50011, USA.

[3] Metals Handbook, Non-destructive Evaluation and Quality Control, Ninth Edition, Vol. 17, 1989.

[4] Using Acoustic Emission in Fatigue and Fracture Materials Research. M. Huang, L. Jiang, P.K. Liaw, C.R. Brooks, R. Seeley, and D.L. Klarstrom, JOM-e, http://www.tms.org/pubs/journals/JOM/9811/Huang/Huang-9811.html.

[5] Physical Acoustic Corporation, 195 Clarksville Road, Princeton Jet, NJ 08550, USA, PAC Internet site.

[6] Fatigue Design Model Based on Damage Mechanisms Revealed by Acoustic Emission Measurements, D. Fang and A. Berkovits, Trans. of ASME, 117, pp. 200–208, 1995.

[7] Non-Destructive Characterization of Hydrogen-embrittlement Cracking by Acoustic Emission Techniques, H.L. Dunegana and A.S. Tetelmanb, Dunegan Research Corporation Livermore, Calif. 94550 U.S.A., University of California at Los Angeles, Los Angeles, Calif. 90024 U.S.A., Available online, 27 February 2003.

[8] Monitoring Hydrogen Embrittlement Cracking using Acoustic Emission Technique, A.K. Bhattacharya, N. Parida, and P.C. Gope, Journal of Materials Science, Vol. 27, No. 6, pp. 1421–1427, 1992.

[9] ASTM Standard E 569-07, Standard Practice for Acoustic Emission Monitoring of Structures during Controlled Stimulation, ASTM website: www.astm.org, Approved July 1, 2007.

[10] Acoustic Emission Response of Adhesive-bonded Joints during Single Lap Shear Testing, A. Ustuner and A. Fahr, NRC Report: LTR-ST-530, 1988.

[11] Health Monitoring of Aerospace Structures with Acoustic Emission and Acousto-Ultrasonic, R.D. Finlayson, M. Friesel, M. Carlos, P. Cole, N. Way, and J.C. Lenain, www.NDTNet.com, 15th World Conference on NDT, 2000.

[12] Application of Acousto-Ultrasonic Techniques to the Characterization of Adhesively Bonded Joints in Fibre Reinforced Composites, A. Fahr, S. Lee, and S. Tanary, NRC Report: LTR-ST-1625, 1987.

[13] ASTM Standard E-1495-02, Guide for Acousto-Ultrasonic Assessment of Composites, Laminates, and Bonded Joints, ASTM website: www.astm.org.

[14] Estimation of Strength in Adhesively Bonded Steel Specimens by Acousto-ultrasonic Technique, A. Fahr, S. Lee, S. Tanary, and Y. Haddad, Materials Evaluation, Vol. 47, No.2, 1989.

[15] Flow Testing of Cytec FM 300 and FM 300-2K Structural Adhesives, R. Vcicka, DSTO-TN-0383 Report, Aeronautical and Maritime Research Laboratory, 506 Lorimer Street, Fisherman's Bend, Victoria 3207, Australia.

[16] Non-destructive Evaluation of Adhesively Bonded Joints using Acousto-Ultrasonic, A. Fahr, S. Tanary, and Y. Haddad, Journal of Acoustic Emission, Vol.8, No.1/2, 1989.

[17] NDE of Adhesively Bonded Joints Using Acousto-Ultrasonic and Pattern Recognition, Y. Youssef, A. Fahr and C. Roy, Proceedings of the 4th International Symposium on Non-destructive Characterization of Materials, Annapolis, MD, 1990.

[18] Adhesive Bond Evaluation Using Acousto-Ultrasonic and Pattern Recognition Analysis, A. Fahr, Y. Youssef, and S. Tanary, Journal of Acoustic Emission, Vol. 12, No. 1/2, 1994.

[19] Acousto-ultrasonic Non-destructive Testing of Fibre Reinforced Plastics, A. Vary, Vol. 2, Elsevier Applied Science Publishers, Essex, England, 1990.

[20] Acousto-ultrasonic Theory and Applications, J.C. Duke, Plenum Press, New York, 1988.

[21] Hydrothermal Aging of Jute-Glass Fibre Hybrid Composites- An Acousto-Ultrasonic Study, K.K. Phani and N.R. Bose, Journal of Materials Science, Vol. 22, 1987.

[22] Non-destructive Inspection of Foam Core Aluminum Panels, A. Fahr, C.E. Chapman, A.M. Charlesworth, S. Tanary, and B. Farahbakhsh, NRC Report: LTR-ST-1655, 1987.

Infrared Thermography

9.1 PRINCIPLES

Thermography involves the measurement of surface temperatures of an object that is subjected to a thermal excitation. In this method, local differences in the surface temperature of a material are related to heat flow that is affected by the material's internal features, discontinuities, and geometry. When an object is uniformly heated, the presence of discontinuities will hinder heat diffusion into the material, causing a region of higher temperature as compared to the rest of the surface. As schematically illustrated in Figure 9.1, point A of the surface that is above the defect will be warmer than point B located on defect-free area. Generally, the closer the discontinuity is to the surface and the larger it is, the greater will be the temperature difference. This temperature difference is detected by an infrared (IR) camera and processed to produce a thermal image of the component. A thermal image displays the object's surface temperature distribution in either gray or color format.

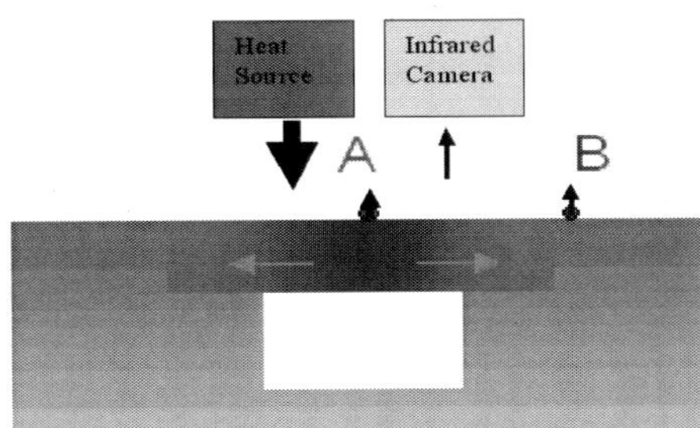

Figure 9.1 Schematic presentation of infrared thermography.

Three mechanisms may contribute to the thermal energy transfer: conduction, convection, and radiation [1]. Conduction mainly occurs in solids, and, to a lesser degree, in fluids, as warmer or more energetic molecules transfer their energy to colder adjacent molecules. In the convection process, warmer and colder molecules actually move and mix. This process only occurs in liquids and gases, such as when stirring or mixing is involved. Radiation is the heat transfer mechanism through electromagnetic energy that needs no medium to flow through and, therefore, can occur even in a vacuum. Electromagnetic radiation is produced when electrons lose energy and fall to a lower energy state. Both the wavelength and intensity of the radiation are directly related to the temperature of the surface molecules or atoms.

In IR thermography NDT, the heat diffusion within a solid test object is primarily by the conduction process, but the transfer of thermal energy from the surface of the object to the remote detector is through radiation or thermal emission. Therefore, thermal diffusivity and thermal emissivity of the test object are important in thermography inspections.

9.1.1 THERMAL DIFFUSION

Thermal diffusivity (α) is the speed at which the heat flows from a region of higher temperature to the surrounding material [2]. Thermal diffusion in a solid material is expressed by the following relationship [3]:

$$\alpha = \frac{k}{\rho \times c_p} \tag{9-1}$$

where k is the thermal conductivity of the material, which is a measure of the heat flow in a given direction when there is a temperature difference in that direction. The unit for k is J/s.m.K (Joule/second.meter.Kelvin) or W/m.K (J/s = Watt). ρ is the material density (kg/m^3) and c_p is the material specific heat capacity (J/kg.K). This parameter is a measure of the amount of heat a material will absorb for a given temperature interval. The thermal diffusion is measured in m^2/s.

The above expression indicates that the heat diffusion in a solid object is directly proportional to the thermal conductivity of its material and inversely related to the density and heat capacity of the material. Table 9.1 provides the thermal diffusivities of some materials at room temperature [4]. Generally, metallic components that have a high thermal diffusion provide better heat penetration and, thus, deeper inspection range than polymeric composite materials that have low heat diffusion capacity. Therefore, thermal inspection of composite parts covers only the near-surface region of the test object and is only effective on thin composite laminates or the front face of a sandwich structure. Near-surface delamination and disbonding up to a few millimeters deep may be identified, depending on the material being tested.

9.1.2 THERMAL EMISSIVITY

Thermal emissivity (ε) is a measure of the efficiency of an object surface in transferring its thermal energy to electromagnetic radiation. Theoretically, the thermal emissivity is defined as the ratio of thermal energy emitted by the test object surface to the energy emitted by a perfect blackbody at the same temperature. A perfect blackbody is an object that absorbs and re-emits all of the thermal energy, having an emissivity of $\varepsilon = 1$. Real objects do not absorb all the thermal energy and emit less radiation than an ideal blackbody, having $\varepsilon < 1$. Emissivity is a dimensionless quantity. Table 9.2 provides the surface emissivity of some materials and substances at 300K.

Thermal emissivity is affected by the radiation angle and wavelength, in addition to the object surface temperature and geometry. The wavelength of thermal radiation lies mostly outside the visible range of the human eye and in the infrared (IR) region (Figure 9.2). Infrared radiation corresponds to the band of the electromagnetic spectrum between 740 nanometers (1 nm $= 10^{-9}$ m) and 10^{-3} m in wavelength. Visible light is between 380 and 750 nm.

Table 9.1 Thermal diffusivity of selected materials and substances (m^2/s) [4].

Pyrolytic graphite, parallel to layers	1.22×10^{-3}
Pyrolytic graphite, normal to layers	3.6×10^{-6}
Gold	1.27×10^{-4}
Copper	1.1234×10^{-4}
Aluminum	8.418×10^{-5}
Aluminum 6061-T6 alloy	6.4×10^{-5}
Steel, stainless 304A	4.2×10^{-6}
Iron	2.3×10^{-5}
Silicon	8.8×10^{-5}
Aluminum oxide (polycrystalline)	1.20×10^{-5}
Glass, window	3.4×10^{-7}
Rubber	1.3×10^{-7}
Nylon	9×10^{-8}
Oil, engine (saturated liquid, 100°C)	7.38×10^{-8}
Water	1.4×10^{-7}
Water vapor (1 atm, 400 K)	2.338×10^{-5}
Air	1.9×10^{-5}

Table 9.2 Surface emissivity of selected materials and substances at 300K [5].

Aluminum, commercial	0.09	Aluminum, oxidized	0.2–0.31
Aluminum, polished	0.039–0.057	Aluminum, anodized	0.77
Black epoxy paint	0.89	Black enamel paint	0.80
Brass, polished	0.03	Brass, oxidized	0.6
Cadmium	0.02	Carbon filament	0.77
Cast iron	0.44	Chromium, polished	0.058
Copper, polished	0.023–0.052	Glass, smooth	0.92–0.94
Gold, polished	0.025	Ice	0.96–0.98
Inconel, oxidized	0.71	Iron polished	0.14–0.38
Iron, rusted red	0.61	Mild steel	0.20–0.32
Nickel, electroplated	0.03	Nickel, polished	0.072
Nickel, oxidized	0.59–0.86	Paint	0.96
Plastics	0.91	Silver, polished	0.02–0.03
Steel, oxidized	0.79	Steel, polished	0.07
Stainless steel, weathered	0.85	Stainless steel, polished	0.075
Titanium, polished	0.19	Tungsten, polished	0.04
Water	0.95–0.963	Wrought iron	0.94

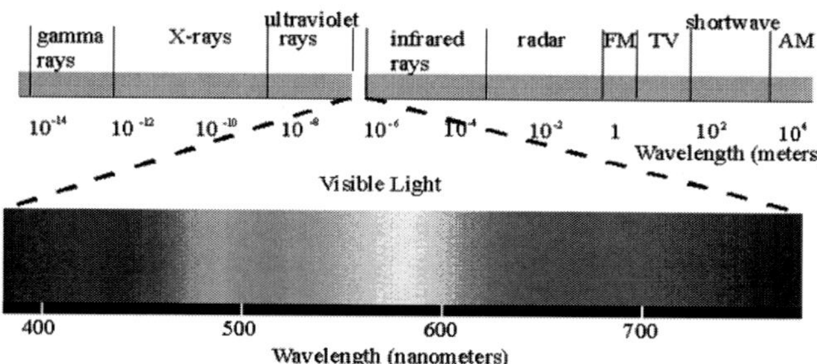

Figure 9.2 Different types of electromagnetic radiation and their wavelengths. The infrared radiation wavelength is between 0.74 and 1000 microns, mostly above the visible light range.

Unlike visible light, IR is emitted from any object with temperatures above the absolute zero. The higher the temperature of the object, the greater the infrared radiation emitted. Radiation energy as a function of wavelength at different surface temperatures illustrates the total radiated energy of a blackbody as a function of wavelength at different temperatures, and shows how the peak wavelength and the amount of radiation vary with temperature. Although this plot shows relatively high temperatures, similar relationships hold true for any temperature down to absolute zero.

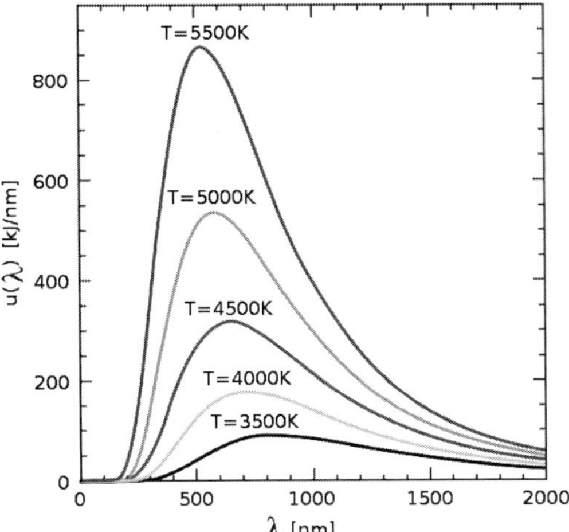

Figure 9.3 Radiation energy as a function of wavelength at different surface temperatures.

9.1.3 IR THERMAL MEASUREMENT

In infrared thermography, the thermal radiation of objects is detected and measured using infrared cameras that are sensitive to temperatures from around –20 to above 500°C. At room temperature, at which most NDT inspections are carried out, with modern infrared cameras, a temperature difference of 0.02°C can be measured with accuracies of ±2%. For NDT applications, in addition to temperature sensitivity, a fast response time of the order of 0.05 second or less may be needed to enable the detection and analysis of transient temperature changes.

In infrared thermal imaging, the surface condition of the test object is an important factor, as the thermal emissivity is affected by surface uniformity, angle, roughness, cleanliness, and shininess. A duller surface that is uniform, free of oil, dust, and dirt will have an emissivity closer to 1. For thermal inspection of polished or reflective surfaces, such as metals that have low emissivity, it is necessary to apply a uniform coat of flat paint that can be removed easily. If an object has low emissivity, IR instruments will indicate a lower temperature than the true surface temperature. Thus, calibration and adjustment of measurement instruments in relation to the emissivity of the object being tested are important.

The American Society for Non-destructive Testing (ASNDT) officially adopted infrared thermography as a standard NDT test method in 1992. Today, a wide variety of thermal measurement equipment is commercially available, and the technology is widely used in many industries, including aerospace. There are a number of ASTM Standards for thermography inspections, including the ASTM E2582-07, "Standard Practice for Infrared Flash Thermography of Composite Panels and Repair Patches used in Aerospace Applications." Continuous improvements are being made in infrared thermal imaging systems that enable new applications.

9.1.4 IR THERMOGRAPHY PROCEDURES

IR thermography tests can be carried by measuring the surface temperature of the object either during its normal use, which is referred to as passive thermography, or by stimulating the part using a heat source, which is known as active thermography. The latter can be carried out in many different ways. In the following sections, the different thermal imaging approaches are described with a few application examples and a summary of their capabilities and limitations.

9.2 PASSIVE THERMOGRAPHY

9.2.1 PRINCIPLES

In passive thermography, an IR camera is simply pointed at the test piece and the surface temperature is monitored during the normal use of the object. If a certain area of the object is warmer (or colder) than the rest of the test part, it will be detected. For example, hot spots in circuit boards can be identified during their normal use by passive thermography, as illustrated in Figure 9.4. In this case, a defective element may exhibit a different thermal radiation as compared to the same component operating properly in another similar circuit board, making it either hotter or colder. In this example, the heat transfer process is in a steady-state condition, and the inspection time is not very crucial. The ASTM Standard E 1934 provides guidelines for examining electrical and mechanical equipment with IR thermography.

Passive thermography is also used when the heat transfer process is transient (unsteady condition). In this case, the timing of the inspection is important since the surface temperature must be monitored as a function of time. The presence of discontinuities is identified by measuring and presenting the difference between the temperature profiles of the defective and sound areas of the same part at a given time sequence (Figure 9.5) or by comparing with an identical object being tested at the same time.

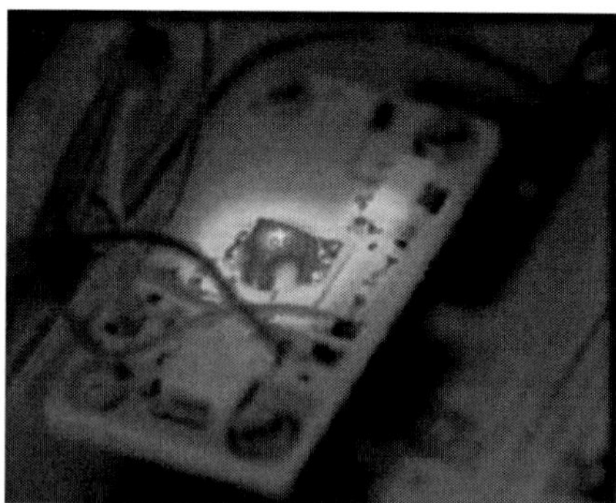

Figure 9.4 Passive thermal image of a circuit board showing hot spots [6].

9.2.2 APPLICATION EXAMPLES

An example of time-based passive thermography is shown in Figure 9.6, which is a thermal image taken after the space shuttle enters the earth's orbit from space. The image indicates that the left wing of the spacecraft is experiencing higher temperatures than the right wing during entry to the earth atmosphere.

A similar approach is used to detect water ingress into honeycomb structures of airplanes. Water penetration through gaps, holes, or broken seals into the interior of critical structures, such as rudders, is a problem for some aircraft and passive thermography provides a useful tool. In this application, the aircraft is moved into a hanger from a colder (sub-zero) outside environment, and the affected area is inspected immediately after. Figure 9.7 shows a thermal image of two rudders of an aircraft taken in a hanger environment. The rudder on the right shows areas of water intrusion into the honeycomb core around the upper hinge, while the rudder

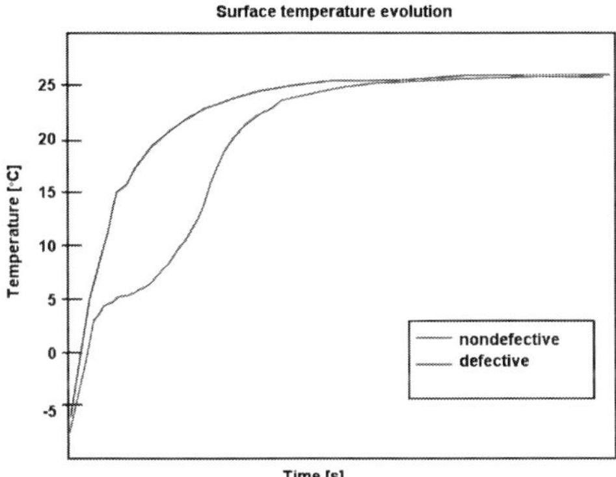

Figure 9.5 Examples of surface temperature (T) as the function of time measured by IR thermography equipment for sound and defective areas. In this case, the defective area is due to water ingress into the structure.

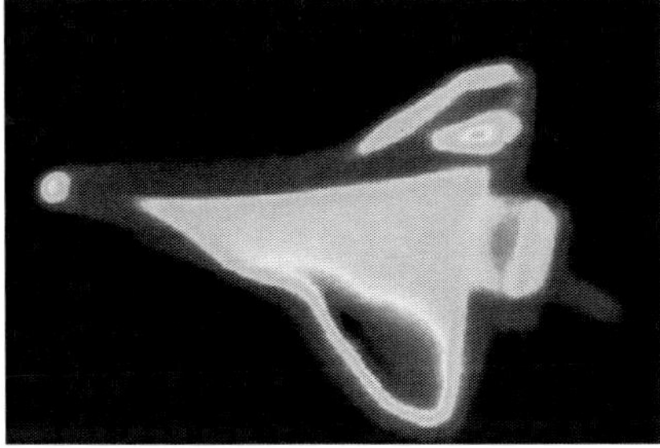

Figure 9.6 Passive IR thermal image of the space shuttle showing areas of the spacecraft that have experienced higher temperatures during re-entry to the earth's atmosphere [7].

on the left only shows thermal differences due to internal features. This approach is only possible when ice has formed at the rear side of the surface being tested.

Passive IR thermography can also be useful for monitoring heat distribution at the aircraft engine exhaust during operation. The change in engine performance, resulting from possible in-service faults, may affect the IR thermal patterns of the exhaust that may be identified. As an example, Figure 9.8 illustrates an IR image of an experimental aircraft engine with damaged turbine inlet guide vanes [9]. Damaged vanes disrupt the gas flow entering the turbine, thus, producing large spatial temperature variations, which result in performance changes in the turbine downstream of the damaged vane affecting the thermal distribution at the exhaust outlet.

Thermal imaging may also be applied to monitor manufacturing processes, such as materials machining, welding, friction stir welding, curing of adhesive bonds, etc. Two examples are provided in Figure 9.9. This figure illustrates IR thermal images obtained during machining of a horizontal spindle as well as during a welding process. Such images help to optimize manufacturing parameters,

Figure 9.7 Passive thermal imaging of aircraft rudders showing water intrusion into the honeycomb core around the upper hinge of the right hand side rudder [8].

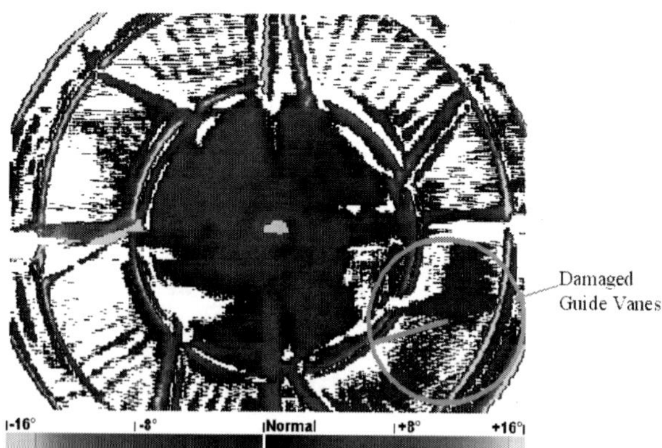

Figure 9.8 An IR thermal image of an experimental aircraft engine showing uneven temperature profile due to damaged vanes [9].

such as machining speed, in order to reduce overheating of tools and, thus, increase their usage life.

As illustrated in the above examples, passive thermography often requires inspection of an identical reference part along with the test part for comparison of thermal images. The reference part enables the inspector to identify the thermal traces of the inherent features of the part from those of the defects and helps in the interpretations of results.

9.2.3 CAPABILITIES AND LIMITATIONS

Capabilities:
- Non-contact inspection of large areas.
- Detects hot or cold spots in real-time.
- No need for external heating source.
- Useful for identifying areas that are experiencing overheating in electrical circuit boards or mechanical instruments.
- Can detect water ingress into honeycomb structures.
- Useful for monitoring heat distribution during machining or other manufacturing processes.

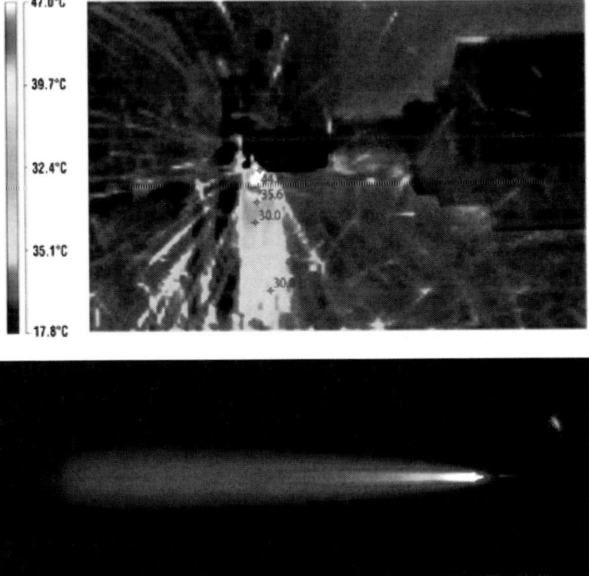

Figure 9.9 IR thermal images obtained during machining of a horizontal spindle (top [10]), and at the end of a welding process (bottom). The coldest areas are black and the hottest are white.

Limitations:

- Low sensitivity and, thus, limited applications for damage detection.
- Differentiation of damage from inherent features requires comparison with an identical damage-free part.
- Hot or cold spots must be close to the surface.
- Inspection timing is important when heat transfer is transient.
- Transient image interpretation is difficult.

9.3 ACTIVE THERMOGRAPHY

9.3.1 *PRINCIPLES*

Active thermography involves heating the object by an external source (optical, thermal, mechanical, or electromagnetic) and observing the surface temperature evolution as a function of time. Internal discontinuities, including inherent features and defective areas, are displayed by mapping the surface temperature distribution or looking at the differences in temperature change rate. A simple active thermography system consists of a heat source (e.g., electric lamp or a vibration device), an infrared camera, and a thermal imaging instrument, as illustrated in Figure 9.10. Optimal heat absorption and emission occur when the surface being tested is perpendicular to the direction of the flow of energy [11]. When there is a significant change of surface slope, points at oblique angles will not absorb or emit as much energy as those that are at normal angle, due to factors such as reflection or scattering of heat waves. Similarly, points farthest away from the heat source will receive and emit less heat. The heat intensity variations caused by surface slopes or geometrical changes of the specimen can lead to incorrect interpretation of results or missing of defects. Indications from defects located below geometrical changes, such as surface slopes or bends, can be easily obscured within the traces of the object shape. If adequate thermal information on the object shape is available, it may be possible to make corrections through signal processing and reduce surface geometry effects. However, this approach is time-consuming and

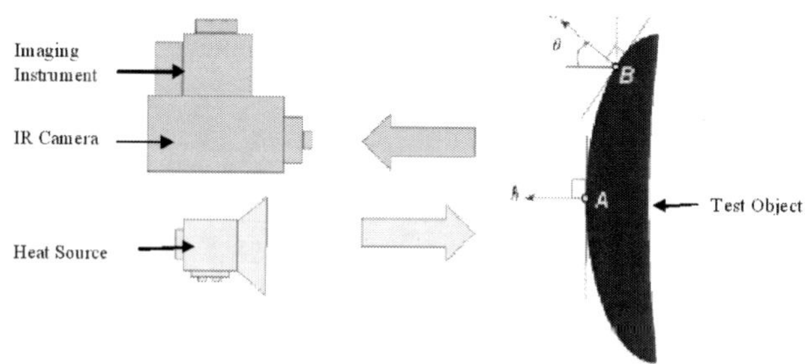

Figure 9.10 Schematic presentation of active infrared thermography.

not always possible. A simpler approach is to compare a series of images taken at different time sequences and consider the earlier ones corresponding to the surface geometry and those of the later times that may be affected by internal defects [9]. The further the defect from the surface, the longer the time will be for the heat waves to reach the defect and create a temperature difference at the surface. This approach is useful when discontinuities are not very close to the surface or too small to allow a reasonable defect contrast to be developed.

There are different active thermography approaches depending on the type of external source used to generate heat in the sample and the way the thermal waves are applied and analyzed. The different approaches are schematically illustrated in Figure 9.11 and described in the following sections. In each section, a few applications are provided that include both metallic and composite reference calibration samples with known internal discontinuities, as well as real service-expired aircraft components. The latter include a piece of airframe aluminum lap-joint with corrosion spots and an aircraft rudder similar to the one examined by passive thermography. The state of this rudder is not clearly known, but, nevertheless, it provides a means for comparative assessment of the different thermography approaches using a real aircraft structure.

9.3.2 PULSED THERMOGRAPHY

In pulsed thermography (PT), a short heat pulse is applied to the test part using flash lamps, laser beams, hot or cold air flow, etc. The heating duration varies from a few milliseconds for good thermal conductors to a few seconds for low-conductivity materials. For aerospace NDT applications, xenon flash lamps are often used. In this case, the heat generated on the surface of the test piece is partly diffused into the object. When the diffused heat is intercepted by internal discontinuities that have a different thermal diffusivity from the surrounding material, that

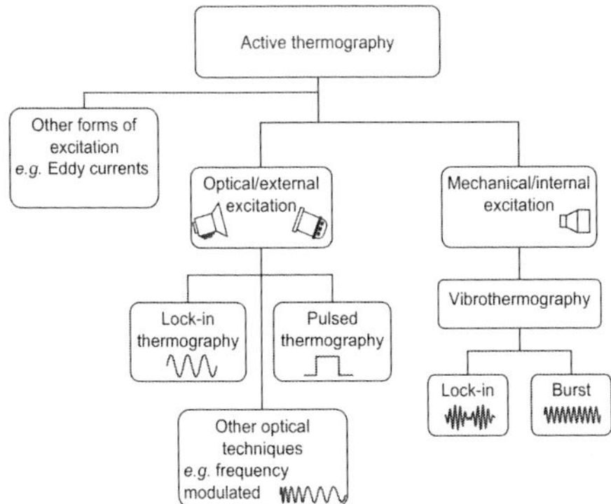

Figure 9.11 Schematic presentation of different active thermography approaches [12].

portion of the thermal energy is dissipated at a slower rate, creating a heat accumulation above the defective region. This heat accumulation makes the surface over the defect slightly warmer than the surroundings, and this can be detected by an infrared camera located in front of the test object. Figure 9.12 shows a schematic diagram illustrating the principles of the pulsed thermography approach.

The pulsed thermography is a time-dependent process, and the time it takes for thermal waves to reach the defect and affect the surface temperature (t) depends on the defect depth from the surface (z) and the material's thermal diffusion rate (α), as below:

$$t \approx \frac{z^2}{\alpha} \tag{9-2}$$

This equation may be used to estimate the defect depth location if the thermal diffusion rate of the test item is known. For inhomogeneous materials or complex parts that contain several materials, it is more practical to use calibration specimens to estimate the flaw depth. In this case, t is measured relative to a calibration sample that contains known discontinuities at different depth levels (also known) from the surface of the part.

Generally, pulsed thermography requires uniform heating of the object surface and any variation in surface conditions, such as brightness, roughness, angle, and paint color or thickness may affect the results. The technique is most effective when defects are close to the surface and have a thermal diffusivity significantly different from that of the host material. Deep flaws (> 4 mm), especially in low conductivity materials, such as CFRP composites, are not easily detectable. Also, inclusions that are similar to the test material in terms of thermal properties may not be detected. Thus, calibration specimens containing foreign material inclusions as artificial defects that are made for ultrasonic testing may not be suitable for thermography tests.

9.3.2.1 APPLICATION EXAMPLES

Pulsed thermography has potential application in a wide range of aerospace materials and components. The approach has been used in many different applications, and a few examples are provided in Figure 9.13 to Figure 9.18 to demonstrate the potential of this approach. More details are available in references [8], [11], [12], [13], [14], and [15].

Figure 9.13 illustrates an aluminum reference sample containing flat-bottom holes of different diameters and depths. This sample was manufactured to establish the capability of an NDT method in characterizing material thickness loss that occurs as a result of corrosion. When the intact surface is facing the NDT sensor (in this case the IR camera), the deeper the hole, the closer it is to the upper surface being inspected, and the more the loss of material thickness. In this figure,

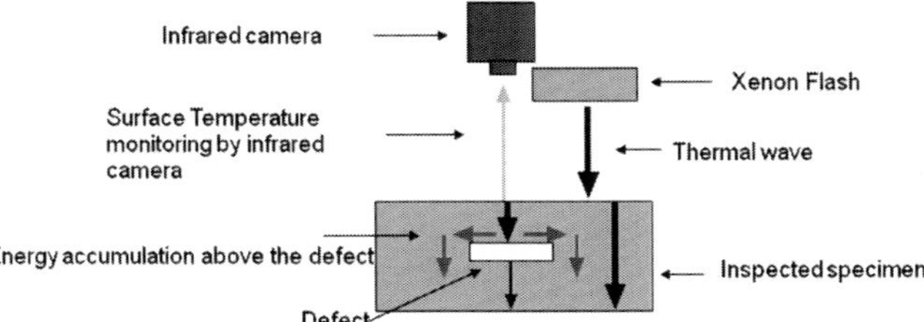

Figure 9.12 Schematic diagram showing the principles of pulsed infrared thermography.

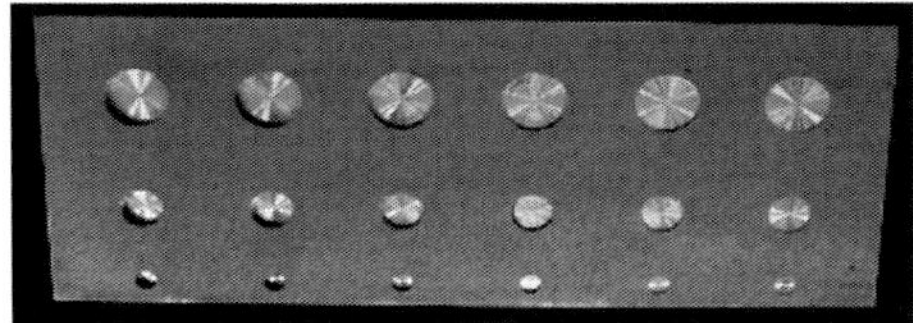

Plate Thickness: 5.35 mm. Hole Diameter: 5 mm, 10 mm. 20 mm
Hole Depth (distance from top surface) in mm:

4.45 (0.9)	3.72 (1.63)	3.05 (2.3)	1.96 (3.39)	0.95 (4.4)	0.44 (4.91)
4.28 (1.07)	3.58 (1.77)	2.83 (2.52)	1.70 (3.65)	0.78 (4.57)	0.45 (4.9)
4.45 (0.9)	3.81 (1.54)	2.90 (2.45)	2.02 (3.33)	0.88 (4.47)	0.42 (4.93)

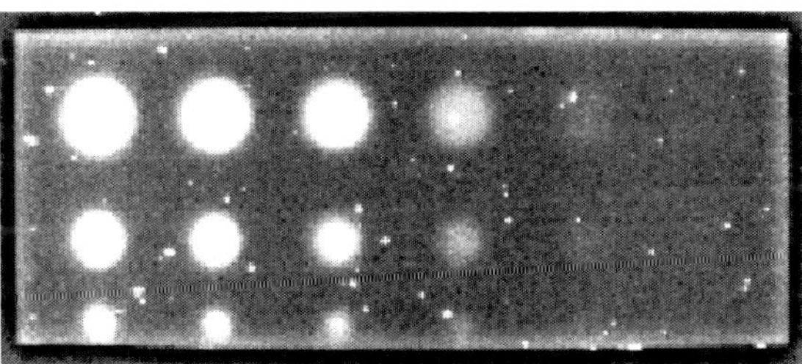

Figure 9.13 (Top): An Al sample with flat bottom holes of different depths simulating thickness loss due to corrosion. The deeper the hole, the closer it is to the surface being tested. (Bottom): a pulsed thermography image of the plate taken when the intact surface is facing the camera. Note that as the discontinuity distance from the inspection surface increases the resolution decreases for all sizes. For Al, discontinuities deeper than 3.65 mm from the top surface are not clearly visible.

the pulse thermography results are also presented, demonstrating the detection capability of this approach with respect to flat-bottom holes in aluminum. For Al, discontinuities deeper than 4.5 mm from the top surface are not clearly visible, especially if they are smaller than 10 mm in diameter.

Figure 9.14 shows a small segment of an aluminum lap-joint cut out of a commercial aircraft fuselage. It contains service-induced corrosion spots, mostly around the fastener holes; some are visible to the naked eye, but many are not

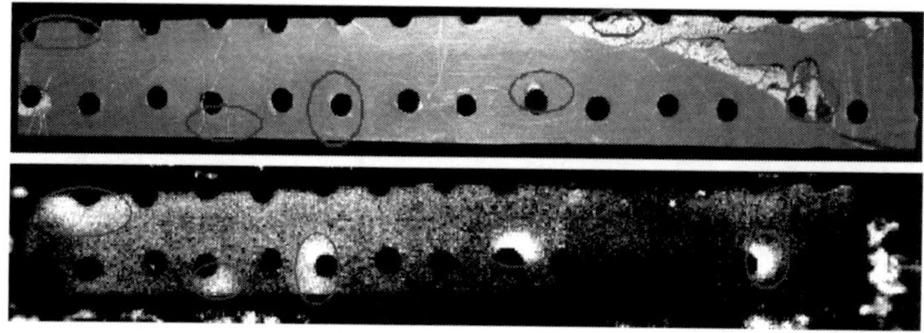

Figure 9.14 (Top): An aircraft part containing corrosion around some fastener holes. (Bottom): Corresponding pulse thermography image identifying corrosion sites. Note that in some areas, corrosion is not visible at the surface, but is detected by thermography due to the different thermal characteristics of the corroded spots as compared to the corrosion-free bare metal.

very visible. However, with the pulsed thermography, the previously invisible corrosion sites become more visible. As mentioned above, this approach is effective when defects bering examined re close to the surface.

Figure 9.15 shows a schematic diagram of a solid laminate composite reference sample containing flat-bottom holes of different diameters and depths. This sample was made to establish the size and depth resolution of thermal images obtained using the pulsed thermography approach on typical CFRP composites. When the hole-free surface is facing the IR camera, the deeper the hole, the closer it is to the surface being inspected. In this figure, the pulsed thermography results obtained at two time sequences are also presented, indicating that the depth resolution is in 2–4 mm range for discontinuities larger than 6 mm in diameter. Beyond that, the sensitivity and resolution drop rapidly as a result of lateral heat diffusion.

Figure 9.16 shows a schematic diagram of a reference sandwich panel made of CFRP skins (10 plies, 2.5 mm thick) and aluminum honeycomb (15.9 mm thick). The panel contains simulated delamination and disbonding, as well as real impact damage introduced using two energy levels. Also, water was injected into honeycomb cells at several locations through small holes that were drilled on the rear face and sealed after injection. The types, sizes, and locations of the discontinuities are identified in the figure caption. The pulsed thermography results, obtained at different time sequences and put together, are also presented in this figure. The thermal image corresponding to the first derivative of the acquired temperature-time signals at 0.73 sec shows all the PTFE inserts in the face sheet and the impact damages, but does not show the other defects. However, at later time-sequences, both the entrapped water and the inserts embedded between the skin and the honeycomb core become visible as heat reaches the deeper regions of the part. Cuts in the core cell walls that are perpendicular to the heat waves could not be seen.

Figure 9-17 shows an aircraft rudder similar to the one tested using passive thermography. This rudder was retired from service due to water intrusion into honeycomb that was identified by ultrasonic and radiography. When the

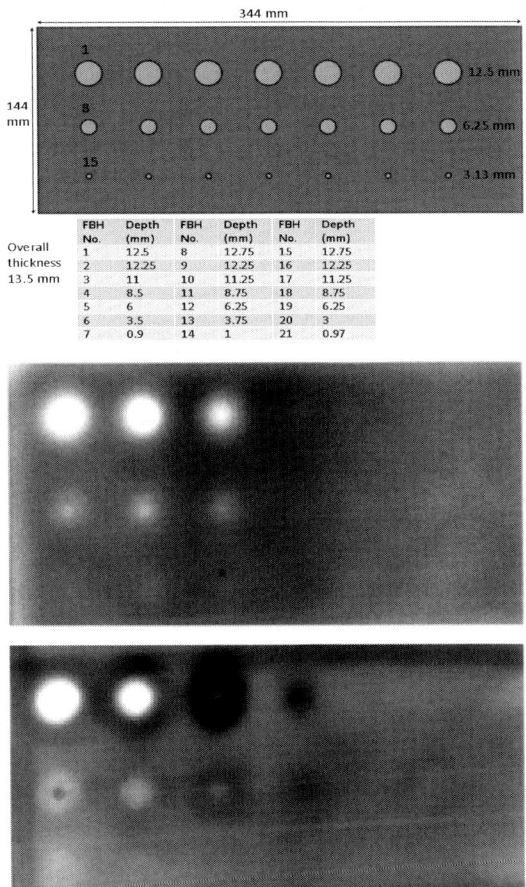

FBH No.	Depth (mm)	FBH No.	Depth (mm)	FBH No.	Depth (mm)
1	12.5	8	12.75	15	12.75
2	12.25	9	12.25	16	12.25
3	11	10	11.25	17	11.25
4	8.5	11	8.75	18	8.75
5	6	12	6.25	19	6.25
6	3.5	13	3.75	20	3
7	0.9	14	1	21	0.97

Figure 9.15 (Top): Schematic diagram of a CFRP solid laminate reference sample containing flat-bottom holes of three different diameters drilled to various depths. (Middle and bottom): Corresponding pulsed thermography images obtained at two time sequences.

rudder was inspected using the pulsed thermography approach, two suspect areas appeared clearly, as shown by B and C.

These were initially thought to be due to entrapped water; however, after further evaluation, it became clear that these indications were the result of thicker paints used in those areas. The PT inspections could not detect any water, possibly due to evaporation, as the rudder was in storage for several months before the thermography tests. However, two additional damage sites were identified that were missed by earlier NDT methods. The damage appeared to be delamination at the edges of composite face sheets that cannot be seen on radiographic images and are not easily detectable by any ultrasonic method, due to edge scattering effects.

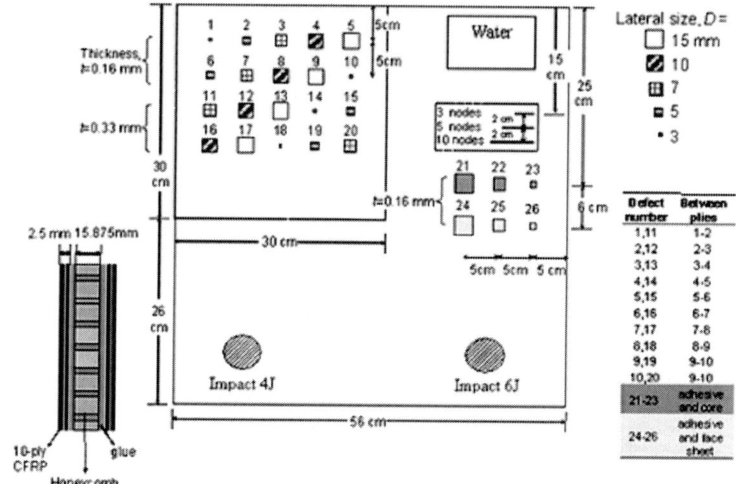

CFRP face sheets- 2.5 mm thick, Al honeycomb core-15.9 mm thick- cell size 4.6 mm. Simulated defects 1-20 are Polytetrafluoroethylene (PTFE) inserts and include single layer (0.16 mm thick) and double-layer (0.33mm thick) square tapes, dimensions 3,5,7,10,15mm. The double-layer tapes entrap air and simulate delamination. 21-23 are single-layer PTFE between adhesive and core and 24-26 are single-layer PTFE between adhesive and face sheet simulating disbonds. Node separation was created by cutting cell walls. Water was injected into several individual cells. Impact damages were introduced in an impact machine at 4 & 6 J. Ply 1 is top face and 10 is next to honeycomb core.

Defect number	Between plies
1,11	1-2
2,12	2-3
3,13	3-4
4,14	4-5
5,15	5-6
6,16	6-7
7,17	7-8
8,18	8-9
9,19	9-10
10,20	9-10
21-23	adhesive and core
24-26	adhesive and face sheet

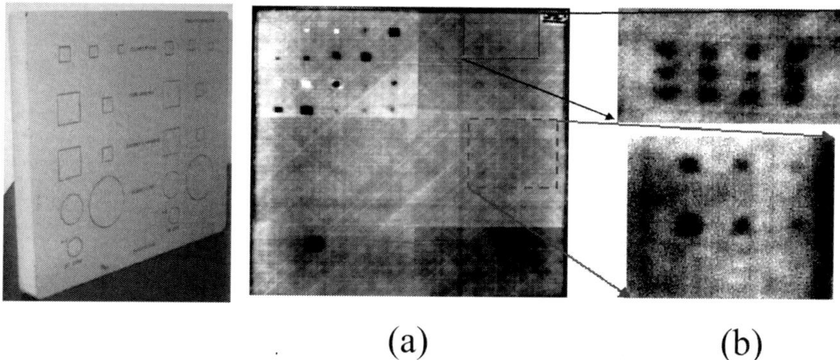

(a) (b)

Figure 9.16 (Top): A reference sandwich panel made of CFRP face sheets and aluminum honeycomb core with implanted defects, including PTFE inserts (single and double layers), cells with water, node disbonds, and impact damage. (Bottom): Corresponding pulsed thermography results showing first derivative of the temperature vs. time (a): at 0.73 sec showing PTFE inserts in the face sheet and impact damages only, but PTFE inserts in the adhesive and water in cells are not visible as they are located deeper. (b): at 8.75 sec both water and simulated defects in the adhesive become more visible as heat reaches the deeper regions at a longer time sequence. Cuts in the core cell walls cannot be seen.

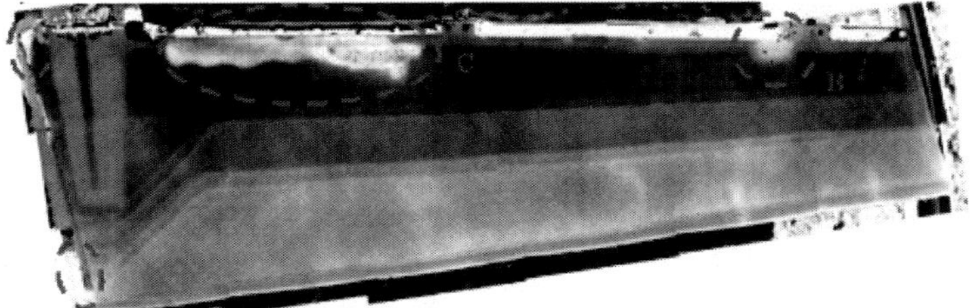

Figure 9.17 Pulsed thermography results of an aircraft rudder similar to the above with areas B and C corresponding to paint thickness differences. The two other red-circled sites are edge delamination. Inherent features, such as skin thickness, changes, and ply drop-offs are also visible. Note that several shots were taken to cover the entire part. Also, several time sequences were used to obtain the optimal image. This image corresponds to the second derivative of PT signals at 0.367 sec.

In addition, the PT image clearly showed the face sheet thickness changes and the locations of the ply drop-offs.

Figure 9-18, a photograph of a helicopter tail rotor half hub is shown along with its PT thermal image. This component is made of CFRP face sheets and fiber-glass filler core attached to metallic fittings and bushings. The part had been retired from service due to the appearance of disbonds at two corners of the central metallic element and the adjacent composite sections. When the part was inspected using PT, other possible damage sites were identified; thus, indicating the usefulness of the approach for inspection of complex components such as this.

9.3.2.2 CAPABILITIES AND LIMITATIONS

Capabilities:
- Non-contact and large area inspection method.
- Relatively fast and inexpensive.
- Near-surface planar defects, such as delamination in composites, can be detected.
- May detect water intrusion in honeycomb structures.
- Metallic plates can be tested for thickness loss if surfaces are not reflective.
- Applicable in the laboratory or field environments.

Limitations:
- Requires uniform heating of the object.
- Surface preparation and coating may be needed.
- Limited depth of penetration; thus, deep defects may not be detectable.
- Requires skilled personnel to perform the tests and interpret the results.
- Verification using other NDT methods may be necessary.

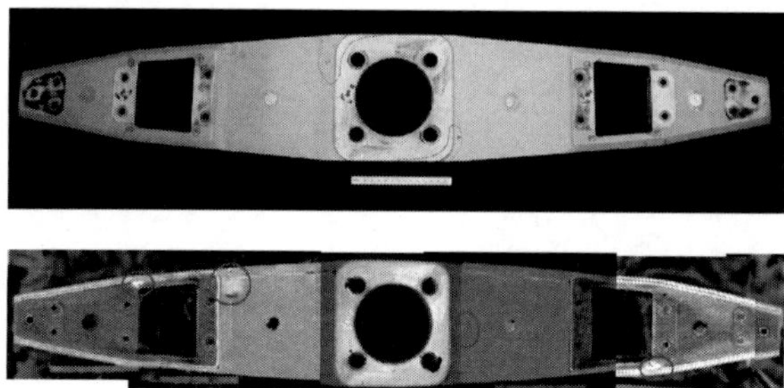

Figure 9.18 (Top): Photograph of a helicopter tail rotor half hub made of carbon fiber strap and fiber-glass filler core attached to metallic fittings and bushings. (Bottom): Corresponding pulsed thermography inspection results showing possible damage sites. Identified damage included disbonding of metal parts from the composite sections.

9.3.3 PULSED PHASE THERMOGRAPHY

Pulsed phase thermography (PPT) is an extension of the regular pulsed thermography that uses discrete Fourier transformation (DFT) to convert the time-domain data to the frequency-domain using:

$$F_n = \sum_{k=0}^{N-1} T(k)e^{-2\pi i k n/N} = \mathrm{Re}_n + i\,\mathrm{Im}_n \tag{9-3}$$

where $T(k)e^{-2\pi i k n/N}$ and F_n represent the temperature signals in time and frequency domains, respectively. In this equation, n designates the frequency increments ($n = 0, 1,...N\text{-}1$), and Re and Im are the real and the imaginary parts of the DFT. Once the data is converted to the Fourier domain, the phase (φ) and amplitude (A) images at different frequencies can be obtained analytically using:

$$A_n = \sqrt{\mathrm{Re}_n^{\,2} + \mathrm{Im}_n^{\,2}} \quad \text{and} \quad \varphi_n = \tan^{-1}\left(\frac{\mathrm{Im}_n}{\mathrm{Re}_n}\right) \tag{9-4}$$

The signal phase usually is less affected by the experimental variations, such as environmental reflections, emissivity changes, heating non-uniformity, and the object geometry or orientation [11]. Thus, this approach is expected to be more suitable for obtaining quantitative information about defects than the conventional time-base pulsed thermography.

9.3.3.1 APPLICATION EXAMPLES

The above-mentioned honeycomb sandwich panel and the aircraft rudder were tested using a pulsed phase thermography system available at Laval University. The test results showed a lower sensitivity to minor discontinuities as compared to the pulsed thermography, both in the reference panel, as well as in the rudder.

The lower sensitivity seen in the PPT results may be due to the differences in thermography system characteristics and capabilities. The pulsed phase tests at Laval University were done using a home-made thermography system equipped with a spectral analysis (i.e., amplitude vs. frequency) capability, as opposed to a commercial system capable of amplitude-time analysis that was employed for PT inspections at IAR. If the same thermography system is used for both PT and PPT, the frequency analysis generally provides additional useful information that is not available through time analysis alone.

9.3.3.2 Capabilities and Limitations

Capabilities:
- Same as pulsed thermography but may be less sensitive to minor discontinuities.
- Images are less affected by the object surface condition or environment.
- Provides additional phase information that may be useful in flaw sizing and depth measurement.

Limitations:
- Same as pulse thermography but requires additional signal analysis.
- Not as fast as pulsed thermography due to additional signal processing.

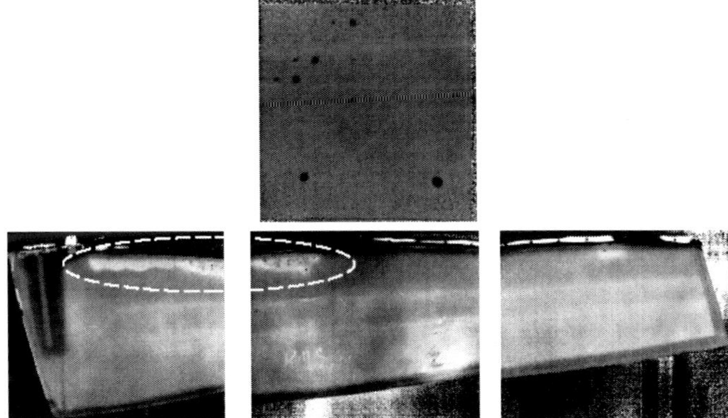

Figure 9.19 Pulsed phased thermography results of the above-mentioned reference panel and the aircraft rudder. Note that the sensitivity, both in the reference sample and the actual rudder, is not as good as the PT. Large folded PTFE inserts and impact damages, as well as areas with different coating thickness are identified with PPT, similar to PT, but other simulated defects or the internal features of the rudder are not clearly visible.

9.3.4 LOCK-IN THERMOGRAPHY

In lock-in thermography (LT), the thermal energy is supplied to the test part using a periodic excitation as opposed to a short pulse. The duration of the thermal excitation varies from a few seconds to a few minutes depending on the material properties and the object thickness. The penetration depth (μ) of the thermal waves is dependent on the excitation period (ω) and the material's thermal diffusion rate (α) according to:

$$\mu = \sqrt{\frac{2 \cdot \alpha}{\omega}} \qquad (9\text{-}5)$$

The thermal waves returning to the surface from the internal features of the component will interact and modify the incident thermal waves. Thus, by monitoring the variation of amplitude and phase between the time-dependent excitation signal and the specimen surface temperature, information about the internal features are obtained. For computing the amplitude and phase variation, several methods can be applied, including the standard lock-in method, Fourier transform, and four-point (or 4-bucket) algorithm [14]. Using the latter, the amplitude (A) and the phase (φ) can be computed using the following equations:

$$\varphi = \mathrm{atan}\,\frac{S1 - S3}{S2 - S4} \quad \text{and} \quad A = \sqrt{(S1 - S3)^2 + (S2 - S4)^2} \qquad (9\text{-}6)$$

where $S1$ to $S4$ are the four equidistant temperature data points, as illustrated in Figure 9.20. If necessary, more points can be used to reduce noise and increase sensitivity. Fourier analysis is then performed at each pixel of the modulated thermal waves, providing the magnitude and phase of the local responses. The magnitude and phase quantities are used to generate the thermal images. The magnitude image is more affected by variations in the illumination, surface absorption, and emissivity than the phase image. Thus, the phase image is particularly useful for

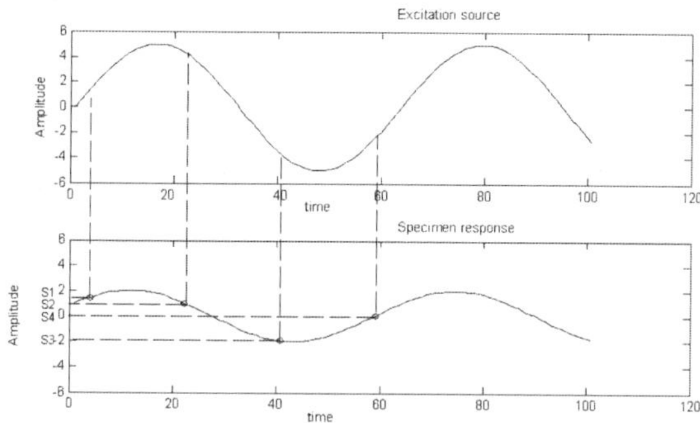

Figure 9.20 Principle of phase and amplitude computation in lock-in thermography.

inspecting parts that are difficult to illuminate uniformly (e.g., large structures) or, have different heat absorption and emissivity (e.g., complex or multi-material parts).

Figure 9.21 illustrates the basic differences between pulsed and lock-in thermography approaches. Like pulsed thermography, heating lamps or laser beams are used as excitation sources; however, the heat waves are periodic and sinusoidal with the excitation time much longer than in pulsed thermography. In terms of signal analysis, in LT, the local changes of the phase angle or phase shift as compared to the input waves are measured for defective and defect-free segments, revealing the defects. Thus, in LT, it is necessary to have data from both the defective and the intact reference points at the same time to enable comparison.

9.3.4.1 Application Examples

Lock-in thermography tests were carried out on the above-mentioned panel and the rudder using the same system as that employed for PPT (Figure 9.16 and 9.17). The results are presented in Figure 9.22, illustrating the lower sensitivity of this approach to embedded discontinuities and defects in honeycomb parts, as compared to both PT and PPT.

9.3.4.2 Capabilities and Limitations

Capabilities:
- Same as pulsed thermography but less sensitive.
- May have better depth of penetration as input energy is less limited.
- Has potential for large structures or complex parts.

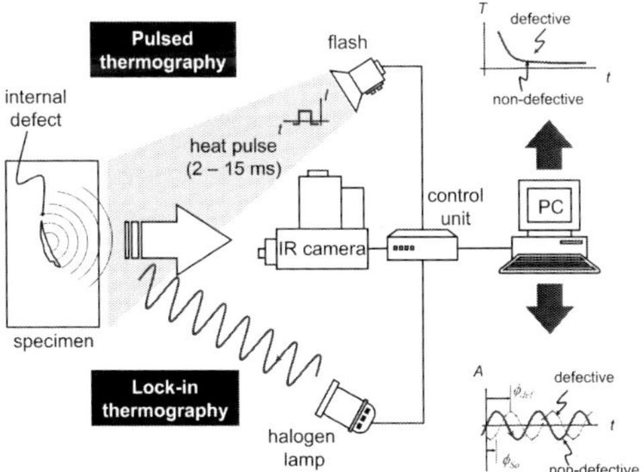

Figure 9.21 Schematic presentation of the differences between pulsed and lock-in thermography approaches, both in the shape of excitation heat waves and the type of processing [12].

Figure 9.22 Lock-in thermography results of the above-mentioned reference panel and the aircraft rudder. Again, large folded PTFE inserts and impact damages are visible, but other simulated defects or inherent features of the rudder are not easily seen.

Limitations:
- Same as pulsed thermography but more complicated signal and image processing.
- An identical reference must be inspected at the same time as the actual defective part for comparison.
- For the same amount of input energy, the sensitivity is less than for both pulsed and pulsed phase thermography methods.

9.3.5 *Vibro-thermography*

Vibro-thermography (VT) or thermosonic thermography uses mechanical vibration to generate heat at crack-type defects. Since defects differ from their surroundings by their mechanical weakness, they are usually sites of stress concentration. In case of cracks in metals, under periodical loading, stress concentrations occur at crack tips, resulting in localized plastic deformation. The energy released during deformation with the friction of the intimate crack surfaces, generates heat at the crack site. The heat generation occurs by the conversion of the mechanical energy to thermal energy. The amount of heat generated is dependent on the material, as well as the vibration frequency and amplitude. Optimal heat generation often occurs at the resonance modes of the test sample. Frequency sweeping is used to find optimal frequencies for a given sample.

Unlike other thermography methods that rely on an external heat source, in VT, the source of the heat is the defect itself, thus, eliminating the effects of non-uniform heating of the part. However, since mechanical vibration is used, there is always the possibility of introducing more damage to the object. Mechanical vibration devices, such as shakers or high-power ultrasonic transducers in the kHz range (such as those used for welding plastics), are often employed to create vibrations. Like other thermography methods, the generated heat is detected using an IR camera. Vibro-thermography is a useful research tool and has been used in a number of applications; examples are provided in the following section.

9.3.5.1 APPLICATION EXAMPLES

Figure 9.23 schematically illustrates the vibro-thermography setup used at IAR. Here the specimen is attached to a mechanical shaker in one end to vibrate it at its resonance frequencies. If an ultrasonic transducer is used, it must be coupled to the sample using a coupling medium (e.g., cork, leather, card board, etc.) in order to provide impedance matching and protect the surface of the sample.

Vibro-thermography was used at IAR in a number of applications, including monitoring of heat generated during vibration of stainless steel compact tension (CT) specimens [15]. This work was carried out to study the mechanism of heat generation in vibro-thermography. In this study, thermo-mechanical finite element (FE) modeling was used to predict heat generation during vibration, and experimental vibro-thermography tests were carried out to verify the modeling results. Examples of the modeling and experimental results on a simple CT specimen containing a fatigue crack are illustrated in Figure 9.24. The specimen was vibrated using an electromagnetic/piezoelectric dual shaker which had a frequency range of 10 Hz to 20 kHz and produced forces of a magnitude up to 1780N.

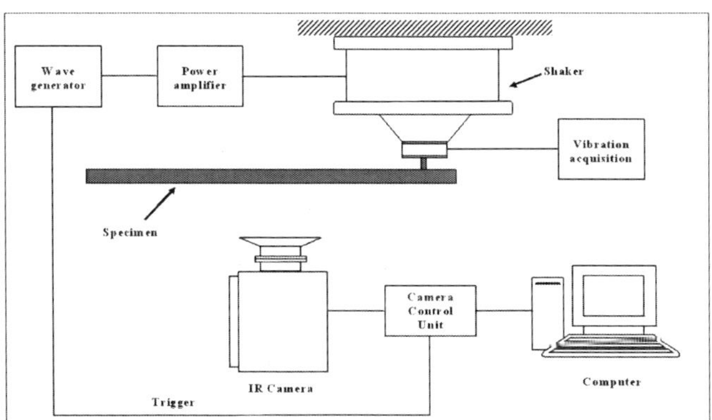

Figure 9.23 Schematic diagram showing a typical vibro-thermography system.

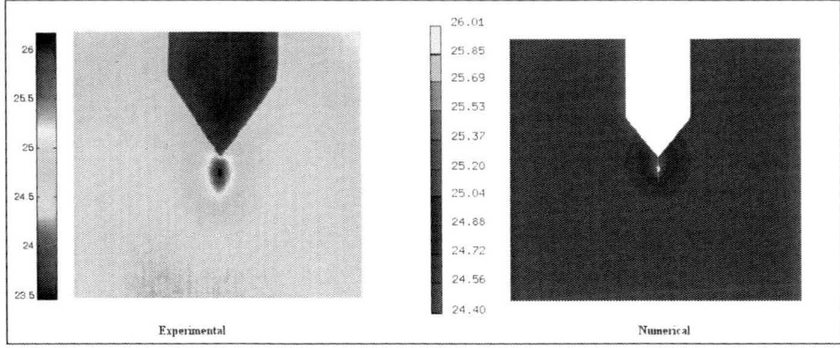

Figure 9.24 Surface temperature distribution in a compact tension steel specimen vibrating at 192 Hz. (Right): Numerical predictions and (left): experimental results.

Both the modeling and experiments revealed that the optimal heat generation was achieved at the specimen natural resonance frequencies. As seen in Figure 9.24, both modeling and experimental measurements clearly show that the maximum heat is generated at the middle of the crack rather than the crack tip. This indicates that at resonance frequencies of the CT specimens, the main source of heat is friction between crack faces, rather than crack tip plastic deformation. The numerically calculated stress distribution in the CT specimens for the experimental conditions employed here indicated that, even at the specimen's natural resonance frequencies and maximum vibration magnitude of the instrument, the stresses at the crack tip for all active frequencies were less than the material's yield stress, making the tests non-destructive. Further details on this work will be provided later, in the chapter entitled NDT Modeling.

The above mentioned reference sample and the rudder were tested using a VT system available at Laval University. Only large (>10 mm × 10 mm) two-layer simulated defects were identified in the test sample, as seen in Figure 9.25. Single layer inserts could not be detected, as there are no free surfaces to rub against each other. Also, smaller double-layer flaws could not be identified due to their small free-surface areas and, thus, lower frictional heat. Generally, the larger the flaw size, the greater the amount of energy that is converted into heat and the higher the likelihood of being detected by VT. Vibro-thermography cannot reveal water ingress into honeycomb since friction cannot be created in the presence of water, which acts like a lubricant. However, this approach identified a few delamination-type damage sites at the edges of the rudder that were missed by the other methods due to edge scattering effects (Figure 9.26).

9.3.5.2 CAPABILITIES AND LIMITATIONS

Capabilities:
- No external heating is necessary since defects act as heat sources.
- Has potential to detect tight cracks that may not be detectable by other NDT methods.
- Is capable of detecting edge delamination that may be missed by other NDT methods.

Limitations:
- Requires vibration of the sample at proper resonance frequencies to create adequate heat.
- Attachment of the vibration transducer to the specimen is necessary.
- There is a possibility of additional damage to the test object if improper vibration forces and frequencies are used.
- Only large discontinuities that have two intimate free surfaces may be identified.
- Foreign material inclusions with no free-surfaces or intrusion of fluids into sandwich parts cannot be identified.

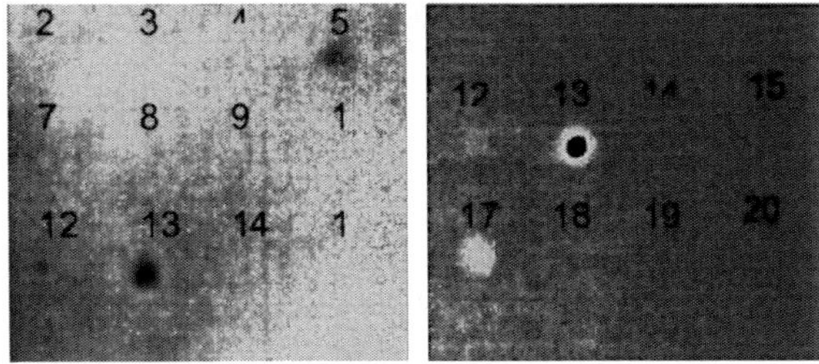

Figure 9.25 Vibro-thermography inspection results of the top left corner area of the above reference sample containing simulated defects. Only defects larger than 10 mm × 10 mm are detected. Different harmonic frequencies were needed to identify these defects.

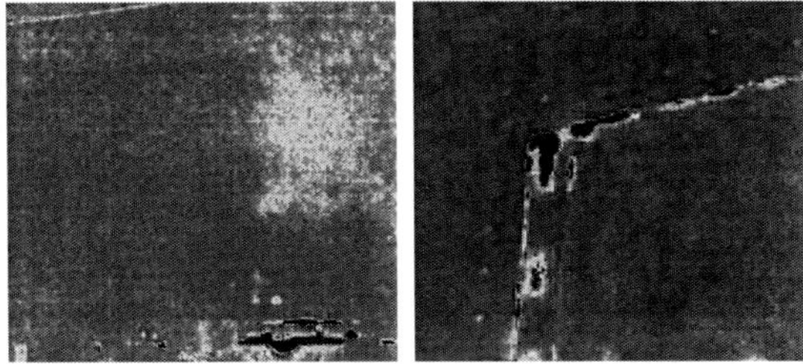

Figure 9.26 Vibro-thermography results of segments of the rudder which contained edge delamination.

9.4 THERMOGRAPHY EQUIPMENT

Typical infrared thermal imaging equipment that can be used for both active and passive thermography may consist of an illumination head, a control system, power supplies, and a computer with appropriate data acquisition, analysis, and graphic software. The illumination head is usually a closed hood in which the heating source (e.g., xenon flash lamps of ~2500 J power) and reflectors are located. An infrared camera is mounted on the top of the hood, and a door is installed to allow access to the hood for sample movements and adjustments. The control system communicates with the computer and the IR camera during the flash heating and thermal measurements. A data acquisition card installed on the computer acquires digital temperature data and a thermal analysis, and graphic software performs analysis, processing, displaying, and recording of the acquired data. A monitor allows real-time viewing of live images taken by the IR camera. A photograph of the IAR thermal imaging equipment (Echo Therm manufactured by Thermal Wave Imaging Inc.) is shown in Figure 9.27 as an example.

The infrared camera installed on this system is a ThermaCAMTM SC3000 IR camera manufactured by Flir Systems Inc. (thermal sensitivity of 20mK at 303K and a spectral response from 8.0 to 8.8 μm). The IAR thermal imaging equipment can be utilized to perform passive or active thermography using pulsed, pulsed phase, or vibro-thermography approaches.

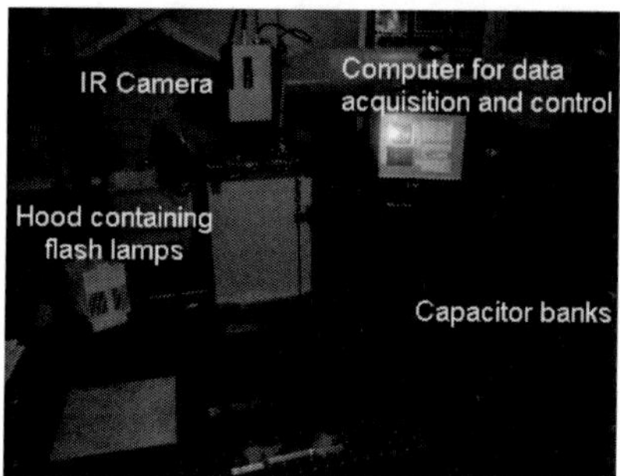

Figure 9.27 A photograph of a commercial thermography system.

9.5 CHAPTER REFERENCES

[1]   Introduction to Non-destructive Testing, NDT Resource Centre, Center for Non-destructive Evaluation, Iowa State University, Ames, Iowa 50011, USA, http://www.cnde.iastate.edu/, 2011.

[2]   Metals Handbook, Non-destructive Evaluation and Quality Control, Ninth Edition, Volume 17, ASM International, Metals Park, OH, USA, 44073, 1989.

[3]   Theory and Practice of Infrared Technology for Nondestructive Testing, X. P.V. Maldague, John Wiley & Sons Inc, New York, NY, 2001.

[4]   Heat Transfer, J.P. Holman, 9th Edition, McGraw-Hill. ISBN 0070296391, 2002.

[5]   The Engineering Tool Box, Emissivity Coefficients of Some Common Materials, www.engineeringtoolbox.com., 2011.

[6]   Infrared Camera Incorporated website: www.infraredcamerasinc.com/images/14c.jpg

[7]   http://www.popsci.com/military-aviation-amp-space/article/2009-09/thermal-imaging-spots-hot-shuttle-landing.

[8]   Comparison of Thermographic Inspection Techniques for Non-destructive Evaluation of CF-18 Rudders, M. Genest, M. Brothers, A. Fahr, A. Bendada, S. Guibert, C. Ibarra-Castanedo, X. Maldague, J-M. Piau, M. Susa, B. Tang. NRC Report: LTR-SMPL-2007-0037, 2007.

[9]   Infrared Thermal Imaging as a Diagnostic Tool for Gas Turbine Engine Faults, J.D. MacLeod, et al. ASME 94-GT-344, 1994.

[10] Machining Compacted Graphite Iron, CIM- Canada's Metalworking and Fabricating Technologies Magazine, September 1, 2010, http://www.cimindustry.com/article/chipping/automotive-work.

[11] Quantitative Subsurface Defect Evaluation by Pulsed Phase Thermography: Depth Retrieval with the Phase, C. Ibarra-Castanedo, Ph.D. Thesis, Laval University, 2005. http://www.theses.ulaval.ca/2005/23016/23016.pdf

[12] Comparative Study of Active Thermography Techniques for the Non-destructive Evaluation of Honeycomb Structures, Short Title: "Active Thermography for the NDE of Honeycomb Structures", C. Ibarra-Castanedo, J-M. Piau, S. Guilbert, N. Avdelidis, C-K Jen, M. Genest, A. Bendada and X. P. V. Maldague, Laval University Publication, Quebec City, Canada. G1K 7P4, 2007. Contact: IbarraC@gel.ulaval.ca

[13] Infrared Thermography Inspection of CH149 Tail Rotor Half Hubs, M. Genest, D.S. Forsyth and A. Fahr, NRC Report: LM-SMPL-2004-0249, 2004.

[14] Thermal Wave Imaging With Phase Sensitive Modulated Thermography, G. Busse, D. Wu, and W. Karpen, Journal of Applied Physics, No.71, Issue 8, April, 1992.

[15] NDE of Foam Core Sandwich Structures by Thermography, M. Genest and D.S. Forsyth, Proceedings of CANCOM conference, Canadian Association of Composite Structures and Materials, 2005.

[16] Frictional Heating Model for Efficient Use of Vibro-thermography, F. Mabrouki, M. Thomas, M. Genest, and A. Fahr, NRC Report: JA-SMPL-2008-0123, 2008.

CHAPTER 10

Radiography

10.1 PRINCIPLES

In radiographic testing (RT), radiation is passed through the test object and is projected onto a recording medium (e.g., film) to produce an image of the interior of the object, as shown in Figure 10.1. Either electromagnetic rays of very short wavelength (x-rays and gamma rays) or particulate radiation (neutrons) are employed. The use of such energy sources requires stringent safety precautions when performing radiographic tests. The radiation is partly absorbed by the test object, and this absorption is affected by changes in section thickness, as well as internal features of the component, such as holes, attachments, fasteners, or welds, as well as defects, such as pores, cavities, inclusions, or cracks, thus, producing differences in the transmitted energy, which are recorded on films or other recording media. All radiographic techniques can perform full-volume inspections for

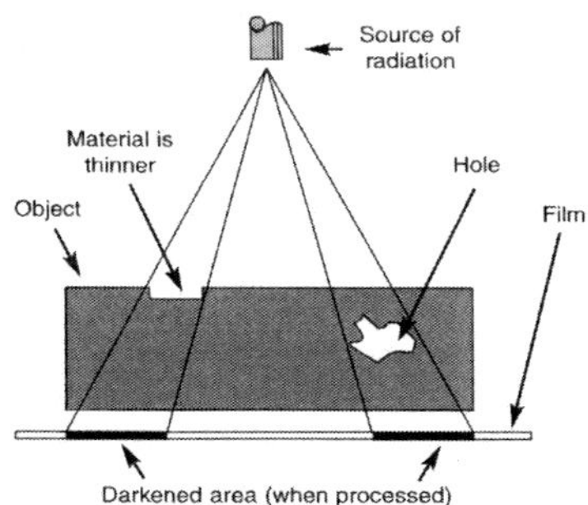

Figure 10.1 Schematic presentation of x-ray film radiography [1].

internal discontinuities that may be inaccessible with other techniques; however, access to both sides of the part is often necessary (except for the Compton back-scattering method).

Radiographic methods can be grouped into different types depending on the source of radiation (e.g., x-rays, gamma rays, neutron particles, etc.), the means of recording radiographic images (e.g., film, paper, digital detector arrays, and real-time radioscopy), or the way the images are presented (e.g., two-dimensional radiographs whose dimensional images are produced by computed tomography). Generally, in radiography, the aim is to produce an image that shows the necessary internal features of the object as clearly as possible. This requires careful control of a number of variables that can affect image quality, as defined in the following section.

10.2 RADIOGRAPHIC IMAGE QUALITY

Since film radiography has a long history and is still being used more than other recording media, the characteristics that determine the image quality are described below in the context of film radiography. However, most of the definitions are generally applicable to other recording media, too.

10.2.1 IMAGE DENSITY

In film radiography, image density or film density is a measure of the degree of film darkening and is related to the logarithm of the ratio of the intensity of light incident on the film and the intensity of light transmitted through the film. A density reading of 2.0 is equivalent to 1% of the incident light passing through the film, while a density of 4.0 corresponds to 0.01% of transmitted light reaching the far side of the film.

Industrial radiography typically requires a radiograph to have a density between 2.0 and 4.0 for acceptable viewing with common film viewers. For a density above 4.0, extremely bright lights are necessary to view the film. Film density is measured with a photo-electric sensor called a densitometer. This device measures the amount of light transmitted through the film when it is placed between the light source and the sensor.

10.2.2 SENSITIVITY

Radiographic sensitivity is a measure of the quality of an image and is defined in terms of the smallest detail or discontinuity that can be detected. Radiographic sensitivity is dependent on the combined effects of two independent sets of variables. One set of variables affects the contrast, and the other set of variables affects the definition of the image.

10.2.3 IMAGE CONTRAST

Radiographic image contrast is defined as the degree of image density difference (e.g., transition from dark to bright in percent) between two regions of a radiograph that makes it possible to distinguish features of interest, such as defects, from the surrounding area. Radiographic image contrast is affected by the wavelength of the radiation, scattering of nearby objects, or the sharp edges of the test piece, and the absorption differences within the test piece, as well as the film type (grain size, density) and processing (chemicals, development time, agitation). The intensity of radiation transmitted by various portions of the test piece depends on the thickness, shape, and composition, as well as the presence of discontinuities.

10.2.4 IMAGE DEFINITION

Radiographic image definition is defined by the abruptness of the change of density from one area to another. Like contrast, definition also makes it possible to see features of interest, such as defects, but in a different way. Image definition is affected by the radiation wavelength, source size, source to film distance, object to film distance, object thickness change, and object movement, as well as the film type and processing.

10.2.5 GEOMETRIC SHARPNESS

In radiography, loss of geometric sharpness, or unsharpness, refers to the loss of definition that may occur as a result of three main factors: source size, source to object distance, and object to detector distance, as schematically illustrated in Figure 10.2. As seen in this figure, when the source size is very small, all of the radiation essentially originates from the same point and very little geometric unsharpness is produced in the image. This is the case in micro-focused radiography. In contrast, when the source size is large, due to the different paths that the radiation can take, the edges of the discontinuities become less defined.

Generally, in industrial radiography, the allowable geometric unsharpness is 1/100 of the thickness of the material up to 1 mm (0.04 in). This is referred to as the degree of penumbra shadow (Ug) in the radiographic image (refer to Figure 10.3). Since it is often difficult to measure the penumbra shadow in a radiograph, it is calculated using the following expression:

$$\mathrm{Ug} = f \cdot b/a \qquad (10\text{-}1)$$

where f is the focal spot size provided by the x-ray source manufacturer, a is the distance from the source to the front face of the object, and b is the distance between the front face of the object and the film. When the film is placed directly

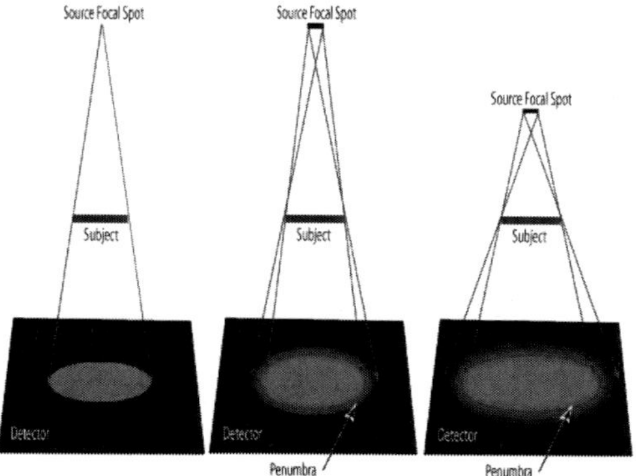

Figure 10.2 Schematic presentation of geometric unsharpness.

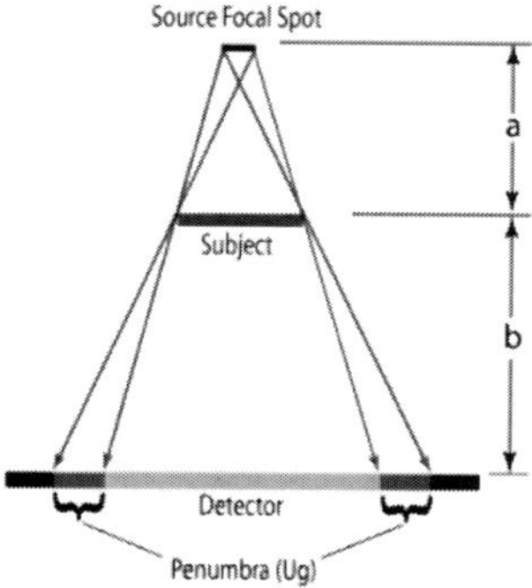

Figure 10.3 Schematic presentation for calculation of geometric unsharpness.

below the object, b is the object thickness. These measurements are made before the equipment is setup and before performing the x-ray inspections.

In addition to the geometric factors mentioned above, the beam angle also affects the image sharpness, as shown graphically in Figure 10.4. Also, any movement of the source, the object, or the detector, as well as abrupt change of the subject thickness will affect sharpness and image definition. Additionally, the source wavelength, detector, or film type (sensitivity) and film processing time are important to achieve a desirable definition.

The image quality is established by exposure of a standard image quality indicator (IQI) along with the test piece during radiographic inspections. The ASTM E747 standard x-ray image quality indicators (also known as penetrameters) consist of a set of six wires arranged in order of increasing diameter and encapsulated between two sheets of clear plastic (see Figure 10.5). They are commercially available in different sizes and materials (aluminum, copper, stainless steel, Inconel, and titanium). For image quality calibration, an appropriate IQI is placed

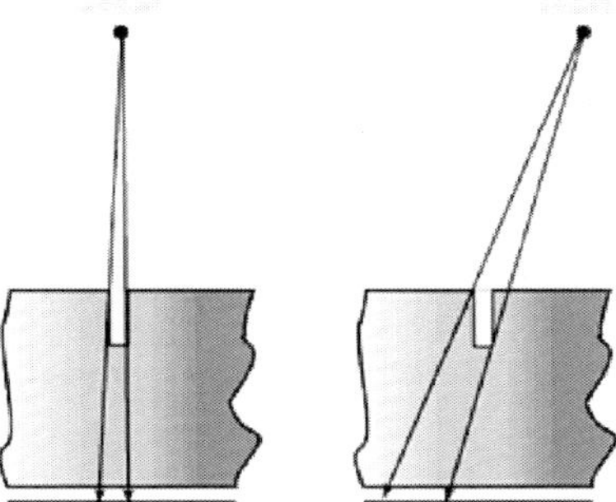

Figure 10.4 Effect of beam angle on radiographic image sharpness. (Left): Right way resulting in a well-defined image. (Right): Improper way resulting in distorted and less-defined image.

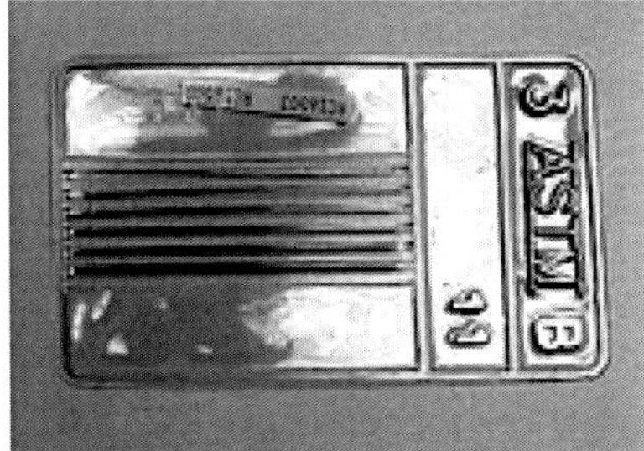

Figure 10.5 The ASTM- E747 standard image quality indicators.

on the source side of the part either over a section that has a uniform thickness or on a block of similar material and thickness to the region of interest.

The image sharpness is improved by using a small focal size source and beams as close to perpendicular to the part as possible. In conventional radiography, the resolution can be improved by using high contrast films placed as close as possible to the component, and by optimizing the orientation of the object with respect to the source.

Scattered radiation is a source of unsharpness and must be taken into consideration when producing a radiograph. Often, radiation is scattered from objects in the immediate area of the test piece or from where the part is resting. Side scatter can be reduced by moving the unnecessary objects away from the film or by placing a collimator at the exit port of the beam, reducing the diverging radiation surrounding the central beam.

In this chapter, the radiographic methods are described on the basis of the radiation source type and the use of recording or imaging approaches with each type.

10.3 X-RAY TECHNIQUES

10.3.1 PRINCIPLES

X-rays are produced when fast-moving electrons collide with a target material (usually tungsten). The essential principle of x-radiography is that radiation penetrates light materials better than dense ones, as the x-ray absorption increases with the materials' atomic number. Generally, most solid materials, except for those with a very high atomic number or density (e.g., lead) and those with a very low density (some plastics), can be inspected by x-rays.

As x-rays pass through a test object containing an internal discontinuity, different amounts of radiation are absorbed locally due to density difference between the discontinuity and the bulk material; this is traced on the recording media. In addition to flaws, internal geometrical features and variations in thickness or composition will produce differences in transmitted radiation, which can be recorded. X-rays have wavelengths in the range of 0.01 to 10 nm and energies in the range 100 eV to 1 MeV [2]. X-rays are shorter in wavelength than UV rays and longer than gamma rays. There are three principle controls to a standard x-ray system: the input current control (mA), the input voltage control (keV), and the time control (minute). The first two determine the radiation intensity and the last controls the exposure time. Changing any of these will affect the characteristics of the radiograph.

10.3.2 EQUIPMENT

10.3.2.1 X-RAY TUBES

X-ray equipment comes in a variety of configurations and sizes. Portable units can be easily moved to the inspection site for field use, while stationary units are mainly intended for use in production or laboratory environments (Figure 10.6). X-ray tubes are available in a wide range of energy levels. For large steel or heavy metal components, systems capable of producing millions of electron volts (MeV) may be necessary in order to penetrate the full thickness of the material. On the other hand, small, lightweight components may require a system capable of producing only a few keV. The major components of an x-ray system are the generator tube, the high voltage generator, the control console, and the cooling system.

The effective focal spot size for typical x-ray tubes is in the range of 1 mm × 1 mm to 4 mm × 4 mm while for micro-focused x-ray systems can be 10 μm × 10 μm to 0.5 mm × 0.5 mm. Micro-focused x-ray tubes with very small focal spot sizes produce high resolution images, but they cannot generate high energies that are often required for most aircraft components. Nevertheless, they are very useful for inspecting small components, such as turbine blades, or electronic devices where image enlargement is necessary. The use of a micro-focused source allows

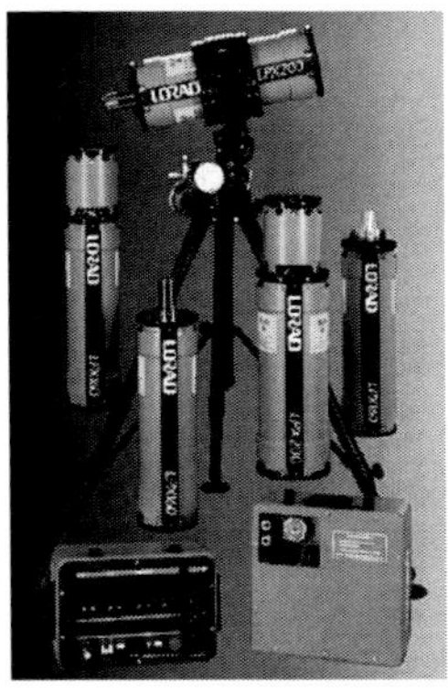

Figure 10.6 Examples of portable and stationary x-ray systems. Pictures curtsey of Lorad Industrial Imaging and PAR Systems.

a greater distance between the object and the detector, resulting in larger images without compromising the resolution.

10.3.2.2 IMAGING MEDIA

Films, radiation sensitive papers, or electronic media may be used to capture images or record the data. When an electronic medium is used, results can be displayed in real time on a monitor.

Film Radiography

In conventional film radiography, photographic films are used for imaging the interior of the test object. X-ray images are variations in film densities that result from the interaction of the radiation photons with the cellulose acetate molecules of the film emulsion. Both the radiation intensity reaching the film or applied exposure and the film characteristics affect the film density.

Film Characteristic Curves

Film Characteristic Curves determine the relationship between the applied exposure and the resulting film density. Film manufacturers commonly characterize their films and provide such curves to users, as different types of radiographic films respond differently to a given amount of exposure. The film characteristic curves are plots of film density in relation to the log of relative exposure (Figure 10.7). Relative exposure is the ratio of two exposures.

For example, if one film is exposed at 100 keV for 6 mA minute and a second film is exposed at the same energy for 3 mA minute, then the relative exposure would be 2. Figure 10.7 shows three film characteristic curves with the log relative exposure plotted on a linear scale. Film characteristic curves are used to adjust the exposure in order to produce a radiograph with a certain density. The curves are also used to relate the exposure produced with one type of film to the exposure needed to produce a radiograph of the same density with another type of film.

Radiographic films generally deliver high-quality spatial resolution, which, in turn, results in high sensitivity. Because of this, film radiography is often used on castings or forged metal parts, such as airfoils, pumps, pump housings, gears, struts, or landing gear components, where very small discontinuities must be identified.

Overall, radiographic films provide high sensitivity and high resolution images and are particularly appropriate for identifying internal discontinuities [3]. In addition, films are lightweight and flexible, making them suitable for in-situ inspections of simple or complex parts that require the film to be in contact with the component surface for improved resolution. However, film radiography requires image development that is time-consuming and involves chemicals potentially harmful to the inspectors or the environment.

Digital Radiography

Digital radiography may be used to speed up the x-ray inspection and save on the cost of films and processing. There are different ways of capturing a digital radiographic image, and the most common is to use a specialized *flat panel* that converts the x-ray quanta directly into electrical charges. The flat panel is a fixed-sized x-ray detector that is positioned in the same place as the film to capture and relay the image directly to a computer. Figure 10.8 shows a typical flat panel and Table 10.1 provides specifications of two of such devices that are available

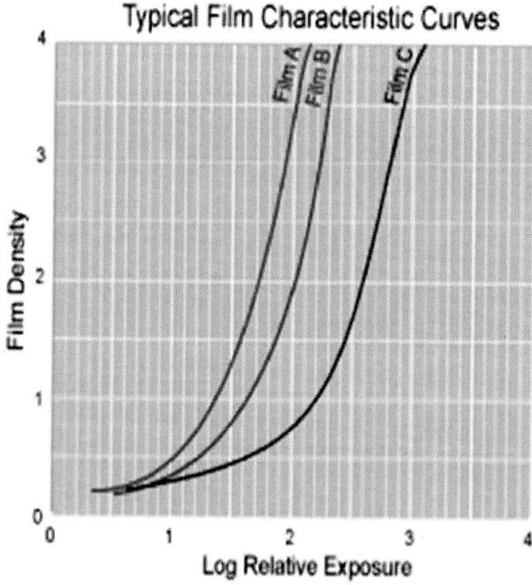

Figure 10.7 Typical film characteristic curves for exposures to achieve certain film density [1].

commercially [4]. More information on digital radiography and details of the related instruments are provided in references [5], [6], and [7].

Briefly, either direct or indirect conversion detectors may be used. The direct conversion detectors have a photoconductor panel that uses *amorphous selenium* (*a-Se*) to convert the absorbed x-ray photons directly into electric charges, which are further processed and displayed as an image immediately after the exposure. Indirect conversion detectors use a *scintillator* screen as an intermittent step for converting the absorbed x-ray energy into visible light, which is later transformed into electric charges by using an *amorphous silicon* (*a-Si*) photodiode array or a charge-coupled device (CCD). Figure 10.9 provides a schematic representation of the direct and indirect conversion mechanisms and the role that thin-film transistor arrays play in both types.

Dynamic Range

Dynamic range is a term used in digital radiography defined as the ratio between the maximum and minimum intensities of x-ray radiation providing identification of geometrical test target with a contrast of 5% [7]. Flat panels provide a larger dynamic range (e.g., 2000/1) as compared to films (e.g., 400/1). This is one of

Figure 10.8 Photograph of a digital X-ray flat panel.

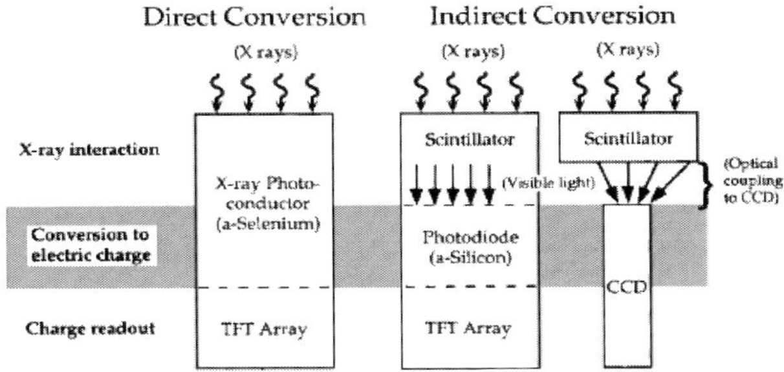

Figure 10.9 Direct and indirect conversion in digital radiography [5].

Table 10.1 Characteristics of two commercial digital X-ray flat panels (Supplied by FlashScan, www.thalesgroup.com, 2010).

Name	Number of Pixels	Pixel Size (μm)	Active Area (mm)	Maximum Spatial Resolution (lines/mm)	Typical Energy Range (keV)	Dynamic Range	Digital Output (bit)	Dimensions (cm)
FlashScan 23	1560 × 1560	143	223 × 223	3.5 @ 4% MTF*	25 to 160	2000:1	14	36 × 33 × 1.3
FlashScan 35	2240 × 3200	127	293 × 406	2 @ 40% MTF*	25 to 160	3500:1	14	50 × 36 × 6.5

*MTF stands for Modulation Transfer Function

the key advantages of digital radiography, as it requires less exposure dosage or shorter exposure times.

Using digital radiography enables one to capture and display the image in real-time and save it in digital format for enhancement or transmission to other sites. A disadvantage is poorer resolution in comparison with fine-grain films. However, the technology is improving with time, and the newer systems provide resolution close to film-radiography. In addition, the real-time display enables multiple exposures of the test object at different positions in a timely manner to achieve optimal sensitivity and resolution. Consequently, major drawbacks are their high initial cost, lower spatial resolution, rigidity, and the fact that they are fragile and cannot take x-ray energies higher than 350kV.

Fluoroscopy

Fluoroscopy is an imaging approach that shows a continuous real-time x-ray image on a monitor. It is also called radioscopy or real-time radiography. In fluoroscopy, an electronic image intensifier (see Figure 10.10) converts the x-ray photons into a visible image, which is then recorded by a video camera. The latter is connected to a frame grabber card that converts the captured analog image into a digital one for real-time display on a monitor. Figure 10.11 shows a schematic representation of x-ray fluoroscopy.

As described in references [8] and [9], the image intensifier is comprised of an "input phosphor scintillation" screen, a "photo cathode," and an "output phosphorous" screen. The x-ray photons strike the input phosphorous screen and are converted into light photons. A photocathode placed behind this screen absorbs the light and gives off electron beams that, after acceleration by a series of electronic lenses, are focused onto an output phosphorous screen. The latter yields an amplified light image, which is detected by a CCD video camera and sent to a monitor for viewing and analysis.

Figure 10.10 Example of a commercial image intensifier available.

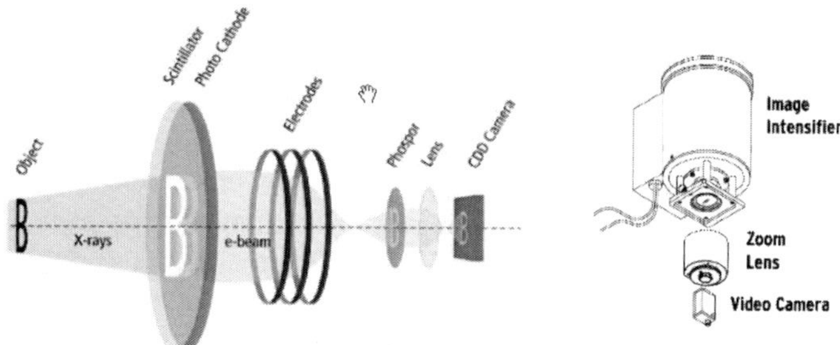

Figure 10.11 Schematic representation of X-ray fluoroscopy (www.phoenix-xray.com, 2010).

Unlike film radiography that provides a negative image of the object, images produced by fluoroscopy are positive, meaning that the brighter areas correspond to higher levels of transmitted radiation reaching the screen. Therefore, lighter or brighter areas in the display image represent thinner sections or areas with lower densities within the test objects (e.g., cavities and porosity). Like digital radiography, the real-time image acquisition and display of fluoroscopy enables the operator to change instrument settings or component position and see the effects immediately in order to optimize the process. While the approach is superior to film radiography in terms of speed and real-time capability, the initial costs associated with the image intensifier, capturing devices, and component manipulation system are high.

Reverse Geometry X-ray (RGX)

Reversed geometry x-ray technology was developed to achieve further improvement in contrast [10]. In this approach, shown schematically in Figure 10.12, the object is placed in front of a large television-like tube that is big enough to provide

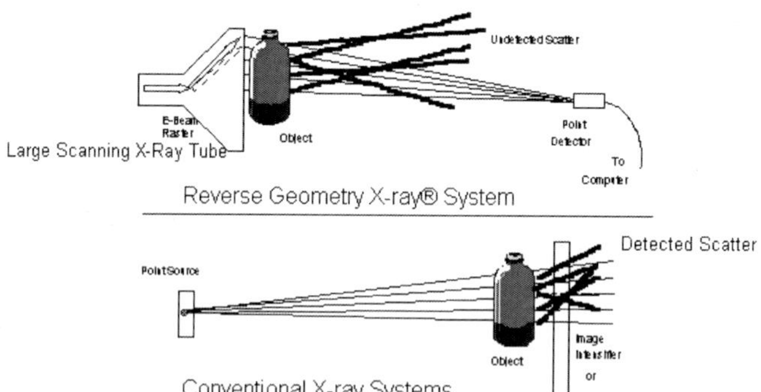

Figure 10.12 Schematic diagrams comparing the reverse geometry to the conventional X-ray.

exposure to the whole object placed very close to it, while a point detector is located several feet away to capture the transmitted rays. The tube contains a raster scanning X-ray source. In this arrangement, scattered rays that cause blurring of the image are bypassed by the point detector, thereby, producing higher resolution images and better contrast.

With reverse geometry x-ray imaging, alignment of the x-ray source, object, and detector are no longer very critical. X-rays are emitted over a wide angle; thus, there is more freedom in detector alignment and tilt. In this case, as long as x-rays can reach the detector, the computer can produce a radiographic image of the object.

10.3.3 APPLICATION EXAMPLES

X-ray radiography is a well-established technique that is widely used in aerospace to detect volumetric flaws in a variety of materials, and there are several ASTM Standards for radiographic inspections and image analysis (e.g., ASTM E 1030, E94, and E1742). In general, material discontinuities that result in an absorption difference of 1% or more can be identified using this method. It is used during manufacturing as a quality control tool to inspect metal castings for flaws, such as voids, cavities, porosity, foreign material inclusions, shrinkage cracks, etc. The inspection is done before the metal casts are cut and machined to avoid defective materials being turned into complex parts at high machining costs.

Reference [1] provides x-ray examples of manufacturing defects using film radiography that are illustrated in Figure 10.13 and Figure 10.14. Porosity and

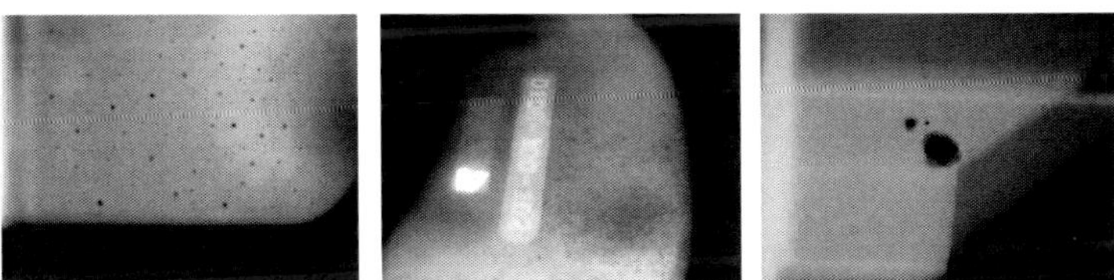

Figure 10.13 Radiographic images of distributed pores (left), localized porosity (middle), and isolated cavities (right) in metal castings.

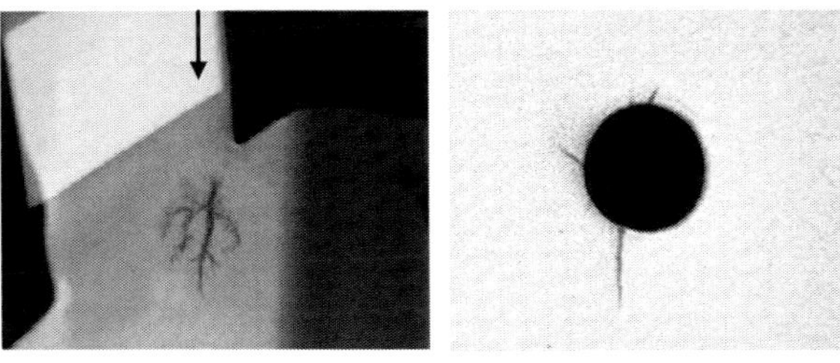

Figure 10.14 Radiographic images of shrinkage cracks in metal castings.

cavities are often caused when gases are entrapped during the casting process. Pores and cavities may also occur in the thicker mid-sections of a casting as a result of metal shrinkage. Shrinkage cavities and cracks occur as a result of the volumetric change of the molten metal as it solidifies to leave isolated cavities that cannot be filled due to the lack of feed of the supplementary melt. Such cavities or cracks are often irregular in shape (Figure 10.14 (left)) or branch out from larger voids (Figure 10.14 (right)). Internal stresses due to thermal gradients that occur during the metal cooling process may also result in casting cracks.

Non-metallic inclusions in metals that are lower or higher in density than the host alloy will appear on the film-radiographs darker or lighter, respectively. Such inclusions are often caused by small pieces of the mold material falling into the molten metal during the casting process or may be due to entrained slag. Figure 10.15 shows an inclusion with a higher density than the metal casting.

Radiography is also used during the service-life of the aircraft to detect damage due to fatigue, wear, tear, impact, corrosion, and other mechanisms. For the detection of fatigue cracks, it is important that the radiation beam is close to parallel to the plane of the crack in order to provide an adequate beam path to create sufficient contrast. Digital radiography and fluoroscopy which provide real-time imaging are useful for rapid identification of the proper component positioning with respect to the source beam. If the component cannot be x-rayed in the direction parallel to the crack plane, radiography should not be used and other NDT methods, such as ultrasonic or eddy current techniques, must be considered. In such a case, radiography and ultrasonic methods are complimentary to each other since the UT is most effective when the beam is normal to the surface of the crack and radiography is sensitive when the beam is parallel with crack plane.

Radiographic inspection of engine parts, such as turbine blades for small cracks or wear of leading edges, requires very high resolution systems. Reverse geometry x-ray is suitable for such applications, and Figure 10.16 provides examples of the RGX used on engine parts. Stress corrosion cracking of aluminum airframe components is also difficult to detect using the conventional x-rays but may be detectable by the RGX approach (e.g., Figure 10.17).

Radiographic inspection of aircraft structures is not possible unless access to both sides of the part is possible. For example, for radiographic inspection of

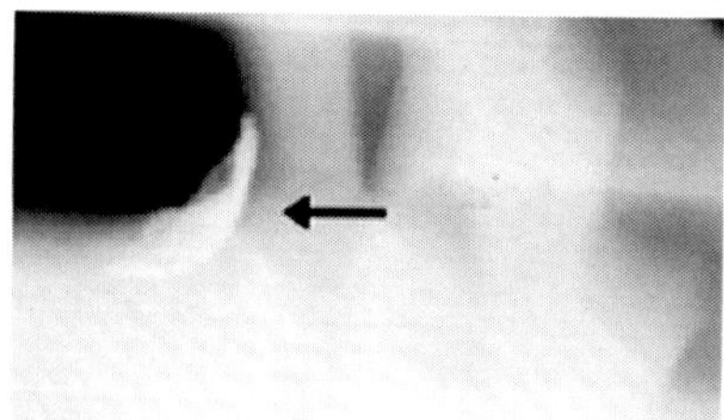

Figure 10.15 A radiographic image of non-metallic inclusion in a metal casting.

a fuselage lap-joint section from the outside of the aircraft, the interior of the aircraft corresponding to that site must be removed to allow access for film placement. Thus, radiography is mostly used in the laboratory or depot environment to inspect disassembled parts. Cracks originating from fastener holes and metal loss due to corrosion in lap-joint structures can be detected (see Figure 10.18 and Figure 10.19).

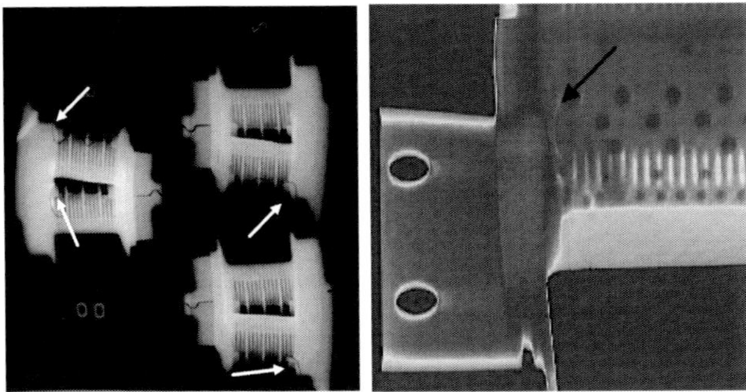

Figure 10.16 Reverse geometry x-ray images of aircraft engine parts showing wear at the edges identified by white arrows (left) and a fatigue crack identified by the black arrow (right) [10].

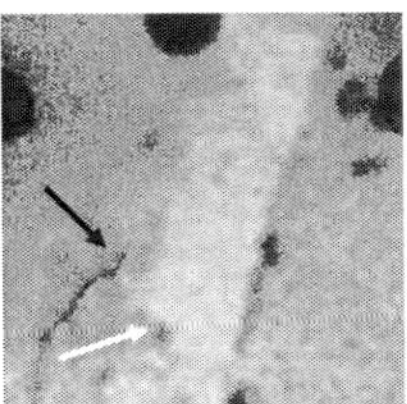

Figure 10.17 Reverse-geometry x-ray image of an aircraft aluminum skin section showing corrosion in blue and stress corrosion cracking identified by an arrow [10].

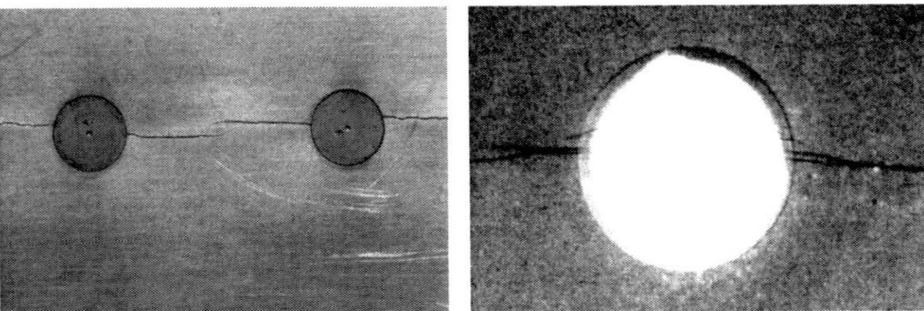

Figure 10.18 A photograph of typical cracks originating from fasteners holes of aluminum airframe lap joints (left) and an x-ray image of similar cracks (right). The steel rivet absorbs all the x-ray energy as indicated by the white circle.

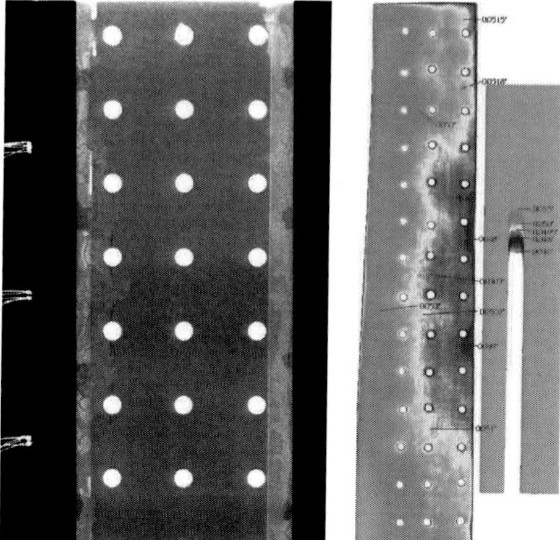

Figure 10.19 An x-ray image of an aluminum lap-joint with corrosion and cracks (left) and an x-ray map of the remaining thickness (right). Note that the thickness is determined by comparing the colors of the corroded site with those of a known wedge-shaped calibration sample that is x-rayed along with the test piece. Each color corresponds to a certain thickness range.

Metal corrosion by itself cannot be easily identified due to the confluence of the radiation absorption by the corrosion products and the remaining metal. However, in aluminum alloys, if a dual-energy source is used, it may be possible to distinguish between aluminum and the aluminum hydroxide that is produced as a result of corrosion. The different energies are properly selected such that each one provides optimal sensitivity to aluminum and its oxides.

The metal loss can be identified and mapped by x-rays after the lap-joint segments are disassembled and the corrosion products are removed. The measurement of the remaining thickness is carried out by comparing the x-ray intensity (represented by different colors) of the corroded site with those of a known wedge-shaped calibration sample that is X-rayed along with the test piece. More detailed description of this approach is provided in a separate chapter on aging aircraft NDT.

In aircraft composite parts, the x-ray method is used to detect impact damage in carbon fiber composite laminates or fiber metal laminates (Figure 10.20). For the x-ray mapping of delamination or disbonding that are a result of impact in these materials, it is necessary to have surface-breaking damage to allow ingress of a radio-opaque fluid (e.g., zinc iodide). If such a fluid is capable of penetrating the damage through the surface-breaking cracks, dents, or gaps, then high contrast images can be produced of planner defects that otherwise would not be detectable due to their small cross-sectional size. With the radio-opaque fluids, a larger density difference between the damage and the rest of the material is achieved. Similarly, water intrusion into honeycomb sandwich structures can be detected

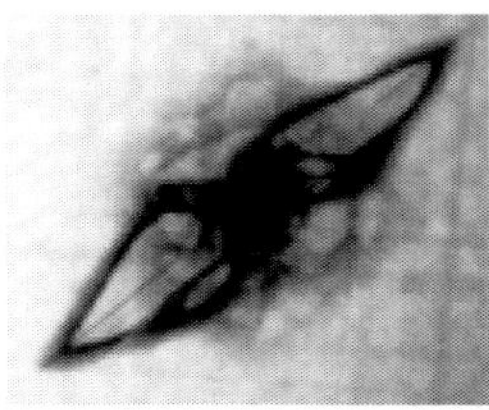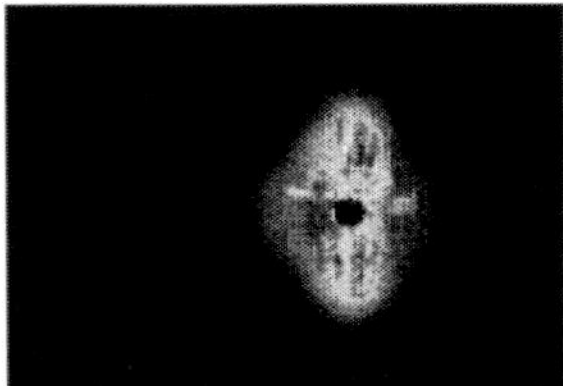

Figure 10.20 Radiographic images of impact damage in a graphite epoxy composite laminate (left), and impact damage in a fiber metal laminate (right).

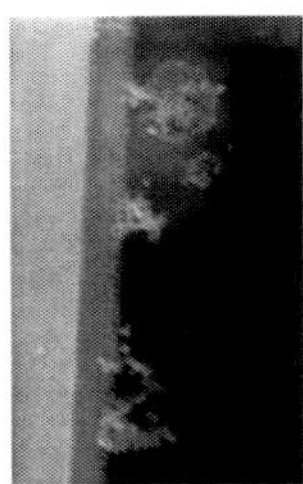

Figure 10.21 X-ray images of water intrusion into honeycomb sandwich parts. (Left): An experimental reference panel. (Right): A helicopter tail rotor blade [3].

using x-radiography, as illustrated in Figure 10.21. In this case, cells with water are identified due to the higher density of water as compared with the intact cells.

Finally, radiography is a useful tool for obtaining information about the internal design or different materials and configurations used in the construction of complex aircraft components. Such information may be needed for proper repair and maintenance. Some examples are provided in Figure 10.22.

10.3.4 Capabilities and Limitations of X-ray Techniques

Capabilities:
- Full volume inspection of a wide variety of materials.
- Detects and locates defects that have different density from the host material.
- Defect size in the plane normal to the beam direction can be obtained.
- Capable of inspecting complex parts for defects or internal information.
- Minimum part preparation is required.
- Provides a permanent image of defects.
- Suitable for after-manufacturing quality control and in-service inspections.

Limitations:
- Hazardous radiation, so human protection from exposure is essential.
- Stringent safety requirements impose economic and operational constraints.

- Relatively expensive in terms of equipment and film costs.
- Access to both sides of the test piece is needed.
- Beam direction may affect the results, and therefore, multiple exposures at different directions may be necessary.
- Cracks cannot be detected unless close to parallel to the beam direction.
- Delamination or disbonds cannot be detected unless they are surface connected and radio-opaque penetrants are used.
- Requires extensive operator training and skills.
- Focal spot size, exposure power, and duration are critical.
- Sensitivity decreases with increased part thickness.

10.4 GAMMA RAYS

10.4.1 PRINCIPLES

Gamma rays are produced by radioactive sources, such as iridium-192 and cobalt-60. These isotopes emit radiation in a few discreet wavelengths. Cobalt-60 emits 1.33 and 1.17 MeV gamma rays and iridium-192 emits 0.31, 0.47, and 0.60 MeV gamma rays. In comparison to an x-ray generator, cobalt-60 produces energies comparable to a 1.25 MeV x-ray system and iridium-192 to a 460 keV x-ray system [1, 11]. These high energies make it possible to penetrate thick materials with a relatively short exposure time; thus, gamma rays are best suited for inspection of thick sections. Like x-rays, the sensitivity of γ- rays varies with the

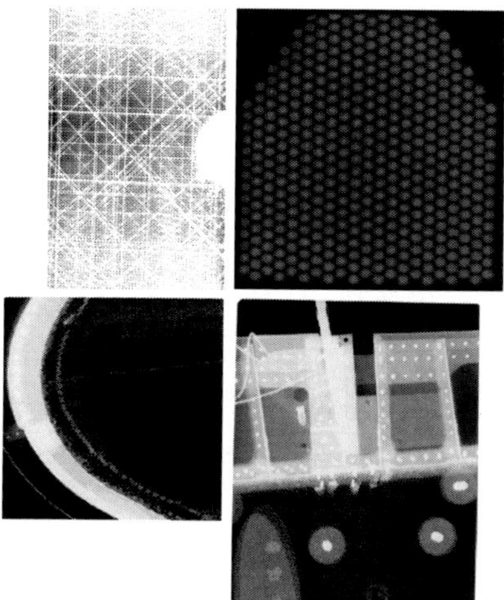

Figure 10.22 X-ray images showing a CFRP composite laminate layup (top left), the internal design of a body armor (top right), a segment of an aircraft rudder with metallic skin and honeycomb core (bottom left), and a segment of an aircraft structure with metallic ribs attached to a honeycomb section (bottom right).

material's atomic number, and, generally, most solid materials, except those of very high density (e.g., lead) or very low density (some plastics), can be inspected by gamma rays.

10.4.2 EQUIPMENT

The physical size of isotope materials varies between manufacturers, but, generally, an isotope material is in the form of a pellet that measures 1.5 mm × 1.5 mm. Depending on the level of activity desired, a pellet or pellets are loaded into a stainless steel capsule and sealed by welding. The capsule is attached to a short flexible cable and a connector that can be attached to a portable source projector and storage housing that shields radiation (Figure 10.23).

In terms of applications, γ-rays provide similar results to x-rays, but gamma rays have more power, while x-rays have a broader spectrum. The high energy of the gamma ray sources along with their small physical size and the fact that they are very portable makes them quite useful for field applications. However, the disadvantage of a radioactive source is that it can never be turned off, and the safe use of the source is very important and a constant responsibility.

10.4.3 CAPABILITIES AND LIMITATIONS OF GAMMA RAYS

Capabilities:
- Similar to x-rays.
- Very portable and useful for field applications.
- Thick sections can be inspected.

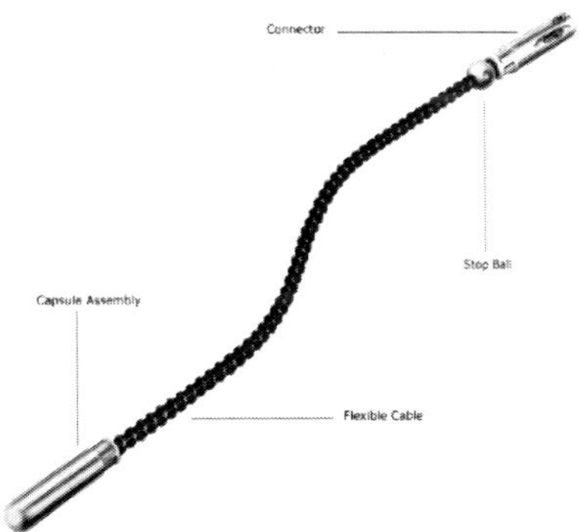

Figure 10.23 A commercial gamma ray capsule (left) and a source projector and storage device (right).

Limitations:

* Same as x-rays.
* Source cannot be turned off.
* Extremely high safety precautions are necessary.
* Source has a certain shelf-life.

10.5 NEUTRON RADIOGRAPHY

10.5.1 Principles

Neutron radiography employs particulate radiation called neutrons to form a radiographic image of an object. Neutrons are produced by nuclear reactors, accelerators, or certain radioactive isotopes and are sensitive to light elements (e.g., hydrogen). Thus, neutron radiography has very few applications in aerospace. A prominent application is the detection of water intrusion into honeycomb structures. However, parts must be removed and taken to the source of neutrons (reactor/accelerator site) for inspection.

Neutron radiography differs from conventional radiography in that the attenuation of neutrons is related to the specific isotopes present in the test material, rather than its density or atomic number. In contrast to x-rays or gamma rays that attenuate more with the increase of the mass number and density, with neutrons, attenuation is random. Certain light elements, such as hydrogen, have high neutron attenuation. Thus, the technique is capable of detecting problems due to hydrogenous media, such as moisture ingress, in aircraft honeycomb structures. Other practical applications include detection of explosives in metallic containers, inspection of rubber o-rings in metal assemblies, detection of residual ceramics in investment-cast turbine blades, and identification of minute amounts of corrosion in lap-joints by detecting hydrogen in the corrosion products.

Reference 12 describes the use of neutron radiography for the detection of moisture ingress in aluminum honeycomb and the associated cell corrosion in the CF-18 flight control surfaces (Figure 10.24). Generally, neutron radiography can be considered as a complementary approach to conventional radiography and is often used when conventional x-rays or other NDE methods are ineffective.

10.5.2 Equipment

The ASTM STP586-EB describes a Californium-based neutron radiography method for corrosion detection in aircraft [13]. This approach is particularly valuable where the corrosion is hidden behind thick metallic structural members. The neutron radiographic technique has been applied successfully to detect corrosion in many aircraft structures, including wing tanks, rear stabilators, aft spars, starboard and port wings, rudders, fuselage skin, rotary blades, tail flaps of helicopters, and landing gear.

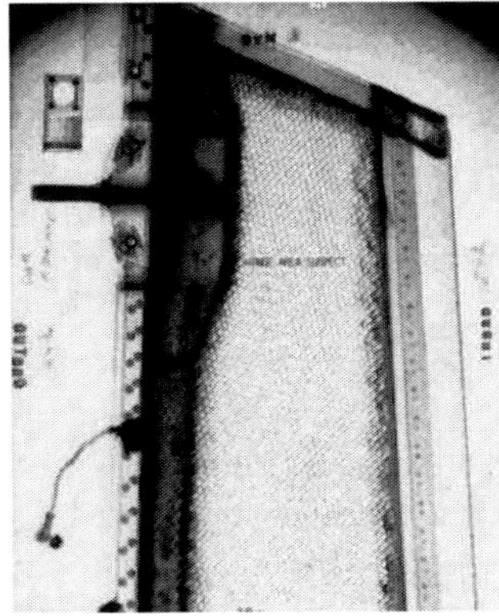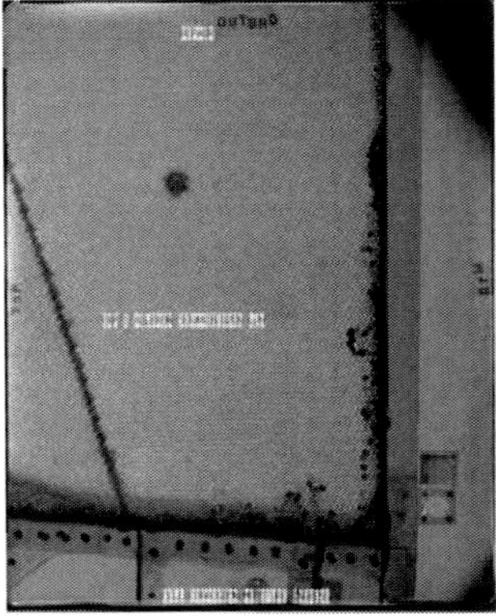

Figure 10.24 Neutron radiographic images of aircraft honeycomb composite structures showing water intrusion near the edges attached to metallic parts (http://www.upei.ca/~phys221/gxie/airplane.htm).

Overall, due to the lack of transportability of neutron radiography and the high cost of using nuclear facilities, the technique is not widely used. However, attempts are underway to design and make transportable neutron radiography systems using radioactive materials along with modern imaging devices to enable the application of this technology in the field (e.g., [14]).

10.5.3 CAPABILITIES AND LIMITATIONS OF NEUTRON RADIOGRAPHY

Capabilities:
- Sensitive to light elements (e.g., hydrogen) that are not detectable by other NDT methods.
- Full-volume inspection of hydrogenous materials.
- Corrosion products in metals may be detectable.
- Water intrusion in honeycomb structures can be detected.
- Detection of some explosive materials is possible.
- Unique applications (e.g., o-rings, plastic parts, and investment-cast turbine blades).

Limitations:
- Neutron sources are not readily available.
- Most discontinuities (e.g., cracks, voids, delamination, inclusions) cannot be detected.
- Application is limited to hydrogenous media.
- Field application is not possible.
- Very high equipment and operational costs.
- Hazardous and requires human protection from exposure.
- Requires highly-specialized people.

10.6 COMPTON BACKSCATTERING

10.6.1 PRINCIPLES

The scattering of photons in a material that results in a decrease in energy of X-rays or gamma rays is called the Compton Effect. The energy shift depends on the angle of scattering and not on the nature of the scattering medium. Since the scattered x-ray photon has less energy, it has a longer wavelength and less penetrating power than the incident photon. Analogous to an optical visual system where the visible light scattered by an object surface is detected by the eye, it is possible to detect the back-scattered radiation and use it to obtain information about the material. This approach is referred to as Compton backscattering [15].

The Compton backscattering based on x-rays uses collimated x-rays aimed at a small region of the object and a collimated radiation detector, located on the same side as the source, to capture a portion of the scattered energy, as illustrated in Figure 10.25. In this case, the inspection volume is the intersection between the collimated detector's field of view and the collimated x-ray beam. By scanning the test object with respect to a collimated source/detector assembly and monitoring the returned signal intensity, a point-by-point measure of the material's density can be obtained [16]. Using this approach, 2-D maps of the density variations as a function of the location of the source/detector assembly are generated.

In contrast with the conventional film radiography that requires access to both sides of the test object, Compton backscattering can be carried out from one side only. However, only areas close to the surface can be inspected as the scattered radiation from deeper sections of the object attenuate before reaching the surface. In addition, the technique is slow and inefficient for routine inspections of aircraft parts. Using a 160 keV constant potential x-ray source, a coverage depth of up to 3 mm, 25 mm, and 50 mm can be achieved in steel, aluminum, and aerospace composites, respectively [17]. The method is likely to be best suited for the detection of volumetric defects like voids, porosity, inclusions, and near-surface damage [18].

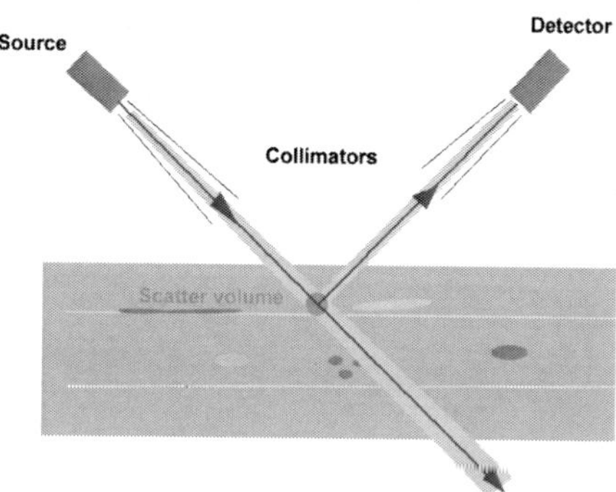

Figure 10.25 Basic principles of Compton backscattering technique.

10.6.2 EQUIPMENT

An x-ray Backscatter Depth Profilometry (BDP) system has been developed at Northwestern University aimed specifically for the inspection of aircraft metallic structures for corrosion detection [19]. To speed up the inspection process, this BDP system captures one-dimensional images, and this makes the data acquisition time on the order of 10 minutes per image. The system is claimed to be able to measure the thickness of a layered aircraft skin with an accuracy of ±0.25 mm, which is about the same as a dial calliper.

The x-ray BDP unit at NW University is relatively large and moves around under its own battery power. It consists of a scan head that can be aligned by a computer-controlled stepper motor system using position sensors which contact the part. In layered materials, such as lap joints, the system, in principle, can provide a density profile of the cross-section of the inspection site. The technique provides a single-sided inspection capable of obtaining a measure of the material density for the volume being inspected [20].

The most important use of x-ray backscattering is in airport security for screening passengers and cargo to detect metallic objects, such as guns. The passenger screening systems use a low energy exposure to take skin-deep radiograph of people while cargo screening systems employ high energies to penetrate deep into the containers.

10.6.3 CAPABILITIES AND LIMITATIONS OF COMPTON BACKSCATTERING

Capabilities:
- Access to only one side is necessary.
- Radiation penetrates only the region near the surface providing unique applications (e.g., near-surface flaw detection in thick sections).
- Minimal influence from surface features.
- Provides a density map similar to x-ray imaging.
- Can identify corrosion in lap-joint structures.

Limitations:
- Very slow and inefficient.
- Limited depth of penetration in high density materials
- Only near-surface flaws are detectable while deeper flaws in thick sections cannot be detected.
- Hazardous and requires human protection from exposure.
- Field application is difficult.
- Equipment cost is high and applications are limited.
- Requires very specialized personnel to conduct the tests.
- It is not an approved approach for aircraft applications.

10.7 COMPUTED TOMOGRAPHY

10.7.1 Principles

Computed tomography (CT) was originally developed for medical diagnosis, but it has also proved to be a valuable NDT tool. Figure 10.26 schematically shows the different components of an x-ray CT system [21]. In this case, the test component is placed on a turntable located between the radiation source and the imaging equipment. The turntable and the imaging equipment are connected to a computer that enables correlation of the collected x-ray images to the position of the test component for subsequent analysis. Electronic detectors are used to collect x-rays transmitted through the object at many angles as the component rotates on the turn table. The detector's readings are then processed using an extensive computing procedure to produce a cross-sectional image of the inside of the object.

Unlike the conventional radiographic imaging that uses a path close to perpendicular to the surface being inspected, in the CT, the x-ray beam and the detector array lie in the same plane as the surface being imaged. Thus, elaborate computing algorithms and specific software packages are required to calculate, locate, and display the point-by-point relative attenuation of the x-ray energy passing through the thin cross-sectional slice of the test piece. The process is repeated to obtain images of other cross-sections until the entire volume of the part is covered. This provides a series of 2-D images that can be put together to produce cross-sectional 3-D images of the test component as if it were being sliced. With this method, the inside of the object can be seen, and its internal characteristics, such as density variations, geometrical changes or internal defects, as well as their dimensions or shape, can be analyzed.

10.7.2 Equipment

The CT images provide both density and geometry information related to volumetric defects, such as voids, porosity, inclusions, and cavities, with very high accuracy (0.025–0.25 mm resolution). However, the application of the x-ray CT in aerospace has been limited to relatively small components, such as turbine blades (e.g., Figure 10.27), due to the extensive processing time and resources that are needed. Internal defects, such as blocked cooling air channels of the blades that are difficult to see using conventional radiography, can be observed. As a laboratory tool, CT can be quite useful for a variety of applications, including precise measurement of thickness of multi-layered components, internal dimensions of integrated parts, and failure analysis of components that cannot be disassembled. However, the high initial cost of the CT systems, as well as the requirement for multiple component exposure and extensive computing resources, prohibit the widespread use of this technology in aerospace.

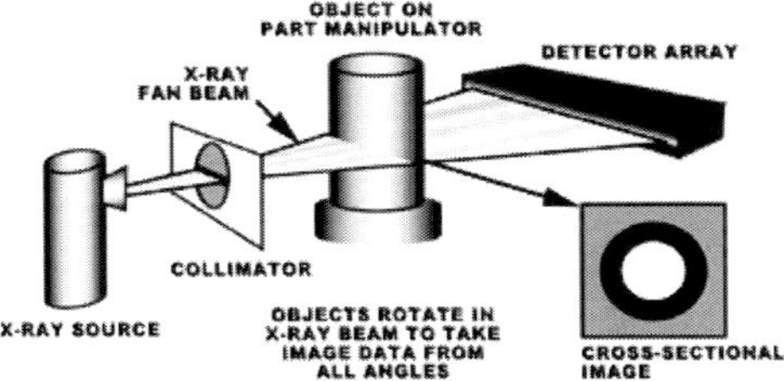

Figure 10.26 A schematic presentation of an x-ray computed tomography system [21].

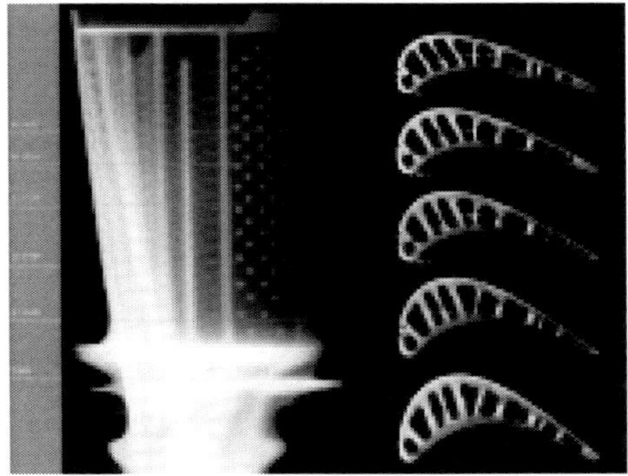

Figure 10.27 CT images of a turbine blade showing internal cooling channels [22].

10.7.3 CAPABILITIES AND LIMITATIONS

Capabilities:
- Applicable to a wide variety of materials and components.
- Provides cross-sectional views of the interior of the part.
- Very versatile in terms of the part geometry or composition.
- Can be used on a variety of part sizes, weights, or densities.
- Can detect small volumetric flaws, such as voids, pores, and inclusions, or measure dimensions.

Limitations:
- High equipment cost and limited throughput for large volumes.
- Limited to high-value small components.
- Access to all sides of the test piece is needed.
- Cracks may not be detected unless close to parallel to the beam direction.
- Requires extensive operator training and skills.
- Not suitable for large parts or for detecting disbonds or delaminations.
- A laboratory technique and cannot be applied in the field.

10.8 CHAPTER REFERENCES

[1]   Introduction to Non-destructive Testing, NDT Resource Centre, Center for Non-destructive Evaluation, Iowa State University, Ames, Iowa 50011, USA.

[2]   Metals Handbook, Non-destructive Evaluation and Quality Control, Vol. 17, 9th Edition, Published by ASM International, Metals Park, OH, USA, 44073, 1989.

[3]   Radiographic Inspection of Aircraft Components, S.A. Mango, Quality Digest Internet Site; http://www.qualitydigest.com/, accessed in 2011.

[4]   Overview of Film and Digital Radiography Systems, M. Khan, NRC Report: LM-SMPL-2010-0032, June 2011.

[5]   Principles of Digital Radiography with Large-Area, Electronically Readable Detectors: A Review of Basics, H.G. Chotas, J.T. Dobbins III, and C.E. Ravin, Radiology, pp 210, 595–599, 1999; Published online December 2, 2002.

[6]   Application Limitations for Digital Radiography, R. Kochakian, Journal of Canadian Society for Non-destructive Testing, Jan/Feb 1999.

[7]   Measurement of Dynamic Range of Digital X-ray Imaging Systems, E.B. Kozlovskii, Biomedical Engineering, Vol. 34, No. 5, pp 257–9, 2000.

[8]   Metal Image Intensifier Tubes, T.T. Thompson, Radiographics, Vol. 4, No. 4 May 1984.

[9]   The AAPM/RSNA Physics Tutorial for Residents, X-ray Image Intensifiers for Fluoroscopy, J. Wang and T.J. Blackburn, Radiographics, 20: pp:1471–1477, September 2000.

[10]   Aerospace NDE, Digitary Digital X-ray Systems Internet Site: www.digitary.com/aircraft/aircraft.htm, accessed in 2011.

[11]   Gamma Radiography, http://duplicate.hubpages.com/hub/gamma-radiography, May 2010.

[12]   Neutron Radiography of Aircraft Composite Flight Control Surfaces, W.J. Lewis, L.G.I. Bennett, T.R. Chalovich, and O. Francescone, www.ndtnet.com, accessed in 2011.

[13]   ASTM-STP-586-EB, Californium-Based Neutron Radiography for Corrosion Detection in Aircraft, 1976, ASTM International, West Conshohocken, PA, USA.

[14] Designing a Transportable Neutron Radiography System, R. Picha, N. Simoroj and W. Sangphet, Kasetsart Journal (Journal of Natural Sciences), No 42, pp: 305–310, 2008.

[15] Compton Scattering Elemental Imaging of a Deep Layer Performed with the Principal Component Analysis, A.Tartari, G. Maino, E. Lodi, C. Bonifazzi, Proceeding of the 2000 World Conference on NDT, also published at: http://www.ndt.net/article/wcndt00/papers/idn261/idn261.htm.

[16] Detection and Thickness Characterization by X-ray Backscatter of Second-layer Corrosion in Aircraft Structures, L.R. Lawson, N. Kim and J.D. Achenbach, Proceedings of the Tri-Service Corrosion Conference, Orlando, Florida, June 21–23, 1994.

[17] ASTM Standard E1931, Standard Guide for X-ray Compton Scatter Tomography, Volume 03.03 Non-destructive Testing, October 2010, ASTM International, West Conshohocken, PA, USA.

[18] NDE of Composites, ESR Technology, http://www.ESRtechnology.com

[19] X-ray Backscatter Depth Profilometry", J.D. Achenbach, Center for Quality Engineering and Failure Prevention, Northwestern University Report No. 1137892, 1998, Northwestern University 633 Clark Street, Evanston, IL 60208.

[20] Compton Scattering Elemental Imaging of a Deep Layer Performed with the Principal Component Analysis, A.Tartari, G. Maino, E. Lodi, C. Bonifazzi, World Conference on NDT, Rome, Italy, 2000.

[21] High Energy X-ray Tomography, M. Crine, http://www2.ulg.ac.be/bioreact/, accessed in 2011.

[22] High Power Micro-focus Computed Tomography, http://www.nikonmetrology.com

CHAPTER 11

NDT of Aerospace Composite Materials and Components

11.1 BACKGROUND

Composite materials generally consist of two or more chemically-distinct constituents: one is the "matrix," which holds everything together, while the other constituents are embedded into the matrix to provide reinforcement. The reinforcing constituents may be fibers, particles, or flakes. Fiber reinforced plastics (FRP) is a term used to describe composite materials that use fibers embedded in a plastic matrix. The most common matrix materials are thermoset polymers (e.g., epoxy, vinyl ester, polyester, or phenolic resins) and thermoplastic polymers (e.g., polyether ether ketone or PEEK, polyphenylene sulphide or PPS, and polyimide). Reinforcing fibers can be carbon, boron, glass, aramid (an aromatic polyamide product developed by the DuPont Company that is trade-named Kevlar®), or mixtures of these. Modern aircraft use significant amounts of FRP composites to achieve lighter weights while maintaining the required strength and stiffness. Generally, composites provide a higher strength-to-weight ratio and better fatigue and corrosion resistance than the counterpart aluminum and titanium alloys. Furthermore, their stiffness and physical properties can be tailored to design requirements. In addition, with composites, it is possible to manufacture large parts using fewer pieces, reducing the need for mechanical fasteners.

Carbon fiber reinforced plastics (CFRP) are most widely used in the load-bearing primary structures of aircraft while composites made of fiber glass or aramid and hybrid materials, like fiber metal laminates (FML), are mostly employed in non-structural areas to take advantage of their unique properties. Sandwich construction consisting of a low-density core sandwiched between two thin plates

(1–2 mm) is very common in aerospace. Honeycomb sandwich panels are made of CFRP skin laminates and either aluminum or Nomex® core (trade-name for a fire-resistant material developed by the DuPont company). Such panels are often used in areas of slight curvature with moderate loads where most loading is due to bending, as they provide very high stiffness-to-weight ratio in bending. Structures, such as horizontal stabilizers, rudders, winglets, control surfaces, and segments of fuselage, may be made out of these materials. Less critical parts may be fabricated using other materials (e.g., glass or Kevlar fibers, foam core, FML). Figure 11.1 shows composite sections in a commercial aircraft (e.g., Boeing 787) and the types of materials used.

The use of composites in aircraft requires NDT during and after manufacturing, as well as in service. The current philosophy for most aircraft composite structures is to perform NDT after fabrication for quality control purposes and to rely mostly on visual inspections during service. More detailed in-service NDT is carried out if damage is detected by visual tests or suspected to have occurred as a result of overloading, accidental impacts, moisture ingress, lightning strikes, high temperature exposures, or fire. In these cases, the choice of the method is dependent on the component material and design, the suspected damage type and location, and the NDT ability to detect and measure the damage.

In this chapter, first, a brief description is provided of composite materials widely used in aerospace and the type of defects or damage that may occur in them. Then, NDT methods that are most suitable for their inspection are mentioned and a few examples of their application are given. Finally, the types of defect detectable in the different materials and configurations and the NDT limits are listed in section summaries.

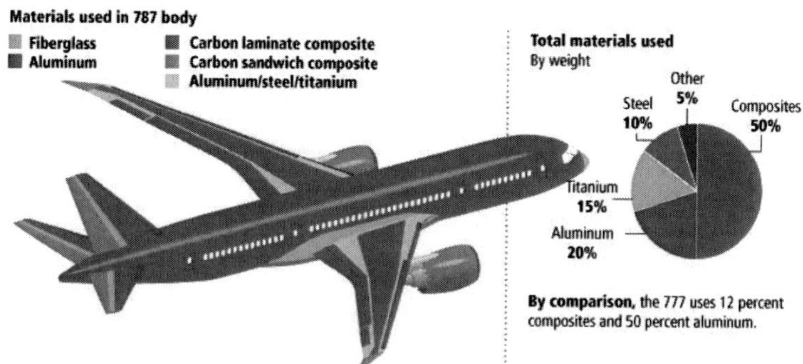

Figure 11.1 Materials used in the Boeing 787 aircraft body (http://metallurgyfordummies com/composite-materials/).

11.2 CARBON FIBER SOLID LAMINATES

Highly loaded structures, such as wing skins, are often made of CFRP solid laminates of several millimeters in thickness with stiffeners similar to those used in metallic structures. Stiffeners can be mechanically joined, adhesively bonded, or made as integral parts of the structure using a co-curing process. CFRP materials for manufacturing of aerospace structures typically come in the form of "prepreg" tapes with resin and fibers already combined. The resin in these "prepreg" tapes has undergone a partial curing operation to produce a tape that has some mild stiffness, so it may be handled. The fibers may be aligned in a single direction to produce a unidirectional tape, or they may be aligned in different directions to provide multi-directional strengthening. And, finally, they may be woven to produce a fabric that again provides strengthening, usually in two directions. The thickness ranges from 0.13 to 0.25 mm for unidirectional tapes and up to 1.5 mm for fabrics. The fibers are also used to repair composite or metallic parts, and repair materials usually come either in the form of a prepreg fabric or a dry fabric that is wetted with liquid resin when the repair is carried out.

Composite sections of an aircraft structure are normally produced by placing multiple layers of the "prepreg" in different orientations in an appropriate mold and curing the stacked material under pressure and heat to produce a solid part that takes the shape of the mold. The fiber orientation and laminate thickness may vary from place-to-place to achieve desirable strength and stiffness values. CFRP solid laminates of up to several millimeters in thickness are attached to inner sub-frames (e.g., doublers, stiffeners, ribs) forming complex structures that require inspections after manufacturing. During service, they may be subjected to impacts, overloading, or hostile environments that may cause damage and necessitate inspections. In this section, the defect and damage types in CFRP solid laminates are first identified and then the NDT procedures commonly used for their detection and characterization are described. This section does not deal with the inspection of adhesive bonds, as this is the subject of an ensuing separate section.

Figure 11.2 Examples of CFRP composite structures. (Left): An aircraft fuselage segment. (Right): An aircraft wing section.

11.2.1 Defect Types

The most commonly-occurring manufacturing defects in composite laminates are cavities that result from entrapment of air or other gases between the individual plies during fabrication (e.g., Figure 11.3). Areas of high porosity content due to lack of adequate resin (resin-starvation), gaps between individual tapes that are filled with excessive resin (resin-rich), material inclusions, such as paper or nylon (peel-plies), that may be inadvertently left behind between the layers during the lay-up process, and misalignment of plies may also occur. Such discontinuities may compromise the mechanical properties of the component and must be detected after manufacturing using NDT methods.

During service, composite parts may be damaged by flying debris, dropped tools, or other accidental impacts that may exceed the design tolerance of the structure. Depending on the impact energy, the damage can be quite visible, barely visible, or invisible on the surface but extensive below the surface, as schematically illustrated in Figure 11.4. At high impact energies, damage often penetrates through the thickness of a laminate,creating a visible hole and extensive internal delamination, cracking, and fracture. At medium energy levels, an impact may create a surface dent and cause internal delamination, fiber fracture, and matrix cracking extending to the rear surface in a pyramid shape. In this case, the initially visible dent may become less visible with time lapse.

Invisible or barely-visible damage may occur at low impact energies, but internal damage may still be significant. Figure 11.5 shows cross-sectional views of CFRP composite samples with typical damage types, while Figure 11.6 shows a typical low-energy impact damage profile. These effects, combined with the generally low impact damage tolerance of the CFRP, require identification of barely-visible impact damage (BVID) by NDT methods. Thus, the ability of NDT methods to identify BVID during service, particularly on large surfaces, is important.

Unlike metallic structures, in composites, fatigue cracking is not a major concern since fibers tend to arrest micro-cracks that may form in the resin matrix as a result of cyclic loading. However, degradation of properties due to exposure to extreme temperatures, such as those due to lightning strikes, fire, and engine

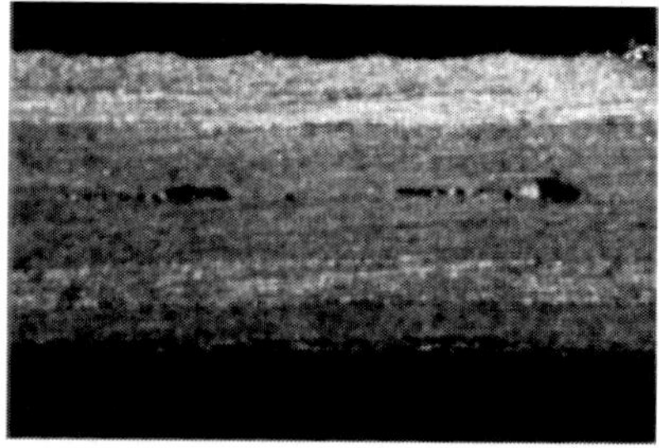

Figure 11.3 Cavities in a typical CFRP laminate.

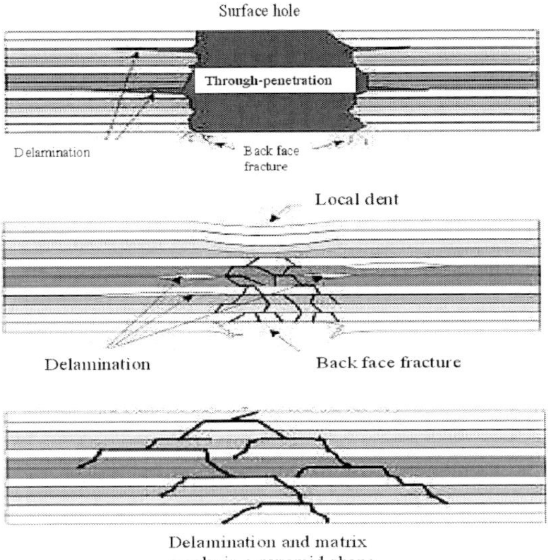

Figure 11.4 Typical impact damage profiles in CFRP laminates at high (top), medium (middle), and low (bottom) energy levels.

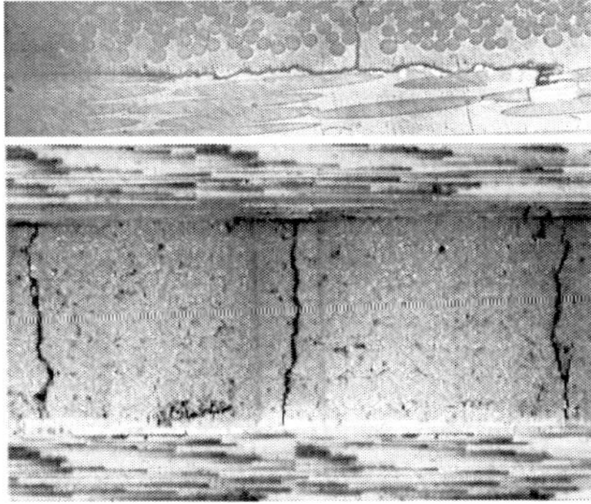

Figure 11.5 Typical damage types in CFRP solid laminates. (Top): Delamination and matrix cracking. (Bottom): Cross-laminate cracks.

exhausts, is of concern, as is degradation due to high humidity or the action of fluids. Composite components exposed to extreme environments may not show visible signs of damage but may be seriously degraded and no longer flightworthy. NDT characterization of this type of damage is complicated by numerous variables inherent to environmental degradation, such as the exposure type (e.g., humidity, fluids, or fire), conditions (e.g., fluid amount or fire temperature), and duration (short, long, or repeated exposure). Also, the degradation mechanism is

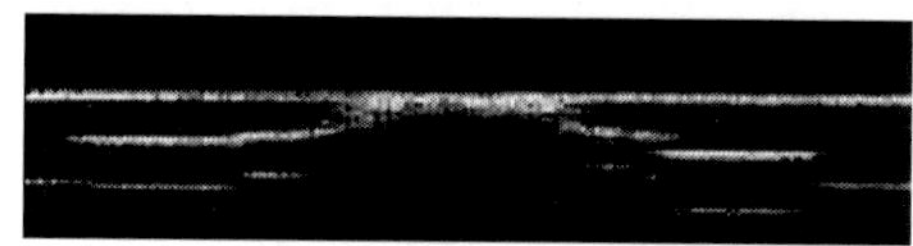

Figure 11.6 A typical low-energy impact damage in CFRP composite laminate showing a pyramid-shaped profile (www.netcomposites.com).

important and varies with the material and the construction of the components, as well as the presence or lack of protective coatings.

11.2.2 NDT METHODS

Ultrasonic methods are widely used to inspect CFRP laminates after fabrication for manufacturing defects. A general introduction to ultrasonic inspection of CFRP composite materials is provided in reference [1]. Automated C-scan systems are commonly employed in order to obtain more reliable inspection results in a timely manner. Flat, slightly curved, and tubular parts are easily inspected using automated water-immersion or squirter systems, often in through-transmission or reflector pulse-echo modes. For inspection of sharp corners or internal surfaces and attachments, the pulse-echo approach using contact probes, phased-arrays, or non-contact laser-ultrasonic may be used. The choice of the approach, sensors, equipment, and test procedures (e.g., pulse-echo, through-transmission) or parameters (frequency, focal size, etc.) is dependent on the test material and component geometry, as well as the discontinuity types and sizes that must be detected.

As described earlier, the ability of an NDT method to identify material defects is often established using reference calibration samples with implanted or real discontinuities. In CFRP composites, Teflon tapes are often used to create simulated defects (e.g., cavities or delamination) by entrapping air between two layers of Teflon, sealing the edges, and embedding them between plies at different depths. Since acoustic properties of Teflon are similar to matrix resins, the entrapped air gap between the two Teflon layers resembles inter-laminar cavities or delamination.

Other materials, such as paper, nylon, and fiber glass cloth, used during fabrication of composites may also be embedded in calibration samples to check their detectability [1]. Generally, foreign material inclusions that have different acoustic impedance from that of the host material are expected to be detected using ultrasonic techniques. When suitable calibration references are not available, lead tapes of different sizes are affixed on the actual part to assess detection sensitivity. Lead has a much higher density than the CFRP and creates an acoustic impedance mismatch that can be easily identified. Reference samples or lead tapes are used to optimize the test setup (procedures and parameters) before performing the actual inspections.

Figure 11.7 shows a schematic diagram of two reference panels, one flat and the other curved (curvature diameter ~60 cm), made of 40 plies of CFRP. The panels contained embedded Teflon tapes of different sizes and depth locations that were folded and sealed at the edges. The panels were tested in an automated ultrasonic immersion C-scan system using the pulse-echo approach. In the case of the curved panel, a contour-following software routine was used to maintain the beam normality to the part during scanning.

The inspection results are presented in Figure 11.8. This Figure shows that, in ~6 mm thick CFRP laminates, discontinuities as small as 2.5 mm × 2.5 mm Teflon inserts can be identified, regardless of the plate curvature. This example demonstrates the sensitivity levels that can be achieved using ultrasonic inspections; however, in real aircraft parts, the size of defects that must be detected is much larger. These images are obtained by mapping the intensity of the returned UT signals that, in addition to locating internal discontinuities, also provide an

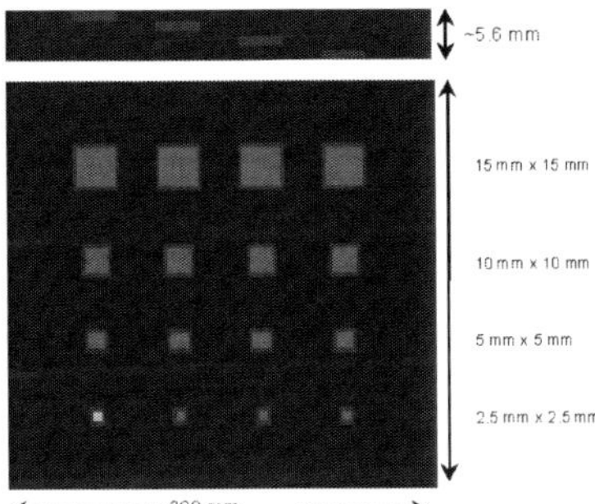

Figure 11.7 A 40-layer CFRP solid laminate $[0°/45°/–45°/90°]_{5S}$ reference panel with embedded discontinuities simulating delamination (folded Teflon inserts) of various sizes and depths. Note that in the schematic image, discontinuities appear much larger than their real size.

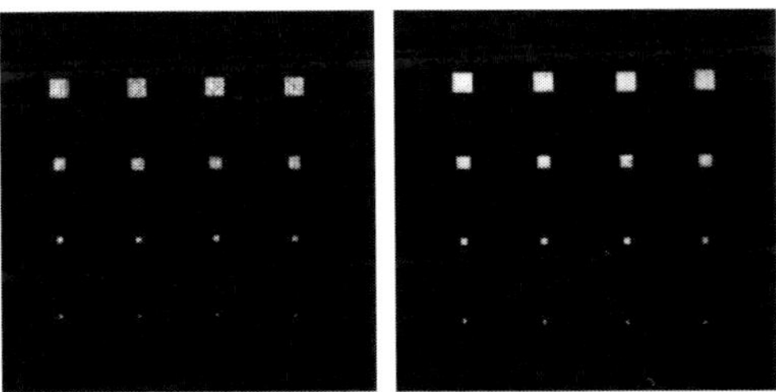

Figure 11.8 C-scan images of the CFRP solid laminate reference panels obtained using the pulse-echo approach. The entire returned signal was captured using a full-wave gate and at the scan completion, only signals between the front and back surfaces were gated to show internal discontinuities. A 5 MHz probe was used. Note that the curved panel was inspected using a contour- following software package to maintain the UT beam perpendicular to the part during scans. (Left): Flat panel. (Right): Curved panel.

estimate of the flaw size. However, they do not indicate how deep the flaws are located; such information is obtained using time-of-flight C-scanning, if needed.

As mentioned earlier, aircraft parts are made in different shapes and sizes, often with additional features, such as stiffeners, ply build-ups (or drop-offs), bends, and corners, that create non-uniformity in the beam path. The additional features make ultrasonic inspections and interpretations difficult, and if they are not visible, they should be known to the inspector so that appropriate inspection procedures may be selected and the results will be properly interpreted. For example, the item shown in Figure 11.9 has regions of ply build-up adjacent to the sharp corners and around the drilled hole. As a result, changes in background color are seen due to the differences in UT attenuation.

Geometrical features such as these are usually identified from their regular shapes and straight border-lines. When the part thickness changes drastically, two or more inspections using different parameters may become necessary. Also, as seen in this image, segments of the part that are parallel to the beam direction and sharp corners are not mapped and may need to be examined separately.

In addition to ultrasonic testing, infrared thermography may be used for identifying manufacturing defects, such as cavities and porosity, as well as internal features of relatively simple geometries that are less than 5 mm in thickness [2–5]. An example is provided in Figure 11.10 where a CFRP near-tubular structure is inspected at IAR using pulse thermography in comparison with the conventional ultrasonic C-scan. Most of the defects (e.g., voids) identified by the ultrasonic method are also seen in the thermal image, except for those that are in the regions where internal ribs are attached. Due to the heat sink into the ribs, thermal contrast is lost in these areas. However, they may become visible at a longer time sequence.

Another example is presented in Figure 11.11 where the C-scan image of a CFRP tube is shown, along with thermal images of a few regions. The C-scan corresponds to the entire tube and is obtained in a single inspection identifying internal features such as thickness transitions and ply build-ups, as well as

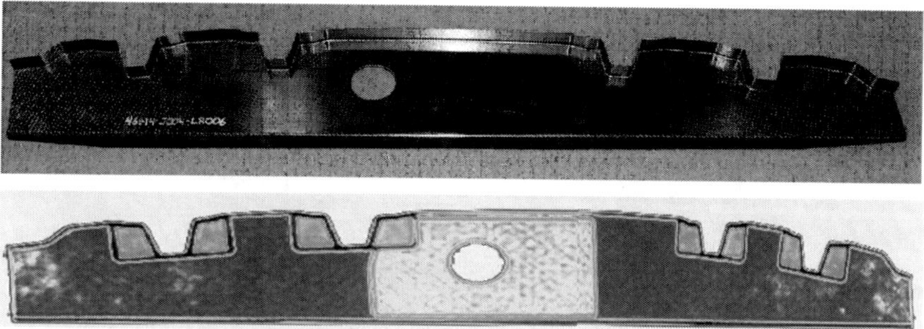

Figure 11.9 A CFRP composite part with the corresponding ultrasonic C-scan image. Change of color corresponds to change of thickness.

Figure 11.10 Schematic of a near-tubular CFRP structures (left) and side-by-side comparison of images obtained using ultrasonic C-scan (middle) and pulsed thermography (right).

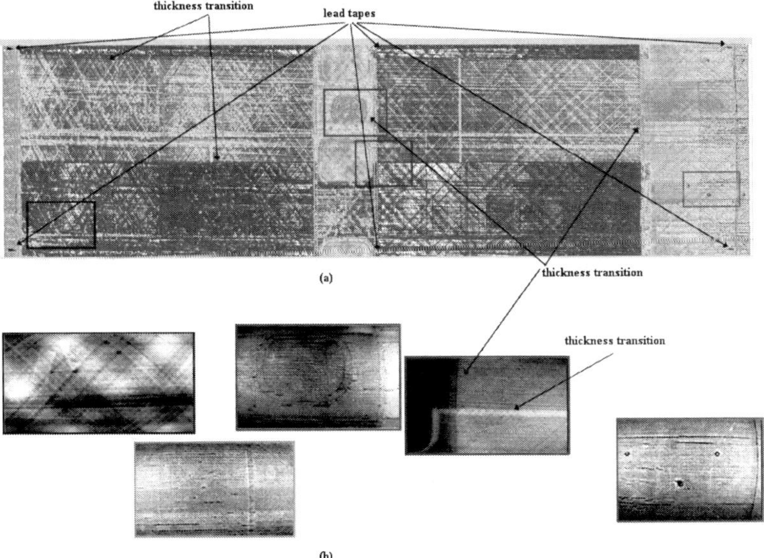

Figure 11.11 Inspection results of a CFRP tubular structure using ultrasonic C-scan (top) and infrared pulsed thermography (bottom). Note that the C-scan image shows the entire tube as it might be opened, identifying thickness transitions and areas of ply build-up as well as discontinuities. The thermal images correspond to selective segments of the tube obtained at different time sequences showing some of the features identified by the UT with a higher resolution.

discontinuities. Thermal images, on the other hand, are taken at different time sequences in order to show the different features. Nevertheless, they show both the internal features, as well as discontinuities that the UT was able to identify. Both methods also show the general pattern of fiber orientation. Thus, several inspections or multiple methods may be necessary to cover all areas of a complex structure. Overall, with ultrasonic C-scanning or thermal imaging, manufacturing defects, such as cavities or regions of high porosity, due to resin-starvation as well as internal features, can be mapped. However, thermography is limited to relatively thin (< 5 mm) CFRP laminates. Also, some indication of ply orientation can be obtained [5].

After completion of inspections and identification of defects, a decision must be made to accept or reject the part based on pre-set criteria. For example, in the IAR inspection of composite arm booms of the Space Station Remote Manipulator System (SSRMS or the Canadarm), the client's accept/reject criterion for C-scan results (example in Figure 11.12) was to reject the part if an isolated single discontinuity larger than 2 mm in diameter was found. In the case of multiple small discontinuities, the criteria considered the discontinuity density. The latter was defined as the ratio of the maximum dimension of the largest indication by the distance to all adjacent ones. If the value exceeded 0.5, then the area was considered to be defective and the part was rejected [7].

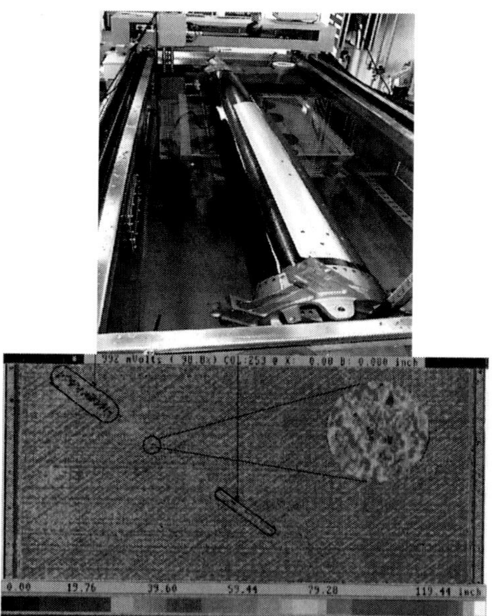

Figure 11.12 (Top): A composite arm boom under inspection at IAR. (Bottom): An ultrasonic C-scan image of a composite arm boom with manufacturing defects exceeding the acceptable limits (encircled areas).

During service, composite structures of aircraft are inspected visually for signs of impact or environmental damage. If visible damage is found, NDT is used to verify the damage and establish its size or severity. Visual inspections are operator dependent and can only identify damage that creates a surface dent (in the case of impact), discoloration (in the case of exposure to high temperatures), or fracture (in the case of lightning strikes). Enhanced optical methods, such as D-Sight or Edge-of-Light, may be used to provide a better visibility, particularly for BVID (e.g., Figure 11.13). These methods are suitable for inspection of large surfaces to identify suspect areas (e.g., Figure 11.14); however, they require verification by more reliable methods, such as ultrasonics. This is due to the fact that most composite structures contain inherent surface distortions resulting from manufacturing or internal features that may be confused with impact damage, creating undesirable false-calls. Laser shearography and infrared thermography are more reliable; however, the damage must be of the delamination-type and very close to the surface to be identified by these methods. Since impact damage in CFRP solid laminates is usually more extensive towards the rear face, the severe portion of the damage may remain beyond the detection limit of these methods, particularly if the composite skin thickness is more than 5 mm. Although the above-mentioned methods provide a fast and economical way of inspecting large surfaces for possible impact damage, none can give a good estimate of the damage extent. For

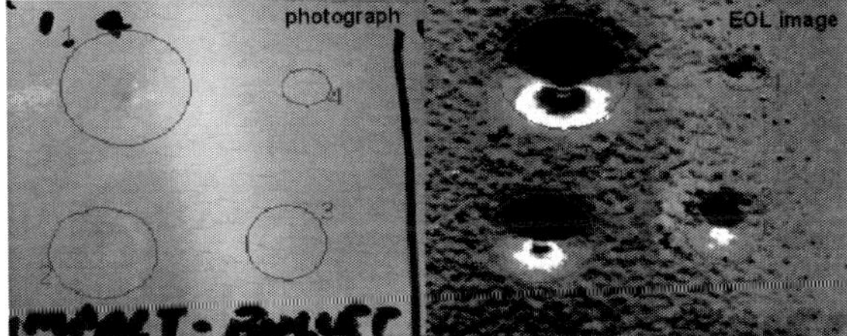

Figure 11.13 A comparison of visual (or photographic) indications of impact damage with those of the Edge-of-Light. Impact damage was introduced at different energy levels, producing visible (1), barely-visible (2 and 3), and invisible (4) dents.

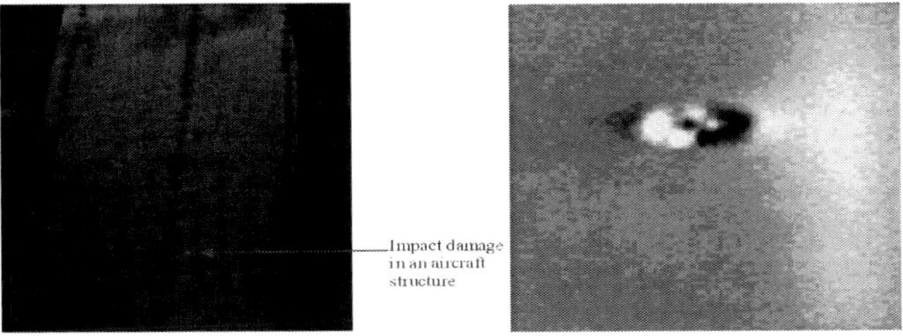

Figure 11.14 (Left): D-Sight inspection of an aircraft vertical stabilizer showing impact damage. (Right): A similar damage identified by laser shearography. Magnifications are different.

more accurate and reliable detection and measurement of impact damage, ultrasonic methods are employed.

The ultrasonic pulse-echo technique is widely used to detect and size impact damage in CFRP laminates. Large area ultrasonic inspections may be carried out using automated scanners or laser-ultrasonic approach. As shown in Figure 11.15, amplitude C-scan provides the overall size of the damage but does not give the damage depth. Time-of-flight (TOF) C-scanning is used to obtain an indication of the damage depth and profile. In the TOF approach, the UT travel time is measured and displayed in colors (or gray scale) that are related to different depths within the sample. By comparison of the colors of the damage indication with a calibrated scale, it is possible to estimate both the size and depth-location of the damage. The TOF approach is mostly sensitive to delamination damage that lies perpendicular to the beam direction, and fiber fracture or matrix cracks are not clearly identified.

Radiography assisted by radio-opaque penetrants may provide additional information in the laboratory tests when access to both sides of the part is available. Damage must be open to the surface so that an enhancing penetrant could enter the defect (e.g., Figure 11.16). This method is also useful for the detection of matrix cracks that may occur in composites due to rapid temperature change or extreme

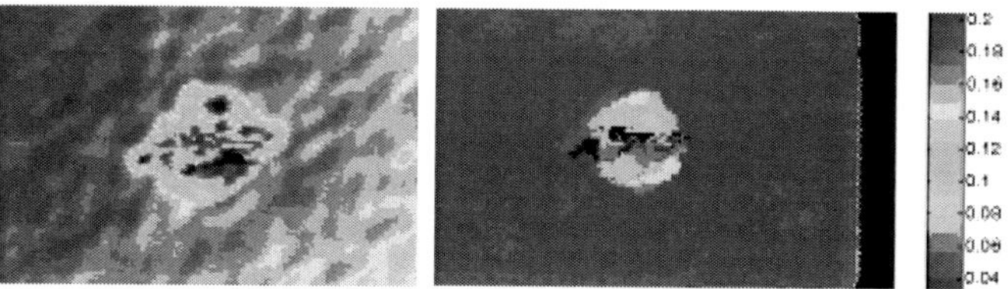

Figure 11.15 Ultrasonic amplitude (left) and time-of-flight (right) C-scan images of an impact damage in a CFRP composite laminate (www.netcomposites.com). Different colors in the image on the right correspond to different depths.

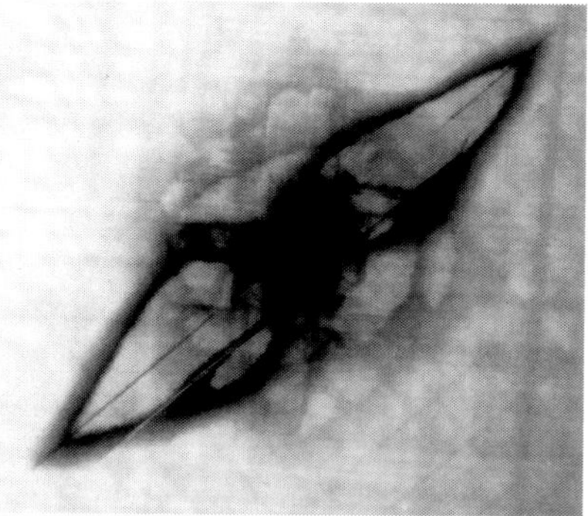

Figure 11.16 Penetrant-enhanced x-ray image of an impact damage in a CFRP composite laminate (www.netcomposites.com).

vibration (e.g., acoustic damage). Such cracks normally run along the fibers and can be easily identified by a florescent penetrant inspection (Figure 11.17).

Lightning and fire damage are often visually detectable by their appearance (burning signs, blistering, discoloration, and physical damage (e.g., Figure 11.18)), but property degradation due to exposure to elevated temperatures, humidity, or harmful fluids is not easily identified using NDT methods. Ultrasonic methods based on guided-waves or Lamb waves, such as those used in the acousto-ultrasonic (AU) technique, have shown potential for detecting diffused microscopic changes that are associated with material degradation [8]. Figure 11.19 shows AU signals in time- and frequency-domains from unidirectional CFRP samples before and after repeated thermal exposures and shows significant changes in both the overall signal intensity and frequency spectra. Figure 11.20 illustrates the change in the shear strength and the transmitted AU energy, measured both in the direction of the fibers, as well as perpendicular to the fibers, with the number of exposures. As seen in this figure, the AU signal energy changes when waves travel through the matrix (normal to fibers) rather than along the fibers, and the change is more

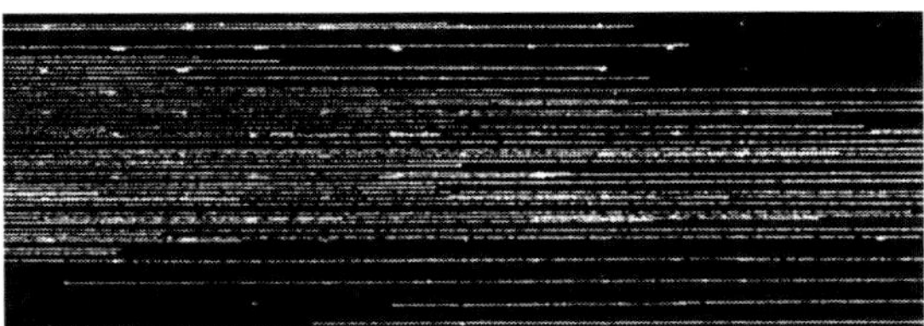

Figure 11.17 Florescent penetrant inspection of a CFRP composite panel subjected to acoustic vibration showing matrix cracks.

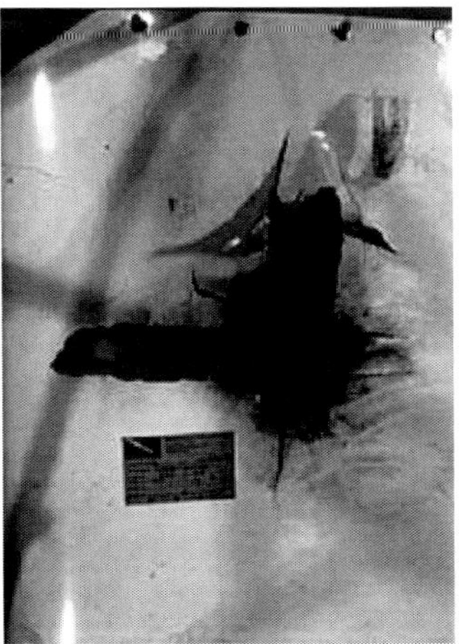

Figure 11.18 Lightning damage to an unprotected aircraft composite panel (http://www. lightningtech.com/ October 2011).

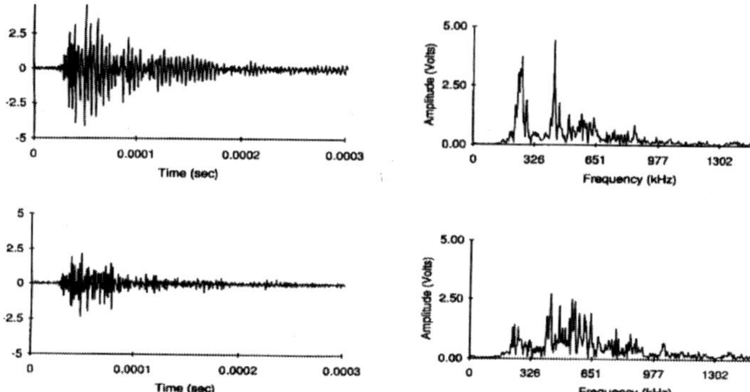

Figure 11.19 Typical acousto-ultrasonic signals in time and frequency domains for unidirectional CFRP samples. (Top): Before and (bottom): after repeated exposure to 255°C (40 times).

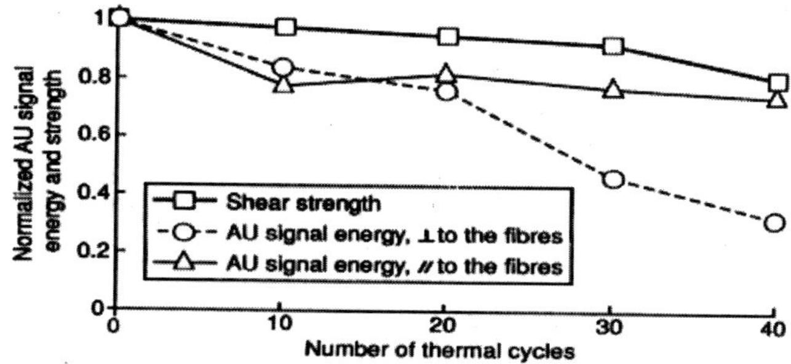

Figure 11.20 Change in the shear strength and the acousto-ultrasonic signal energy of unidirectional CFRP samples with repeated exposures (up to 40 cycles) to 255°C.

pronounced with increasing thermal cycles. In comparison, the material's shear strength that is a measure of the matrix degradation is less affected by the repeated exposures to 255°C, indicating that minute changes in the material can be identified by this approach before they affect the strength. It must be noted that the AU tests were performed in a laboratory under well-controlled conditions, and their potential for field use must be established.

11.2.3 SECTION SUMMARY

*   Manufacturing flaws, such as cavities, porosity, and some inclusions, in composite solid laminates can be detected by automated ultrasonic C-scanning. Several scans or different approaches (pulse-echo or through-transmission) may be needed for full coverage of large or complex parts.

- CFRP composite laminates several millimeters in thickness can be easily inspected using ultrasonic methods. For thicker parts, the through-transmission approach using low frequencies is suitable; however, flaw detection sensitivity and resolution will decrease with increasing thickness.

- Flaws at sharp corners or internal surfaces may not be identified by automated C-scanning, but pulse-echo contact probes or transducer arrays may be more suitable. Alternatively, laser-ultrasonic method may be used if available.

- Rapid, large area inspection of composites for impact damage may be carried out using visual or enhanced optical methods (e.g., D-Sight and Edge-of-Light); however, damage sites must be re-examined using an ultrasonic technique for verification and size estimation.

- Impact damage size and delamination profile can be obtained with good accuracy using the ultrasonic time-of-flight approach, but it does not clearly identify matrix cracks or fiber fracture.

- X-radiography assisted by radio-opaque penetrants may provide additional information on the damage; however, access to both sides of the sample must be available.

- Infrared thermography or laser shearography may be used to identify near-surface delamination in thin laminates (< 5 mm in thickness); however, these methods cannot provide the damage size accurately.

- Localized degradation of CFRP composites due to exposure to elevated temperatures may be detectable by an ultrasonic guided-wave approach, such as the acousto-ultrasonic technique; however, damage must be severe enough to create changes in the material's microstructure.

11.3 KEVLAR LAMINATES

Kevlar® is a DuPont registered trademark for an aromatic polyamide product that has good impact and fatigue resistance combined with desirable high temperature endurance, chemical protection, and electrical insulation properties. It often comes in a fabric form and may be used in aircraft applications where a balanced combination of the above properties is needed. For example, it is used in the Airbus A380 aircraft in flooring and electronic equipment casings, as well as wing flaps. When used in structural applications like CFRP composites, it requires NDT inspection, both during manufacturing and in-service.

11.3.1 Defect Types

The types of manufacturing defects that may occur in Kevlar composites are similar to CFRP; the most important are cavities and areas of resin-starvation or high porosity. Impact damage and property degradation due to moisture intake are typical damage types that may occur during service. However, Kevlar absorbs the impact energy much better than the CFRP, and at low energies, often no visible dent is seen. At high energies, Kevlar strands may break, creating visible damage and internal delamination.

11.3.2 NDT METHODS

Ultrasonic inspection of composites made of Kevlar fabric is very difficult due to the high attenuation of this material (e.g., netted strands absorb and scatter UT waves). Only Kevlar laminates less than 4 mm in thickness can be inspected using the ultrasonic through-transmission approach, while the pulse-echo method is not easily applied. An example is provided in Figure 11.21, which shows a C-scan image of a Kevlar plate with internal defects, mostly in the central region and between the strands of the fabric [9]. However, the image resolution is not sufficient to accurately indicate the defect borders. In contrast, the pulsed thermography inspection of the same sample provides a much better resolution, clearly showing the small internal voids scattered within the panel. Unlike ultrasonic waves that are attenuated rapidly by this material, the heat diffusion rate within the Kevlar laminate appears to be very uniform, being affected mostly by internal discontinuities rather than by the fabric of the material.

Pulsed thermography is equally sensitive to delamination damage, as shown in Figure 11.22. This Figure shows a photograph of a Kevlar specimen that contained impact damage at the center and edge-delamination in one corner, along with its thermal image that identifies the location and extent of internal delamination. Similar to CFRP, delamination in Kevlar laminates may occur at different depth levels, creating a change of thermal contrast as is shown in the thermal image. The optical examination of the sample's cross-section (Figure 11.23) confirmed that the edge-delamination was at different depths.

Figure 11.24 shows an Airbus A380 avionic casing inspected using pulsed-thermography after fabrication. In the thermal image, which is also shown in this figure, a defective region has been identified. Thermal inspection of large and complex components, such as this one, requires several thermal shots to cover all areas and analysis at different time sequences to identify possible defects at different depth levels.

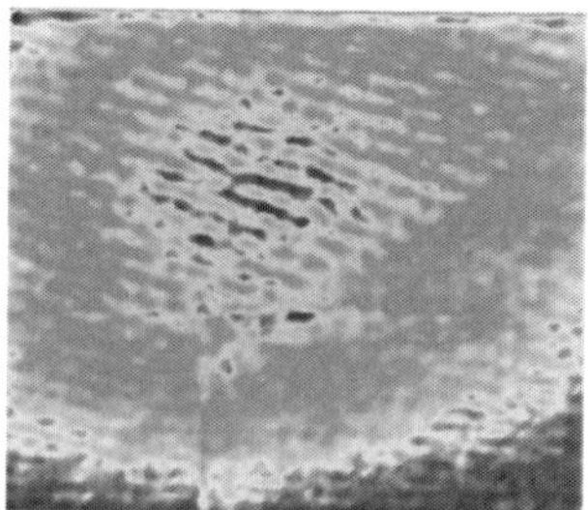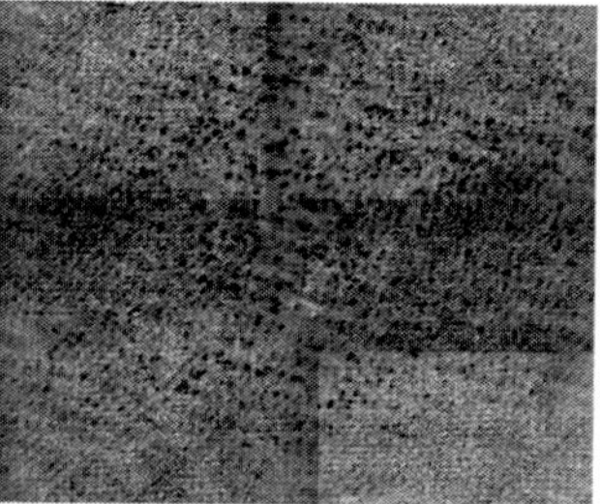

Figure 11.21 (Left): Ultrasonic through-transmission C-Scan and (right): Infrared thermal image of a Kevlar panel containing internal voids.

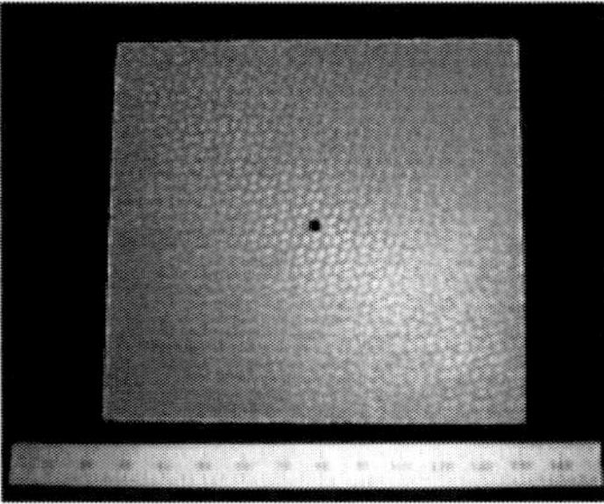 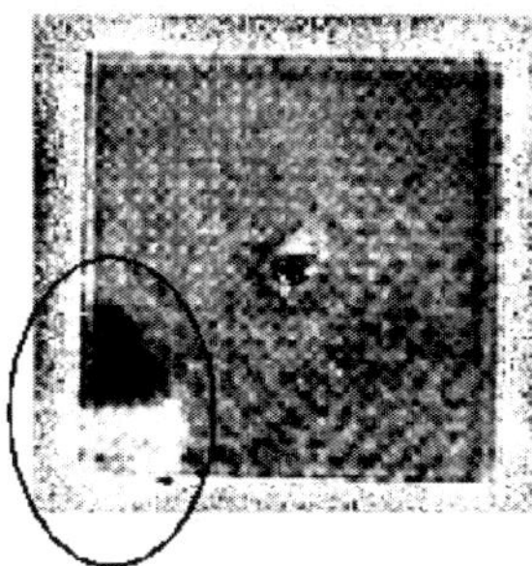

Figure 11.22 (Left): A photograph a Kevlar composite panel with impact damage at the center and edge delamination at one corner. (Right): Corresponding thermal image showing the internal damage caused by impact and delamination at the edge, located at different depths as indicated by change of contrast.

Figure 11.23 Cross-sectional view of the above-mentioned edge delamination confirming the pulsed thermography results.

Figure 11.24 (Left): Avionic casing of an A380 aircraft. (Right): Corresponding thermal image of portion of the casing identifying a defective area.

11.3.3 SECTION SUMMARY

- Kevlar composite laminates are often difficult to inspect using ultrasonic techniques.
- Infrared pulsed thermography has potential for parts less than 5 mm in thickness.
- Manufacturing defects, such as small voids and service damage, or delamination, can be identified using pulsed thermography.
- Thermal inspection of large or complex aircraft parts will require several shots to cover all areas and analysis at different time sequences to identify defects at different thicknesses.

11.4 FIBER-METAL LAMINATES

Fiber-metal laminates (FML) are made of thin aluminum sheets stacked together with fiber glass or Kevlar inter-layers [12]. Glare is a commercial FML material made of 2024-T3 aluminum and unidirectional S2 glass fibers impregnated with Hexcel F185 epoxy that comes in different thicknesses and configurations (e.g., Figure 11.25). FML materials are approximately 15 to 20% lower in density than aluminum and have significantly higher strength in the fiber direction. In addition, they have superior fatigue resistance, damage tolerance, fire resistance, and lower sensitivity to impact damage than most composite materials. FML has been recently introduced in parts of the fuselage of some new aircraft (e.g., Airbus A380).

As a part of a research program on impact damage tolerance and durability of FML materials, NDT procedures were used to detect and characterize different damage types [13]. The study provided indication of the capabilities and limitations of the IAR NDT methods when used for the inspection of FML materials, as described in the following sections.

11.4.1 DEFECT TYPES

Potential manufacturing defects in FML may include inadequate resin impregnation of composite inter-layers and un-bonded areas between the aluminum and composite sheets. Foreign material inclusion between the layers is also possible. Damage types that may occur in service are delamination and disbonding of composite layers, as well as cracking and corrosion of thin aluminum sheets.

11.4.2 NDT METHODS

A number of NDT methods are applicable to FML materials; a few are assessed and presented here. For this assessment a step-shaped reference sample was made which contained simulated delaminations or disbonds using Teflon inserts of different sizes imbedded in different layers, as shown in Figure 11.26. A peel-ply made of Nylon was also embedded to establish its detectability.

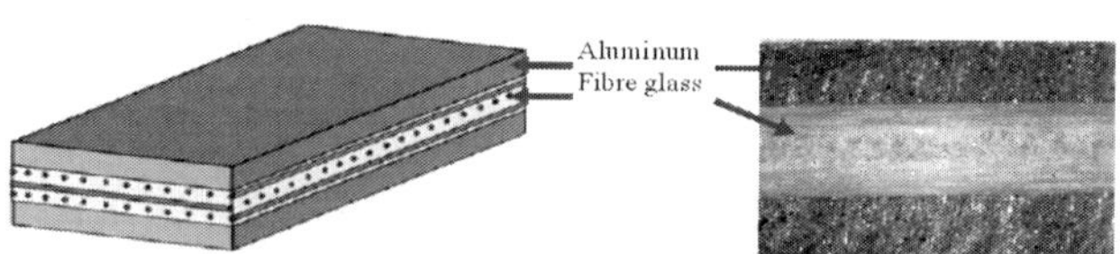

Figure 11.25 (Left): Schematic diagram showing the construction of a commercial FML material (GLARE 4-2/1, 4 mm thickness). (Right): Cross-sectional view of the GLARE under microscope.

Ultrasonic through-transmission C-scanning and infrared pulsed-thermography that are commonly employed for testing composite materials were used on the reference sample containing artificial defects. The results are illustrated in Figure 11.27, indicating that the ultrasonic method is capable of detecting the embedded flaws, both Teflon and Nylon. Only one scan was necessary to identify the inserts regardless of their depth location. However, when inserts overlapped, only the larger ones could be identified. In contrast, the pulsed-thermography required analysis at different time sequences to reveal flaws located in various layers. In addition, the resolution of thermal images was not as good as in C-scans. Thus, the ultrasonic method was considered to be more suitable for FML inspection than pulsed-thermography.

In addition, intact FML plates were subjected to impact damage at various energy levels and inspected using different NDT methods. Foreign object impact, even at low energies, creates surface deformation on FML that can be easily identified using a visual approach or an enhanced optical method, such as D-Sight (e.g., Figure 11.28). However, delamination caused by impact may extend beyond the visible dent and can be mapped using ultrasonic C-scan, as also seen in this

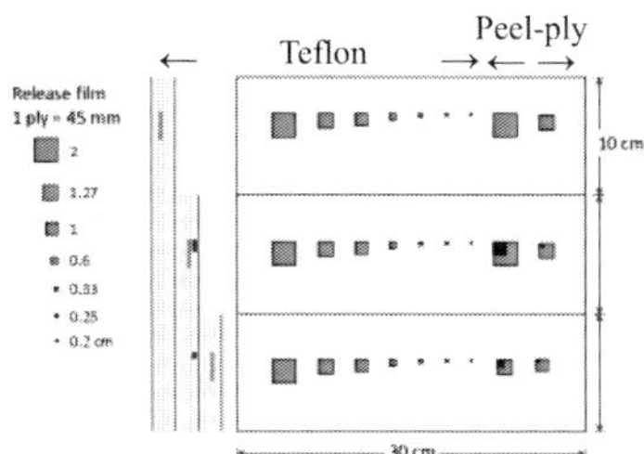

Figure 11.26 A test sample made of fiber metal laminate plates containing embedded flaws (Teflon and release film) of different sizes at different layers and the corresponding ultrasonic C-scan and infrared thermal images. The white spots on thermal images are artifacts.

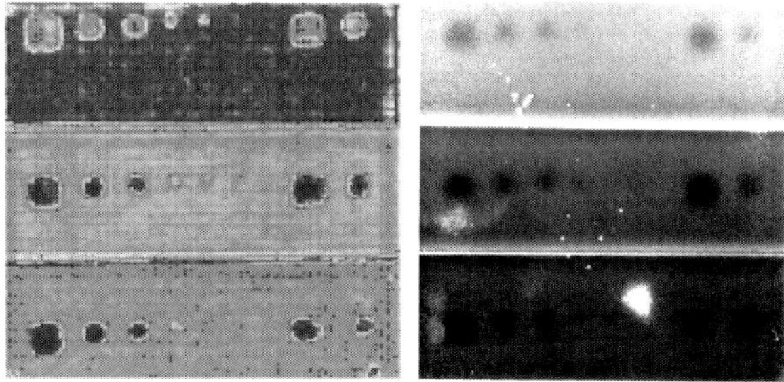

Figure 11.27 Ultrasonic C-scan (left) and infrared thermography (right) images of the sample shown in Figure 11.26.

figure. The blue and red indications are due to the surface deformation, and the surrounding green trace indicates the extent of delamination. The C-scans provide an estimate of the size and shape of delamination within the FML.

X-radiography aided by a radio-opaque penetrant was also used to map the damage. In this case, a small hole had to be drilled in the damage site to allow the penetrant to seep into the material. Using this procedure, it was possible to obtain high resolution images of the internal damage without interference from the surface dent. X-ray images of three different FML samples subjected to the same impact energy level are provided in Figure 11.29, indicating that the shape and extent of internal damage is dependent on the type of the FML material. The black circles are the small holes drilled into the sample.

As in conventional aluminum fuselage structures, FML sheets may have to be attached together using fasteners in the form of lap-joint components. If this type of construction is considered for aircraft applications, fatigue, corrosion, and disbanding, the FML must be evaluated. The work described in reference [12] presents the results of cyclic fatigue testing of FML lap-joint samples (Glare 3) containing three rows of rivets. NDT was used to identify cracks and disbonding of the FML plies, and examples are presented in Figure 11.30.

The NDT methods employed in this case were visual, eddy current scanning, and radiography with penetrant enhancement. Cracks developed on the top and bottom aluminum sheets could be easily identified visually. The eddy current technique also identified the surface-breaking cracks but could not detect internal ones, as the composite inter-layer between the aluminum plates is electrically non-conductive and prevents the electromagnetic field from penetrating below the

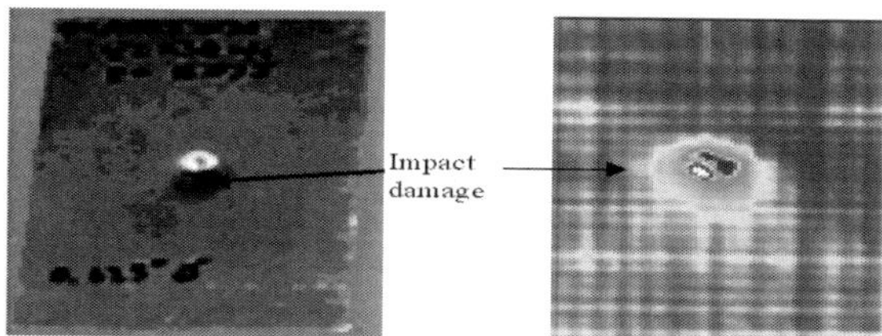

Figure 11.28 A photograph (left) and an ultrasonic C-scan image (right) of an impact-damaged FML specimen. The surface dent is indicated by blue and red colors, and the internal delamination is shown in green.

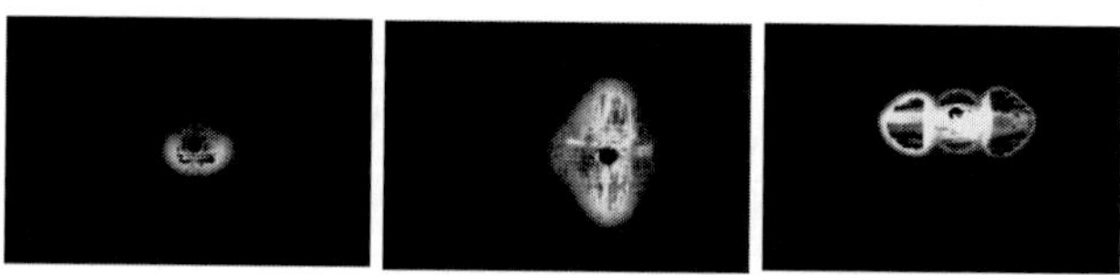

Figure 11.29 Radiographic images of impact damage in different Glare materials showing the internal delamination profiles. A radio-opaque penetrant was employed to map the damage.

surface layers. However, all cracks as well as delamination of FML plies could be identified using a penetrant-enhanced x-ray approach, as seen in Figure 11.30. The penetrant was applied to the specimen and penetrated through the cracks and the exposed edges of the samples.

11.4.3 SECTION SUMMARY

- Ultrasonic methods can be used to detect delamination and disbond in FML.
- Surface dents due to impact affect the UT results, and the extent of delamination can be mapped if it extends beyond the deformed area.
- Surface cracks are easily detectable using visual methods as well as the eddy current technique. However, neither approach can identify cracks in the internal aluminum layers.
- X-rays may be used with a radio-opaque penetrant to reveal internal disbonding as well as cracks if access is available to both sides and damage is open to the surface so that a penetrant fluid can be applied.

11.5 HONEYCOMB SANDWICH PANELS

Honeycomb sandwich materials are hybrid composites that are made of a honeycomb core sandwiched between two CFRP solid laminates. The core is either aluminum or Nomex® that is bonded to the face sheets using adhesives. They are widely used in modern aircraft in components, such as control panels, rudders, blades, access doors, and even sections of fuselage. Figure 11.31 shows the basic construction of a honeycomb panel, along with photographs of a Nomex® honeycomb specimen, and a typical aircraft part made of honeycomb materials. As with composite solid laminates, the use of honeycomb sandwich materials in the primary structures of aircraft requires inspections after manufacturing and during service.

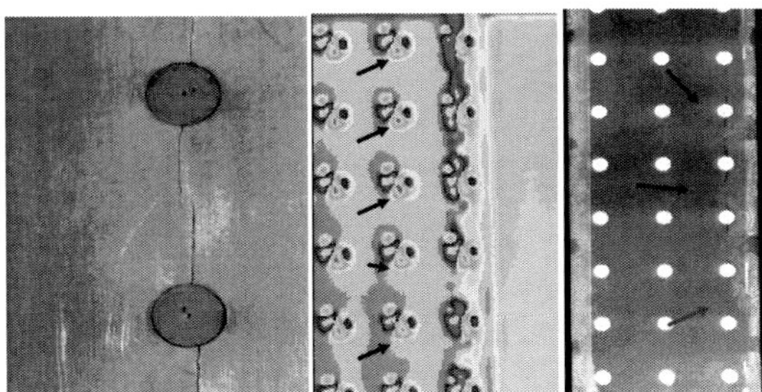

Figure 11.30 (Left): Close-up photograph of fatigue cracks between rivet-holes in FML lap-joint sample. (Middle): Eddy current image of the same sample showing traces of fastener and cracks (black arrows). (Right): Enhanced x-ray image of the same sample showing cracks (black arrows) and internal disbonding (red arrows).

11.5.1 DEFECT TYPES

The types of manufacturing defects in honeycomb face sheets are the same as those described in the section on CFRP solid laminates and include cavities, porosity, resin starvation, and foreign material inclusions. In addition, abnormalities, such as un-bonded areas due to lack of adhesive between the face sheets and the core, excessive adhesive inside the core cells, and gaps between the core sheets or between the composite and metallic parts, may exist. Damage types that may occur during service could include impact damage, disbonding of the face sheets from the core, and environmental degradation due to water intrusion into honeycomb or exposure to fire and corrosive fluids.

Impact damage is an issue as the CFRP face sheets of honeycomb structures are generally very thin (< 2 mm in thickness) and are less resistant to impact than the thick-section solid laminates (> 4 mm in thickness). Impact, even at low energies, may create a combination of skin delamination and fiber breakage as well as skin-core disbonding and core damage. These are schematically illustrated in Figure 11.32. Typical impact damage in aluminum and Nomex® honeycomb panels are presented in Figure 11.33. In some cases, the surface dent may be invisible or may become invisible over time, making it difficult to detect.

11.5.2 NDT METHODS

Manufactured honeycomb sandwich parts are typically inspected for processing defects in an automated ultrasonic C-scan system using the through-transmission approach. The part edges must be sealed to avoid water penetration into the core. The UT energy entering through the first face sheet is transmitted by the core

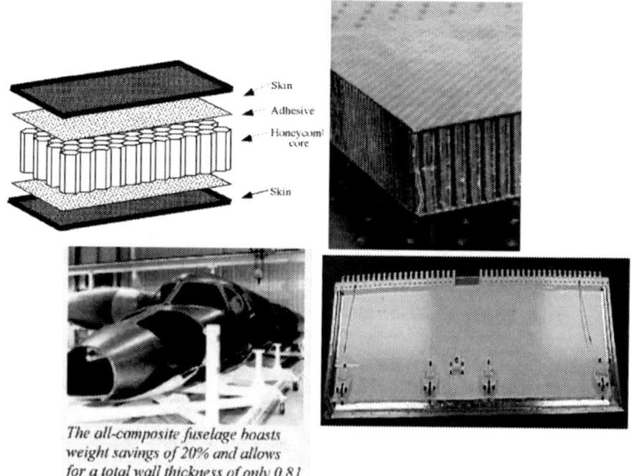

Figure 11.31 (Top-left): Schematic of a honeycomb material construction. (Top-right): A sample made of CFRP face sheets and Nomex® core. (Bottom-left): Honeycomb aircraft fuselage. (Bottom-right): Honeycomb control panel with metallic attachments.

walls, and after passing through the second face, the sheet is picked up by the receiving sensor. The core material and thickness are important factors that affect the inspection. Aluminum core can easily transmit ultrasonic energy along the cell-walls, while Nomex® is less efficient in this regard. Generally, frequencies less than 5MHz are used for Al honeycomb parts, while for Nomex® parts even lower frequencies may be necessary. Figure 11.34 shows an example of a composite honeycomb panel representing a typical access door of an aircraft. The panel is made of CFRP face-sheets and Nomex® core that is tapered down close to its edges where the two skins are joined together, enclosing the honeycomb core.

The corresponding ultrasonic through-transmission C-scan image is presented in Figure 11.35. In this figure, the pink color represents the surrounding solid laminate, and the red shows the honeycomb sandwich section. A change in the thickness of the surrounding solid laminate and a few other indications are identified, as shown in this figure. The cross lines indicate that the honeycomb core consists of four individual segments, and there are gaps between them. Also, there

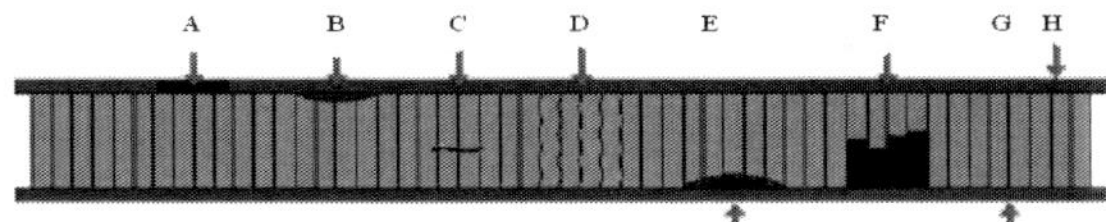

Figure 11.32 Schematic illustration of typical defects in honeycomb panels. A: Front face delamination. B: Front face-core disbonding. C: Core cracks. D: Core walls disbonding. E: Rear-face disbonding. F: Crushed core. G: Rear face. H: Front face.

Figure 11.33 Typical impact damage in aluminum and Nomex® honeycomb parts. Note that the surface dent is invisible, particularly in the Nomex® sample, but internal damage is significant.

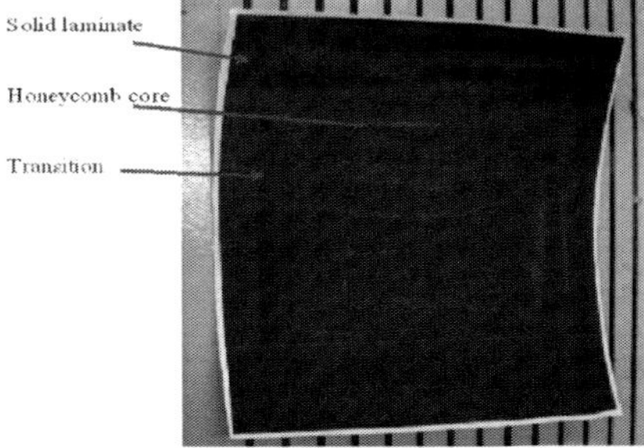

Figure 11.34 Photograph of the inner surface of a honeycomb sandwich panel.

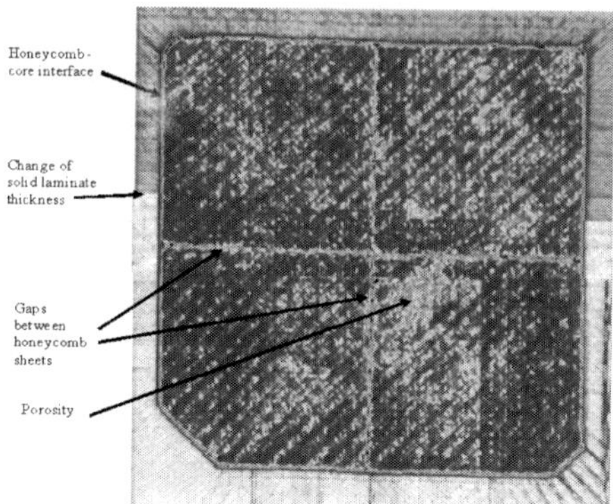

Figure 11.35 Ultrasonic through-transmission C-scan inspection result of the above-mentioned panel.

are indications of scattered pores in the face sheets that are more pronounced in one area near the middle of the panel.

The through-transmission method enables the detection of flaws in both skins and the core in a single inspection; however, it cannot provide the exact through-the-thickness location of the flaws. Also, in large aircraft parts, such as fuselage sections, access to both sides may not be possible, or attachments may prohibit the use of the through-transmission approach. Other methods, such as the pulse-echo, pulsed thermography, or laser-ultrasonic option, may be used to inspect internal surfaces, but these methods cannot provide much information about the core.

Figure 11.36 shows a pulsed thermography image of the same panel taken from the inner surface. This image shows features, such as ply drop-offs, at one edge, and ply overlaps that were not quite visible in the ultrasonic through-transmission test. However, it does not show the thickness transition zone and the scattered pores seen in the through-transmission C-scan. Overall, more than one NDT method may be necessary to inspect honeycomb parts and identify different types of manufacturing defects that may exist.

As in composite solid laminates, honeycomb parts are inspected for signs of impact damage or environmental degradations using visual and aided visual methods. However, visual inspections are not very reliable and do not provide information about the extent of damage inside the structure. Tap-testing is also used to identify disbonding of bonded honeycomb structures of aircraft using either a simple coin-tapping or automated devices. An experienced inspector may recognize the disbond from the duller sound at damage sites as compared to the surrounding undamaged areas. Sonic methods, such as acoustic resonance or mechanical impedance analysis (MIA), are also employed. In the resonance approach, the probe is coupled to the test part with a coupling fluid, and a resonance condition is created on damage-free sections of the part by frequency sweeping. The presence of a disbond is detected by the change in the resonance

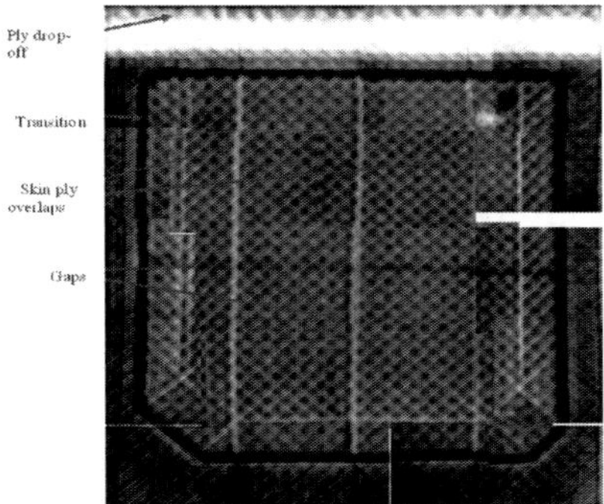

Figure 11.36 Pulsed thermography inspection result of the above-mentioned panel taken from the inner surface- second derivative at 2.733s.

frequency when the probe is moved from a sound area to the damaged site. The MIA approach uses a dual-element probe, which is dry-coupled to the part and measures the local stiffness. An optimum drive frequency is first set on a sound area (or reference sample) and then, testing is performed at the fixed optimal frequency. The presence of a disbond is identified by measuring changes in the local stiffness (mechanical impedance) as the probe is moved from damage-free regions of the part to the disbond site. Skin disbonds and core damage that affect the local stiffness can be identified.

The above mentioned approaches work well when the disbond is large and the damage site is not highly constrained. However, with these methods it is difficult to size the damage accurately. For more precise damage characterization, ultra-sonic tests are carried out by scanning the impact site using hand-held contact probes or portable automated x-y scanners. Ultrasonic methods use high frequencies (1–10 MHz range) and have better sensitivity than the sonic tests that operate in the kHz range. In the following paragraphs, ultrasonic tests (both pulse-echo and through-transmission) on an impact-damaged honeycomb panel are assessed by comparing the UT results with actual damage size and profile observed under a microscope. The assessment also includes two other methods, namely infrared pulsed thermography and laser shearography, that were available at IAR.

Figure 11.37 shows a CFRP-Nomex® honeycomb panel containing barely-visible impact damage. The panel was inspected using ultrasonic pulse-echo, as well as through-transmission methods. The latter was carried out in an immersion tank, but the sample edges were properly sealed to avoid water intrusion into the core. As shown in this figure, the damage size provided by the pulse-echo approach was smaller than that indicated by the through-transmission method. Shortly after the through-transmission test, infrared pulsed thermography was performed. The thermal image revealed an area of high thermal difference in the middle that was close in size to the pulse-echo estimate surrounded by a larger trace of lower contrast that was similar in size to the through-transmission result

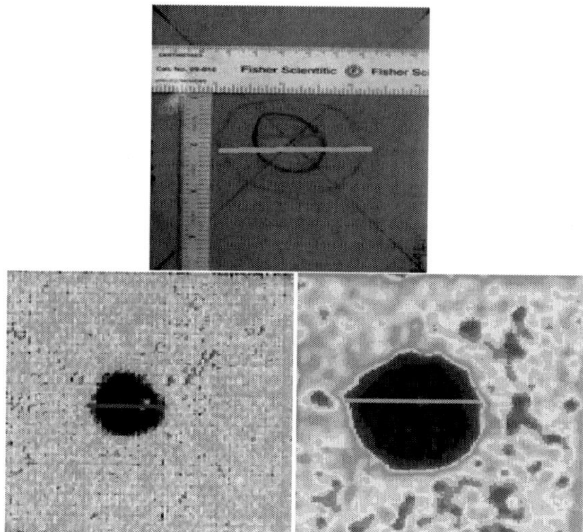

Figure 11.37 (Top): A CFRP-Nomex honeycomb panel with low-energy impact damage in the middle. (Bottom-left): Ultrasonic pulse-echo C-scan. (Bottom-right): Ultrasonic through-transmission C-scan. Horizontal lines in the top photograph correspond to damage sizes provided by pulse-echo and through-transmission approaches, respectively.

(Figure 11.38, left). Then, laser shearography was applied using a vacuum, and the damage trace was estimated to be similar to that given by through-transmission UT (Figure 11.38, right).

Finally, after completion of NDT tests, the panel was cut through the damage diameter, and its cross-section was examined under a microscope to verify the NDT results and correlate the NDT size estimates with the actual damage size and profile (Figure 11.39).

The comparison indicated that the largest delamination in the top skin exceeded the size of the visible dent on the surface by a factor of two. The damage size provided by the pulse-echo was very close to the largest delamination size in the top face, while the larger indication given by the through-transmission method correlated well with the core damage that extended laterally beyond the skin delamination. The results of the infrared thermography were interesting as the high thermal difference in the middle matched well with the top skin delamination size, while the surrounding lesser thermal contrast correlated with the core damage. The damage indication produced by shearography was close in size to that of the through-transmission UT image.

This example shows that the pulse-echo approach provides a good estimate of the front-skin delamination but does not identify the core damage. In contrast, ultrasonic through-transmission gives the overall damage size, including the core fracture as the beam passes through the entire thickness of the material, and is affected by the broken cell-walls. However, it is difficult to apply this approach in

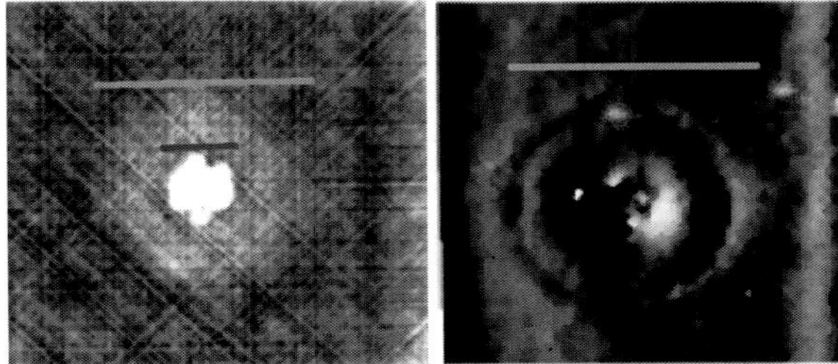

Figure 11.38 (Left): Infrared thermal image of the panel taken from the damaged side shortly after ultrasonic inspection in an immersion tank. (Right): Vacuum shearography image of the above panel taken from the damaged side.

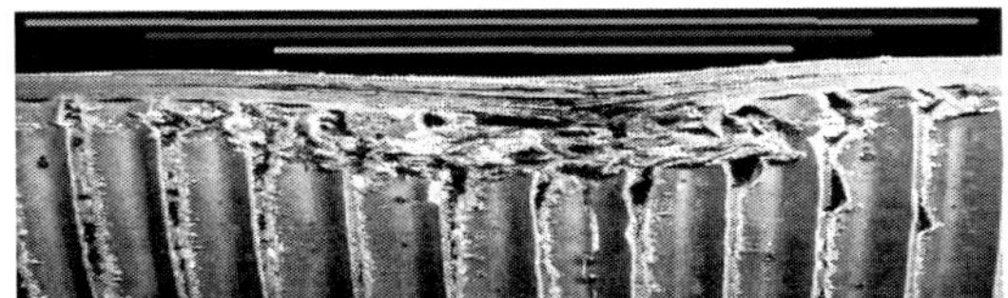

Figure 11.39 A micrograph of the cross-section of the damage showing surface dent, skin delamination and fracture, and core damage. The maximum sizes of each damage type are identified by blue for dent diameter, red for skin delamination, and green for core damage.

the field. The thermography results led to the view that, during ultrasonic immersion tests, water had penetrated through the skin cracks into the delaminated layers and the broken cells; thus, heat diffuses at different rates in these areas as compared to the intact region. In fact, when thermography tests were repeated after a few weeks, the core damage could not be identified since water had evaporated. In the shearography test, the vacuum causes the skin to deform out-of-plane where cell-walls are broken, and, thus, the results can be related to the core damage.

Overall, if skin delamination is of concern, tap-testing or resonance methods followed by ultrasonic pulse-echo or infrared thermography are adequate to identify and size the damage. However, if damage to the core is considered important, then inspection using MIA followed by either through-transmission UT, if possible, or vacuum shearography may be needed. It must be pointed out that the above-mentioned observations are specific to the CFRP-Nomex® material evaluated here and require validation for other honeycomb materials or configurations.

As seen in the above example, water can penetrate through surface cracks or gaps into honeycomb cells and over time cause deterioration of Nomex® or corrosion of aluminum core. Water intrusion into honeycomb can be identified using ultrasonic through-transmission or radiography (e.g., Figure 11.40); however, in either case, access to both sides of the part is necessary. Alternatively, infrared thermography can be carried out from one side. However, the entrapped water must be in contact with the skin laminate that is being inspected.

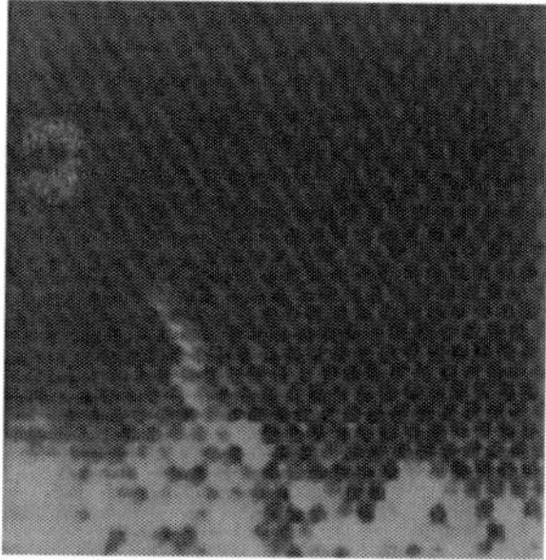

Figure 11.40 A reverse geometry X-ray image of a composite honeycomb panel (Nomex® core with Kevlar skin) with water entrapment (Digiray® Digital X-ray Systems Inc.).

11.5.3 SECTION SUMMARY

- Manufacturing defects, such as skin voids and porosity, skin-core unbonded areas, and core gaps in honeycomb sandwich parts can be detected using the ultrasonic through-transmission technique; however, the edges must be properly sealed to prevent water ingress.
- The pulse-echo approach may be used to identify defects in skin laminates only.
- Like pulse-echo UT, pulsed thermography is best suited for inspection of skin laminates and may be employed as a complimentary approach to ultrasonic tests.
- Visible impact damage and BVID can be identified using visual or enhanced optical methods but damage size or profile cannot be inferred.
- BVID may disappear with time, especially in CFRP-Nomex® honeycomb parts, making the optical methods less effective.
- Tap-testing or the sonic resonance approach can be used to identify skin delamination or disbonding; however, damage size cannot be easily determined.
- Either pulse-echo UT or pulsed thermography can be used to map the extent of delamination of the top skin, but core damage cannot be identified using these procedures.
- Mechanical impedance analysis may provide indications of core disbonding and damage, but quantitative information cannot be easily obtained.
- If access is available to both sides of the part, through-transmission UT can provide information about the overall damage size (skin delamination and core damage). However, in-situ application may be difficult or impossible.
- Laser shearography is much easier to apply in the field to reveal the entire damage in honeycomb parts; however, the material's stiffness and the size and location of the damage are important considerations.

- Water intrusion into the honeycomb core can be identified using ultrasonic through-transmission or radiography; however, access to both sides of the part is necessary.
- Infrared thermography may be used to identify water ingress into the honeycomb; however, the entrapped water must be in contact with the skin laminate that is being inspected.

11.6 FOAM-CORE SANDWICH PANELS

Foam-core sandwich materials are used in some small aircraft in areas where the strength requirement is low but a high stiffness is needed (e.g., flight control surfaces). Such components are typically made of a very low-density, rigid foam core covered by thin layers of high-strength materials (e.g., Al, CFRP, or glass fiber epoxy) as face sheets. The examples provided here are panels that are made of E-glass/epoxy face sheets and PVC closed-cell foam core.

11.6.1 DEFECT TYPES

Manufacturing of the foam-core structures varies depending on the face-sheet and core materials and includes processes, such as adhesive bonding of the face sheets to the core or resin-infusion through the fiber-glass sheets laid on the foam core. Thus, the defect types may vary, too. Missing foam or cavities and gaps, as well as resin starvation of the face sheets or lack of bond between the face-sheet and the core, may occur as a result of poor manufacturing processes. During service, as with other sandwich structures, damage by impact and material degradation due to intrusion of fluids or exposure to hostile environments (e.g., extreme heat) may occur.

11.6.2 NDT METHODS

The conventional sonic and ultrasonic methods used for testing of the majority of composite materials and sandwich parts may not be applicable for the inspection of foam-core sandwich structures due to the high acoustic attenuation of the foam. For example, in Figure 11.41, a 5-mm thick PVC foam-core with E-glass face sheets is shown where segments of the foam have been removed (3 mm and 4 mm in depth) to simulate large cavities or missing foam. The sample was tested using low frequency sonic and ultrasonic methods. As seen in this figure, even using low (< 1MHz) frequency through-transmission approach, which is often recommended for materials with high acoustic attenuation, did not reveal the simulated cavities. However, infrared pulsed-thermography, which was also performed in the through-transmission mode, provided some indication of the internal defects [14]. The foam core, being a porous and soft material, completely absorbs the sound energy even when low frequencies are used, while heat is diffused at different rates depending on the foam condition, creating minute temperature differences on the surface that are measured using a sensitive infrared camera.

In another example, infrared thermography was carried out on a panel made of E-glass/ epoxy face sheets and foam core. A resin-infusion process had been used to manufacture the part. Figure 11.42 shows a photograph of the sample along with its thermal image obtained from one side only. Both images clearly show the resin-infusion sites (darker dots), as well as the areas that resin has not reached (resin-dry sites). Since the glass/epoxy face sheet is thin and transparent, the dry areas can be seen visually, thus, supporting a correlation with thermal maps. As seen in these examples, with infrared thermography, it may be possible to identify manufacturing defects, such as resin starvation, of the face sheets or large core cavities.

Similar samples were also subjected to impact damage at different energies and evaluated visually, as well as by pulse-thermography. The transparency of the glass/epoxy face sheet was slightly changed as a result of impact damage, and it was possible to see the damaged areas visually. However, visual examination was not very conclusive for surfaces that were coated with a non-transparent latex paint. Thus, thermography tests were carried out, and the results are shown in Figure 11.43, along with their photographs. While the impact sites could be visually identified

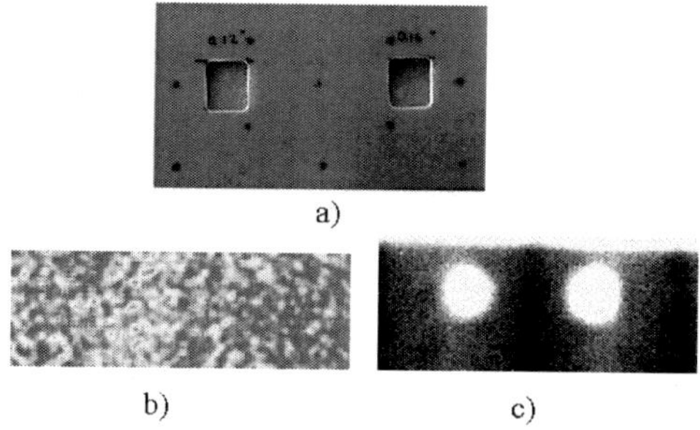

Figure 11.41 A photograph of a PVC foam core with void (a) and ultrasonic C-scan (b) and pulse-thermography (c) images both obtained in the through-transmission modes.

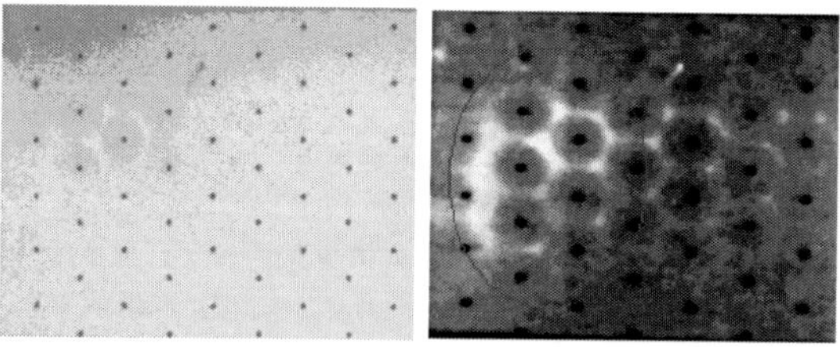

Figure 11.42 (Left): Photograph of a foam-core sample made of E-glass/epoxy face sheets and PVC core using a resin infusion process. Glass/epoxy face sheet is transparent and resin-dry areas (lighter colors) and resin-rich infusion sites (darker spots) are visible. (Right): Pulsed-thermography image of the same sample showing the resin-starved area.

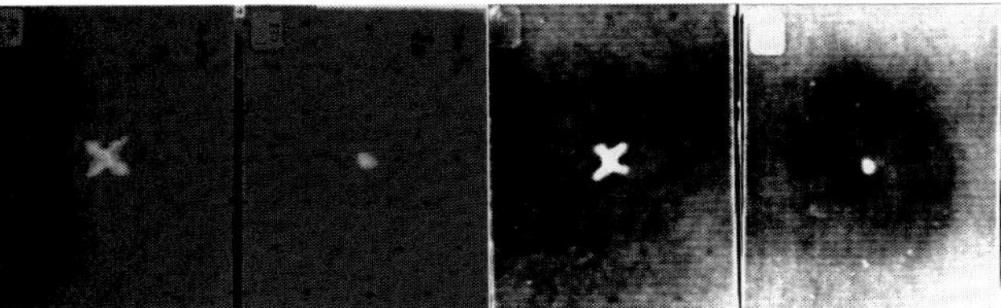

Figure 11.43 (1st and 2nd Left): Photographs of foam-core panels subjected to impact damage at two energy levels. Damage sites and shapes are visible. (Right): Corresponding thermal images showing the damage sites and shapes.

from the surface dents they caused, the thermal maps provided a better indication of the shape and extent of the damage. A comparison of the resulting images indicated that skin delamination and disbonding did not exceed much beyond the visible dent, and the foam material was able to mostly contain the damage in the impact site. The differences in contrast around the damage sites in thermal images are due to the surface reflectivity variations and are not related to damage.

11.6.3 SECTION SUMMARY

- Foam-core sandwich parts are difficult to inspect using conventional sonic and ultrasonic techniques.
- Infrared thermography may provide some indication of manufacturing defects, such as resin starvation of skins or large core cavities.
- Impact damage may be identified by visual examinations and mapped using pulsed-thermography.
- In the glass/epoxy foam-core panels examined, the impact damage was mostly contained within the boundaries of the visible surface dent.
- Core damage was not easily detectable using thermography or sonic/ ultrasonic methods.

11.7 INSPECTION OF ADHESIVE BONDS

Adhesively-bonded joints are extensively employed in the construction of aircraft structures. Adhesives may be used to attach two or more parts together in a variety of configurations, as schematically illustrated in Figure 11.44. The parts or adherends may be metallic, composite, or other materials. In many respects, bonded joints have advantages over mechanical fastening methods for aerospace applications. They allow greater design flexibility and provide superior aerodynamic surface finish, more uniform stress distribution, higher strength-to-weight ratio, better fatigue behavior, and lower cost. However, there are a number of disadvantages that limit their use in primary structures. A major problem is that the ultimate bond strength is very dependent on the preparation of the adherend surfaces,

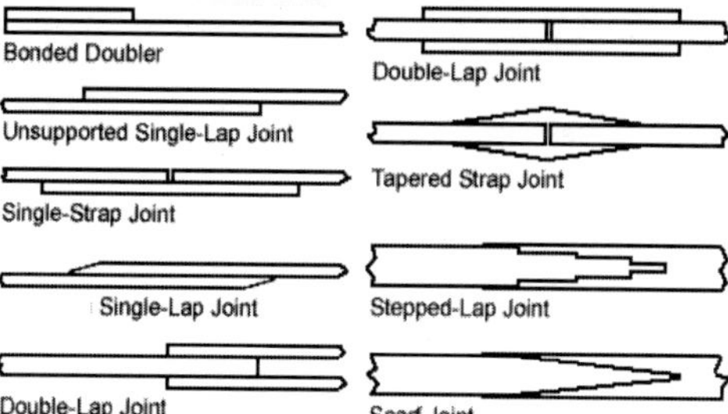

Figure 11.44 Schematic presentation of adhesive bonding configurations used in aerospace (http://www.vircon-composites.com/2_3_2.asp).

properties of the adhesive, and its curing conditions. Furthermore, it is difficult to obtain an accurate measure of the bond quality or integrity non-destructively.

The bond quality after processing is influenced by physical discontinuities within the adhesive layer, such as air-gaps (voids), porosity, and foreign material inclusions, as well as contamination of the adherent surfaces. The latter is very difficult to detect using NDT. Also, during service, the bond integrity may be affected by disbonding of the adherends as a result of overloading, impact, or vibration, as well as by deterioration of the adhesive layer as a consequence of exposure to high temperatures, humidity, or hostile fluids. Again, deterioration of adhesive cannot be easily quantified by NDT.

In this section, first, defect and damage types that may occur in adhesively-bonded structures of aircraft will be mentioned, and then the NDT methods employed for identifying physical discontinuities, as well as those that have shown potential for bond quality or integrity assessment will be described. The bonded configurations covered in this section include mostly lap joints and strap joints that are used in the construction of both metallic and composite airframe structures, as well as in composite patch repairs. Sandwich structures that use adhesives to bond face-sheets to the internal core material are not covered in this section as they have been described earlier.

11.7.1 Defect Types

Lack of adhesive or air-gaps between the two parts that are being attached are referred to as un-bond areas or voids and are the most common processing defects in bonded joints. Excessive porosity in the adhesive layer and lack of adequate adhesion of the adherents or poor bonds may also exist in a bonded joint. Poor bond quality is known as "kissing bond" and is defined as a condition where the two bonded surfaces are in intimate contact, but there is little bond strength. Kissing bond may be caused by adhesive aging, improper surface preparation, or contamination. During service, overloading, vibration, and foreign object impact may cause disbonding of bonded parts, while prolonged or repeated exposure to humidity, hostile fluids, or extreme temperatures may degrade the bond quality.

11.7.2 NDT Methods

NDT detection of physical air-gaps, excessive porosity in the adhesive layer and disbonding of the joint is possible; however, the adherent material, component geometry, bond configuration, and access to the site are important considerations. Sonic or ultrasonic methods, radiography, infrared thermography, and shearography may be used to identify processing defects. For example, as illustrated schematically in Figure 11.45, ultrasonic pulse-echo or through-transmission methods are employed to detect voids, porosity, and un-bonded areas in the adhesive layer after joining the two metallic or composite parts together.

Figure 11.46 shows an example of an ultrasonic C-scan inspection of an adhesively-bonded aluminum lap-joint specimen. Large lap-joint segments of aircraft structures are inspected using automated ultrasonic C-scan systems while small patch-repaired sections are tested manually by hand-held probes or transducer arrays. Depending on the component geometry, bond configuration, and access, either pulse-echo or through-transmission approaches may be used.

Disbonding of bonded parts during service is identified using sonic or ultrasonic devices (e.g., tap-testing, resonance, MIA, or contact pulse-echo). Similar equipment to that used for sandwich parts is utilized for lap-joint structures; however, procedures may vary, depending on the materials and geometry. As mentioned earlier, tap-testing is often used, as it is simple and inexpensive; however, the approach is highly operator-dependent and lacks reliability, especially when the disbond is small or the structure is constrained and stiff. Sonic resonance and mechanical impedance analysis provide better reliability, but these methods are qualitative and do not provide an accurate estimate of the disbond size. Ultrasonic pulse-echo tests carried out from the front surface are suitable for bonded composite laminates, lap-joints, and patch repairs. Adhesively-bonded patch repairs are widely used on aging aircraft to provide reinforcement, slow down crack growth, and fix impact or battle damage. Figure 11.47 (left) shows an example of composite patch repair of an aluminum aircraft structure used at a critical location

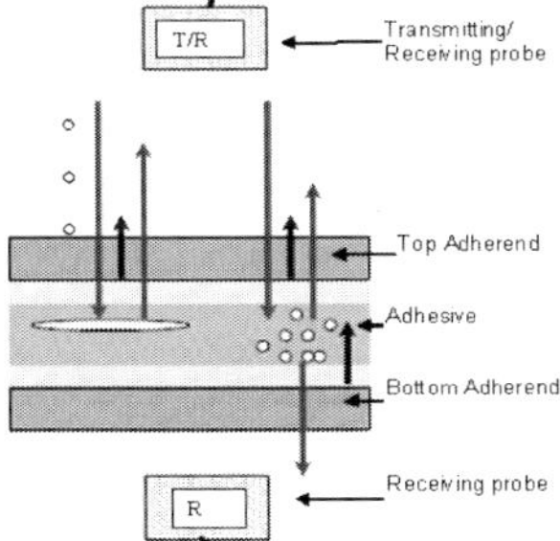

Figure 11.45 Schematic representation of ultrasonic inspection of adhesively-bonded joints using either a single transmitting/receiving probe in pulse-echo mode or two probes in through-transmission. The latter approach will require access to both sides of the joint.
http://www.adhesives.org/StructuralDesign/JointTesting/NonDestructiveEvaluation.aspx

Figure 11.46 Ultrasonic C-scan image of an adhesively-bonded aluminum lap-joint showing an un-bonded area.

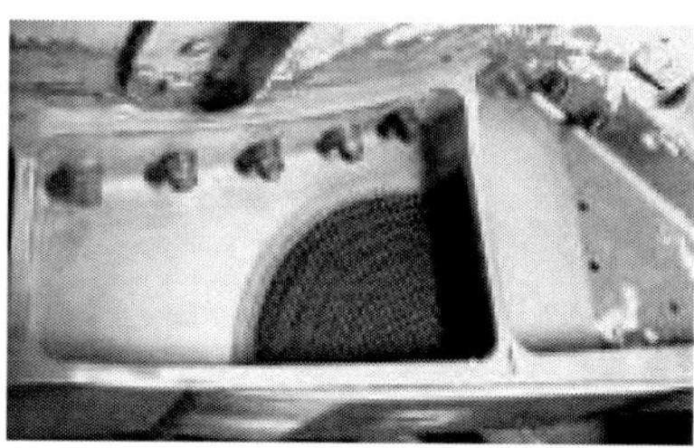

Figure 11.47 Example of composite patch repair applied to a critical location of an aluminum aircraft structure.

to reinforce and prevent crack growth. In addition to sonic and ultrasonic methods, infrared thermal imaging has shown potential for mapping disbonding of composite patches applied to metallic substrates.

In a study described in reference [14], pulsed thermography was employed to detect and follow disbonding during cyclic loading of a 1.5-mm-thick CFRP patch applied to rectangular plates of Al 2024-T3. Figure 11.48 shows one of the specimens, along with pulsed thermography images obtained at progressively increasing fatigue cycles. The disbond progress during the test is clearly shown, and the disbond size is estimated. This approach was also used to check the above-mentioned patch-repaired structure for possible disbonds. Figure 11.49 shows the infrared thermal image.

While NDT detection of physical defects, such as voids, un-bonded regions, and disbonding of bonded parts, is relatively easy, identification of kissing bonds or degradation of bond quality is very challenging. Only a few NDT methods have shown potential, and mostly in laboratory tests carried out on simple specimen geometries. One such technique is the acousto-ultrasonic approach, which uses guided-waves in pitch-catch configuration, as illustrated in Figure 11.50. Studies on single lap-joint specimens made from either metallic or composite adherends and commercial adhesives indicated that degraded bonds were less effective in transmitting the AU waves, as compared to good quality bonds [15,16, 17]. Subsequent shear tests of the specimens resulted in shear strength values that correlated well with the AU signal characteristics that represented the efficiency of the bond line in transmitting the guided waves.

Laser shearography using vacuum loading has also shown potential for detecting a weak bond; however, in this case, the applied vacuum has to separate the

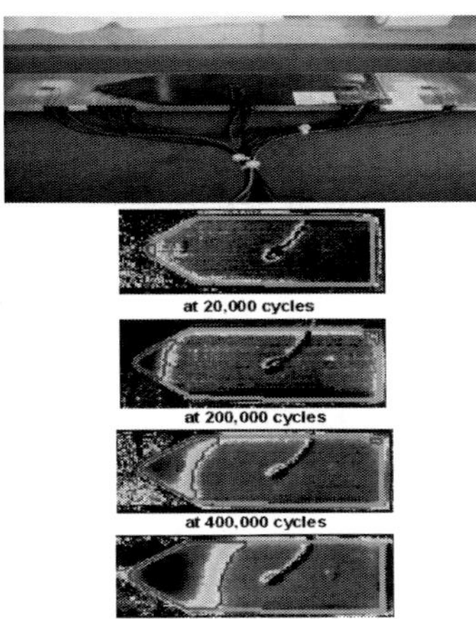

Figure 11.48 (Top): An experimental composite patch bonded to aluminum for fatigue testing. (Bottom): Infrared thermal images obtained at different fatigue cycles showing the progress of disbond from the tip of the patch.

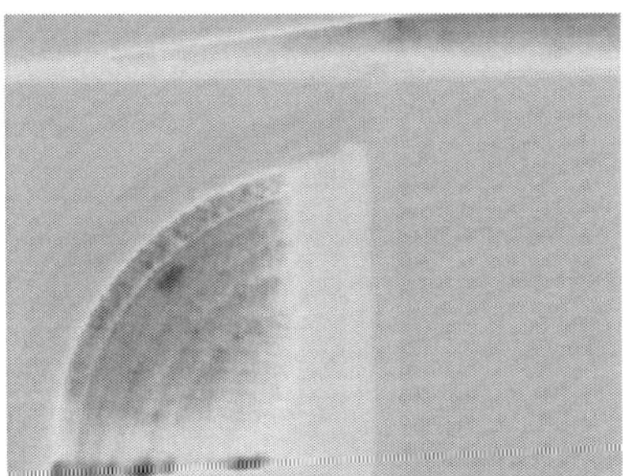

Figure 11.49 A composite patch bonded adhesively to an aircraft structure and the corresponding infrared thermal image showing areas with possible defects.

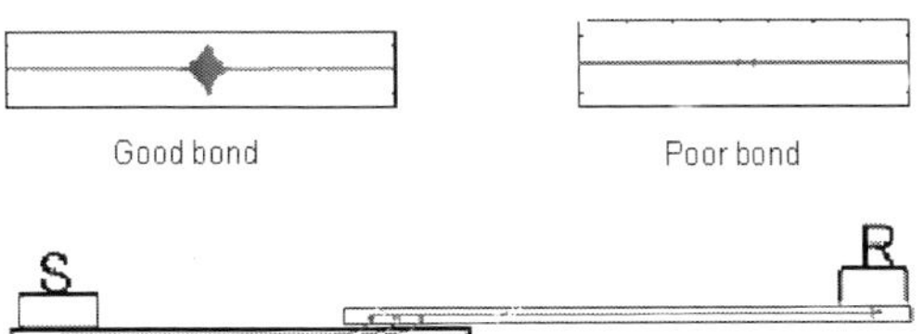

Figure 11.50 Schematic illustration of acousto-ultrasonic evaluation of bond quality in single lap-joint specimens. S is the wave sending probe and R is the receiving probe. Poor bond quality is identified by the AU overall signal attenuation.

bond line and create an out-of plane deformation of the surface in order to be detected. A similar approach has been developed that uses a laser to create disbonding at the weak bonds that may occur due to thermoplastic effects [18]. The out-of-plane deformation at the disbonded area is then detected by a laser interferometer. Since, in these approaches, the bond-line must be detached using either vacuum force or thermal shock in order to create out-of-plane surface deformation, they may not be considered to be non-destructive.

11.7.3 Section Summary

- Tap-testing is a simple approach for identifying large disbonds in sandwich structures but is not very reliable.
- Automated tap-testing may increase reliability but cannot provide size estimates.
- Sonic resonance and mechanical impedance analyses are more suitable for sandwiched parts; however, they are slower and more costly.
- Ultrasonic methods are the most reliable and can measure the disbond size, both in bonded laminates and sandwich components.
- Infrared thermography has shown potential for disbond detection in composite patches as well as skin delamination in sandwich structures.
- Vacuum shearography and laser tapping may have potential to identify weak bonds and ultrasonic guided waves may provide some estimate of bond quality in a laboratory environment.

11.8 CHAPTER REFERENCES

[1]   An Introduction to the Ultrasonic C-scan Inspection of Advanced Composite Materials, A. Fahr and A.M. Charlesworth, NRC Report: LTR-ST-1602, 1986.

[2]   A Comparison of Pulsed Thermography and Ultrasonic C-scanning for Inspection of Aircraft Composite Structures, A. Fahr, M. Genest, M. Brothers, and R. Rutledge, NRC Report: CPR-SMPL-2009-0003, 2009. Also presented at NDT in Canada Conference, London, Ontario, Canada, August 2009.

[3]   Ultrasonic and Thermographic Non-destructive Inspection of Composite Laminate Structures, A. Fahr, M. Genest, M. Brothers, D. Djokic, J. Laliberté, and A. Yousefpour, NRC Report: CPR-SMPL-2007-0142, 2007. Also presented at the 7th Canada-Japan Composite Workshop, Kanagawa, Japan, July 28–31, 2008.

[4] Pulsed Thermography of Non-Planar Surface for Non-destructive Evaluation, M. Genest, X. Maldague, E. Grinzato, P.G. Bison, and S. Marinetti, Proceedings of the CASI 2005 Conference, Toronto, Ontario, April 26–27, 2005.

[5] Ultrasonic and Thermographic Non-destructive Inspection of Composite Tubular Structures, A. Fahr, M. Genest, M. Brothers, D. Djokic, J. Laliberté, A. Yousefpour, and C. Marsden, NRC Report: CPR-SMPL-2007-0142, 2007.

[6] Determination of Fiber Orientation in Composite Laminates Using Polar C-scan Imaging of Scattered and Leaky Ultrasonic Waves, A. Fahr and C. E. Chapman, NRC Report: LTR-ST-1606, 1987.

[7] Ultrasonic and Mechanical Qualification of Composite Arm Booms for the Space Station Remote Manipulator System (SSRMS), A. Fahr, C.E. Chapman and C. Poon, Journal of the Canadian Institute for Non-destructive Evaluation, Vol. 22, No. 5, September/October 2001.

[8] Damage Assessment in Carbon Fiber Epoxy Composites using NDE Techniques, A. Fahr, N.C. Bellinger, C.E. Chapman, S. Beland and C. Poon, NRC Report: CPR-SMPL-1993-0083 published in Proceedings of the Canadian Association for Composite Structures and Materials (CACSMA) Conference, 1993.

[9] Non-destructive Evaluation of Adhesively Bonded Joints in Graphite/ Epoxy Composites using Acousto Ultrasonics, A. Fahr, S. Tanary, S. Lee and Y. Haddad, Journal of Pressure Vessel Technology, Transactions of the ASME, Vol.114, No. 3, August 1992.

[10] Thermographic NDE Inspection of Kevlar Panels and Kevlar Box, M. Genest, NRC Report: LM-SMPL-2005-0171, 2005.

[11] Thermographic Inspection of Kevlar Enclosures SPDB and SPEDC, M. Genest, NRC Report: LM-SMPL-2005-0199, 2005.

[12] Application of Fiber Metal Laminates in Airframe Structures, J.F. Laliberte, C. Poon, P.V. Straznicky and A. Fahr, Polymer Composites, SPE Topical Journal, Vol. 21, No.4, August 2000.

[13] Non-destructive Evaluation of Fiber-Metal Laminates, A. Fahr, C.E. Chapman, J.F. Laliberte and C. Poon, NRC Report: CPR-SMPL-2000-0072, 2000. Also in the Proceedings of the 7th International Conference on Composite Engineering, Denver, Co, July 2–8, 2000.

[14] Pulsed Thermography for Non-destructive Evaluation and Damage Growth Monitoring of Bonded Repairs, M. Genest, M. Martinez, N. Mrad, G. Renaud, A. Fahr, Composite Structures, Volume 88, Issue 1, pp. 112–120, March 2009.

[15] Review of IAR NDI Research in Support of Ageing Aircraft, A. Fahr, J.P. Komorowski, D.S. Forsyth, C.E. Chapman, NDT Net, Vol. 4, No.1, January 1999.

[16] Non-destructive Evaluation of Adhesive-Bonded Joints, A. Fahr and S. Tanary, Proceedings of the AGARD 69th Meeting on The Impact of Emerging NDE/NDI Methods on Aircraft Design, Manufacture and Maintenance, Brussels, 1–6 October. 1989.

[17] Non-destructive Evaluation of Adhesively Bonded Joints using Acousto-Ultrasonics, A. Fahr, S. Tanary, and Y. Haddad, Journal of Acoustic Emission, Vol.8, No.1/2, Jan.–June 1989.

[18] Application of Laser Tapping and Laser Ultrasonic to Aerospace Composite Structures, A. Blouin, C. Neron, B. Champagne and J-P Monchalin, Insight, Vol. 52, No. 3, pp. 130–133, March 2010.

CHAPTER 12

NDT of Corrosion in Aluminum Airframe Structures

12.1 BACKGROUND

Corrosion and fatigue cracking are prime concerns in aging airframes made of aluminum alloys, and there is an increasing need to detect and characterize this damage at early stages. Current aircraft structural integrity processes primarily consider fatigue cracking and do not fully address time-related degradation due to corrosion and corrosion-fatigue. Aircraft life prediction based on a comprehensive damage tolerance analysis considers all variables affecting the residual strength. Such analysis requires NDT limits and metrics for cracks and corrosion. Cracks can be considered as two-dimensional discontinuities that are measured in terms of length or area. Corrosion, on the other hand, is a three-dimensional discontinuity, which often has an irregular profile, and cannot be easily characterized by a single unit. Often, several metrics of corrosion are needed, and this may require the very judicious use of several techniques.

Airframe multi-layer aluminum lap-joints are primary structural elements of aircraft that have received considerable attention (e.g., [1]), and they will be the focus of this chapter. Corrosion of aluminum lap-joints results in the loss of material that, in turn, reduces the residual strength and fatigue life of the remaining structure. The loss of a material's thickness as a result of corrosion is the metric that is widely used to measure corrosion. Ten percent thickness loss is the maximum level that is generally tolerated in large transport aircraft; beyond that, the replacement or repair of the part is needed. NDT is used to assess the condition of a structure in terms of the remaining material thickness.

This chapter briefly describes the most common corrosion types in aging aircraft and provides a brief review of NDT techniques that have been applied for the detection and measurement of corrosion in airframe components. The review is based on a previous IAR publication [2] and describes the capabilities, limitations, and key measurement metrics of the potential NDT methods. It also provides information on their ability to characterize corrosion with relevant examples of application. As before, most examples are taken from various studies carried out at the NRC Institute for Aerospace Research. The IAR NDT results are compared with destructive verification tests employed at IAR for lap-joint corrosion characterization and are presented here.

12.2 CORROSION TYPES

A description of different types of corrosion is provided in references [3] and [4], and those that are most prominent in aluminum lap-joint structures are mentioned below, with a few examples shown in Figure 12.1:

Crevice corrosion occurs when a corrosive medium is trapped in between metal parts. For example, water entrapment between faying surfaces of lap-joints could result in crevice corrosion, which may lead to pitting and other types of degradation.

Pitting corrosion is a highly localized attack that may lead to the formation of deep and narrow cavities. This type of corrosion can occur on any metal but the effect is dependent on the alloy and the environment. Pitting is a common form of corrosion seen on aircraft parts. In lap-joint splices, pitting often starts when the cladding of the aluminum is damaged. Pits have been identified as crack nucleation sites in aircraft structures.

Intergranular corrosion is a localized form of attack that affects the grain boundary regions in a polycrystalline material. This can be a particularly insidious form of corrosion, as it may have little effect on the exposed surface where it started. In rolled products like aluminum plates, severe intergranular corrosion may produce delamination-like effects that are referred to as **exfoliation** corrosion and frequently cause a visible deformation of the surface.

Galvanic corrosion occurs when metals of different electrochemical potential are in contact in a corrosive medium. In practice, this often occurs in lap-joint structures when fasteners are made of a different metal than the aluminum substructure.

Filiform corrosion often starts as a pit but spreads sideways. It can be difficult to detect until it has progressed enough to create visual evidence, such as blisters. This type of corrosion is unusual in typical aircraft materials.

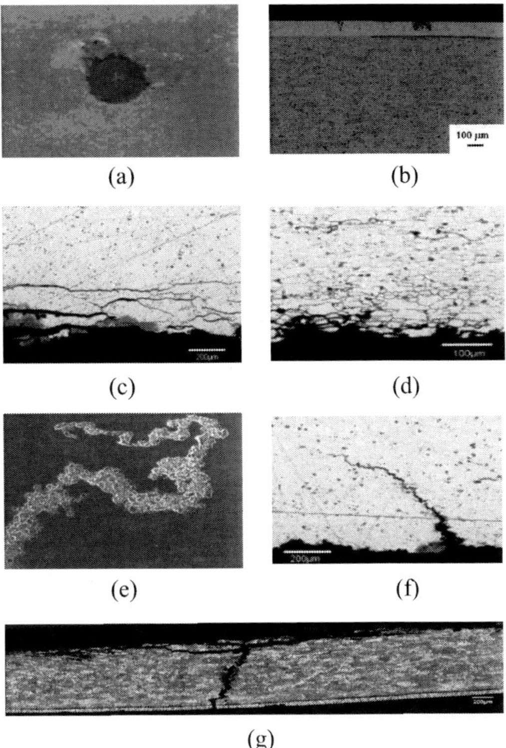

Figure 12.1 Examples of corrosion in aircraft aluminum alloys. (a): Beginning of corrosion on the surface near a fastener. (b): Corrosion pits in the aluminum cladding. (c): Exfoliation corrosion. (d): Intergranular corrosion. (e): Filiform corrosion. (f): Crack originating from a deep corrosion pit and propagating along an intergranular path. (g): Corrosion of the faying surface in an aluminum lap-joint and a through-the-thickness crack.

Fretting corrosion occurs when a corrosive attack is combined with fretting wear resulting from very small-amplitude cyclic movement of two surfaces relative to each other. This type of corrosion is often in the form of pitting and is often seen at fastener holes of lap-splice joints.

Stress corrosion is a term employed when sustained stresses acting in a corrosive environment increase the susceptibility of metals to corrosion and accelerate the nucleation and growth of cracks. Cracks resulting from these conditions are referred to as **stress corrosion cracks (SCC).**

Hydrogen embrittlement refers to material degradation caused by diffusion of hydrogen into metals. Hydrogen embrittlement could reduce crack nucleation times and increase crack growth rates. This is not considered a common problem in the materials used in aircraft structures, but some studies have suggested it can contribute to stress corrosion cracking in the 7000 series of Al alloys. In contrast, hydrogen embrittlement is commonly found in landing gear components and other parts made from high-strength steels.

12.3 ALUMINUM LAP-JOINT CORROSION

Lap-joints consisting of two or more thin aluminum sheets or plates (thickness 1 to 3.75 mm) fastened to internal sub-frames (e.g., Figure 12.2) are common structural elements of aircraft wings and fuselages. Many transport aircraft manufactured in the 1950s (e.g., KC135, B52, and B707) frequently used spot welding with no joining compounds in lap-joints. By the 1960s, lap-joints were assembled using rivets and either sealed (DC 9) or bonded using an adhesive (e.g., B737) to delay the onset of corrosion. Water ingress from the edges of lap splices or doublers often contribute to corrosion between the thin layers of aluminum fuselage skins. The corrosion is occasionally visible on the surface (e.g., Figure 12.1, a), but, most of the time, it is not visible or easily detectable. Lap-joint corrosion is the most common problem with aging aircraft made of aluminum alloys.

Corrosion creates pitting and roughness of the surfaces and produces aluminum oxide and hydroxide that have much higher volume than the initial aluminum alloy from which they form. When corrosion occurs between the layers of lap-joints, the higher volume of the resulting oxides and hydroxides causes out-of-plane deformation of the skin between fasteners that is referred to as pillowing.

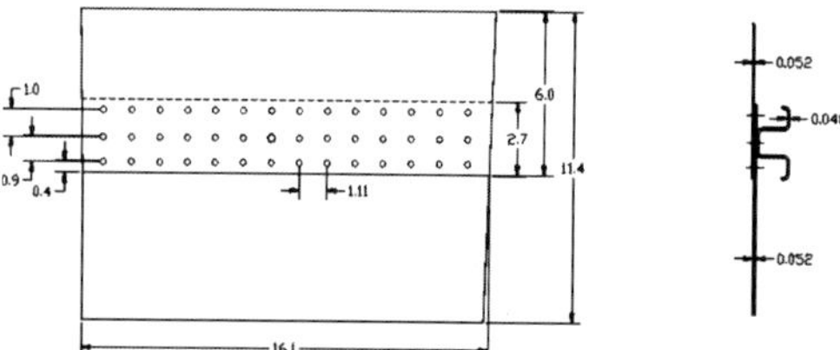

Figure 12.2 A typical fuselage lap-joint design. Dimensions are nominal and are in inches.

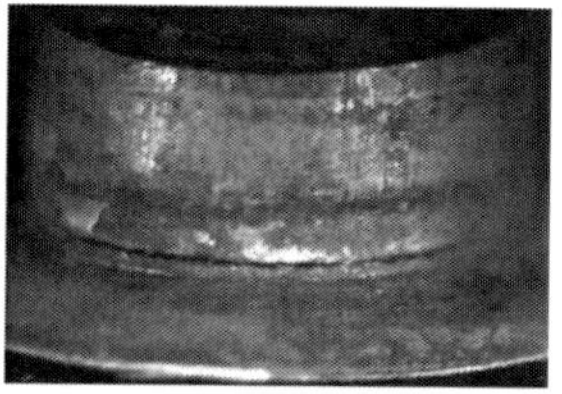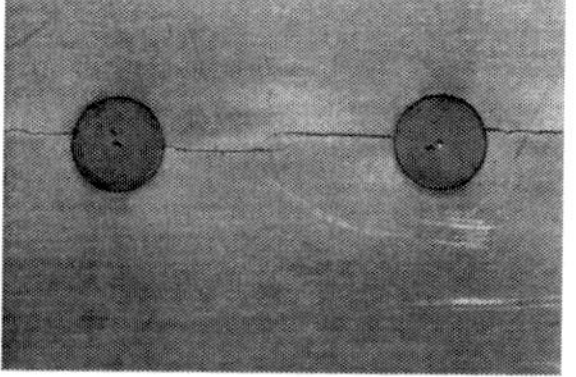

Figure 12.3 (Left): A service-induced crack at the neck of a fastener. (Right): Fatigue cracks emanating from fastener holes in the aluminum plate.

Increased pressure may lead to the separation of the top skin from fasteners or may cause fasteners to fail in a tensile overload mode (Figure 12.3, left). The extra pressure on the faying surface, along with the thinning of the skin, may also cause cracks to form in the surrounding skin. Such cracks occurring around consecutive fastener holes, such as those shown in Figure 12.3 (right), are referred to as multi-site damage (MSD) and could cause failure of the affected region. If the corrosion damage is detected and found to be beyond the manufacturer's tolerances, the corroded area is cut out and the damage repaired by patching; otherwise, the corrosion is removed by chemical or mechanical means, leaving thinner skins. In the case of multi-layer riveted joints, in addition to the material thickness loss by layer, pit size and distribution, pillowing deformation, crack size, location, and orientation may be required to fully characterize the damage created by corrosion.

Thick aluminum sections, such as wing skins (thickness 4.75 to 6.25 mm), are less vulnerable to the loss of small amounts of material to corrosion. However, they are susceptible, in many cases, to exfoliation corrosion starting in fastener holes. The fasteners are often made of a different material, such as cadmium plated steel, and when moisture penetrates between the fastener and the aluminum skin, galvanic corrosion and fretting may occur at the point of contact between the dissimilar metals. The intergranular progress of such corrosion along Al grain boundaries, which run parallel to the surface of the wing skin, will lead to delamination of the thin layers of aluminum, and this is known as exfoliation corrosion.

It is desirable to detect corrosion pits at the surface of the holes or intergranular corrosion before they progress to the exfoliation stage. However, this is not an easy task and requires removal of the fasteners. In the case of thick sections, for full damage tolerance analysis, corrosion characteristics, or metrics, such as the extent and distribution of intergranular corrosion, pit sizes on the faying surfaces of bolt holes, extent and number of layers of exfoliation, depth location of exfoliated layers with respect to the surface, and direction of exfoliation, may be needed. Corrosion of other structural components, such as hinges, may involve a combination of the above-mentioned types and require similar metrics. Further information on the various types of corrosion and how each one may impact on the structural integrity and residual strength/life of aircraft structures is provided in reference [4]. This publication provides numerous examples of how corrosion damage can be incorporated quantitatively in aircraft structural integrity assessments, and it elaborates on many of the forms of corrosion that are mentioned in the present publication, as well as some that are not dealt with here, such as hydrogen embrittlement.

12.4 NDT METHODS FOR LAP-JOINT CORROSION

A number of NDT methods are available for inspection of airframe structures for signs of corrosion; however, no single technique is capable of inspecting every different component or providing all the above-mentioned metrics that may be required. The following review considers a wide range of NDT techniques with particular attention to their capabilities or limitations in quantifying corrosion in terms of metrics that are required for life assessment analysis.

12.4.1 OPTICAL METHODS

Visual inspections of aircraft are generally carried out to detect bulging, pillowing, paint flaking, or other signs of damage that might be indicative of underlying corrosion. However, visible signs usually appear when corrosion is already at an advanced stage. There are a number of commercial devices that can be used to magnify and improve the visual detection and documentation of surface features. Examples include the various types of magnifying devices and aids, such as boroscopes, cameras, and microscopes, as well as techniques, such as optical metrology, holography, shearography, shadow moiré, and interferometry, that are sensitive to localized defects or stresses. Enhanced optical techniques, such as D-Sight and Edge-of-Light (EOL), can detect surface corrosion or the effects of corrosion, such as pillowing, at early stages [5, 6]. These methods are sensitive to surface deformation on a small scale. They are inexpensive, fast, and easy to use and, thus, attractive for corrosion that affects exposed surfaces. However, their widespread application has been limited by the difficulty in interpreting images and recognizing corrosion pillowing from the normal deformation of the skin, which occasionally lead to false indications. Figure 12.4 shows a fuselage lap-splice joint cut from a retired commercial aircraft, and Figure 12.5 presents the corresponding EOL and D-Sight images. These techniques are sensitive to small amounts of pillowing deformation that may be related to underlying corrosion; however, they cannot easily quantify the extent of the damage. The D-sight and EOL methods would be best used for rapid large-area inspection to identify potential corrosion sites so that they can be examined more rigorously using other NDT techniques offering reliable and quantitative information.

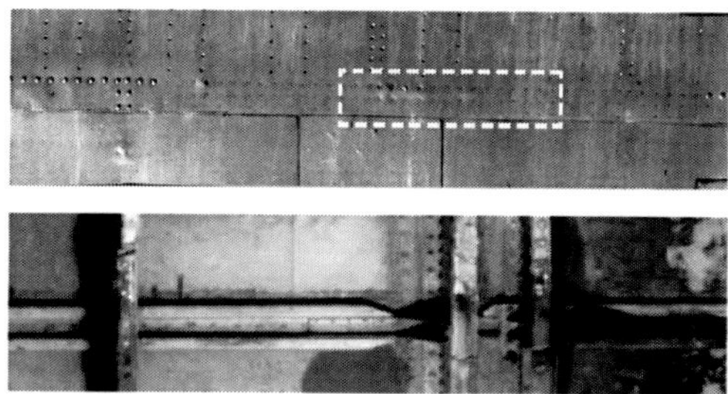

Figure 12.4 An aluminum lap-joint cut from a commercial aircraft. (Top): Front view showing the area that is inspected using different NDT methods. (Bottom): Rear view showing the inside surface and the stiffeners.

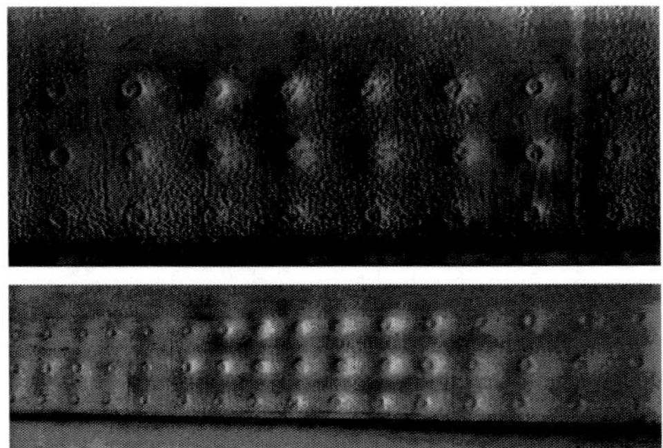

Figure 12.5 A lap-joint sample viewed by two different enhanced optical methods showing surface pillowing, which is more pronounced in the middle. (Top): Edge-of-light. (Bottom): D-Sight. (These methods are described earlier in this publication).

12.4.2 LIQUID PENETRANT INSPECTION

Liquid penetrant inspection (LPI) is commonly used to detect surface-breaking cracks and, occasionally, surface pitting caused by corrosion. The ease of use and total surface coverage in a relatively short time make the LPI a method of choice for the first line inspection of aircraft parts after general visual inspection. Surface cleaning is often necessary to increase the chance of identifying small cracks. The detection capability of LPI methods is largely dependent on the surface cleaning process employed, as well as the fluid types and the size, location, and thickness of cracks. Generally, cracks with a surface length larger than 2 mm can be detected under favorable conditions by experienced inspectors using fluorescent liquid penetrants; although, probability of detection (POD) can vary from one situation to another [7]. Liquid penetrant inspection is also used during corrosion repair to ensure complete removal of corrosion products. Even a small amount of corrosion product can lead to rapid re-initiation of corrosion.

12.4.3 EDDY CURRENT

Eddy current (EC) techniques are commonly used in the inspection of specific critical components of aircraft and to verify the visual test results [8]. These methods require access to one side only and can detect cracks as well as material thinning due to corrosion. At frequencies of less than 10 kHz, eddy currents can penetrate relatively thick multi-layer structures and can detect approximately 10% material loss [9]. Figure 12.6 shows an example of an eddy current scan of an aluminum lap-joint showing areas of corrosion between the two layers [10]. Areas of corrosion damage would need to be larger than the dimensions of the probe. Typical probe dimensions for the EC thickness loss measurements are in the order of 1 cm. The thinning of the material affects the electromagnetic response of the eddy current sensor that is presented in grey-scale or color maps and related to the thickness loss by prior calibrations.

Multi-frequency techniques can be implemented using conventional EC equipment to find material loss in either first or second layers. Using a dual-frequency mixing EC method, corrosion greater than 10% in the second layer of two-layer structures can be found [11]. The multi-frequency techniques have a number of practical advantages in that they use commercially available equipment in a familiar manner. However, set-up and calibration for these methods are time-consuming and complex. Also, corrosion on the bottom of the first layer cannot be distinguished from corrosion on the top of the second layer.

Eddy current inspections are performed by manually or automatically scanning a probe over the area of interest in a raster mode. Thus, inspection speed is slower than with the enhanced optical methods, but they provide more reliable results. Automation of inspection improves the performance of the eddy current approach in terms of speed and detection capability. Newer probes, such as super-conducting quantum interference devices (SQUID) [12], giant magneto-resistive (GMR) [13], and self-nulling coil sensors, have also been tried and found useful for specific application. For example, GMR is reported to be sensitive to exfoliation and stress corrosion cracks.

Transient or pulsed eddy current techniques are the most promising methods available for quantification of corrosion in multi-layer structures [14, 15]. This approach is used to map corrosion of the same aircraft sample shown before (see Figure 12-6), and the results are presented in Figure 12.7. The results are similar to those of the single frequency eddy current method; however, this approach may be able to identify the specific faying surface that contains corrosion. Also, with PEC method, it is possible to quantify corrosion independent of lift-off variations.

The remote field eddy current (RFEC) method is based on the measurement of the differences in the energy flow patterns in the near-field and remote-field regions. The energy induced by the excitation coil traverses the component thickness twice before reaching the pickup coil. The RFEC method is characterized by its equal sensitivity to a flaw irrespective of its through-thickness location. Due to this characteristic, the technique has potential for detecting corrosion located

Figure 12.6 An eddy current inspection map of the lap-joint specimen at 8 kHz, corrosion severity increases from light green to red.

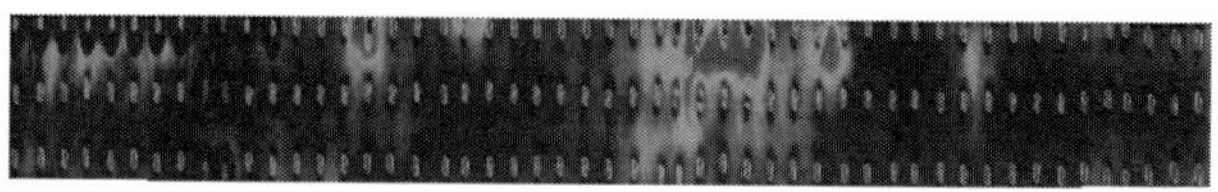

Figure 12.7 A pulsed eddy current inspection map of the lap joint sample, corrosion severity increases from light green to red.

several skin-depths away from the excitation source. The remote field eddy current method has been shown to be capable of detecting anomalies located in the second and third layers up to 25.4 mm in thickness [16]. The magneto-optic imaging (MOI) method uses a magnetically activated crystal to image the magnetic field induced by either an eddy current probe or by direct induction. The MOI shows the presence of corrosion in lap-joints as areas of different brightness. Because the technique provides a visible trace of the resulting magnetic field, with the adjustment of the frequency of excitation, the inspector can make reasonable estimates of the depth or percentage of corrosion detected [17].

12.4.4 ULTRASONIC TECHNIQUES

Ultrasonic testing (UT) is based on the transmission of high frequency (0.5–2 MHz) sound waves into the test area and measurement of the amplitude, frequency, or time of arrival of the returned echoes that provide information about the nature and position of flaws. Conventional UT uses piezoelectric probes to generate and detect sound waves and a coupling fluid to transfer the energy between the probe and the material. The need for a coupling fluid is a disadvantage as it reduces the inspection speed. False indications could also be caused if the coupling is lost. There are many different ways that UT inspections can be carried out. Those that have been used for aircraft parts or in-service inspections are mentioned below.

The pulse-echo technique is commonly used for ultrasonic inspection of aircraft structures, such as lap-joints or wing skins. An example is provided in Figure 12.7 where the same specimen as examined by eddy current method is inspected using a portable ultrasonic scanning system. If a joining compound has not been employed between the layers, the inspection range will be limited to the outer skin only. The technique can indicate first layer thickness loss by measuring the time of travel of sound waves. Small commercial thickness gauges that are very inexpensive can be used, giving a resolution of about 2% in typical Al aircraft skins [18]. Individual pits cannot be detected, but surface roughness of the corroded parts can affect the return signal and reduce the accuracy of this approach. If a joining compound has been used, then it may be possible to measure the amplitude of multiple reflections to produce C-scan maps of internal interfaces. However, C-scan inspections require more elaborate and expensive automated scanning frames. Also, careful interpretation of the signals is necessary, as the C-scan does not identify different interfaces. To differentiate the interfaces, time-of-flight measurements and B-scan presentations are used.

Figure 12.8 Ultrasonic amplitude C-scan inspection map of the lap-joint specimen at 30 MHz, corrosion sites are shown in blue.

Normal- and oblique-angle pulse-echo UT techniques coupled to the part with a coupling fluid in immersion mode have been used to detect exfoliation corrosion around fastener holes in thick sections without removing the fasteners [19]. At the normal angle, when a transducer passes over the corrosion-free areas of the first layer, back-surface multiple reflections will be seen. However, when it passes over the exfoliation corrosion where sound waves cannot penetrate, the back reflections that provide indications on the C-scan maps will attenuate or disappear (see Figure 12.9, left). With this method, exfoliation extending beyond the fastener heads can be detected.

When an oblique angle is used and the time-of-travel (TOT) of the UT beam is measured, a TOT profile of the entire depth of the hole is generated. If there is no corrosion, the TOF profile will show a constant slope of the countersink without abrupt change of time, but if exfoliation corrosion is present, an additional blip will be seen (Figure 12.9, right). The width of this blip can be used to estimate the exfoliation size using simple geometrical relations. Figure 12.9 also shows a micrograph of the cross-section of an actual aircraft sample with exfoliation corrosion at a fastener hole.

These tests are carried out in the laboratory environment using the immersion approach and a focused transducer; however, similar tests may be possible using contact probes. The latter approach is more useful for field applications, but the results may be affected by the coupling fluid, transducer alignment, pressure applied

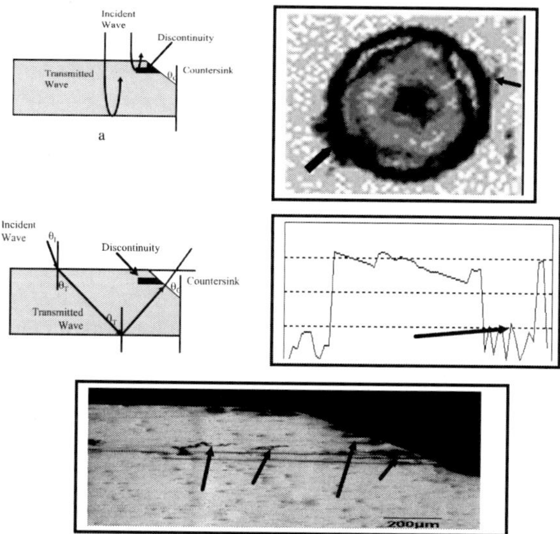

Figure 12.9 Ultrasonic pulse-echo inspection of a sample with exfoliation corrosion. (Top left): Normal angle beam approach and (top right): corrosion area shown by an arrow. (Middle left). Oblique angle beam approach and (middle right): corresponding time-of-travel profile showing corrosion area by an arrow. (Bottom): A micrograph of the cross-section of the specimen near the fastener hole showing actual exfoliation corrosion and cracks.

on the transducer, paint condition or thickness, and surface finish of the part. To minimize these variables, automated scanners and closed-cycle water-coupling systems have been developed for use on aircraft in the field.

One such system is known as the "Drip-less Bubbler" proposed by Barnard and Hsu[1]. This device uses a specially-designed fluid containment and circulating mechanism that prevents water leakage while allowing automatic scanning of the probe. The main advantage of the automated approach over the manual contact UT is the ability to use higher frequencies and focused immersion transducers that result in better spatial resolution and reproducibility. Consistent coupling, ability to scan over surface protrusions, and conforming to curvature with flexible frames are some of the advantages. C-scan images, however, pertain to the first prominent reflection from an exfoliation crack and deeper cracks cannot be imaged if they are in the shadow of a shallower exfoliation. The drip-less bubbler has been adapted to a number of commercial scanners for in-service inspections.

Ultrasonic spectroscopy involves analysis of the frequency spectra of the reflected echoes from a particular interface and may be used to provide information about the roughness of surfaces [20]. Small surface pits may be detected by measuring changes in high frequency components of the spectra. However, in this approach, the full waveform must be digitized and processed, and, thus, it is slow and only suitable for small area inspections in laboratory environments.

Guided waves are generated in thin plates using two angled-beam transducers in a pitch-catch configuration. The velocity of guided waves varies as a function of the frequency-thickness product. If frequency is kept constant, in principle, the thickness loss due to corrosion can be identified by measuring the change in the arrival times of the signals at the receiving probe. In this case, the differences in velocities within the lap-joint with and without material loss may be large enough to distinguish between the two configurations. However, the field applications require a stable scanning mechanism to ensure proper transducer coupling and alignment. If a single wedge transducer is used, obliquely back-scattered ultrasonic signals (OBUS) are generated that have been shown to detect and characterize corrosion under paint in metallic panels [21]. In the referenced study, OBUS have been measured through the paint and used to produce C-scan images of corrosion damage located on both top and bottom faces of test panels.

Ultrasonic surface waves are created when the incident angle approaches the critical angle of the material. Surface waves travel on or near the surface with a penetration depth of about one wavelength and, thus, are sensitive to small surface-open cracks. The feasibility of using surface waves for the detection of cracks under rivet heads in aluminum lap-joints has been studied at IAR using both contact and immersion approaches [22]. The contact method using a 5MHz wedge probe can detect hidden cracks in the range of 0.25–0.75mm in the laboratory

---

1   D. J. Barnard and D. K. Hsu, Center for Non-destructive Evaluation, Iowa State University, Ames, IA 50011.

environment. The technique is only suitable for inspection of small areas, as it is very labor-intensive, but can be easily used in the field.

Similarly, the potential of surface waves to detect cracks around bolt-holes in samples simulating wing skins has been studied using the immersion approach. A sample made of two aluminum plates (2024-T3 Al, 1 mm thick) with three rows of rivet-holes containing EDM notches of different sizes was used. The first row contains type A rivets (7 mm head diameter), the second row contains smaller type B rivets (5.75 mm head diameter), and the third row has no rivets. Figure 12.10 shows the specimen and the corresponding C-scan images using back-scattered surface waves. In the butterfly-shaped indications, the body is the bolt-hole and the wings are cracks. As seen in the C-scan traces, EDM notches as small as 0.75 mm under smaller rivets and 1.25 mm under larger rivets are detectable. However, this approach is time consuming for field applications.

Non-contact techniques greatly increase the capability of testing large areas and eliminate the transducer coupling problems associated with conventional ultrasonic techniques. Several non-contact UT techniques are available, including electromagnetic acoustic transducers (EMAT) [23], laser-ultrasonic [24], and air(gas)-coupled ultrasonic methods [25]. However, while these approaches are generally effective in laboratory environments, they are less useful in field applications.

12.4.5 RADIOGRAPHY

Conventional X-radiography uses a point source on one side at a distance from the structure to illuminate the region of interest and an area detector (e.g., film) located on the opposite side to capture the transmitted rays. High-resolution, real-time

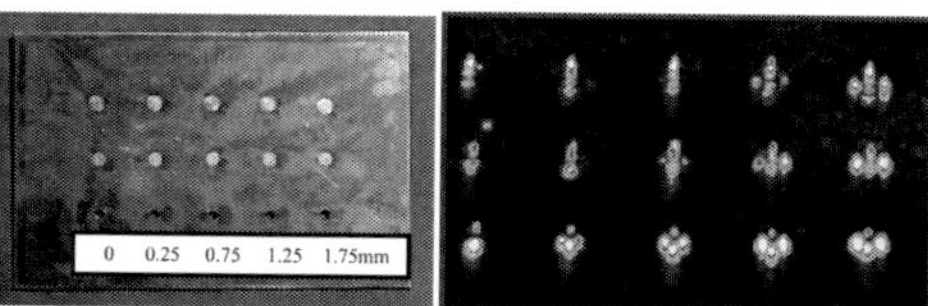

Figure 12.10 A lap joint specimen with EDM notches in holes and under rivets and the corresponding ultrasonic immersion C-scan images obtained using back-scattered surface waves.

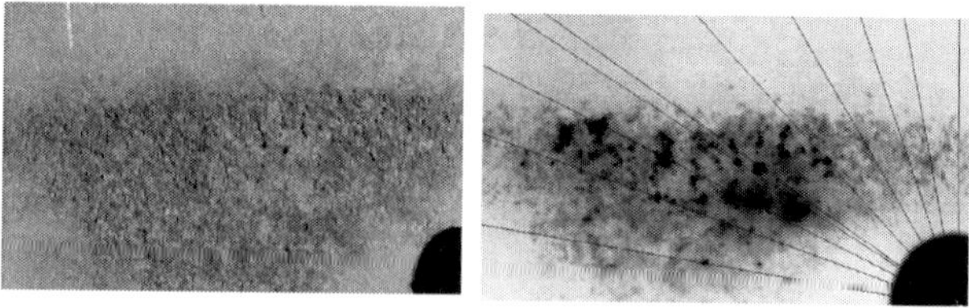

Figure 12.11 (Left): Pitting corrosion in aluminum. (Right): Corresponding x-ray image.

systems have been developed for on-aircraft inspections for cracks and corrosion [26]. Exfoliation and pitting corrosion can be detected (e.g., Figure 12.11), but the remaining metal thickness cannot be determined due to the confluence of the absorption of the metal and the corrosion product. A dual-energy method has been developed, which can distinguish between aluminum hydroxide and aluminum, and has been used to make measurements of metal thickness in an artificially-corroded specimen in a laboratory. However, this method is not practical, as it requires access to both sides of the part, which is often not available [27]. Compton x-ray imaging is based on measurement of the back-scattered photons that yields a point-by-point measurement of material density [28]. The technique is a single-sided inspection capable of penetrating about 4 cm in aluminum. However, this approach is extremely slow and inefficient for routine inspections.

X-ray computer tomography (CT) produces a cross-sectional image of the inside of the object providing both density and geometry information. However, the application of CT in the aerospace industry has been limited to relatively small components that can be brought to the scanner. Nonetheless, it can be used to measure internal dimensions, including wall thickness of complex components. Similarly, neutron radiography is performed on small samples that can be brought to nuclear reactors and, for this reason, is not used very often. However, this technique is perhaps the only approach that may provide indications of hydrogen embrittlement or identify minute amounts of corrosion by detecting the presence of hydrogen molecules. Consequently, there is no obvious advantage in using this approach when corrosion is large enough to be detected using other NDT methods [29].

12.4.6 THERMAL METHODS

Thermography uses an infrared camera and detects the presence of corrosion from the differences in thermal diffusion rates when the object is subjected to a thermal excitation. It is a large area technique but generally suffers from limited resolution and difficulty of interpretation of thermal images. To perform measurements on aluminum, a layer of high emissivity coating must be applied to the surface. Painted aircraft may require no special treatment, but, usually, a water-soluble black paint is applied to surfaces before inspection to increase infrared thermal emissivity. Significant work has been carried out to develop thermal methods for aging aircraft problems (e.g., [30]). Generally, thermal techniques are best suited for large-area inspections to identify defects close to the external surfaces where precise size measurement is not required. However, quantitative measurements on multi-layer specimens is difficult, requiring separation of thermal response of aluminum from that of corrosion products, adhesive or sealant, and air gaps. In Figure 12.12, a cut-out segment of an aircraft stringer made of 7075-T6 aluminum alloy with stress corrosion is shown, along with its pulsed-thermography image

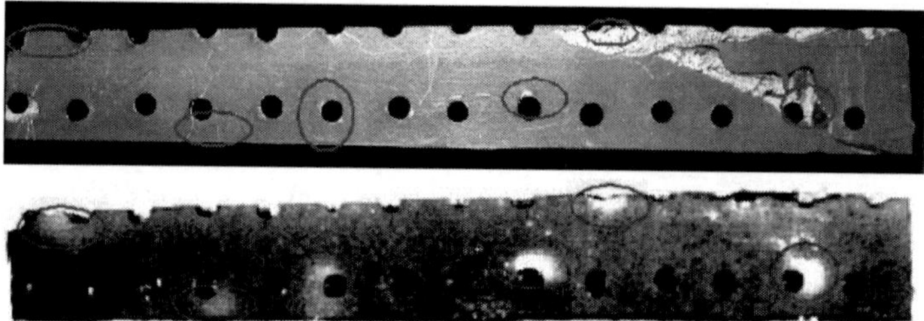

Figure 12.12 (Top): An aircraft stringer made of 7075-T6 aluminum alloy with stress corrosion. (Bottom): The corresponding pulsed-thermography image showing both visible and hidden corroded regions.

[31]. The thermal map shows both the visible corrosion sites as well as areas of hidden corrosion that are not visible on the surface.

12.5 CORROSION VERIFICATION TESTS

After detection of corrosion or cracks using NDT techniques, it is often necessary to verify the data and to measure the actual size of the damage. For surface-breaking flaws, visual examinations at high magnification are often used. Microscopes or boroscopes can be used to measure the size of surface-open flaws. If these tools cannot be used on-site, a replica of the affected area can be made using commercially available replication tapes, which is then examined under a microscope.

This approach is particularly useful for determining the size or distribution of surface pits or cracks. Verification of hidden corrosion in lap-joints requires disassembling or the pry opening of the layers to reach the corrosion sites. Visual examinations of corrosion are often followed by mechanical or chemical cleaning of the affected areas to enable accurate measurement of the remaining thickness.

For measurement of the remaining thickness in aluminum lap-joints, a technique has been developed at IAR that is applicable in the laboratory environment. This approach is described in reference [32] and is based on the fact that the intensity of the x-ray transmitted through a homogeneous solid body of uniform density is a linear function of the thickness of the material. In practice, after disassembly and removal of corrosion products, the corroded panels are x-rayed, along with a calibration piece.

The calibration piece is basically an aluminum wedge with a progressively-increasing thickness from zero to just above the thickness of the part that is being assessed. The transmitted x-ray intensity is recorded on a sensitive film, which, after processing, is digitized and analyzed using an image processing software package. Different shades (or colors) of x-ray intensity obtained for the calibration wedge are related to different known thicknesses.

Once the relationship is obtained, the thickness of the unknown specimens is calculated by comparison with the calibration wedge. If the test specimen is not exactly the same alloy as the calibration wedge, physical calliper measurements on a few locations of the specimen are used to provide a correction factor.

Figure 12.13 shows a photograph of the disassembled lap-joint specimen used in the eddy current and ultrasonic inspections shown earlier in Figure 12.7 and Figure 12.8 before cleaning. White areas are corrosion products at the faying surfaces. Figure 12.14 shows the x-ray maps of the same plates taken after cleaning and gives the remaining thickness in different colors. The thickness changes are displayed in color for better viewing and correlations with images provided by NDT.

Optical and SEM micrographs of the corroded surfaces are shown in Figure 12.15 and Figure 12.16. Examples of thickness profiles measured using this technique in comparison with metallographic sectioning and confocal microscopy

Figure 12.13 Photographs of the faying surfaces of the specimen after disassembling and before removal of corrosion products and sealant residuals.

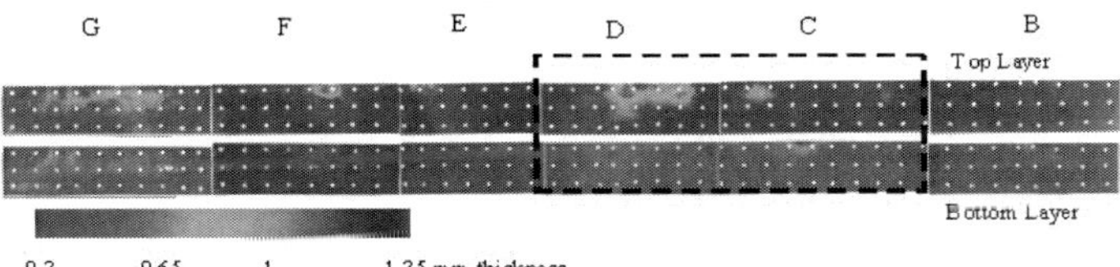

Figure 12.14 Digital X-ray images of the above-mentioned disassembled and cleaned lap-joint samples. Remaining thickness is displayed and measured by comparison of the corresponding colors with those of the calibration ramp below the image.

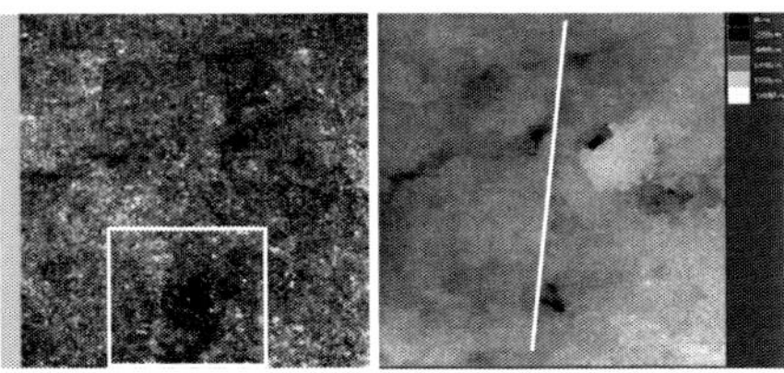

Figure 12.15 Confocal micrograph and a close-up from a retired aircraft lap-joint sample. The dynamic range of the image is 1 mm from black to white. The line marks a surface profile, which was measured using confocal microscopy and radiography, and compared to metallographic sectioning in Figure 12.17.

are shown in Figure 12.17. It must be noted that, since the radiographic technique is actually measuring remaining thickness, it is sensitive to the topography of both top and bottom surfaces, as well as the subsurface features.

On the other hand, the profiles obtained using confocal microscopy or metallography are only sensitive to the top surface features, such as pitting cavities. Even with these differences, good correlations exist between the thickness profiles obtained using different approaches. With the x-ray mapping, spatial resolution of 0.01 mm and thickness resolution of 0.02 mm are possible.

12.6 NDT METRICS FOR CORROSION

Damage tolerance life assessments require metrics for corrosion that are sufficiently representative of the overall damage and are realistically measurable by

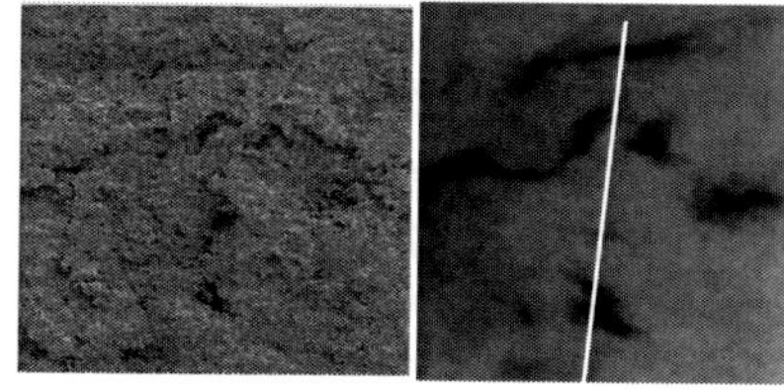

Figure 12.16 The SEM micrograph (left) and the radiography thickness map of the same area as the confocal microscopy image (right).

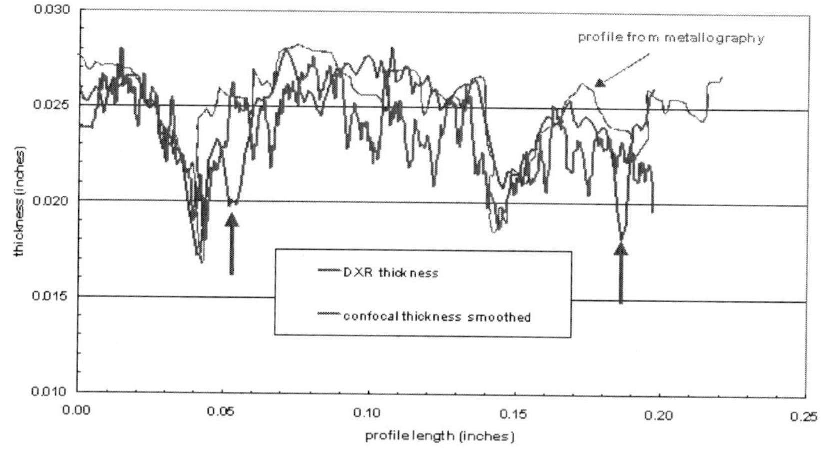

Figure 12.17 Surface profiles of an Al 2023-T3 lap joint specimen from a retired aircraft obtained using three surface topography measurement techniques. The profiles are from the lines noted on Figure 12.15 and Figure 12.16 and show corrosion pittings and the remaining thickness of the plate. DXR corresponds to digital x-ray.

NDT methods. In the IAR publication [2], an attempt has been made to define and estimate NDT metrics that may be used to characterize corrosion in the critical aircraft structural elements, including multi-layer lap-joints of fuselage and wing sections, as well as other critical structural members. While corrosion in multi-layer thin structures of a fuselage is often characterized by average thickness loss, measurement of corrosion pitting and cracks, as well as surface pillowing, may be necessary, too. On the other hand, corrosion of wing sections is predominantly of an exfoliation type and estimates of the extent, depth, and number of layers of exfoliation are important. Corrosion in other structural elements may contain one or a combination of the above-mentioned types, depending on the environment and stress conditions they are subjected to, and, thus, the same metrics may be used for characterization.

12.6.1 AVERAGE THICKNESS LOSS

Average thickness loss can be estimated by NDT using eddy current, thermography, or ultrasonic methods. In multi-layer structures, measurement of the first layer thickness is oftenpossible. To date, single frequency eddy current has demonstrated a sensitivity of about 10% material loss on fuselage multi-layer specimens with a precision of 2% of the total thickness. The spatial resolution is limited by probe diameter, which is typically about 10 mm. Ultrasonic pulse-echo and pulsed-thermography may provide slightly better results (about 5% material loss with 2% precision) if carried out carefully; however, they are less reliable. Multiple frequency and pulsed eddy current techniques can also identify 5% thickness loss per skin with an accuracy of 1%.

12.6.2 SURFACE PITTING

Corrosion pits occurring on hidden surfaces are extremely difficult to measure. A few methods, such as advanced eddy current, Compton back-scattering, and backscattered ultrasonic spectroscopy, have shown potential to provide indications of the overall surface roughness due to pitting, but their accuracy has not been documented. Generally, the typical sizes of corrosion pits are much smaller than the probes used in most NDT techniques. It may be possible to infer pitting depths from material studies in combination with measurements of surface roughness. That is, given a certain level of surface roughness due to pitting in a certain material, it may be possible to infer a distribution of pit depths. This information would have to be generated empirically. Individual pits on exposed surfaces could be measured using optical devices, such as boroscopes or microscopes, at appropriate magnifications. A sensitivity of 200–500 microns pit size is achievable on cleaned surfaces under favorable conditions. Replication can also be used for precise measurement of the dimensions under a scanning electron microscope.

Replication is useful for tight and hard-to-reach areas, such as inside surfaces of bolt-holes. Individual pits as small as 0.02 mm can be measured using these methods. The spatial distribution of pitting on an exposed surface can be easily obtained using fluorescent liquid penetrant as well as optical devices, such as EOL or boroscopes. Also, replication in combination with microscopic examination may be used.

12.6.3 CRACKS

All types of cracks are of concern to aging aircraft; however, those that originate at fastener or rivet holes are hard to detect because they are hidden by the fasteners. Crack detection by conventional NDT is generally limited to situations where cracks have progressed beyond the fastener head, except for a few laboratory techniques, such as ultrasonic surface waves. In the laboratory environment, this method can detect hidden cracks 0.25–0.75 mm long in the first layer of lap splices. However, it is expected that in the field, its sensitivity would be less and comparable to the eddy current methods (1 mm). Cracks under fasteners in the second layer of multi-layer structures are not easily detectable with the current NDT technology. EC methods may give indications of second layer cracks when they are beyond the fastener heads by about 2 to 3 mm. The measurement of surface-breaking cracks is much easier for NDT, and cracks as small as 0.1 mm may be detectable by eddy current or high-frequency ultrasonic methods under favorable conditions. When fasteners are removed from bolt or rivet holes, surface-breaking cracks in the hole can be detected using eddy current rotating probes, and a sensitivity of about 1 mm is achievable.

12.6.4 SURFACE PILLOWING

Surface pillowing of fuselage lap-joints can be measured using enhanced optical techniques, such as D-Sight and Edge-of-Light, optical metrology, or mechanical measurements. Mechanisms other than corrosion, such as clamping stresses due to the manufacturing process or mechanical overloading during operation, can also cause deformation of riveted joints that may be confused with corrosion pillowing. These effects reduce the ability to infer material loss data from pillowing measurements. Both D-Sight and EOL techniques are sensitive to extremely small changes in surface topography (~0.05 mm out-of-plane deformation with a precision of 0.025 mm). However, the quantitative capabilities of these techniques are not well defined, and their reliability needs to be improved. Optical metrology methods, such as laser triangulation, can make fast, accurate three-dimensional measurement of pillowing with very good precision, but they are difficult to set up. In contrast, mechanical measurements of deformation are easily done but provide less accurate results than optical measurements.

12.6.5 EXFOLIATION

Layers of exfoliation corrosion occurring in thick-section wing structures are very difficult to detect. The pulse-echo ultrasonic approach is the most suitable NDT method for identifying exfoliation corrosion that extends beyond the fastener head, but will only measure the top layer and any layers beneath, which extend beyond the top layer of exfoliation. More advanced UT methods use surface waves or reflections to interrogate the volume obscured by fasteners and can detect smaller areas of exfoliation. These methods are still subject to the same limitation of the top or bottom layers, obscuring exfoliation occurring between them.

The depth location of the first layer of exfoliation that extends beyond fasteners is relatively easy to determine using the ultrasonic pulse-echo technique. In thick-sections, accuracies in the order of 0.1 mm can be achieved. More advanced UT techniques can interrogate the bottom of the fastener hole and locate the bottom layer exfoliation with less accuracy. The extent of exfoliation extending beyond the fastener head can be estimated using the pulse-echo approach, and an accuracy of about 1 mm is achievable.

12.7 CHAPTER SUMMARY

There is a great need for cost-effective and reliable NDT techniques that can be used to ensure the safe operation of aging aircraft. A large number of NDT methods have been applied to the detection and characterization of corrosion in simulated or real aircraft parts. This chapter summarizes the application areas, capabilities, limitations, and metrics for each technique (see Table 12.1). However, while some of the methods mentioned above have been applied in the laboratory environment, there is little information on their field use. Therefore, the NDT metrics quoted here are best estimates under the most favorable conditions. Each NDT method has certain capabilities and limitations, and, often, more than one technique is needed to cover the various components and corrosion types encountered in aging aircraft.

For NDT of aluminum airframe structures, such as lap-joints and wing skins, a two-step inspection may be necessary. The first pass inspection may involve the use of an enhanced visual method as a fast approach to identify suspect areas. This is followed by a second inspection using either eddy current or ultrasonic methods to verify the visual inspections and quantify the damage. A combination of techniques may be needed for different flaw types. The choice of NDT methods for internal components is dependent on the material, shape, location, and accessibility of the component, as well as the nature of the problem. Usually, visual inspections followed by a more detailed NDT using either eddy current or ultrasonic techniques are necessary.

In terms of corrosion metrics, surface bulging or pillowing can be measured using enhanced optical methods, such as D-Sight, Edge-of-Light, or optical

Table 12.1 Summary of NDT Techniques for Aging Aircraft Made of Aluminum Alloys.

TECHNIQUE	ADVANTAGES	LIMITATIONS	METRICS, RESOLUTION
Eddy Current	*One side access can be automated*	*Slow, correlation difficult*	*Multi-layer corrosion, material loss, cracks*
- Low frequency	Equipment available	Specific depth	10% loss, 1 mm crack
- Multi-frequency	Various depths	Time consuming	<10% loss
- Pulsed/transient	Depth location	At experimental stage	5% loss, cracks
- Remote field	Deep flaws	No through-thickness	>1 mm cracks
- Magneto-optic	Fast imaging of small areas	Flat surfaces	> 10% loss
Ultrasonic	*Well developed, reliable, automation*	*Coupling fluid needed Slow, first layer only*	*Thickness, cracks, exfoliation, disbond*
- Pulse-echo	Easy, inexpensive	Only 1st layer	2–5% loss, 2 mm crack
- Through-transmission	Less attenuation	Access to both sides	Disbond, delamination
- Spectroscopy	Small flaws detectable	Very slow	Roughness, pits, 8mic.
- Guided waves	Large area coverage	Poor reproducibility	Disbond, corrosion
- Surface waves	Very sensitive, simple	Small area, slow	
- Non-contact	No coupling needed	Experimental stage	Small cracks, 0.3 mm
- Acoustic emission	For hidden parts	Poor reliability	Active flaws
Optical	*Large area coverage, easy*	*Surface features only, Interpretation difficult*	*Surface dents, corrosion, pillowing*
- Visual	Fast, inexpensive	Not reliable	Paint flaking, dents
- Enhanced visual	Provides image	Indirect measurement	3% loss, surface pits
- Boroscopes	Precise dimension	Time consuming	Surface pits, 0.25 mm
- Liquid penetrant	Total coverage	Cleaning needed	Open cracks, 2 mm
Thermography	*Large areas, fast*	*Interpretation difficult, poor resolution, must paint surface*	*>5% thickness loss on one layer, disbond*
Radiography	*Full volume inspection*	*Safety precautions*	*Density profiles, dimension, internal corrosion*
- X-rays	Visible records	Access to both sides needed	Cracks, exfoliation
- Compton	One side access only	Inefficient, experimental	Roughness, 0.25 mm
- Tomography	Cross-sectional view	Small parts, expensive	Internal dimensions, 0.025 mm
- Neutron	Sensitive to hydrogen	Not portable, laboratory tool	Stress, hydrogen embrittlement

metrology, with good precision. Calibration with respect to known references is needed for these and many other NDT methods. Surface pitting is also detectable after cleaning of the part using enhanced visual inspections and can be measured in terms of surface dimensions with a good accuracy using boroscopes or by producing a replica and examining it under a microscope. Pit depth measurements are more difficult, but pit distribution can be mapped using enhanced optical methods or by fluorescent liquid penetrant inspection. The latter is also useful during cleaning of corrosion to ensure removal of corrosion pits and products.

Thickness loss due to corrosion is best determined using the ultrasonic pulse-echo method as it provides a direct and accurate measure of the first layer thickness. However, surface roughness can affect the results, and the method is not applicable beyond the first layer. This method is also suitable for locating exfoliation cracks around fastener holes of thick wing skins. The frequency analysis of ultrasonic echoes may provide indications of overall roughness due to pitting. Thickness loss can also be measured using the single frequency eddy current method; however, the induced field is affected, not only by the corrosion, but also by geometrical changes, such as thickness or gap variations, as well as the presence of fasteners and substructures. Transient or pulsed eddy current and multi-frequency EC inspections offer better prospects for improvements.

Cracks can be measured using a variety of NDT approaches, including LPI, eddy current, x-ray, and ultrasonic methods, although the detection sensitivity and measurement accuracy are different. LPI is only suitable for surface-breaking cracks larger than about 2 mm, while eddy currents can achieve much smaller detection levels at a higher reliability. Hidden cracks under fasteners are problematic for the eddy current method, but ultrasonic surface waves have shown potential in detecting such cracks with a detectability of about 0.75 mm in the laboratory environment. Finally, all NDT methods require representative samples and flaws for calibration, and availability of appropriate reference standards is essential for success of inspections.

12.8 CHAPTER REFERENCES

[1]	Overview of NRCC Research into the Interaction of Corrosion and Fatigue in Aircraft Structure for the USAF/Lockheed Martin Aeronautics CFSD Phase 2 Project, G. F. Eastaugh, N. C. Bellinger, A.A Merati, D.S. Forsyth, and B.A. Lepine, NRC Report: LTR-SMPL-2003-0191, 2003.

[2]	Survey of Non-destructive Evaluation Techniques for Corrosion in Aging Aircraft, A. Fahr, D.S. Forsyth and C.E. Chapman, NRC Report: LTR-ST-2238, October 1999.

[3]   AGARD Corrosion Handbook Volume 1, Aircraft Corrosion: Causes and Case Histories, W. Wallace, D.W. Hoeppner, and P.V. Kandachar, AGARD-AG-278 Volume 1, 1985.

[4]   Corrosion Fatigue and Environmentally Assisted Cracking in Aging Military Vehicles, NATO Research and Technology Organization Scientific Report Number: RTO-AG-AVT-140, (AC/323(AVT-140) TP/346), March 2011.

[5]   Corrosion Detection in Aircraft Lap Joints: Proposed Approach to Development of POD Data," Presented at RTO Workshop on Airframe Inspection Reliability under Field/Depot Conditions, J. P. Komorowski, D.S. Forsyth, D.L. Simpson, and R.W. Gould, 13–14 May 1998, Brussels RTO-MP-10, AC/323(AVT)TP/2, pp. 8-1 to 8-8.

[6]   Quantification of Corrosion in Aircraft Structures with Double Pass Retroreflection, J.P. Komorowski, N.C. Bellinger, R.W. Gould, A. Marincak, and R. Reynolds, Canadian Aeronautics and Space Journal, Vol. 42, No. 2, pp. 76–82, June 1996.

[7]   POD Assessment of NDI Procedures Using a Round Robin Test, A. Fahr, D. S. Forsyth, M. Bullock, W. Wallace, A. Ankara, L. Kompotiatis and H.F.N. Goncalo, AGARD-R-809, January 1995.

[8]   The ABC's of NDT Development and Adoption for Aging Aircraft, D.S. Wilson and D.J. Hagemaier, Materials Evaluation, pp. 336–346, March 1999.

[9]   An Investigation of Non-destructive Inspection Equipment: Detecting Hidden Corrosion on USAF Aircraft, J. Alcott, Materials Evaluation, pp. 64–73, January 1994.

[10]   NDT Research at IAR in Support of Aging Aircraft, A. Fahr, D. Forsyth, J.P. Komorowski and C.E. Chapman, Proceedings of the 1st Pan American Conference for NDT, Toronto, Sept 1998.

[11]   Dual Frequency Eddy Current Testing for Second-layer Corrosion, W. D. Chevalier, Presented at ATA 1993 NDT Forum, Scottsdale, Arizona, September 1993.

[12]   Evaluation of High-sensitivity Magnetic Imaging for Hidden Corrosion, A. D. Hibbs, R. Chung, J.S. Pence, Q.C.C. Talley and T.A. Sage, Proceedings of Non-destructive Inspection of Aging Aircraft, San Diego, CA, SPIE Vol. 2001, pp. 200–208, 1993.

[13] Low-frequency Magnetoresistive Sensors for NDE of Aging Aircraft," E.S. Boltz, Q.W. Cutler, T.C. Tiernan, Non-destructive Evaluation of Aging Aircraft, Airports, and Aerospace Hardware II, San Antonio Texas, 31 March–02 April 1998, SPIE Proceedings Vol. 3397, pp. 39–49.

[14] Quantitative Characterisation of Corrosion Using Scanning Pulsed Eddy Currents, J.C. Moulder and J.A. Bieber, The First Joint DoD/FAA/NASA Conference on Aging Aircraft, pp. 1022–1034, 8–10 July 1997.

[15] Pulsed Eddy Current Method Developments for Hidden Corrosion in Aircraft Structures, B.A. Lepine, D.S. Forsyth, B.P. Wallace and A. Wyglinski, The First Pan-American Conference for Non-destructive Testing, pp. 107–117, September 1998.

[16] Recent Advances in Remote Field Eddy Current NDE Techniques and their Application in the Detection, Characterization and Monitoring of Deeply Hidden Corrosion in Aircraft Structures, Y. Sun, T. Ouyang and S. Udpa, SPIE Proceedings, Vol. 3586, pp. 200–210, 1999.

[17] Aircraft Corrosion and Crack Inspection using Advanced MOI Technology, D.K. Thome, G.L. Fitzpatrick and R.L. Skaugset, Non-destructive Evaluation of Aging Aircraft, Airports, and Aerospace Hardware, Scottsdale, AZ, SPIE Proceedings Vol. 2945, pp. 365–373, 03–05 December 1996.

[18] Non-destructive Evaluation of Corrosion in Aging Aircraft, Part 1, Introduction, Ultrasonic and Eddy Current Methods, R.A. Smith, INSIGHT, Vol. 37, No 10, October 1995.

[19] Emerging Technologies for the Detection and Characterization of Exfoliation Corrosion in Aluminum Aircraft, A. Fahr, B. Rogé and V. Lefebvre, Final Report of the Technical Cooperation Program- Materials and Processes Technology Group, Report Number TTCP/MAT-TP-5/4/2005, 2005.

[20] Non-destructive Evaluation of Corrosion in Aging Aircraft, Part 1, Introduction, Ultrasonic and Eddy Current Methods, R.A. Smith, INSIGHT, Vol. 37, No 10, October 1995.

[21] NDE of Hidden Flaws in Aging Aircraft Structures using Obliquely Backscattered Ultrasonic Signals (OBUS)", Y. Bar-Cohen, A.K. Mal and M. Lasser, SPIE Proceedings Vol. 3586, pp. 347–353, 1999.

[22] Non-destructive and Destructive Inspections of Aircraft Lap Joint Specimens for TTCP-MAT-TP5, A. Fahr, D.S. Forsyth, R.W. Gould and C.E. Chapman, Operating Assignment 17, NRC Report: LTR-SMPL-2000-0216, August 2000.

[23] Non-contact Ultrasonic NDE Systems for Aging Aircraft, R.E. Green, B.B. Djordjevic, Proceedings of 2nd Joint NASA/FAA/DoD Conference on Aging Aircraft, Williamsburg, VA. 31 Aug.–03 Sept. 1998, NASA/CP-1999-208982/PART1, pp. 244–249, 1999.

[24] Laser Ultrasonic Inspection and Characterization of Aeronautical Materials, J-P. Monchalin, C. Neron, M. Choquet, D. Drolet and M. Viens, NDT.Net, Vol. 3, No. 11, November 1998.

[25] Air-coupled Ultrasonic Measurements of Adhesively Bonded Multi-Layer Structures, D.W. Schindel, Ultrasonics, Vol. 37, No. 3, pp. 185–200, March 1999.

[26] Corrosion Detection and Characterization using High Resolution Real-time Radiography, C. Bueno, M.D. Barker, R.C. Barry and R.A. Betz, 41st International SAMPE Symposium and Exhibition, March 25–28, 1996.

[27] Using Energy Dispersive X-ray Measurements for Quantitative Determination of Material Loss due to Corrosion," J. Ting, T. Jensen and J.N. Gray, Review of Progress in Quantitative NDE, Vol. 12, D.O. Thompson, D.E. Chimenti, eds. Plenum Press, New York, pp. 1963–1969, 1993.

[28] X-ray Backscatter Depth Profilometry, J.D. Achenbach, Center for Quality Engineering and Failure Prevention, Northwestern University, Report No. 1137892, 1998.

[29] The Use of Neutron Tomographic Techniques for the Detection of Corrosion Damage in Aircraft Structures, R.C. Lanza, ASME International Mechanical Engineering Congress and Exposition, San Francisco, AD Vol. 47. Structural Integrity in Aging Aircraft, ASME, pp. 253–259, November 1995.

[30] Thermographic Detection of Corrosion in Aircraft Skin, H.I. Syed, W.P. Winfree, K.E. Kramer and P.A. Howell, Review of Progress in Quantitative NDE, Vol. 12, D. O. Thompson, D. E. Chimenti, eds. Plenum Press, New York, pp. 2035–2041, 1993.

[31] Infrared Thermography Inspection of Parts of a CP140 Tail Horizontal Stabilizer Stringer, M. Genest, NRC Report: LM-SMPL-2005-0054, April 2006.

[32] Corrosion Detection and Thickness Mapping of Aging Aircraft Lap Joint Specimens Using Conventional Radiographic Techniques and Digital Imaging, C.E. Chapman, and A. Marincak, NRC Report: LTR-SMPL-1996-0067, 1996.

CHAPTER 13

NDT Reliability

13.1 BACKGROUND

The use of the damage tolerance approach in aircraft design and maintenance requires reliable NDT methods, both during component manufacture as well as during the service life, and it requires that the NDT reliability be established quantitatively. The reliability of an inspection is basically the degree an NDT process or system is capable of achieving its purpose; for example, the requirement to identify defects larger than a given size in a particular part. When NDT is used for flaw detection, the reliability is quantified by probability of detection (POD), which is typically presented in the form of a curve similar to the one illustrated Figure 13.1. This figure shows that the probability of detection increases with increasing flaw size. A POD curve provides the percentage of flaws expected to be identified as a function of the characteristic flaw size (e.g., length or area). This is often referred to as the "inspection limit" and is defined as the flaw size detectable at 90% probability of detection with 95% confidence. This 90/95% value is used in fracture mechanics calculations to determine the component life limits and to establish the periodic safe inspection intervals (SII). In addition, the POD data may be employed for many other purposes including: validation of NDT procedures, assessing the ability of the NDT personnel, comparison of the capability of different NDT processes or equipment, and qualification of NDT systems.

The NDT POD can be determined in many ways including:

- Experimental demonstration of NDT reliability.
- Use of modeling to estimate NDT outcome (model-assisted POD).

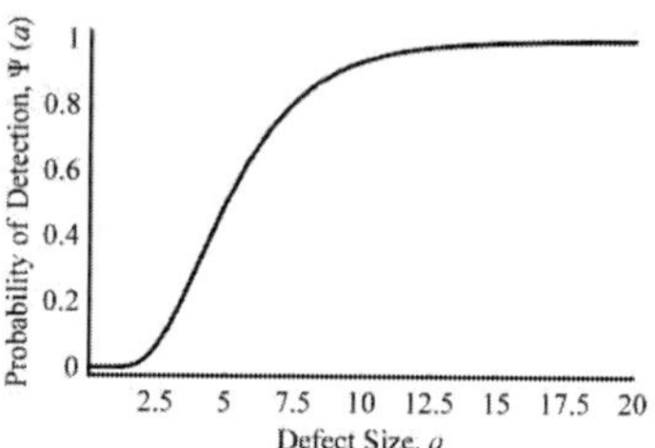

Figure 13.1 Typical probability of detection curve.

Over the past three decades, many studies have been carried out to determine NDT POD. The earliest POD studies were carried out by NASA in support of the Space Shuttle Program [1–3]. Soon after, the United States Air Force (USAF) initiated a program and performed a large NDT POD qualification experiment known as "Have Cracks, Will Travel" [4]. The USAF program resulted in the publication of the Military Handbook MIL-HDBK-1823 [5] and the Military Standard MIL-STD-1783 [6].

At the same time, the US Department of Transportation and the Federal Aviation Administration (FAA) sponsored inspection reliability studies for aging transport aircraft (e.g., [7–9]). Concurrently in Canada, the NRC Institute for Aerospace Research (IAR) carried out a number of NDT POD studies on military gas turbine engine components (e.g., [10–13]) and later on helicopter rotor blades [14] and airframe structures ([15, 16]). The majority of these studies used experimental demonstrations to establish NDT reliability in terms of POD. The data generated under these studied have been compiled by the Non-destructive Testing Information Analysis Center of the United States Department of Defense to produce a data base that is available in the public domain [17].

The experimental demonstration of NDT reliability requires testing of a large number of representative specimens or real parts under situations replicating the actual inspections and containing flaws that have a wide range of sizes. Such tests often involve evaluation of a myriad of variables that influence the NDT outcome, such as component condition or position, NDT equipment and settings, inspection environment, change of operators, etc. The time and cost associated with such studies can be quite significant; thus, model-assisted POD provides an attractive alternative, especially when the conventional experimental POD assessments are not possible or perhaps cost-prohibitive.

In recent years, the capability of models to simulate NDT tests has increased dramatically, and there are several commercial software packages for most commonly-used NDT methods. In model-assisted POD, modeling is employed to quantify the effects of test variables on inspection results, thus, reducing the time and cost of POD studies. The Model-Assisted Probability of Detection (MAPOD) Working Group was established in 2004 by the US Air Force Research Laboratory (AFRL) in cooperation with the FAA and NASA [18]. The goal of the MAPOD is to increase the understanding, development, and implementation of modeling methodologies for POD estimation. Modeling tools may be used along with statistical analysis to determine simulation-based POD data (e.g., [19]).

In this chapter, first, the experimental demonstration of NDT reliability based on the above-mentioned recommended practices and standards is described as well as the model-assisted POD analysis methods. Then, a few examples of NDT POD studies carried out at the NRC Institute for Aerospace Research are provided. The examples include a myriad of NDT tests carried out on life-expired components with artificial or service-induced cracks, representative samples with simulated flaws, specimens with damage, and generic coupons with artificial or real defects. Gas turbine engine parts, helicopter rotor blades, and airframe structural parts are among the components used in the IAR investigations. These examples provide several possible scenarios and the related procedures that may be used in the design and performance of similar NDT POD assessments.

13.2 EXPERIMENTAL DEMONSTRATION OF NDT RELIABILITY

The experimental approach generally involves testing of a statistically-valid number of representative samples with and without flaws, using NDT procedures that duplicate the real component inspections, and, finally, determining the probability of detection and confidence level. The test samples must represent the actual part in terms of material, geometry, surface conditions, and flaw characteristics. The MIL-HDBK-1823 handbook entitled: Non-destructive Evaluation System Reliability Assessment provides guidelines for performing POD experiments [5]. This handbook provides testing and evaluation procedures for assessing NDT system capability for the determination of POD that involves the following steps:

- Define the NDT system that requires assessment
- Design appropriate experiments
- Carry out demonstration tests
- Analyze results using suitable statistical procedures
- Document the results.

13.2.1 NDT System Definition

The first step of an NDT capability assessment is to precisely define the NDT system that is being evaluated in terms of operational parameters and range of application. Also, it is necessary to demonstrate that the system remains stable throughout the assessment and will not change with time.

Many factors may affect the outcome of an inspection; some are controllable while others are not. The controllable factors include the capability of the

NDT technique and the equipment used, the parameters or settings employed, the knowledge, experience, and the ability of the inspector (human factors), and the inspection environment (light, temperature, etc.). The uncontrollable factors may include the component material and condition, access to the inspection site, the discontinuity characteristics (type, size, and location). Important process parameters and factors that contribute to the system variability need to be defined prior to conducting the experiments. These form the NDT system "variables." In addition to the physical attributes of the NDT system, planned statistical assessments of the factors responsible for the system variability may be included in the tests.

13.2.2 DESIGN OF EXPERIMENTS

The experimental design identifies the NDT system variables that may influence the flaw detectability but cannot be controlled in the real inspection environment. The uncertainty in detectability of flaws is caused by both the physical attributes of the flaw (e.g., size) as well as the NDT process variables. The uncertainty caused by different flaw sizes is accounted for by using representative specimens with a range of known flaw sizes in the experimental assessments. The uncertainty caused by the NDT process is accounted for by a test matrix of different inspections that need to be carried out to investigate the variables that are considered important. The ultimate objective is to determine the POD versus flaw size relationship, which defines the capability of the NDT system under representative application conditions. Thus, an experimental design comprises the following steps:

- Define the process variables that may influence flaw detectability but cannot be controlled in the real inspection environment (e.g., component material and geometry, inspection site, etc.).
- Specify a test matrix of conditions that represent the real inspection environment taking into account the variables that influence flaw detectability (e.g., equipment settings, inspection conditions, surface preparation, etc.).
- Specify the order of performing the experiments included in the test matrix and the number of flawed and unflawed inspection sites that must be considered.

The general guidelines for the design of experiments to assess POD are provided in the previously-mentioned references [5] and [6]. However, the proper identification of the NDT process variables, the development of a suitable test matrix, representative test specimens, and flaws may require assistance of an expert in NDT POD analysis. These aspects of the experimental design are described in more detail in the following paragraphs.

13.2.2.1 Test Variables

Test variables that may influence the flaw detectability must be well-represented in the experimental demonstration tests. Process variables may include both the characteristics of the part (e.g., material, geometry, surface condition, cleanliness, etc.) as well as the inspection process. The latter is dependent on the specific NDT method that is being assessed. For example, in liquid penetrant inspections process variables may include the LPI fluid sensitivity, dwell time, or viewing conditions while in ultrasonic inspection the sensor frequency, focal length, beam direction, threshold level, scan rate, or the inspector training, physical ability, etc. will be important.

13.2.2.2 Test Matrix

Once the important process variables that are expected to influence the NDT system capability are defined, a test matrix is generated for their experimental evaluation. The test matrix is a list of conditions that collectively define experiments necessary to assess the NDT system capability. Individual experiments that assess the effect of a single variable at a time are preferable; however, they may be time consuming or expensive. Alternatively, factorial experiments, which consider the influence of several influencing variables simultaneously, are used. More details on factorial experimental design are provided in the above-mentioned reference [5].

Finally, a document is developed that specifies the experimental design for the inspections, the specimens that need to be inspected, the procedures that need to be followed to perform inspections, the way the data are recorded, and the process for ensuring the inspection system is under control.

13.2.2.3 Test Samples

Test samples must represent the real part in terms of material, geometry, surface condition, and flaw characteristics. Life-expired components of aircraft, such as engine parts or structural elements of retired airframes, are ideal, if adequate inspection sites and representative flaws are available to perform NDT reliability tests. Aircraft parts containing service-induced damage provide valuable opportunities for such experiments. A number of examples of IAR POD studies utilizing life-expired aircraft engine components with service-induced cracks are provided later in this chapter. However, such parts are very rare, particularly for airframe structural elements. In the latter case, it may be possible to introduce representative flaws in adequate numbers on a service-retired structure and perform inspections under field conditions to establish POD. For example, this approach can be used to determine the detectability of impact damage caused by runway debris in aircraft structures using visual, enhanced optical, or other NDT methods. An IAR example of such tests carried out on helicopter rotor blades is provided later in this chapter. For situations where the above-mentioned approaches are not possible, samples representing the actual part are manufactured for POD assessments.

Manufacturing of representative samples with realistic flaws in adequate numbers is very time-consuming, difficult, and costly. Generic samples provide a more economical alternative, especially for simulating airframe structural components, such as lap-joints [15]. Generic samples may be designed in such a way that several component configurations (e.g., single or double-lap) or different flaw locations (e.g., cracks in the first-layer or the second-layer) are evaluated by using the same sample set but assembled in different configurations. Later in this chapter, a generic POD study carried out at IAR on airframe lap-joint structures is described to provide an example of this type of study. The characteristics of the test samples and the flaws are established based on the actual component for which the NDT POD data is required. However, the inspector must not be aware of the presence or absence of flaws or their characteristics. Specimens that become familiar to the inspectors will bias the resulting POD and must not be used.

13.2.2.4 FLAWS

The accuracy of the POD estimates is dependent on the number of inspection sites with and without flaws, the size of flaws at the inspection sites, and the basic nature of the inspection results. Unflawed sites are necessary to insure integrity of results and to determine the rate of false-calls. The effects of false-calls on POD will be described later. Reference [5] specifies at least 60 flawed sites for the NDT methods that provide hit/miss data (e.g., LPI) and 40 flawed sites for technique that provide a signal response (e.g., eddy current). This reference also recommends at least three-times as many unflawed sites.

The flaw sizes must cover the expected range and must be uniformly distributed. Too many large flaws, which are always detected, or too many small flaws, which often are missed, will bias the POD. The majority of flaws must be in the mid-region of the POD curve where the curve changes rapidly.

13.2.3 DEMONSTRATION TESTS

Demonstration tests involve inspection of the sample sets under normal conditions used in the NDT facilities or in the field and following the documented experimental design and test procedures provided to the inspectors. The document on the experimental design and test procedures provides the inspectors with information and instructions that they must follow before, during, and after the tests, such as information on necessary standards and calibrations, process parameters, data analysis techniques, accept/reject criteria, and data recording.

Three categories of experiments may be used to evaluate the NDT capability [20].

13.2.3.1 DEMONSTRATION AT ONE DISCONTINUITY SIZE

Historically, this type of experiment has been performed to establish the ability of an NDT system by using multiple test samples with the same size flaw. Based on statistical sampling theory, 29 successes in 29 trials at one discontinuity size give a 90% confidence that the same size defects will be found every time [5]. This category is used mainly to satisfy regulatory concerns and provides much less information about the inspection system than when a range of discontinuity sizes are assessed.

13.2.3.2 DEMONSTRATION OF A SINGLE INSPECTION

This type of experiment is performed to qualify the POD of a specific NDT process applied to a particular problem. This approach is widely used and involves either using test coupons with simulated discontinuities representing the problem at hand or actual components with service-induced defects. A properly designed experiment with a sufficient number of discontinuities and a well-represented distribution of sizes will allow the generation of valid POD data. Another way of performing this type of experiment is to analyze NDT data from fleet inspections [21]. In this approach, the results of the inspection of a single subject, for example one fastener hole, are recorded until a flaw is detected and repair or replacement takes place. Crack growth data is then used to estimate the size of the flaw at the times of previous inspections that may have missed this flaw. If the same inspection has been performed at intervals of time on the same subject, there may be a set of NDT results for different crack sizes. Given a large enough sample population, it may be possible to estimate a statistically valid POD.

13.2.3.3 DEMONSTRATION OF MULTIPLE INSPECTIONS

An experiment of this type is essentially the same as performing multiple Category 2 experiments by using more than one NDT process on the same subject. It is expected that, for a given discontinuity type, performing multiple inspections will increase the chance of detecting discontinuities. However, reference [22] demonstrates a marginal or no benefit in performing multiple independent inspections.

13.2.4 STATISTICAL ANALYSIS

The results of the demonstration tests are analyzed using appropriate statistical functions to develop probability of detection (POD) curves. The choice of the statistical functions is dependent on the type of the inspection data. Generally, the response of an inspection process may be in the form of a trace (e.g., in the case of liquid penetrant inspection), a signal output (e.g., in the case of manual ultrasonic tests), or an image (e.g., in the case of radiographic film). Often a threshold level is set above the background noise (e.g., noise from the test material, NDT process, equipment, or environment) and traces, signals, or images that exceed the threshold are considered as being due to defects. Defect identification is done either by a human operator or by the NDT instrument. The possible outcome of an inspection could be one of the following:

- A defect exists and is detected. This is referred to as "true positive (TP)" or "hit."
- A defect does not exist and none is detected. This is referred to as "true negative (TN)."
- A defect exists but is not detected. This is referred to as "false negative (FN)" or "miss."
- A defect does not exist but the inspection falsely identifies one. This is referred to as "false positive (FP)" or "false-call."

The NDT data produced in this manner can be grouped into two categories: those that only provide the inspection outcome in either "hit" or "miss," and those that provide additional information related to the "flaw size." The "hit/miss" or "pass/fail" inspection output is often qualitative and is presented in a binary format; "1" for a defect found and "0" for a defect missed. The majority of manual inspections carried out at the depot level or in the field are of this category and are analyzed using a binary regression technique, such as a "logistic" model. The analysis requires a minimum of 60 flawed sites and three times as many flaw-free inspection sites in order to be valid [5].

Most modern NDT instruments provide a signal response (often in digital format) from a discontinuity that can be related to the flaw size. The NDT response to a discontinuity of size "a" is referred to as "$\hat{a}$" and is measured in different ways depending on the NDT method and instrument being used. The quantity "$\hat{a}$" is the NDT signal feature of interest, such as signal amplitude (e.g., ultrasonic), phase-angle (e.g., eddy current), or the number of events (e.g., acoustic emission). Although this type of data is considered more rigorous, from the inspector's perspective, it may be more strenuous to obtain than the previous one as the inspector needs to carefully measure the signal feature of interest. For this type of data, a threshold is selected above the background noise level, and after elimination of data below the threshold level, called "left censored" and those above the signal saturation level, called "right censored," a linear regression analysis is carried out. In this case, the recommended number of inspection opportunities are 40 for flawed and 120 for flaw-free sites [5].

Following reference [23], POD at a discontinuity of size "a" is defined to be the average probability of detection of all discontinuities at the size "a." This definition reflects the fact that the detectability of discontinuities is dependent on many factors, including, but not limited to, the flaw size. Therefore, the mean POD curve is derived from the assumption that each flaw has a range of probability of detection as expressed below:

$$\text{POD}(a) = \int_0^1 p \cdot f_a(p) \, dp \tag{13-1}$$

where p is the proportion of flaws detected and f_a is its density function. A confidence bound is often associated with the POD curve that reflects the fact that the curve is calculated using a sample population. The higher the number of samples used to obtain the POD, the higher will be the confidence level. If sufficient "hit/miss" and flaw size or "â vs. a" data are available, then a POD-flaw-size relationship can be obtained using an appropriate statistical function. Different statistical functions have been suggested for calculating POD and confidence curves depending on the type of the inspection data. Two models are often used for the POD analysis of inspection data depending on whether the outcome of the systems is "hit/miss" or "â vs. a". These are described in the ensuing paragraphs.

13.2.4.1 THE LOG-LOGISTIC (LOG-ODDS) MODEL

Berens and Hovey examined various methods of modeling NDT data to determine POD curves [24]. They concluded that the log-logistic (or log-odds) model was the most consistent distribution for determining a POD curve as a function of defect size, a_i. The functional form of the log-logistic distribution is as follows:

$$P_i = \frac{\exp[\alpha + \beta \ln(a_i)]}{1 + \exp[\alpha + \beta \ln(a_i)]} \tag{13-2}$$

where P_i is the probability of detection for defect i, a_i is the size of defect i, α and β are constant parameters that define the curve. α and β are referred to as location and slope parameters, respectively. They presented two approaches for estimating these constants, the Range Interval Method (RIM), which is also known as Regression Analysis, and the method of Maximum Likelihood Estimators (MLE). They also suggested the calculation of confidence bounds on the log-logistic POD curve by assuming that, for large sample sizes, the estimates of the mean POD curve will be normally distributed about the true POD curve.

13.2.4.2 (I) RANGE INTERVAL METHOD (RIM)

The range interval method has been used to estimate the parameters α and β required to define a log-logistic POD curve based on equation (13-2). It is assumed that the variability of POD within a small defect size range or interval is small and the detection within that range follows a binomial distribution [24]. To implement the range interval method, the defect data is divided into t intervals of equal length. The probability of detection is calculated for each interval as being the ratio of defects detected to the total number of flaws in that interval. This gives t data points. The t data pairs of POD and flaw size are transformed into a linear domain and a linear regression is performed on the data pairs in order to obtain the intercept and slope parameters, α and β, of the log-logistic function. The reverse transformation gives the POD curve.

The data points are transformed into a domain where the POD relationship is linear. Then linear regression analysis is performed using the following transformations on the log-logistic distribution function:

$$Y_i = \ln\left(\frac{P_i}{1-P_i}\right), \quad X_i = \ln(a_i) \tag{13-3}$$

where P_i is the proportion of flaws detected and a_i is the size of defect i. For ease of analysis, defects are grouped into size-intervals where a_i is the average defect size in interval i, and n is the total number of intervals.

The result of the transformation on equation (13-3) is a set of points, which are fitted with the line:

$$Y = \alpha + \beta X \tag{13-4}$$

The parameters α and β can be substituted into equation (2) and used to calculate a POD curve for a range of flaw sizes. It should be noted that if the estimated POD value for an interval is 0 or 1, this transformation is undefined. For a POD of 0, a value of $1/(n+1)$ is used, and for a POD of 1, a value of $n/(n+1)$ is chosen. This approximation overestimates the POD at small flaw sizes where the POD in an interval is zero, and underestimates the POD at large flaw sizes where the POD in an interval is 1.

The 95% confidence bound, $Y_{95\%}$, on the sample mean can be found in the transformed domain using:

$$Y_{95\%} = Y_{(MEAN)} - 1.96 S_{Y/X} \tag{13-5}$$

where 1.96 is the standard normal variate for 95% of the area for the normal distribution curve, and $S_{Y/X}$ is related to the variance and covariance of the transformed variables.

13.2.4.3 (II) MAXIMUM LIKELIHOOD ESTIMATORS (MLE)

The maximum likelihood method also has been used to estimate the parameters α and β in equation (13-2) that maximize the probability of obtaining the observed data. The likelihood L for a single observation is:

$$L(P_i; a_i, x_i) = P_i^{x_i} \bullet (1 - P_i)^{1 - x_i} \tag{13-6}$$

where P_i is the probability of detecting flaw i of size a_i and x_i is the inspection outcome, 0 for a "miss," and 1 for a "hit."

The overall likelihood of observing all the data is the product of their individual likelihoods. Thus, the likelihood of a series of independent inspections is the multiplication of the individual observations:

$$L(P; a, x) = \left[\prod_{i=1}^{h} P_i \right] \left[\prod_{j=1}^{n-h} (1 - P_j) \right] \tag{13-7}$$

where L is the overall likelihood, n the total number of flaws, h is the number of flaws detected, and n-h is the number of flaws missed. Note that probability of a miss = 1 − probability of a hit. The hit and miss values are established by destructive verification tests or other acceptable means. By taking the logarithm of equation (13-7), the series of products becomes a series of sums, as illustrated by equation (13-8) [6].

$$\ln L(P; a, x) = \sum_{i=1}^{h} \ln P_i + \sum_{j=1}^{n-h} \ln P_j \tag{13-8}$$

Now, the values of α and β in equation (13-2) can be selected to maximize the likelihood of equation (13-8). Since the logarithm is a monotonic function, the maximum of the log likelihood is the same as the maximum of the likelihood itself [6]. Thus, equation (13-8) can be differentiated with respect to parameters α and β of equation (13-2). The maximum likelihood of α and β are denoted as α' and β' and found by maximizing the log likelihood with respect to α and β [25]. The derivatives of the log likelihood are taken and set equal to zero:

$$\sum_{i}^{n} \frac{\exp[\alpha' + \beta' \ln(a_i)]}{1 + \exp[\alpha' + \beta' \ln(a_i)]} - \sum_{i}^{n} x_i = 0 \tag{13-9}$$

The variable x_i is as defined previously, 1 for a hit and 0 for a miss. The sum is obtained over a set of flaws numbered from 1 to n. Similarly, the derivative of the log likelihood is taken with respect to β and set to zero;

$$\sum_{i}^{n} \ln(a_i) \cdot \frac{\exp[\alpha' + \beta' \ln(a_i)]}{1 + \exp[\alpha' + \beta' \ln(a_i)]} - \sum_{i}^{n} x_i \cdot \ln(a_i) = 0 \tag{13-10}$$

In a similar way, the 95% confidence bound given by equation (13-5) is created using an assumption of a normal distribution.

13.2.4.4   LOG-NORMAL MODEL

The cumulative log normal distribution has been suggested in reference [5] for modeling POD data. The cumulative log normal distribution is expressed as:

$$P_i = 1 - Q(z_i)$$

$$\text{for } z_i = \frac{\ln(a_i) - \mu}{\sigma} \qquad (13\text{-}11)$$

where $Q(z)$ is the standard normal survivor function, z_i is the standard normal variant, and μ and σ are the location and scale parameters of the POD curve. The maximum likelihood method can be used to find the values for the location and scale parameters. The same likelihood function as in equation (13-8) applies to both the log-logistic and log normal distributions. Equation (13-8) is differentiated with respect to μ and σ, derivatives set to zero, and the resulting simultaneous equations are solved to provide estimates of μ and σ. The method of determining the confidence bound on the log-normal POD curve is derived by Cheng and Iles as described in reference [26].

The "hit/miss" data can be analyzed using both log-normal and log-logistic algorithms along with the above-mentioned Range Interval Method (RIM) or Maximum Likelihood Estimators (MLE) for the estimation of the location and scale parameters. A comparison of the different statistical approaches will be made later in this chapter. Briefly, the log-logistic distribution is easy to implement on large "hit/miss" sample sizes and provides conservative POD results. In contrast, the log-normal distribution is more complicated in terms of mathematics and slower to calculate but can be implemented using common spreadsheet programs. For the estimation of the parameters required for the above distributions, the RIM approach is highly dependent on the interval size and the approximation for intervals with POD of 0 or 1. In contrast, the MLE method does not require any information other than the actual inspection data. Therefore, the log-normal along with MLE provides more realistic POD curves and is preferable for modeling "hit/miss" type NDT data.

For â versus a type data, it has been noted in many NDT methods that the logarithms of â and a are linearly related, with residuals ε normally distributed with mean zero and standard deviation δ^2; thus, one can write:

$$\ln \hat{a} = \beta_0 + \beta_1 \ln a + \varepsilon \qquad (13\text{-}12)$$

and, therefore:

$$POD(a) = 1 - Q\left[\frac{\ln a - \mu}{\sigma}\right] \quad \text{where } \mu = \frac{\ln(y_{th}) - \beta_0}{\beta_1} \text{ and } \sigma = \frac{\delta}{\beta_1} \quad (13\text{-}13)$$

where Q is the standard normal survivor function, and y_{th} is the value of the signal â at the decision threshold.

The choice of the model of equation (13-10) implies particular meanings for the coefficients β_0, β_1, and ε. In particular, ε captures all the random variances in the NDT measurement; be they due to electronic noise, variation in discontinuities of the same characteristic size, manual probe manipulation, inspector performance, or any other random source of noise. The coefficients β_0 and β_1 are simply those of a linear fit to the data (in the log domain). The most common method to estimate all of the coefficients is to use maximum likelihood estimators, as described in MIL-HDBK-1823 [5].

13.2.4.5 FALSE-CALLS

The above-mentioned functions do not take into account the false-calls. As explained before, a false-call occurs when a non-existing flaw is identified as a flaw. Too many false calls in inspection data can bias a POD curve because when false calls are high, a portion of the detected flaws are likely to be coincidental identifications. Reference [27] considers the effects of false-calls by using probability of indication (POI) and provides mathematical expressions to calculate POI. The calculation is based on the assumption that the recorded indications are the sum of indications that result from true identification of flaws, plus the indications that result from false identifications, minus the overlap where noise signals (i.e., false calls) are identified as flaw signals on specimens that are actually flawed. A false call rate of more than 5% is recommended to be taken into account to ensure an accurate modeling of the true POD.

Later, an extension to the curve-fitting method of estimating POD was suggested by Spencer that includes terms to account for false-calls and miss rates independent of flaw size [28]. The Spencer model can be written as:

$$POD(a) = p_h + (1 - (p_m + p_h)) \cdot F(a\,;\mu,\sigma) \tag{13-14}$$

where $POD(a)$ is the probability of detection at the flaw size a, p_h is the false call probability, p_m is the probability of missing a flaw independent of flaw size, and $F(a;\mu,\sigma)$ is the two-parameter distribution used to fit the data. The distribution F is usually modelled by a log-normal or log-logistic curve. This model overcomes the key problems associated with previous curve-fitting methods, by providing reasonable values at very small and very large flaw sizes and reducing sensitivity to outliers. All of the parameters in equation (13-14) can be found using a maximum likelihood estimation method.

Large values of either p_h or p_m will indicate that an inspection procedure is not well controlled due to problems, such as large variability in signal at the decision threshold level, poor correlation of the discontinuity size with the POD, and insufficient number or distribution of discontinuity sizes. A comparison of two or more POD curves can be made using statistical methods, such as analysis of variance (ANOVA) and multivariate analysis of variance (MANOVA), which provide a test of hypothesis that the curves under consideration are the same. In statistics, ANOVA referred to a collection of statistical models and procedures in which the observed variance in a particular variable is partitioned into components attributable to different sources of variation. MANOVA is a statistical test procedure for comparing multivariate (population) means of several groups that unlike ANOVA uses the variance-covariance between variables in testing the statistical significance of the mean differences. The algorithms used for the ANOVA/MANOVA comparisons between POD curves are standard methods as described in references [5] and [28]. ANOVA provides a comparison of one POD curve to a POD curve that is selected as the "standard." MANOVA provides a test as to whether two or more POD curves are statistically equivalent (i.e., differences are small enough to be expected in a random sampling).

The effective POD seen in practice will be due to both the actual true detection as well as false-calls. In many inspection methods, decision thresholds can be varied, which may result in a different POD and false-call rate for each threshold. This information can be shown graphically using a Relative Operating Characteristic (ROC) curve. The ROC curve addresses the issue of total accuracy (hit calls made on both flawed and unflawed sites) by incorporating false-call data into the analysis. The use of ROC and theory of signal detection in NDT applications has been discussed elsewhere [29]. Generally, ROC curve is a graphical plot that illustrates the performance of a binary classifier system as its discrimination threshold is varied. It is created by plotting the fraction of true positives out of all the positives vs. the fraction of false positives (or false calls) out of all the negatives at various threshold settings. A typical ROC curve is a plot of the probability of detecting a true flaw against the probability of a false-call as the decision-making threshold is varied. An example of a ROC curve is presented in Figure 13.2. In an accurate inspection, lowering the threshold and increasing the false call rate should not

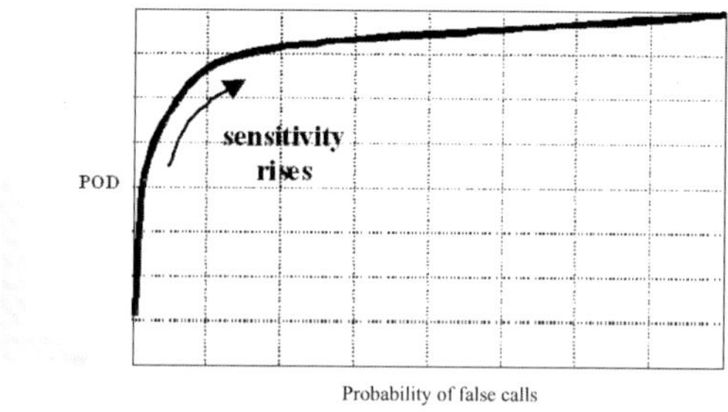

Figure 13.2 Example of ROC curve.

affect the probability of making a true call. A poor inspection on the other hand is one where flaws cannot be discriminated from noise, and the probability of detection rises with the probability of false-calls. ROC curves can be useful in deciding what inspection threshold to use in combination with POD information and cost-benefit analyses. They can also be used to rank inspector performance where NDT data is inspector-dependent.

13.2.5 DOCUMENTATION OF RESULTS

Finally, the outcome of an NDT reliability demonstration is documented describing the details of the NDT systems being evaluated, the design of the experiments, individual test results, statistical methods used, POD and confidence curves obtained, and a summary of the results. Proper documentation is important for extrapolating the results obtained for a specific NDT system and component to other similar NDT systems or components. Usually, POD data developed for one NDT system and component cannot be used for others unless all the factors that have a major influence on the POD results are accounted for and necessary corrections made. For this, a full knowledge of all the test variables and their effects on POD is necessary. Examples of extrapolating existing POD data to new situations will be described later in this chapter under the section entitled "Examples of NDT Reliability Studies."

13.3 MODEL-ASSISTED POD

R. B. Thomson defines model-assisted POD (MAPOD) as methodologies that estimate or predict probability of detection of NDT using models [30]. MAPOD is used to augment or replace traditional experimental POD demonstrations in order to reduce the cost and time needed for such studies. In addition, POD data generated using experimental demonstrations are component-specific and only applicable to the exact conditions under which the inspections are carried out. Any broader application of the POD data to other components or inspection conditions relies on qualitative engineering judgement and further verifications that the changes will not reduce the POD.

As mentioned before, many factors may influence the outcome of an NDT test and POD is determined to reflect the variability of an NDT system. Some of the influencing factors are controlled by well-understood physical phenomenon and can be addressed by physics-based models. Others, such as the inspector variability, cannot be established by modeling and must be quantified by empirical experiments. Thus, MAPOD may be carried out in different ways, and these are described in the following sections.

One approach is to rely primarily on physics-based models to predict the NDT response due to influencing variables that are well-understood by physical phenomenon and use empirical data only for those that are less well-defined (e.g., operator, system variations). This is referred to as the "full model" approach [31]. Commercial modeling software packages are available for some NDT techniques, such as eddy current or ultrasonic inspection, that may be used for this purpose.

Another approach is to use transfer functions to convert data measured for one specific application and a set of conditions to another related application or set of conditions. For example, POD curves developed using empirical demonstrations for an eddy current inspection on an aircraft engine part may be used to generate new POD data for the same method and component inspected using a different instrument, different settings (frequency, gain, threshold), by a different operator, or in different conditions.

Likewise, existing POD data may be transferred to another part of different material, different geometry, or different flaw type. In such cases, the transfer functions relating the old and the new situations are established. In this case, generic POD curves can be developed for a given component geometry or configuration and used as a basis along with modeling to obtain new POD data. Both approaches may require limited experiments to verify the modeling results. A combination of modeling and experiments that uses transfer functions from numerically simulated NDT signals and the POD data generated for a generic well-defined situation may be the most cost-effective way of developing NDT reliability data. In either case, the use of modeling to estimate POD is a practical and economical alternative to experimental approaches.

In practice, a combination of both the full model and the transfer function approaches is likely to be used as illustrated in the MIL-HDBK-1823A (2007) schematic of Figure 13.3.

In this figure, the theoretical and experimental assessments are combined to obtain POD. The protocol for model-assisted POD determination based on this handbook consists of the following steps:

1. Define the scope of the POD study.
2. Identify factors that control signal and noise.
3. Identify a subset of factors that can best be assessed empirically.
4. Prepare samples and conduct the empirical tests.
5. Analyze the results and obtain the mathematical model relating flaw response to flaw size relationship (i.e., â vs. a).
6. Determine if experiments or physics-based models are to be used to assess the influence of the factors.
7. Conduct the evaluation of the factors using either empirical experimentation or physical modeling.
8. Analyze the results and determine adjustments (e.g., experimental transfer functions or theoretical models).

9. Update the â vs. a relationship obtained in step 5 with the necessary adjustments resulting from step 8.
10. Compute the new POD.

The above-mentioned protocol is a conceptual overview and the details may vary depending on the specific situations. In reference [31], a few examples of the MAPOD applications are provided, and the different approaches are compared.

13.4 EXAMPLES OF NDT RELIABILITY STUDIES

As mentioned before, NDT reliability may be determined experimentally or estimated using model-assisted POD or by a combination of the two approaches. For the experimental development of NDT POD, the commonly accepted approach is to use test samples with artificial defects that simulate real parts and real flaws as described in references [5] and [6]. A large number of flawed and flaw-free samples are needed to generate statistically valid POD. In many cases, the cost and time associated with the preparation of representative samples and performing NDT tests duplicating those of the actual field inspections may become very significant. Thus, alternative and more cost-effective ways of developing POD are desirable.

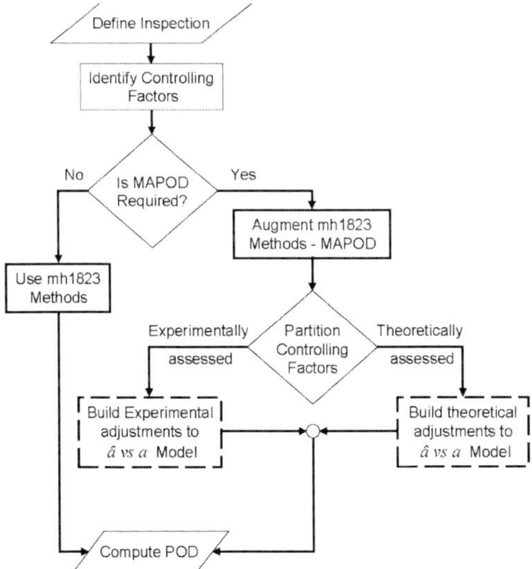

Figure 13.3 MIL-HDBK-1832A approach for model-assisted POD determination [5].

With the above-referenced recommended practices and standards in mind, over the years, the NRC Institute for Aerospace Research (IAR) has developed and performed POD assessments in many different ways including the following:

1. POD development using actual life-expired parts with service-induced damage.
2. POD development using actual parts with artificially-introduced damage.
3. POD development using representative samples with artificially-induced defects.
4. POD development using generic test blocks representing several structural configurations with artificially-induced defects.
5. POD estimation using data generated during an NDT procedure development.
6. POD estimation using field inspection data.
7. POD estimation using modeling and verification by limited experiments.

In the following sections, the above-mentioned approaches will be described, and the related observations or issues as well as advantages and limitations will be mentioned.

13.4.1 POD Development using Actual Life-expired Parts with Service-induced Cracks

Several NDT reliability demonstrations on actual parts that contained service-induced damage have been carried out [11], [12], and [32], one example is provided below that used retired compressor disks of a military aircraft engine (J85-CAN-40).

13.4.1.1 Methodology

Background: For the development of valid NDT POD, components with a large number damaged and damage-free sites are needed. The parts are then inspected using NDT procedures and conditions that duplicate the real field inspections, and the results are recorded. After completion of inspections, the test sites are carefully examined using destructive tests to verify the presence or absence of defects and measure the size of flaws identified. The recorded NDT data are compared with the verification test results to establish the number of flaws detected or missed as well as the true and false-calls on flaw-free sites. The data are then analyzed statistically to determine POD and the confidence curves. This approach is schematically illustrated in Figure 13.4 and described in references [11], [12], and [32].

Test Parts: The components were service-expired compressor disks of an aircraft engine (J85-CAN-40) shown in Figure 13.5. Although the disks were

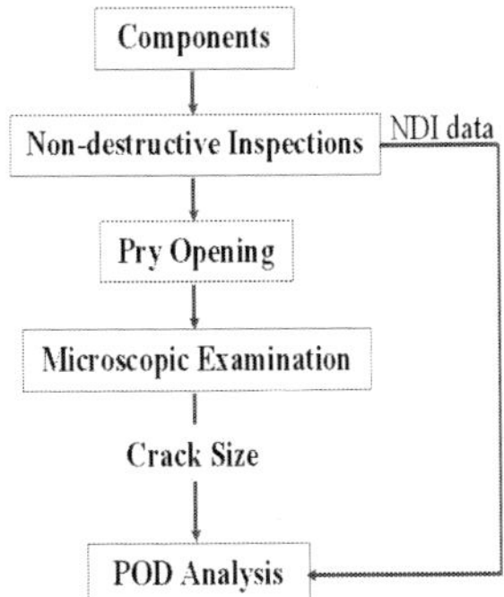

Figure 13.4 Steps used in the J85-CAN-40 POD assessment.

Figure 13.5 The J85-CAN-40 compressor disc and spacer.

considered as safe-life parts, they had developed low-cycle fatigue cracks in some of their fastener bolt-holes towards the end of their service life. The disks were made of precipitation-hardened martensitic stainless steel (AM355). Each disk had 40 bolt holes, and each spacer had two integral segments with 40 bolt holes in each segment. The diameter of the bolt-holes in both the discs and spacers was identical (4.8 mm or 0.188 in). Also, the thickness of the disks and spacers in the bolt-hole region was the same (i.e., 1.9 mm).

Calibration Specimens: Forty test specimens were prepared from actual compressor disks by removing material from between the holes. Half of the specimens were used for the creation of fatigue cracks in the laboratory. In these nearly rectangular specimens, a 2.8mm hole was drilled, and a 1.4 mm EDM notch was cut in the hole. These specimens were then loaded under cyclic tension-tension conditions in the range of 67–670 kg at 10Hz. Cracks occurred at about 5000 cycles and grew at about 0.1 mm per 1000 cycles. Cracks from 0.2 mm to 2 mm, as measured using optical microscopy, were grown in the specimens. After loading of each specimen was completed, the EDM notch and the 2.8 mm hole were drilled out to a 4.6 mm diameter hole that is the same size as the actual bolt holes.

In addition, whole compressor disks were also used to fabricate inspection test sites by drilling new holes, 4.6 mm in diameter, between existing holes in four disks, providing 160 virgin holes. EDM notches of various sizes (0.2 to 2 mm long) were placed in 15 of these holes. EDM notches were also rectangular in shape and had a width of 0.17 mm open on both surfaces.

NDT Methods: Several NDT methods were used by four different organizations (A, B, C, and D) to examine the bolt-holes of the test components as described in Appendices of reference [12]. The techniques included the liquid penetrant inspection (LPI) and magnetic particle inspection (MPI), which, at the time of this investigation, were widely used in Canada by the aircraft engine maintenance facilities. Also, eddy current inspections (ECI) with manual probe insertion by the operator as well as automated probe insertion were used. All used established NDT standards and test pieces containing laboratory-grown fatigue cracks of various sizes to set the instrument settings and perform necessary calibrations.

The automated EC inspections were carried out at two organizations using different systems. One (A) used an experimental system consisting of a probe-spinning device attached to an automated scanner with three linear axes and a turntable. The system was equipped with a commercial pattern recognition software package (Tektrend ICEPACK0- described in reference [33]). At this facility, the ECI signal interpretation was carried out either by a certified operator or by using pattern recognition. The other organization (B) employed a commercial system with five axes of movement capable of inspecting a variety of component geometries. The latter performed inspections at different settings (e.g., threshold: 0.5, 0.6, 0.7, and 0.8 of maximum based on calibration).

In addition, two new approaches developed at IAR were explored in this assessment to establish their potential. One approach used ultrasonic surface waves generated in immersion mode (leaky surface waves) and looked at their scattering by surface-breaking cracks [34]. The other was based on an enhanced optical method that looked for surface-breaking cracks at oblique angles by a narrow band of light (Edge of Light) [35].

Verification Tests: After completion of all NDT tests, the existence of cracks in the bolt-holes was verified by destructive testing as schematically illustrated in Figure 13.6. First, 2 cm × 3 cm samples were cut from the region surrounding each bolt-hole using a laser cutting tool. Each sample was then sectioned into two pieces along the diameter of the bolt-holes. In the inward piece (the larger section), a notch was introduced in the side opposite to the bolt-hole, and the sample was then pried open by closing the notch in a vice. The second piece was loaded in three-point bending until failure. In both cases, crack faces were under tension during loading. During pry opening, the sections were examined periodically under an optical microscope to check the crack opening and to determine if multiple cracks existed.

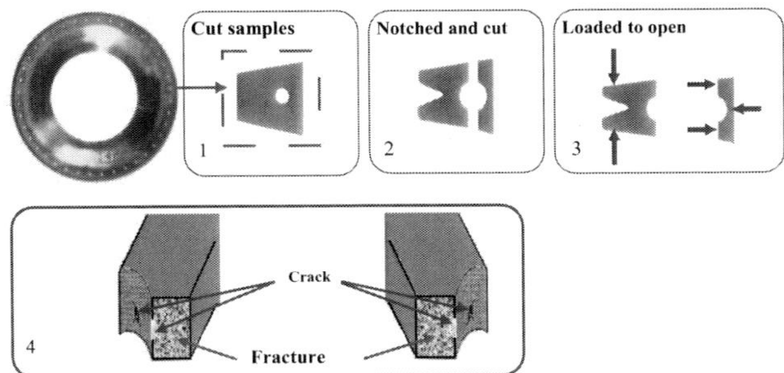

Figure 13.6 Schematic illustration of crack verification procedures.

After pry opening, the fracture surfaces were examined under an optical or scanning electron microscope, as required, depending on the crack size. Under the microscope, the service-induced fatigue cracks were easily distinguishable from the induced rupture failures due to their smooth and oxidized surfaces and crack sizes were measured. The specimens that did not reveal a crack on the fracture surfaces were further examined on the bolt-hole surface to make sure a crack was not missed.

Statistical Analysis: The inspection data generated in this study were of the form of "true positive or "hit", "true negative," "false negative or "miss," and "false positive" or "false-call." The hit, miss, and false-call rates were found for each procedure, and the POD as a function of crack size was determined as well as the lower 95% confidence bound on each POD curve. An earlier comparison of the log-normal and log-logistic analysis of "hit/miss" data showed that the log-normal function using the maximum likelihood estimation approach provided more realistic POD results [15] and, thus, was used for the ensuing POD calculations.

13.4.1.2 RESULTS

Microscopic Observation: Examples of different crack types seen in the compressor discs are provided in Figure 13.7. Most holes had a single radial crack, either open to both top and bottom surfaces (through cracks), or open only to one surface (corner cracks), or open to neither surface (middle or internal cracks). Most cracks were small (<1 mm) corner cracks, but some had grown to larger through cracks. There were also a few middle cracks; many of these initiated at what appeared to be voids or inclusions in the material. The crack faces were smoother than the ruptured surfaces and were covered with an oxide layer of different color that made it possible to recognize and measure them under a microscope. It should be obvious that it is not possible to make artificial cracks simulating the different

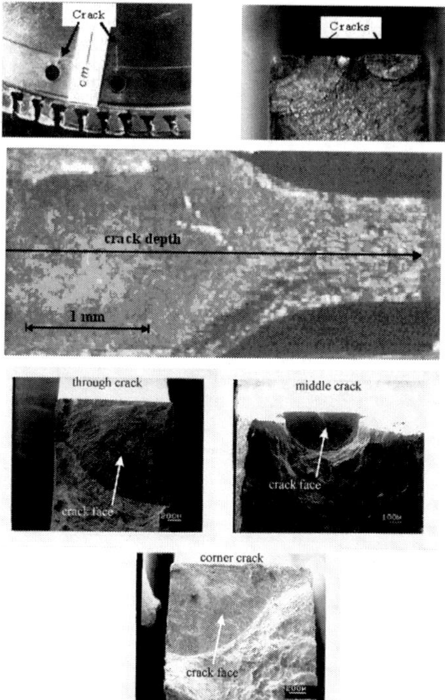

Figure 13.7 (Top left): Photograph of a section of a J85-CAN-40 disc showing bolt-hole cracks as identified by an LPI approach. (Top right): Service-induced fatigue cracks as seen under a scanning electron microscope. (Middle): A 3.5 mm through crack found in one disk. (Bottom): Typical crack types seen in the J85-CAN-40 disks.

shapes, sizes, locations, surface textures, and combinations of different types that were seen in real parts used in this investigation.

Service-Induced vs. Artificial Cracks: As mentioned before, laboratory-grown fatigue cracks or EDM notches are commonly employed for instrument calibration or POD development. As seen above, service-induced cracks in engine components are often covered with an oxide layer and might be very tight due to residual stresses. These characteristics influence the inspection results for many NDT methods. In this study, a comparison of eddy current response between artificial and real cracks was made under the same experimental conditions, and the EC signal magnitude was then plotted as a function of the crack area. Despite scatter in the data for real cracks due to variations in the shape, type, location, or orientation as well as coil position with respect to the crack and the bolt hole, a general trend appeared to exist.

As shown in Figure 13.8 of service induced cracks due to the wider gap between the two faces that provides a higher electrical resistance. Similarly, the oxide layer present in real cracks acts as an insulator resulting in larger (30–40%

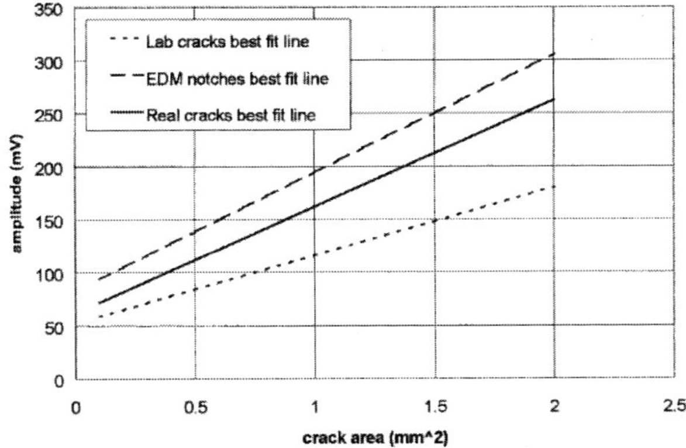

Figure 13.8 Eddy current signal amplitude for different flaw types.

in this case) magnitude EC signals as compared to the laboratory grown cracks. Therefore, if EDM notches or fatigue cracks are employed for setting a detection limit for the NDT systems, these variations should be taken into account in the final estimation of POD. Figure 13.8 also shows that the EC signal amplitude increases with increasing crack size for the three types of discontinuities in the size range investigated.

Comparison of the NDT Methods: Table 13.1 provides a summary of results, and Figure 13.9 gives the overall POD curves for the different NDT methods

Table 13.1 Summary of the NDT reliability assessments carried out at four organizations (A,B,C, and D) on bolt-holes of J85-CAN-40 compressor disks using different NDT procedures including liquid penetrant inspection (LPI), magnetic particle inspection (MPI), manual eddy current inspection (ECI-M), automated eddy current inspection (ECI-A), automated eddy current inspection and pattern recognition analysis (ECI-AP), ultrasonic inspection using leaky surface waves (UTI), and edge of light inspection (EOL).

Technique	LPI	MPI	UTI	ECI-M		ECI-A		ECI-AP	EOL
Organization	A	D	A	A	C	A	B - 0.8	A	A
holes	296	372	372	381	381	381	381	381	328
cracks	246	311	311	320	320	320	320	320	281
hits	63	75	99	183	145	194	252	184	113
rate (%)	26%	24%	32%	57%	45%	61%	79%	58%	40%
misses	183	236	212	137	175	126	68	136	168
rate (%)	74%	76%	68%	43%	55%	39%	21%	43%	60%
false calls	2	8	1	2	0	1	9	1	2
rate (%)	4%	13%	2%	3%	0%	2%	15%	2%	4%
90/95 length (mm)	2.59	2.76	1.90	0.84	1.11	0.65	0.51	0.87	1.37
largest missed (mm)	2.10	2.04	1.30	0.99	1.00	0.89	4.69	1.00	1.26

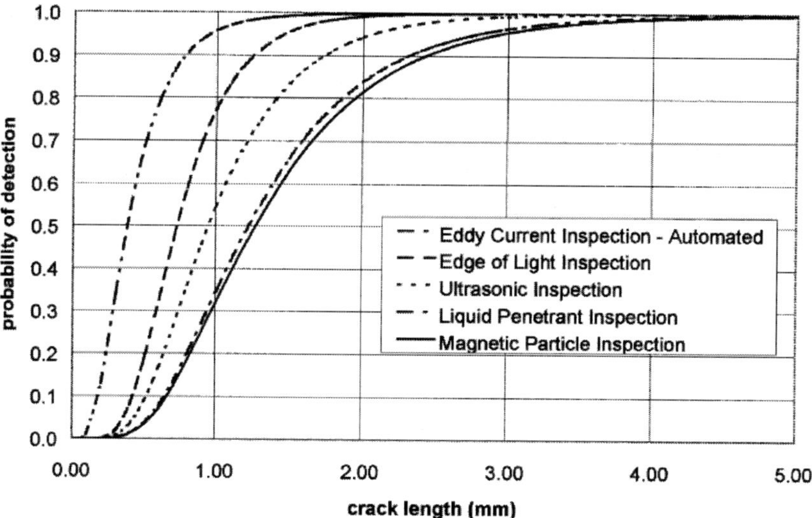

Figure 13.9 Comparison of the overall POD of several NDT methods applied to bolt-hole cracks in J85-CAN-40 compressor disks.

evaluated in this study. Despite the fact that the liquid penetrant inspections were carried out using the standard military procedures, significant variations were observed between the results from different inspection sites as revealed in Table 13.1. Such a variation is expected since the LPI method generally involves several steps that are very dependent on the operator. More importantly, the detection rate for all the LPI techniques investigated was alarmingly low for a technique most commonly used by the aircraft engine maintenance facilities.

The crack tightness and oxidation as well as component surface condition may have contributed to missing a large number of small cracks, and, therefore, the usefulness of this method for such applications is questionable. The 90/95% values for the LPI methods also varied widely and were the largest among the techniques investigated. Magnetic particle inspection at one facility achieved a detection rate and 90/95% value similar to the LPI and was consistent with the results obtained in an earlier study, which involved two other organizations [10].

Ultrasonic surface wave inspections were carried out using two different approaches. In one case, a contact wedge probe and manual inspection were used, and, in the other case, an automated immersion C-scan system using leaky waves were employed. The overall detection rates and the 90/95% values for both ultrasonic procedures were better than for the LPI and MPI, and the largest crack that was missed was smaller. Also, the new edge of light (EOL) technique performed better than the conventional LPI and MPI and was similar to the ultrasonic tests.

The detection rate for the manual eddy current inspection (ECI-M) was better than the LPI, MPI, and ultrasonic techniques, but, clearly, the automated eddy current method (ECI-A) provided the best POD results among the techniques investigated. The use of pattern recognition (ECI-AP) did not seem to improve the results. Since the eddy current method provided the best POD for the components studied, the data for this approach was further analyzed. In practical applications, EC signals are affected by a wide range of factors, and two such factors were studied here: (a) eddy current system and (b) crack type.

POD of Different Eddy Current Inspections: The manual eddy current inspections performed at two different facilities (A and C) and by two different inspectors provided different 90/95% results (A = 0.84 mm and C = 1.11 mm). This is expected as the human factor cannot be controlled easily due to human factors, such as differences in the training, experience, attitude, and ability of people. The difference for automated eddy current systems at facilities A and B was much less with 90/95% values of: A = 0.65 mm and B = 0.51 mm. This demonstrates the importance of using automated inspections for improved probability of detection.

Although the data are not included here, it was also apparent that the instrument settings, such as the reject threshold level, can substantially change the outcome of the automated systems. The use of high thresholds caused one inspection to miss a large crack (4.69 mm) while at low thresholds the false call rates were high. Thus, the instrument settings, especially the reject threshold level, must be selected with care.

POD of Different Crack Types: In the eddy current inspection of bolt-hole cracks using an automated probe-spinning and insertion device, as Figure 13.10 illustrates, through-cracks provided the best POD since they extended from one surface to another and, therefore, provided more resistance to the induced currents.

On the other hand, corner or middle cracks provide partial resistance to the currents and, thus, result in smaller-magnitude signals and poorer POD. Therefore, POD not only varies with the crack size, but it is also dependent on the crack type. Artificial cracks made in the laboratory often simulate through cracks that generally have higher POD than corner or middle cracks. However, in reality, the majority of service-induced cracks are corner-cracks.

Safe Inspection Intervals (SII): The SII for the compressor disks were calculated using a probabilistic fracture mechanics analysis based on the work by Koul et al. [36] as described in Chapter 2 of this publication. The SII was taken to be half the number of cycles required to grow a crack from its assumed initial size to the dysfunction size that was taken as half the critical size.

In order to calculate a realistic SII for a given component of a specific engine, statistically-significant information must be available on the engine mission profile, operational loads on the component, crack propagation rate under the most severe operational conditions, dysfunction or critical crack size, and POD of NDT methods that are used during maintenance inspections. Except for the latter, all the necessary information was taken from the above-referenced previous work on the same types of components.

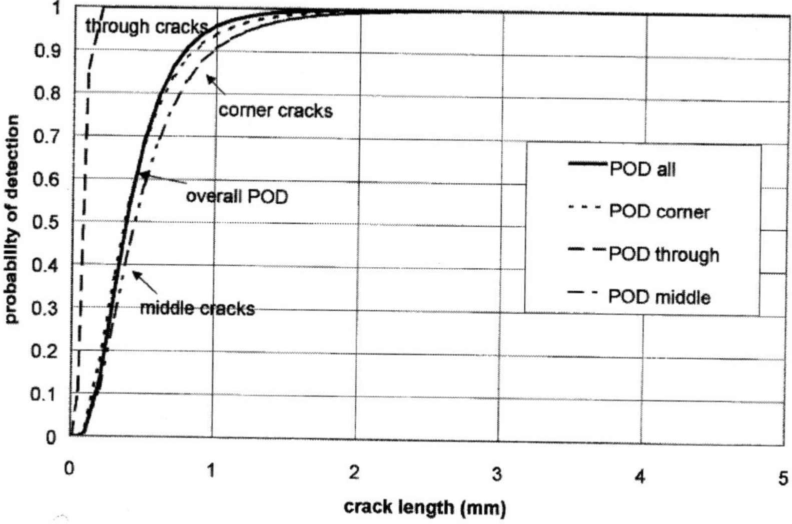

Figure 13.10 POD for various crack types resulting from automated eddy current technique.

The crack sizes detectable at 90% POD with 95% confidence were used in the calculations, and the SII values obtained using the two statistical curve fitting models are given in Table 13.2. The SII values obtained based on the log-normal distribution for the eddy current procedures are considerably larger than the minimum required (4000 cycles) for a cost-effective application of the damage tolerance approach to the J85-CAN-40 engine compressor disks and spacers. Thus, for damage-tolerance-based maintenance of these engine parts, it is essential to inspect the parts every 4000 cycles using the automated eddy current technique.

The results also show that the more sensitive and the more reliable the inspection method, the greater the operating interval between inspections, which translates to lower operating costs and increased availability of the engines.

13.4.1.3 OBSERVATIONS

Of all the NDT methods evaluated, an automated eddy current approach provided the highest POD in detecting fatigue cracks in these specific engine components. However, using very low thresholds resulted in high false calls while at a very high threshold a large crack was missed. Manual eddy current inspection results were not as good as those of the automated systems but better than all the other techniques investigated. The use of pattern recognition did not improve the detection rate or the 90/95% value.

Table 13.2 Safe inspection intervals calculated using probabilistic fracture mechanics.

Technique / Organization	90/95 crack length (mm)	correlation coefficient	SII (cycles)
LPI / A	2.57	0.985	3854
MPI / D	2.67	0.963	3740
UTI / A	1.90	0.987	4198
ECI-M / A	0.84	0.994	4971
ECI-A / A	0.65	0.997	5083
ECI-A / B 0.5	0.23	0.999	5561
ECI-A / B 0.8	0.48	0.997	4984
ECI-AP / A	0.87	0.995	4949
ELI / A	1.36	0.995	4435

The commonly-used LPI and MPI methods produced similar POD results on these components and were inferior to other methods. NDT response from artificial cracks can be quite different from that of serviceinduced cracks probably due to shape, surface texture, and tightness. Through-cracks are easier to detect than corner or middle-cracks and, therefore, give higher POD. Artificial cracks, in most cases, resemble the through-cracks. Most cracks in the investigated parts were corner-cracks that are difficult to simulate and detect. For successful application of damage tolerance maintenance approach to engine components of the type investigated here, the use of automated eddy current systems is necessary.

13.4.1.4 Advantages and Limitations

Development of POD data using actual parts with service-induced damage has the following advantages and limitations.

Advantages:
- Can be performed as a round-robin demonstration involving several facilities and include a number of NDT methods, instruments, systems, settings, or personnel.
- Does not require manufacturing of a large number of test specimens or artificial damage.
- Can be carried out in a timely and cost-effective manner.
- Includes test variables that are difficult to consider (e.g., damage features due to operational environment).
- Significant amount of information can be generated.
- Provides realistic POD data that can be used for component life estimation.
- Most suitable for short-life or disposable parts (e.g., engine blades, discs, etc.)
- POD data may be used for qualification or improvement of inspection methods, or comparison of different systems, facilities, or inspectors.

Limitations:

- Requires a large number of actual parts with service-induced damage that are rare to find.
- POD data are specific to components and damage types studied.
- Requires elaborate damage verification tests.

13.4.2 POD DEVELOPMENT USING ACTUAL PARTS WITH ARTIFICIAL DAMAGE

Aircraft components containing service-induced defects in large numbers that are required for POD studies are seldom available. Thus, artificial defects are introduced on retired parts to enable NDT reliability tests. A work carried out on tail-rotor blades of a military helicopter (CH-146) is briefly described here as an example of this type of POD development. Details are available in reference [37].

13.4.2.1 METHODOLOGY

Background: The CH-146 helicopter tail rotor blades are susceptible to certain small surface scratches, dents, and nicks that may occur during service operations due to airborne sand particles, runway gravel, or other debris. There are different allowable sizes for each of these damage types, depending on their location on the blade.

Objectives: The purpose of this study was to establish the reliability of a visual inspection used at the Canadian Forces (CF) maintenance depots to find the above-mentioned damage types on the CH-146 tail rotor blades and to determine important variables that may affect the POD results.

Test Parts: For this purpose, four CH-146 tail rotor blades were first cleaned and repainted with the same color schemes as those of the CF squadrons operating the helicopters. Two blades were painted in dark green (denoted as g1 and g2), and the other two blades were painted in white (denoted as w1 and w2). Discontinuities were introduced on these parts to simulate the damage types that occur in service and to provide a set of representative discontinuities appropriate for use in a reliability experiment. There were three categories of discontinuities as listed in the CF Operator's Maintenance Manual for this component. The discontinuities had been termed as "nick/scratch," "sharp dent," and "non-sharp dent." The allowable size limits had been specified for each discontinuity type and position on the blade and are given in Table 13.3. Also, the tail rotor blades in service suffered from varying levels of paint abrasion, solely on the inboard side that was most severe at the tip due to take-off and landing on non-paved sites. Sandblasting was used to mimic the in-service paint ablation.

Artificial Defects: For valid POD, the artificial defects must mimic the types of damage that occur in service. It had been determined that due to the nature of the operation of the tail rotor blades, impacts by objects, such as gravel, at high angles of attack with respect to the surface were the main cause of damage that result in scratches of very high aspect ratio (i.e., damage surface length to the maximum depth). A scratch of 1.5 mm in length with a maximum depth of 0.2 mm had been analytically identified to be the primary crack nucleation site in these components. Accidental impact damage that may be caused while the aircraft is on the ground would not necessarily exhibit the high aspect ratio of an oblique impact on a rotating blade.

Multiple methods were evaluated, and those that provided the most realistic discontinuities were used to produce the desired damage types. Since the severity of the damage had been described only in terms of depth, discontinuities of different depths and lengths were made in the test blades. Some penetrated through the paint into the aluminum skin while others did not penetrate beyond the paint. Figure 13.11 shows one of the test parts identifying the simulated damage locations on a transparency placed on the blade. Replicas of all the discontinuities were made using a silicon-rubber replication approach, and their lengths, widths, and maximum depths were measured.

Field Tests: The blades were then mounted on helicopters and inspected by a number of qualified NDT personnel under the same conditions as those used during routine inspections (Figure 13.12). The inspectors had no prior knowledge of the damage locations or sizes in these parts. The inspection results were collected

Table 13.3 Damage characteristics and limits specified for the CH-146 tail rotor blades.

location	orientation	nick/scratch maximum allowable depth, mm (inches)	sharp dents maximum allowable depth, mm (inches)	non-sharp dents maximum allowable depth, mm (inches)
from hub to ~2/3 of the length (station 30)	Span-wise +/- 15°	0.125 (0.005)	0.125 (0.005)	0 375 (0.015)
from hub to ~2/3 of the length	Chord-wise +/- 75°	0.075 (0.003)	0.125 (0.005)	0.375 (0.015)
outboard of station 30	any	0.125 (0.005)	0.375 (0.015)	0.75 (0.030)

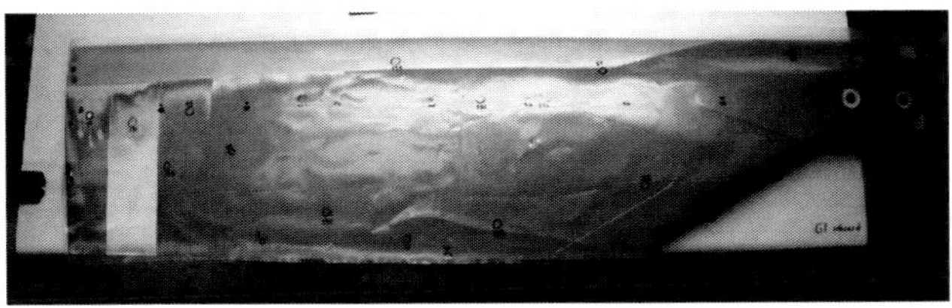

Figure 13.11 Inboard surface of a tail rotor blade showing abraded paint coating close to the tip. A transparency is placed on the blade and the locations of artificial defects are marked on it.

and analyzed to determine POD and establish if the CF visual inspection approach was adequate or not.

In order to preserve the normal routine that is followed by the inspectors, the tests were carried out in a hangar under the same conditions as those used in normal visual inspections. Inspectors were asked to identify all discontinuities on the test blades and make a note describing the type of damage detected and whether or not they exceeded the allowable limits.

13.4.2.2 RESULTS

Since the allowable damage sizes are different for the three damage types, the POD results are presented separately for each category.

Nick/Scratch Type: For this type of discontinuity, the detection rate increased with increasing length and was independent of the damage depth indicating that the inspectors could not get a realistic estimate of the depth in practice. As a result, the "hit/miss" data could not be properly analyzed to determine POD as a function of the damage depth that was required based on the specified allowable limits. In contrast, from the same data, when analyzed as a function of the surface length of scratches or nicks, very reasonable POD plots were obtained as shown in Figure 13.13.

In this figure, two POD curves are displayed. The blue curve was obtained using all the detection data that were obtained when the inspectors identified all discontinuities correctly without classifying them into the different categories. The red curve, on the other hand, corresponds to the data when the inspectors not only detected the discontinuities correctly but also classified them into the appropriate type. The classification into a correct category (i.e., nick/scratch, nonsharp dent, or sharp dent) was important because the allowable limits for each of the damage types were different. In the latter case, a "hit" was when damage was correctly detected and rightly classified, and a "miss" was when a discontinuity

Figure 13.12 Depot aided visual inspection of a helicopter tail rotor blade.

was not identified or was detected but wrongly classified. A "false-call" was when a damage-free site was considered as damaged.

Figure 13.14 shows the overall POD as compared to those obtained using only the data obtained from the green-painted blades, or those of white-painted blades, or those corresponding to discontinuities located close to the edge of the blades. These curves indicate that nick/scratch-type defects were more visible and detectable when the blades were painted in green as compared to white; i.e., a darker color shows scratches or nicks better. However, when the nicks and scratches were close to the edges where the paint coat had been abraded, they were not easily detectable. Unfortunately, using the visual inspection data, POD curves could not be obtained as a function of the damage depth for this type of discontinuity to enable an assessment with respect to the allowable sizes.

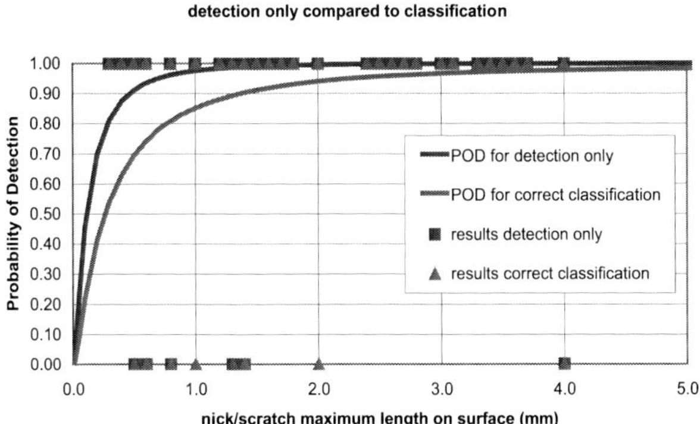

Figure 13.13 Probability of detection of nick/scratch type discontinuities as a function of their maximum surface length.

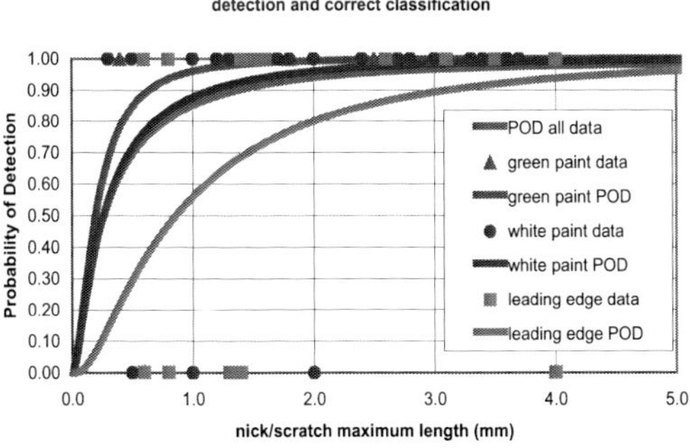

Figure 13.14 Probability of detection of nick/scratch type discontinuities as a function of their maximum surface length and the blade color as well as flaw location on the blade.

Sharp Dents: As in the previous case, here also the POD plots were obtained for correct detection of discontinuities only as well as for correct detection and classification into the right categories (Figure 13.15). The red vertical lines at 0.005 in (0.125 mm) and 0.015 in (0.375 mm) are the allowable limits, which are different according to the position on the blade. Based on this figure, the probability of detection for sharp dents of 0.125 mm (0.005 in) in depth was only 50% and of 0.375 mm (0.015 in) in depth was close to 80%; both are considered inadequate for the safe operation of the aircraft.

Non-sharp Dents: For this type of damage, the correct classification into the right categories was less important than the previous cases because the blunt dents were considered as the least severe of the three types, and a misclassification was only an economic burden (i.e., resulting in unnecessary repair actions). Thus, POD curves were obtained for the correct detection of this type of damage without taking into account their classification (Figure 13.16). This figure indicates that the probability of detection for the non-sharp dents of a size equivalent to or larger than the respective allowable limits was better than 90%, and this was considered to be adequate.

13.4.2.3 OBSERVATIONS

The study showed that the visual inspection approach required for the CH-146 tail rotor blades could not provide adequate information about the depth of nicks or scratches to enable comparison with their respective allowable limits. Additional approaches, such as replication of the flaw sites followed by detailed microscopy,

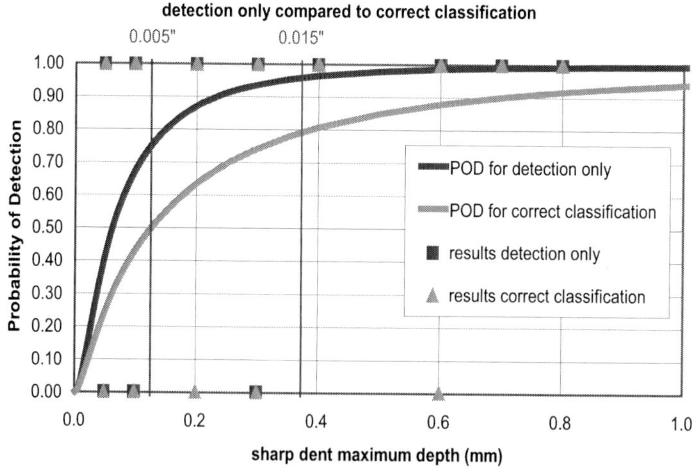

Figure 13.15 Probability of detection of sharp discontinuities as a function of their maximum depth for detection only and detection and correct classification.

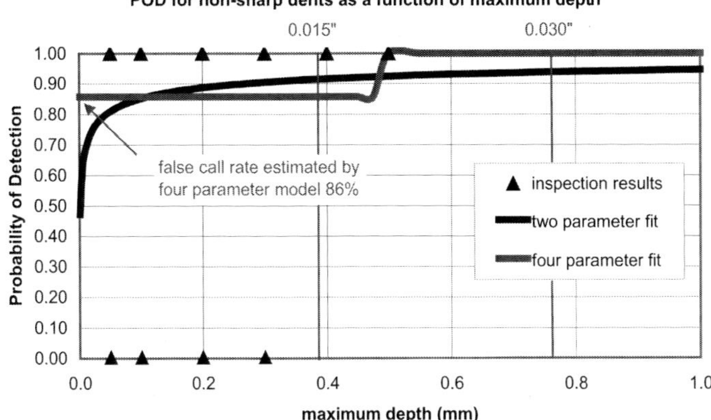

Figure 13.16 Probability of detection of non-sharp dents as a function of their maximum depth. The CFTO repair limits are either 0.015 in or 0.030 in depending on the location of the damage on the blade. These are shown as red lines.

was necessary for depth estimation for this type of damage. For sharp dents, although it was possible to make an estimate of the damage depth using visual tests, the probability of detection was not quite adequate (i.e., 50% for 0.125 mm deep flaws and 80% for 0.375 mm deep flaws). In contrast, the probability of detection for the non-sharp dents of a size equivalent to or larger than the respective allowable limits was higher than 90%, and this was considered to be adequate. Generally, small discontinuities, such as scratches or nicks, were more visible on the green-painted blades than those painted in white, but those located near the edge or in the paint-ablated areas were more difficult to detect.

13.4.2.4 Advantages and Limitations

Development of POD data using actual parts with artificial damage has the following advantages and limitations.

Advantages:
- Uses actual components that are not needed (e.g., service-expired parts).
- Manufacturing of a large number of test specimens is not necessary since only representative artificial discontinuities must be created.
- Provides the ability to perform inspections at maintenance facilities; thus, realistic POD data can be obtained in a cost-effective manner.
- Suitable for large components or structures that cannot be represented by small specimens (e.g., helicopter blades, airframe structures).
- POD data may be used to identify shortcomings of the NDT tests and help to make improvements.

Limitations:

- Requires an adequate number of undamaged real parts.
- Creation of realistic artificial discontinuities of various sizes may be problematic.
- Artificial discontinuities must be characterized by other means (e.g., microscopic examinations).
- POD data are specific to the component, sites, and damage types studied.

13.4.3 POD DEVELOPMENT USING REPRESENTATIVE SPECIMENS WITH ARTIFICIAL DISCONTINUITIES

In many cases, it is not possible to introduce artificial discontinuities in large numbers that are required for POD studies in the actual components. Thus, test specimens representative of the actual part in terms of material, geometry, surface finish, and other characteristics are manufactured. Artificial discontinuities simulating the type of damage that occurs in service are introduced on the specimens for NDT reliability tests. The IAR work carried out on a Nene 10 impeller is described here as an example of this type of POD development. Details are available in reference [38].

13.4.3.1 METHODOLOGY

Background: The original Nene 10 engine design dates back to 1943 and the manufacturer no longer provides replacement parts. In order to continue with the safe use of the engine beyond the initial design life, a damage tolerance analysis had to be carried out, and NDT POD data was needed for critical components to enable such analysis. Under the DT approach, the engine may remain in service for a given life interval based on a crack-free inspection of the critical components.

One of the problems identified as life-limiting for the Nene Mk 10 engine by the Canadian Department of National Defence (DND) was fatigue cracking in the engine impeller. The cracks originate on the surface of the bore and grow into the material. The aspect ratio of the cracks (surface length to maximum depth) was about 4 to 1. It was determined by DND, through analysis, that the dysfunction flaw size was a crack of 2 mm in surface length and 0.5 mm in depth. Based on the deterministic damage tolerance analysis, a probability of failure of < 0.001 is demonstrated if three inspections are performed between the time the flaw reached the 90/95% POD size and the dysfunction size.

Objectives: Eddy current techniques are commonly used for the detection of bore-hole cracks. With this in mind, the objectives of this study were: (I) to develop an eddy current-based approach for the inspection of bore-holes of the Nene 10 impellers and (II) to determine the 90/95% POD crack size. This section primarily deals with the second objective. Having the crack size at 90/95% POD and the crack growth rate for the impeller material and operational conditions, safe inspection intervals can be calculated, and inspection scheduling can be done to achieve the desired risk levels.

Eddy Current Procedure: Inspection of bore-holes for fatigue cracks using eddy current rotating probes is a common practice. However, the specifics of the probe design and equipment setting depend on the material properties, component geometry, and desired flaw sizes to be detected. The probe coil and the rotating head must be designed to satisfy these constraints, and the instrument must be adjusted to match the probe characteristics to achieve optimal signal to noise ratio.

The Zetec MIZ22 instrument equipped with a probe-rotating device (gun) and a specialized custom-built probe were used in this study (Figure 13.17). The blue "gun" has a motor to rotate the probe and is connected to the eddy current instrument via a cable. A photograph of a Nene Mk 10 impeller undergoing an eddy current inspection of the bore is shown in Figure 13.18.

Specimens and Artificial Discontinuities: For valid POD, specimens and artificial discontinuities must represent the actual inspection site and the defect types that occur in service. In this study, hollow cylindrical specimens were made of 2618 T61 aluminum alloy that is similar to the impeller material in terms of composition and conductivity. Two sets of specimens were made, in one set EDM notches were introduced, and in the other set laboratory-grown fatigue cracks were produced.

EDM notches were introduced in six cylindrical specimens that were machined to the same internal diameter as the impeller. The length of these specimens was 225 mm (9 in) and was restricted by the access required to embed EDM notches inside the bore. Since service-induced cracks in the Nene 10 impellers were open to the surface, narrow EDM notches with 4/1 aspect ratio introduced on the surface of the hollow cylinders were thought to be good representations of surface-breaking cracks for use during eddy current probe and procedure development as

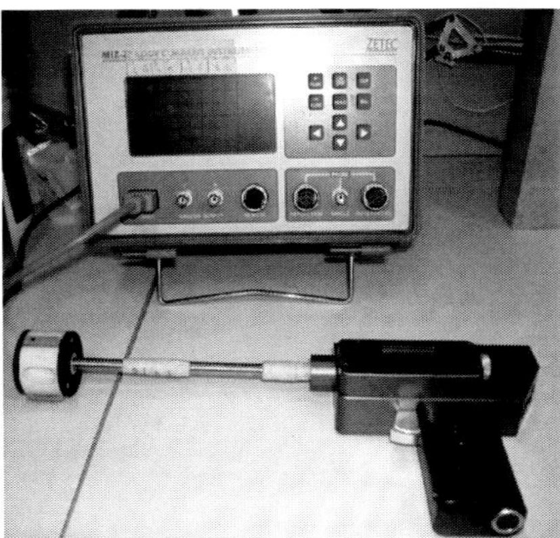

Figure 13.17 Photograph of the Zetec MIZ-22 eddy current instrument and the probe-rotating device with a custom-built probe.

well as for POD study. Figure 13.19 shows different views of the specimens simulating the bore-hole region of the Nene 10 impeller. The internal surfaces were mapped so that there were 16 inspection opportunities on each of the cylinders, yielding 96 total inspection opportunities. In order to estimate false-call rates, 48 sites were reserved with no discontinuity. The other 48 sites had EDM notches embedded. After the notches were introduced, the bore surface was painted with a low gloss black paint to ensure that the notches were not visible.

In addition, three hollow cylinders were manufactured with a slightly smaller internal diameter. These were used to produce fatigue cracks by capping both ends and pressurizing the volume at relatively high frequencies. The maximum pressure and number of pressurization cycles required to generate cracks were determined analytically. The internal surface of each specimen was divided into sections, 4 radial (denoted by 1, 2, 3, and 4) and 3 axial (denoted by A, B, and C), and an EDM starter notch was placed in some sections. By cyclic pressurization

Figure 13.18 Photograph of a Nene 10 impeller with an eddy current probe inserted into the bore. A probe-rotating device is used to cover the entire surface of the bore during inspection.

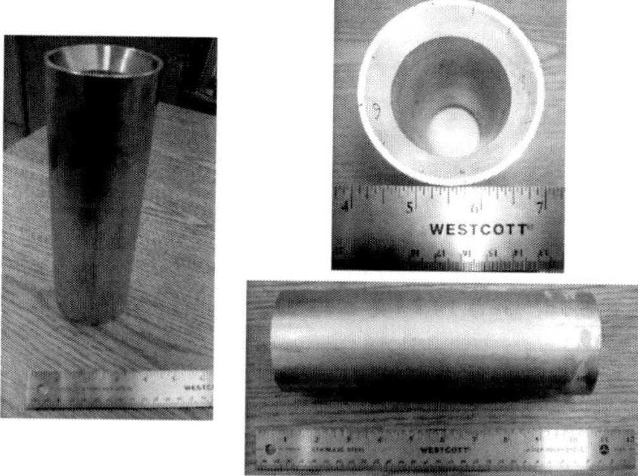

Figure 13.19 Several views of test specimens simulating the bore of the Nene 10 impeller.

of specimens at a load surpassing the endurance strength of the material, it was possible to generate cracks originating from EDM notches. Once the desired crack sizes had been attained, the internal diameter was milled out to eliminate the EDM starter notches and achieve the same diameter as the impeller bore. Table 13.4 provides the details of fatigue cracks.

Inspections: Inspections of the specimens were carried out in a laboratory environment by several DND and IAR certified NDT practitioners and two engineers with over 10 years NDT experience. The inspectors were given a demonstration of the technique and reporting procedures, and introduced to a test specimen (not used in the POD data) for training and practice. After sufficient familiarization tests, actual tests were performed using specimens with EDM notches and fatigue cracks. The inspection results are summarized in the following section.

POD Analysis: Since the above-mentioned eddy current instrument provides a signal for each discontinuity, â vs. a analysis was carried out where â is the signal magnitude recorded, and a is the discontinuity size (e.g., surface length). The key assumption of the â vs. a analysis is that the data can be modelled by the following equation:

$$\ln(\text{signal magnitude}) = \beta_0 + \beta_1 \times \ln(\text{flaw size}) + \varepsilon \qquad (13\text{-}15)$$

where "ln" refers to the natural logarithm, and residual ε is approximately normally distributed with zero mean and variance δ^2. The Maximum Likelihood Estimation method was used to find the parameters of this model from the experimental data. If this model holds, the POD at any flaw size a is the value of the two-parameter log-normal distribution with mean μ and variance σ at a, where:

$$\mu - \frac{y_{th} - \beta_0}{\beta_1}, \text{ and } \sigma - \frac{\delta}{\beta_1} \qquad (13\text{-}16)$$

y_{th} is the signal amplitude at the decision threshold. Signals above this amplitude are called flaws and signals below this amplitude are considered flaw-free. False calls and miss rates are accounted for using the Spencer model, which includes terms to account for false call and miss rates that are independent of flaw size. The model can be written as:

$$POD(a) = p_h + (1 - (p_m + p_h)) \cdot F(a; \mu, \sigma) \qquad (13\text{-}17)$$

where $POD(a)$ is the probability of detection at the flaw size a, p_h is the false call probability, p_m is the probability of missing a flaw independent of flaw size, and $F(a;\mu,\sigma)$ is the two-parameter cumulative log normal distribution used to fit the data. In order to use the preceding analysis with confidence, it must be demonstrated that the experimental data does support the assumptions.

13.4.3.2 RESULTS

The combination of data from all inspectors on EDM-notched specimens resulted in the relationship between the eddy current signal amplitude and flaw size as shown in Figure 13.20. A plot of the error between the individual data points and the straight line fit obtained by maximum likelihood estimation is shown in Figure 13.21. The POD data analysis assumes that these residuals are normally distributed, and if that is true, then they should fall on the straight line plotted on

Table 13.4 Fatigue cracks grown in cylindrical specimens.

specimen	location on radius	location on axis	depth (mm)	opening (mm)	aspect ratio
1	1	A	6.67	16.00	2.40
1	2	A	8.43	17.84	2.12
1	3	A	3.81	8.39	2.20
1	4	A	THRU	THRU	N/A
1	1	B	5.11	10.95	2.14
1	2	B	1.54	3.37	2.19
1	3	B	2.12	4.58	2.16
1	4	B	THRU	THRU	N/A
1	1	C	3.95	8.72	2.21
1	2	C	7.59	15.47	2.04
1	3	C	4.27	9.00	2.11
1	4	C	THRU	THRU	N/A
2	1	A	0.00	0.00	N/A
2	2	A	2.49	2.49	1.00
2	3	A	2.72	2.72	1.00
2	1	B	1.65	3.38	2.05
2	2	B	6.33	3.07	2.06
2	3	B	2.29	4.60	2.01
2	1	C	0.79	1.80	2.29
2	2	C	1.12	2.16	1.93
2	3	C	0.00	0.00	N/A
3	1	A	0.86	1.71	1.99
3	2	A	0.48	1.45	3.00
3	3	A	1.22	2.49	2.04
3	1	B	2.71	5.31	1.96
3	2	B	2.01	4.18	2.08
3	3	B	2.26	3.90	1.72
3	1	C	0.88	1.97	2.25
3	2	C	1.27	2.53	1.99
3	3	C	0.00	0.00	N/A

the chart. It can be seen that the data are closely distributed about this line, and the assumption of normally distributed residuals is good in this case.

Using the analysis described above and the data shown in Figure 13.20, along with an appropriate threshold level, POD curves can be generated. The threshold was chosen to be just above the observable noise level of the instrument, but low enough to provide high detectability at the critical flaw size of 2 mm in surface length. The POD results for all the data obtained on EDM-notched specimens combined and separated for the individual inspectors are shown in Figure 13.22.

Since EDM notches are not physically the same as fatigue cracks, the relationship in terms of detectability between the two must be established if the above POD curves are to be used for damage tolerance life estimation. The detectability of a discontinuity is related to its signal magnitude above the selected threshold. In Figure 13.23, the relationship between eddy current signal amplitude and flaw

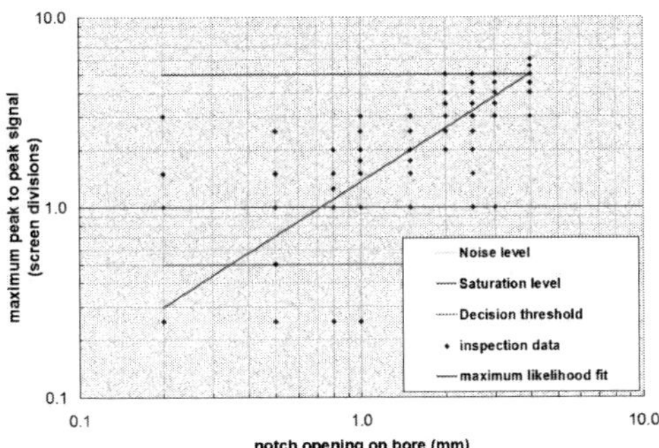

Figure 13.20 Relationship between the eddy current signal amplitude and the EDM notch surface length.

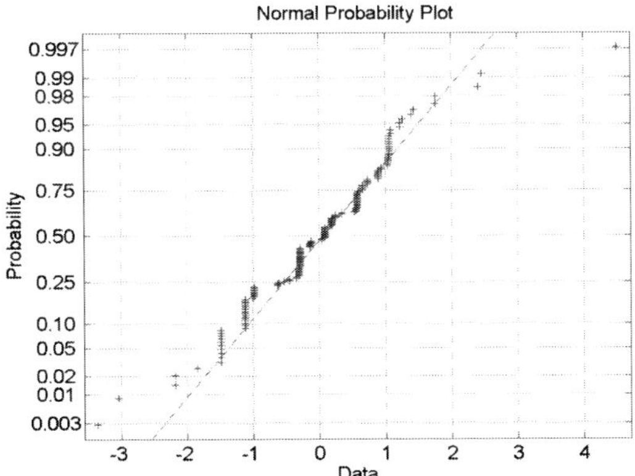

Figure 13.21 A plot showing the difference between the inspection data and the straight line (i.e., residual ε) fit to the POD relationship.

size for the EDM notches is plotted with the results from the fatigue cracks. The response from the fatigue cracks is higher than the mean response for all EDM notches but within the variation of results for the individual inspectors. This is a good indication that the detectability of the EDM notches is a close approximation to the detectability of the fatigue cracks in this case.

Two inspectors performed inspections on specimens containing fatigue cracks using the same threshold level as that used for the analysis of specimens with EDM notches. The difference between the two inspector's data for each discontinuity type was very small, and there was no difference in the individual POD curves. Also, there was no discernible difference in the POD curves for the EDM notches and fatigue cracks as shown in Figure 13.24.

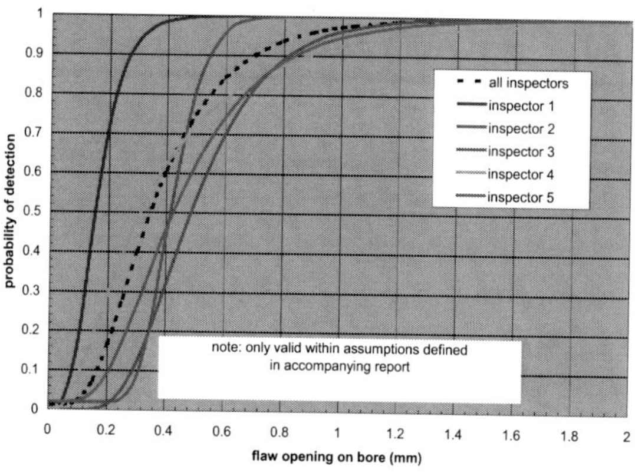

Figure 13.22 POD curves for all inspectors, and individual inspectors, on the EDM-notched specimens.

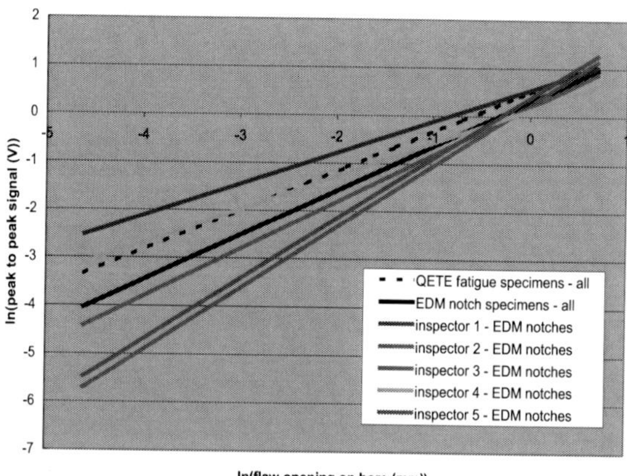

Figure 13.23 Maximum likelihood fits to the relationship between the eddy current signal magnitude (peak to peak) and the discontinuity surface length for EDM notches and fatigue cracks.

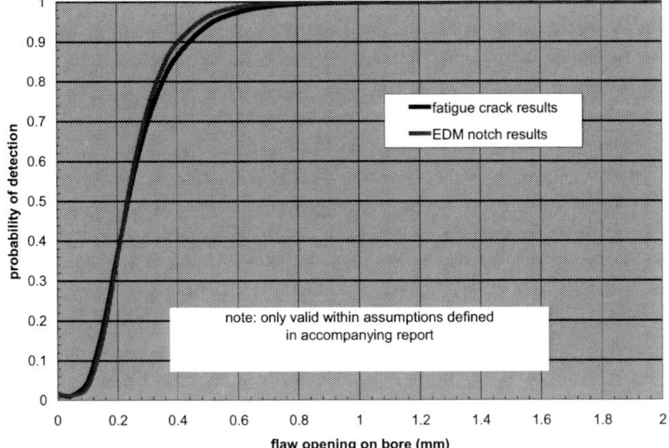

Figure 13.24 Comparison of POD curves obtained from the inspection of specimens containing fatigue cracks and inspection of specimens containing EDM notches. The same two individuals performed inspections on the two discontinuity types.

13.4.3.3　OBSERVATIONS

The study demonstrated the capability and limitations of an eddy current approach based on commercial eddy current instruments and a custom-built rotating probe in the detection of fatigue cracks in the Nene 10 impeller bore. POD data developed using representative specimens and two types of artificial defects (i.e., EDM notches and laboratory-grown fatigue cracks) indicated that the inspection requirement of identifying cracks of 2 mm in surface length and 0.5 mm in depth can be easily met by all the inspectors using this approach. In all cases, the individual POD or group POD was better than 90% for a crack of 1 mm in surface length. It was also shown that the POD data developed on simulated specimens containing EDM notches was not statistically different from that of similar specimens that contained fatigue cracks, indicating that it is satisfactory to use EDM notches to represent fatigue cracks in eddy current tests.

13.4.3.4　ADVANTAGES AND LIMITATIONS

Development of POD data using representative specimens with artificial discontinuities has the following advantages and limitations.

Advantages:
- This is the standard approach for which recommended practices are readily available.
- Specimens and tests can be designed to allow investigation of many variables.

- Does not require actual components or service-induced defects.
- Provides the ability to perform tests in the laboratory or in maintenance facilities.
- POD data may be used for many purposes (e.g., life estimation of new components, qualification of NDT systems, operators, facilities, etc.).

Limitations:
- Preparation of a large number of specimens and artificial defects can be costly or time-consuming.
- Specimens or artificial defects may not fully represent the actual part or service-induced damage.
- Some damage characteristics may be difficult to simulate (e.g., shape, surface oxidation).
- Extensive analytical and experimental work may be necessary to generate realistic POD.

13.4.4 POD DEVELOPMENT USING GENERIC TEST BLOCKS WITH ARTIFICIAL DEFECTS

The work on "Generic Bolt Hole Eddy Current Testing Probability of Detection" provided in reference [16] is an example of this type of POD study and is briefly described in the following paragraphs.

13.4.4.1 METHODOLOGY

Background: POD data generated experimentally using a specific sample/component geometry or discontinuity type/location or inspection procedures/equipment setting often cannot be used for other geometries or flaw configurations when changes are made in the inspections. This is problematic especially for complex airframe structures that are made of several parts of similar design but different geometries or configurations. Examples of such structures are riveted or fastened aluminum lap joints that are widely used in the construction of both wings and fuselage skins. However, the number of aluminum layers and their geometry (shape and thickness) may vary depending on the specific design requirements. These differences may influence the NDT POD and must be evaluated. The present study was undertaken with the support of the Canadian Department of National Defence with the following objectives.

Objectives: The main purpose was to generate the data required to estimate probability of detection of fatigue cracks originating in the fastener holes of aluminum wing and airframe skin structures typical of the DND CC130 and CP 140 aircraft (examples in Figure 13.25) using bolt-hole eddy current procedures. In doing so, the following four objectives were to be achieved:

(i)   Validate the current smallest detectable discontinuity size in fastener bolt holes of aircraft wing box structures when inspected using the bolt-hole eddy current procedure.

(ii)   Generate generic POD data that could be used for future estimates of POD for a range of typical constructions when inspected using the bolt hole eddy current technique.

(iii) Verify a commercially-available POD model so that the POD of discontinuities in similar structures can be interpolated.

(iv) Prepare a handbook to assist future users in the use of modeling along with the generic POD data to estimate POD for similar structures and assess variability.

In this chapter, only the work towards the first and the second objectives is described.

Specimens: The wing components were simulated by manufacturing simple two-layer test coupons, named inspection sets, with a centrally located bolt-hole (Figure 13.26, left). The inspection sets were positioned in a box behind an aluminum covering plate so that only the bolt-holes were accessible through larger holes that were milled-out in the covering plate (Figure 13.26, right). Inspection sets consisted of two plates (coupons) of either 7.8 mm (0.313 in) or 2.25 mm (0.090 in) in thickness with a central hole of 4.5 mm (0.182 in) diameter. A total of 1,434 coupons were made of aluminum alloy AA7075-T6. Of these, 306 coupons contained laboratory-grown (LG) fatigue cracks, 180 coupons had EDM notches, and the remaining 948 were kept blank.

These coupons were used in four different configurations representing those used in the wing structures of these aircraft,;each configuration having 468 inspection sets. A more detailed description of the manufacturing of the test coupons and inspection sets is provided in Appendix B of the above-mentioned report [16]. Figure 13.27 shows the different configurations of the inspection sets chosen for this study.

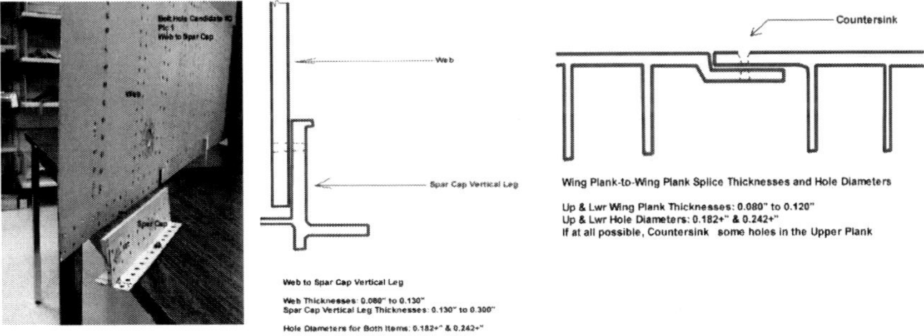

Figure 13.25 A portion of a CP-140 wing and typical configurations used in the CP130 and CP140 wing structures.

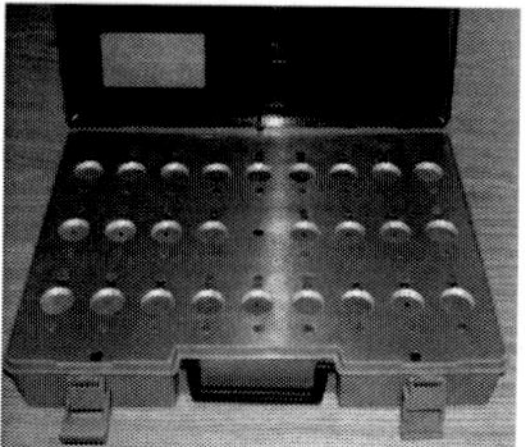

Figure 13.26 (Left): A generic bolt-hole specimen that can be assembled to create different configurations. (Right) Inspection box containing 27 sets of generic specimens placed behind an aluminum plate with 27 disc-shaped milled-out areas that provide access to the inspection sites, i.e., bolt-holes in the middle of the milled-out areas.

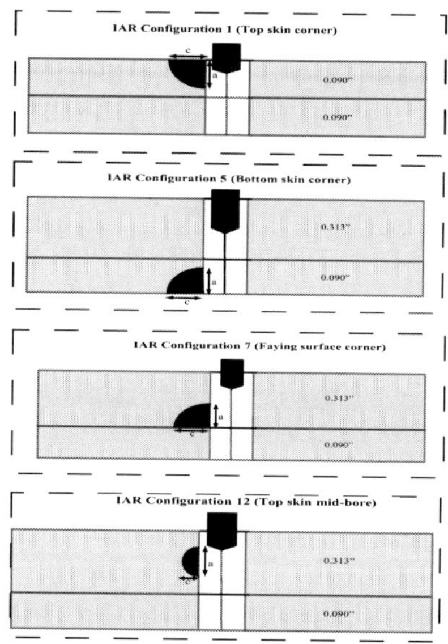

Figure 13.27 Different specimen configurations used in this study. The discontinuities in the aluminum plates and the eddy current probe in the bolt-holes are shown in black.

Discontinuities: Each configuration contained two flaw types: laboratory-grown fatigue cracks and Electrical Discharge Machined (EDM) notches of different sizes either in the corners of the bolt-holes or in mid-bore. There were also inspection sets with no discontinuity (blank coupons) in every configuration. In total, 468 inspection sets were tested for each configuration. The discontinuities were characterized in terms of their length (dimension "a") and depth (dimension "c") by a silicone rubber replica and eddy current methods as well as using modeling. Also, real aircraft parts from a retired Lockheed P-3 (CP140) aircraft were used. The parts did not have service-induced cracks, thus, artificial discontinuities were introduced. For this purpose, wing plank lap-joints were disassembled, de-painted, and EDM notches or cracks were introduced in selected fastener holes for eddy current inspections. For economic reasons, artificial cracks were induced by over-loading rather than fatigue cycling despite knowing that the two loading approaches provide cracks of different characteristics (e.g., crack face smoothness, tightness, residual stresses). However, these differences were not considered significant for eddy current tests. The crack and notch dimensions were estimated using optical and eddy current methods.

NDT Tests: All the coupons and real parts were marked with hidden identification numbers, placed in boxes, and then inspected at DND bases and at the NRC using the bolt-hole eddy current inspection procedure. Inspections were performed by several inspectors, some being certified in eddy current with different levels of certification and experience, while others were trainees. The inspectors had no prior knowledge of the presence or lack of discontinuities in the inspection sets and were instructed to perform inspections in the same manner as they would routinely on aircraft structures. Inspection results were recorded in terms of "hit-miss" data as well as the ET signal amplitudes "â." A few inspectors performed inspections of the real aircraft parts before and after flex honing of the bolt-holes (a cleaning process) to determine if this action improves inspection results.

Verification Tests: After eddy current inspections, fractography was carried out on selected holes to verify the inspection results and measure the dimensions of discontinuities found. The fractographic examination involved opening the holes at specified angles and measuring the flaw dimensions. The crack or notch sizes estimated by the replica, eddy current, and modeling, were compared with those of fractography to establish correlations and accuracy of the methods employed. An example of the correlation between replica and fractography measurements is shown in Figure 13-28. The results were then tabulated to perform statistical analysis in order to estimate inspection POD.

POD Analysis: The flaw sizes obtained using fractography, replica or modeling were used in the same order to calculate POD as a function of flaw size. Both the mean POD and the 95% confidence relationships as a function of discontinuity size, "a," as well as the flaw sizes at 90% POD with and without the 95% confidence (90/95% and 90/mean values) were established. These values are required for component life prediction using damage tolerance methodology. The POD curves and the 90/95% or 90/mean values were used to study factors affecting the POD and make comparisons between different POD models, inspectors, facilities, discontinuity types, locations, etc.

13.4.4.2 RESULTS

Statistical Approach: The different statistical models used for POD calculations provided different POD curves and 90/95% or 90/mean values, with the difference being less pronounced on the mean POD curves than the 95% confidence

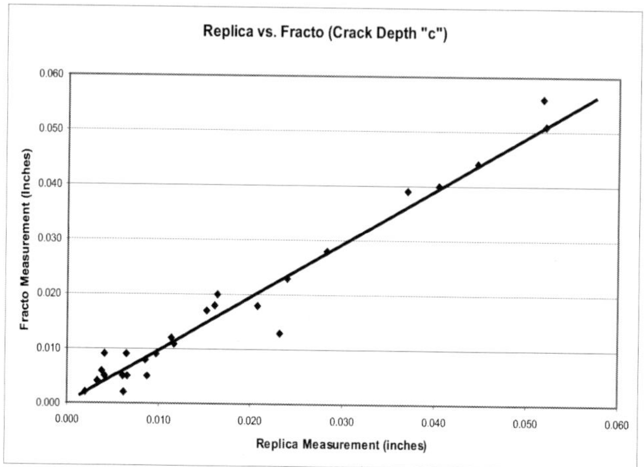

Figure 13.28 Correlation between replica and fractography estimates of dimension "c" for corner cracks.

curves. For the bolt-hole eddy current inspection investigated here, POD relationships determined using the "hit-miss" data were less dependent on the individual inspectors than those using the signal amplitudes ("â vs. a" data) as it was easier to call a "hit" or a "miss" than to measure and record eddy current signals. Most inspectors reported the signal height only when they thought a flaw existed. Also, the log-logistic model generally provided conservative estimates of POD as compared to the log-normal algorithm.

Discontinuity type: From the ET inspections it became evident that the detection rates for the EDM notches smaller than 1 mm (0.040 in) were generally lower than for the fatigue cracks of similar size, for both corner and mid-bore cases. This is associated with the rounded roots of small EDM notches (manufactured using a 0.1 mm-thick shim), which have a less pronounced effect on the eddy current field than the sharp crack tips. At larger sizes, the total area of the cracks/EDM notches became an influencing factor in addition to the crack tip or EDM notch root. The lower detection rates resulted in larger 90/95% sizes for the EDM notches in comparison with the fatigue cracks. EDM notches are commonly used for the calibration and adjustment of eddy current instruments with the assumption that cracks larger than the calibration EDM will be detectable. This study established that, in the case of bolt-holes, the POD for small cracks (<1 mm) is better than for small EDM notches, which is encouraging. Some previous studies have reported that the eddy current response is dependent on the current path and orientation of cracks [39] while others have related the dependence on probe geometry, crack size, material conductivity, and the instrument frequency [41]. Since, in this study, the material, probe, and instrument settings were kept constant, the geometrical differences between cracks and EDM notches are believed to be the contributing factor to the variation in the ET signal.

Inspectors: The years of experience since the original training of the inspectors or the location where the inspections took place (different CF Bases or laboratories) did not affect the outcome of the inspections to a noticeable degree. When test coupons were inspected, all inspectors who participated in the study met the current assumed 1.25 mm (0.050 in) minimal detectable crack depth and reported less than 5% false calls. However, when real aircraft specimens with over-load cracks were inspected, some inspectors fell short of the requirements and reported a significantly higher number of false calls.

Specimen Configurations: Using the "hit-miss" data of the test coupons and their corresponding POD results obtained by the log-logistic method, the four different configurations studied resulted in the following combined (all results combined) and maximum (single worst case) 90/95% crack depth values:

(a) Configuration #1: combined 0.43 mm (0.017 in) / maximum 0.77 mm (0.031 in).
(b) Configuration #5: combined 0.30 mm (0.012 in) / maximum 0.55 mm (0.022 in).
(c) Configuration #7: combined 0.4 mm (0.016 in) / maximum 0.87 mm (0.035 in).
(d) Configuration #12: combined 0.63 mm (0.025 in) / maximum 0.83 mm (0.033 in).

As examples, the actual POD curves obtained for fatigue cracks in the different configurations are shown in Figure 13.29. In these plots, discontinuity sizes detectable at 90% POD with 50%, 90%, and 95% confidence are also presented. The combined and maximum 90/95% crack depths for the four different configurations studied were all below the current assumed value of 1.27 mm (0.050 in), and there were no misses of cracks larger than 1.4 mm (0.056 in) in depth by any inspector who participated in this study. The higher 90/95% values for the mid-bore cracks were related partly to the inaccuracy of replica measurement of mid-bore crack depth and partly to the uneven distribution of the crack sizes in the population (few small cracks and many large cracks) that skewed the POD curve. The combined and maximum 90/95% data for the different configurations obtained using "â vs. a" as well as those of EDM notches are also available in the above-referenced report [16].

Actual Parts: For the actual aircraft parts, based on the limited data available on over-load cracks, the average and maximum 90/95% values obtained using the same log-logistic model were considerably larger (average 2.5 mm or 0.102 in and maximum 4.4 mm or 0.177 in). This may be due the differences in the conditions of the bolt-holes in real service-retired aircraft and those of the specimens made in the laboratory. Fretting between the fasteners and the surface of the bolt-holes is a source of additional noise that affects the EC signals. Also, over-load cracks may produce different EC response as compared to fatigue cracks that are usually tighter and have smoother faces. These effects must be evaluated in order

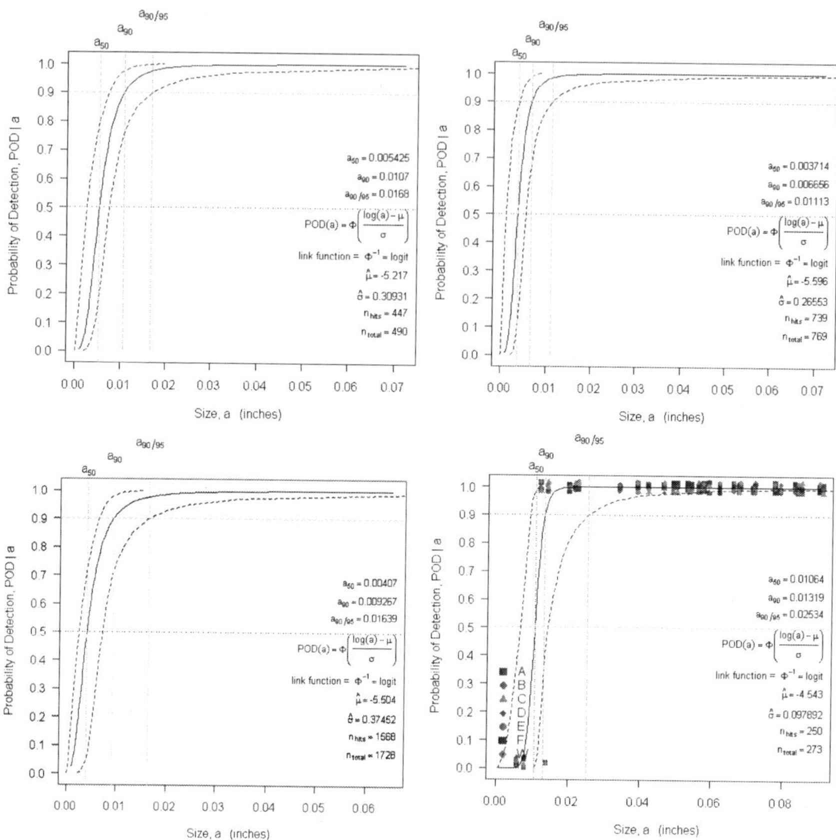

Figure 13.29 Eddy current POD curves obtained for the different specimen configurations when the data sets from all inspectors who tested were combined. (Top-left): Conf. 1. (Top-right): Conf. 5. (Bottom-left): Conf. 7. (Bottom-right): Conf. 12.

to be able to use the over-load crack POD to deduce fatigue crack POD. However, in the real aircraft specimens, some inspectors fell short of the requirements. The tests indicated that by flex honing of the bolt holes of real aircraft parts, some improvements in inspection results can be realized.

In the lack of an adequate number of real parts with fatigue cracks for realistic POD assessment, the material noise associated with the eddy current tests of a few real aircraft parts measured by five inspectors was mathematically incorporated into the POD data of the same inspectors. This data was obtained on test coupons containing fatigue cracks and was used to estimate the POD for real parts. In this case, the POD results were much better than those of the over-load cracks. The generated POD data provided a 90/95% value of 0.87 mm (0.035 in) for the bolt-hole eddy current procedure studied, which was less than the 1.25 mm (0.050 in) value used before.

13.4.4.3 OBSERVATIONS

This study demonstrated that EDM notches larger than 1 mm deep in the bolt-holes of Al lap-joints provide similar eddy current signals to those of fatigue cracks of the same depth. However, smaller notches generated smaller ET signals than cracks of similar sizes due to their round roots. In contrast, the results of different inspectors or facilities obtained under similar conditions were almost identical. The POD curves determined using "hit-miss" data were less dependent on the individual inspectors than "â vs. a" POD. Variability in probe positioning relative to discontinuities as well as in the interpretation and measurement of the signals might be contributing factors. Generally, it is easier for an operator to call a "hit" or a "miss" than to record signal amplitude while inserting the probe into the bolt-hole.

Test coupons of different configurations provided different POD and 90/95% values. With no misses larger than the currently assumed 1.25 mm (0.050 in) minimal detectable crack size (in terms of depth) and very few false calls, all the inspectors participating in this POD study performed quite well on the coupons. The 90/95% values for fatigue cracks in test coupons were all smaller than 0.87 mm (0.035 in). However, the results on real parts with over-load cracks were not as impressive for some inspectors who missed some large cracks or had high false call rates, perhaps due to the bolt-hole surface noise. However, when the noise level was measured on real aircraft parts and mathematically included in their POD of coupons with fatigue cracks, the estimated POD results were much better than those of over-load cracks with the 90/95% values smaller than 1.25 mm. Flex honing of the bolt holes before inspections helped to improve the detection rate and reduce false calls.

13.4.4.4 ADVANTAGES AND LIMITATIONS

Development of POD data using generic test coupons with artificial discontinuities has the following advantages and limitations.

Advantages:
- Test coupons and experiments can be designed to allow investigation of many different component configurations and test variables.
- Does not require actual components or service-induced defects.
- Provides the ability to perform tests in a laboratory or in maintenance facilities.
- POD data may be used for many purposes (e.g., life estimation of new components, qualification of NDT systems, operators, facilities, etc.).
- Test coupons are usable for future investigations when inspection changes occur (e.g., change of probes, systems, operator, etc.).
- Destructive verification tests are not necessary.

Limitations:

- Preparation of a large number of test coupons and artificial defects can be costly or time-consuming.
- Test coupons or artificial defects may not fully represent the actual parts or service-induced damage.
- Some damage characteristics may be difficult to simulate (e.g., shape, surface profile).

13.4.5 POD Estimation from Data Generated during NDT Procedure Development

Extensive analytical and experimental work may be necessary to generate realistic POD. The report described in reference [42] provides a way to estimate POD from the data generated during the development of a typical NDT process.

13.4.5.1 Methodology

Background: As described in the previous sections, the majority of NDT POD studies involve dedicated inspection demonstrations whereby the same samples or parts are inspected by NDT technicians in a laboratory or in service conditions. The results generated in such programs are sometimes challenged on the basis that the inspection conditions may not be quite representative of those seen in the field in terms of environment, access, or human factors. In addition, such investigations are time-consuming, expensive, and require access to many inspectors, NDT equipment, and representative test articles. On the other hand, often NDT techniques are developed and implemented without ever being subject to a full-scale POD verification. The developer of a technique is usually given a target discontinuity size that must be detected or is asked to do the best possible and estimate the resulting "detection threshold." In this work, an attempt has been made to use inspection data that is typically available from an NDT technique development process to estimate the detection thresholds or POD.

Objective: This work was aimed at providing an estimate of NDT POD using data generated during a typical inspection development process. To achieve this objective, IAR carried out a project in collaboration with the DND Aerospace and Telecommunications Engineering Support Squadron (ATESS). The purpose was to provide data on a set of specimens that represent a specific fleet problem, replicating an NDT procedure development process, and also to provide sufficient data to enable POD analysis using the conventional approach. By doing this, it was possible to compare the POD estimate from NDT procedure development to the actual POD of the same technique.

Specimens: A set of specimens was manufactured with fatigue cracks that represented a structural component from an area of the CF-116 fleet known as the "golden triangle." A drawing of the specimens is shown in Figure 13.30. In the actual parts, cracks originate from the corner of fastener holes on the faying surface; thus, cracks of various sizes were introduced in all specimens similar to those seen in service and illustrated in Figure 13-30. There were seven specimens with three holes per specimen and two cracks per hole, for a total of 42 cracks.

NDT procedure development: Generally, the NDT developer is asked to provide a procedure capable of detecting discontinuities greater than a certain size

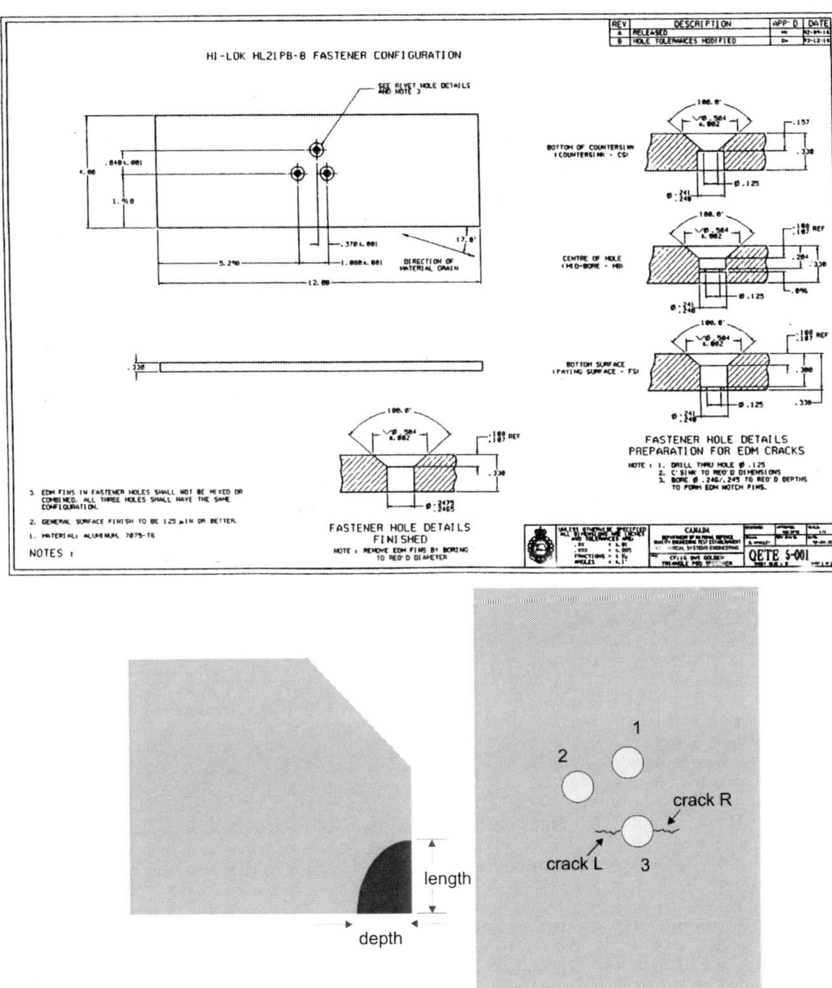

Figure 13.30 (Top) A drawing of the "golden triangle" specimens used for NDT procedure development. (Bottom): Crack type and dimensions as well as labelling key for the specimens.

for a particular application. To achieve this goal, the NDT developer identifies the most appropriate technique (e.g., eddy current or ultrasonic) based on his knowledge of the capabilities and limitations of NDT methods for the required application. Then, suitable equipment is selected and a test specimen (or calibration piece) representing the actual part is made. Artificial discontinuities of different sizes (EDM notches and/or laboratory-grown cracks) covering the desired range of detection are introduced in locations that require inspections. The inspection procedure is developed and the related variables, such as equipment settings (e.g., frequency, gain, filters, and threshold), and the inspection process (e.g., probe scanning, signal analysis) are assessed and optimized to achieve detection of the desired discontinuity size, or the smallest possible size. At this point, the procedure development is completed and fixed and details are recorded to be followed during inspection of the actual parts.

In this study, an eddy current procedure was developed at ATESS using three cracks from the specimen set. As per normal ATESS practice, the equipment settings (e.g., gain, phase, etc.) were set before inspection on a single calibration notch. Then the same procedure and settings were used by three inspectors on the entire specimen set, and signals from individual cracks were recorded and provided to IAR for analysis. The data was of the â vs. a type.

POD Analysis: As described earlier in this chapter, for â vs. a type data, the logarithms of â and a are linearly related with residuals e normally distributed with mean zero and standard deviation δ^2. The relationship is repeated below once again for convenience:

$$\ln \hat{a} = \beta_0 + \beta_1 \ln a + \varepsilon \tag{13-18}$$

To obtain a POD curve, the coefficients β_0, β_1, and ε of the above equation must be estimated. Reference [42] describes in detail the effects of the controllable and uncontrollable NDT variables on these coefficients and how they could change the POD curves. These coefficients can be easily obtained from the â vs. a data plotted on a log scale by drawing the best fit lines through the data. The intercept of the best fit line with the Y-axis provides an estimate of β_0, and the slope of the line gives an estimate of β_1.

Any data which deviates from the straight line indicates that there is noise in the system, and, thus, a non-zero ε. To estimate ε and set the appropriate decision threshold, two important factors must be considered: (i) discontinuity detection requirements and (ii) false call rate requirements.

As described in [42], when β_0 and β_1 are kept constant and ε is changed, the mean discontinuity size at the signal threshold has a POD of 50% (and this size is denoted a_{50}). This is an inevitable fact not an adjustable parameter. Often the discontinuity size at 90% POD, with 95% confidence (denoted $a_{90/95}$), is used as a threshold value in deterministic risk assessments. Because the 95% confidence interval is based on the sample size used to determine POD, it is not a function of the NDT-specimen system; thus, the size at 90% POD (denoted a_{90}) is considered in this analysis.

13.4.5.2 RESULTS

The ATESS inspection results for all the 42 cracks are plotted in terms of the percent of the screen full-scale height (FSH) as a function of the crack depth on a log-log chart (Figure 13.31). This figure shows a typical â vs. a data type that is adequate for POD analysis using the conventional approaches. However, since usually limited data is available from an NDT procedure development process, the data from only three specimens that had a "small," a "medium," and a "large" crack were chosen since only these were used for the procedure development. The results on these cracks obtained by the three inspectors are referred to as "calibration" and shown in Figure 13.32.

To estimate a POD from an â vs. a plot, the decision threshold must be defined. A decision threshold of 30% FSH was selected which roughly corresponds to a crack length of 1 mm in Figure 13.32. Using this threshold, estimates of POD curves were made for the three inspectors using the individual and combined data on 42 cracks in all the 7 specimens (Figure 13.33). The curves in this figure represent the individual and overall POD resulting from a conventional NDT reliability demonstration. Then, POD estimates from individual and combined inspector results for the three "calibration" specimens were obtained and compared with the overall POD curve of the previous figure (see Figure 13.34). The calibration POD curves in this figure represent those that would result from an NDT procedure development process.

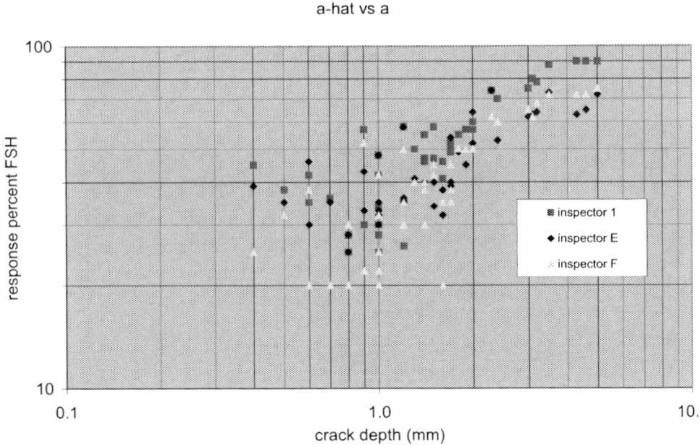

Figure 13.31 Eddy current response in terms of percent of screen full scale height (FSH) as a function of crack depth for three different inspectors using the same procedures and settings (all 42 cracks).

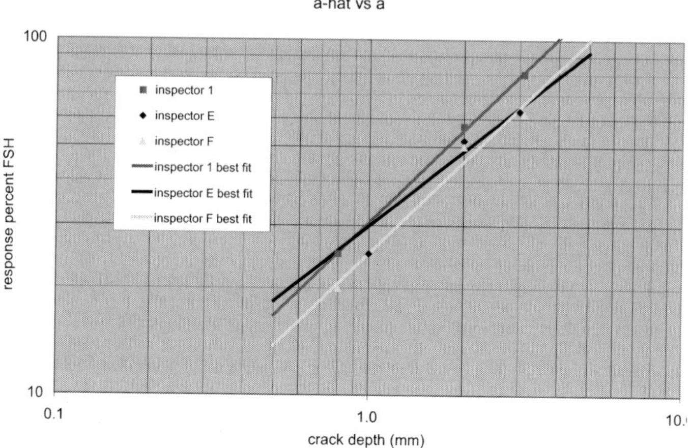

Figure 13.32 Eddy current response in terms of a percent of screen full scale height (FSH) as a function of crack depth for three calibration specimens. The solid lines are the linear best fits to the data.

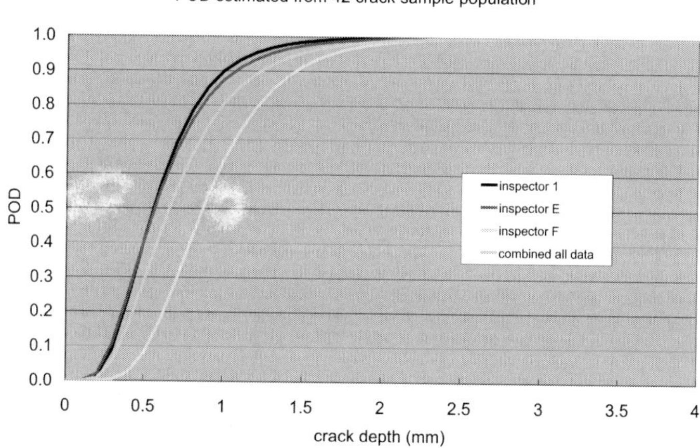

Figure 13.33 A plot of the POD estimates from individual and combined inspector results on 42 cracks in all the 7 specimens.

13.4.5.3 OBSERVATIONS

This study demonstrated a number of important points that are mentioned below. First, even with a set of 42 cracks, there was a significant difference between inspector performance. Only one inspector achieved 90% POD at a 1 mm crack size corresponding to the detection threshold. In the worst case, the POD of the same size crack was about 65% considering all 42 cracks. However, when the

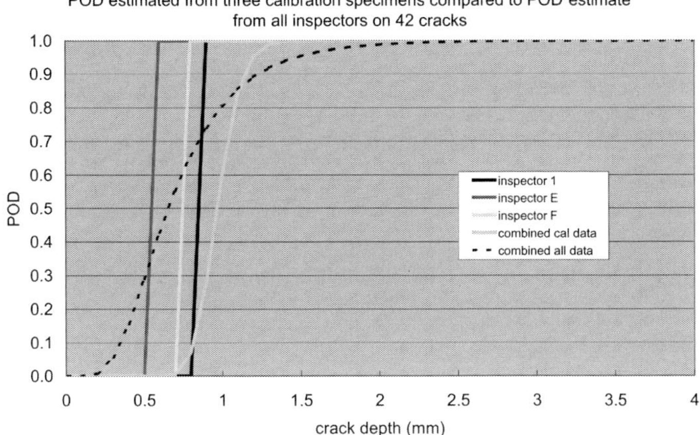

Figure 13.34 A plot of the POD estimates from individual and combined inspector results using calibration specimens with only three cracks representing those used in NDT procedure development in comparison with combined results of the previous figure on all 42 cracks.

combined POD curve obtained on all specimens with 42 cracks was compared with that generated using three calibration specimens with only three cracks, the POD was about 10% less. The variability in the performance of an NDT technique is potentially underestimated if technique development is made on a few specimens only. As seen in Figure 13.34, using a larger sample size provides more accurate POD, especially for smaller crack sizes as one would expect. The latter figure also shows that the few data generated by individual inspectors on calibration samples resulted in erroneous POD curves. This indicates that if the requirement of the inspection is to identify very small discontinuities, a larger number of samples and inspectors are needed for more accurate POD estimation using data generated during an NDT procedure development process.

13.4.5.4 Advantages and Limitations

POD estimation using data generated during an NDT procedure development process provides the following advantages and limitations.

Advantages:
- Estimation of POD from data generated during an NDT procedure development is more cost-effective than the conventional approaches.
- Requires only a few test samples with artificial discontinuities representing the actual parts.

- The approach improves the procedure development process and provides additional useful information.
- The data may be used for evaluation of the performance of operators.

Limitations:
- POD data are not as accurate as those generated using conventional approaches.
- POD analysis requires a good knowledge of the developed NDT procedure and variables that may affect reliability.
- At least three inspectors are needed to perform calibration tests on few samples.
- Safety factors must be incorporated to offset inherent inaccuracies.

13.4.6 POD ESTIMATION USING FIELD INSPECTION DATA

This section is based on reference [43] describing the concept of estimating POD from in-service NDT data, and the associated problems when this approach is tried.

13.4.6.1 METHODOLOGY

Background: The conventional approach for determining NDT POD is practical but can be very expensive. A more economical approach may be to use actual field inspection data as proposed in [21]. This approach is based on back-projection that is estimating the size of discontinuities missed during previous field inspections after a flaw is found. The approach is particularly attractive for airframe inspection techniques as such structures cannot be easily simulated. There are, however, a number of difficulties with this approach. First, there is usually a very limited amount of field data, and those available are often not well-defined. This may require special statistical treatments. Second, crack growth data must exist to allow the estimation of flaw sizes at inspection sites before the flaws were found. These difficulties have prevented the widespread use of field data for POD estimation; however, a few studies have incorporated elements of this methodology [44, 45]. In this section, problems associated with POD development from small data sets are described and inspection data from a full-scale test are analyzed with the view of deducing POD data. The difficulties encountered in the analysis of the field data are discussed.

Approach: In this approach, inspection data at a particular site are recorded over time or operational cycles until a flaw is found. This gives an NDT signal or a hit. At this stage, an estimate of the flaw size can be made either from the inspection result itself or by performing a secondary inspection, or by disassembly and verification tests. The preceding inspections performed at this site may be used to predict miss points by estimating the flaw size at earlier inspections before the flaw was found. This is a complex procedure that requires knowledge of initial flaw sizes, flaw growth rates, and the loading on the inspection site as well as information on inspection and sizing methods, which result in several uncertainties. However, if the uncertainties are fully identified and analyzed, it may be possible to use the field data for POD estimation.

Analysis of uncertainties: The first inherent uncertainty is in the use of crack growth models to estimate the size of a crack at a time before the crack has reached a known size "a". Based on fracture mechanics, the fatigue crack growth rate is stated by the "Paris Law" as:

$$\frac{\partial a}{\partial N} = C(\Delta K)^m \qquad (13\text{-}19)$$

where α is the crack size, N is the number of fatigue cycles, C and m are material constants, and ΔK is the stress intensity range. This equation can be written as:

$$a_{n+1} = a_n + C[\Delta K(a_n)]^m \qquad (13\text{-}20)$$

It is apparent that calculating previous flaw sizes from a given flaw size requires knowledge of several material parameters and local stress concentration factors that are not always available. Leemans [46] showed that while iterative approaches are most accurate, a simple approximation using the slope at cycle n to back-cast to cycle $n-1$ is adequate. However, in back-casting the crack size, parameters C and m of the Paris equation have been shown to be random variables [46]. This results in additional uncertainty in calculating flaw sizes at inspection times before the flaw has been detected.

There is also uncertainty in the measurement of the flaw size when it is detected in service, as service-induced cracks are often blended out, holes are re-drilled, or other maintenance actions are taken, which destroy or change the characteristics of the flaw. Leemans [46] showed that the level of uncertainty in the final flaw size determination is reflected in determination of previous flaw sizes, with standard deviations of similar magnitude in relation to the mean flaw size.

Also, there is uncertainty in operational conditions that is not easy to resolve. Often aircraft are not used in the same role as assumed in the original design, and little information may be available about the actual loading conditions. Uncertainty in loading can have a great impact on calculation of flaw sizes at times before flaw detection. In addition, crack growth rates in different materials can be affected by factors including temperature, humidity, and corrosion. If at least some information about the loading does not exist, it will be impossible to use the field NDT data for POD determination unless cracks are allowed to exist and crack growth is monitored under the aircraft normal operating conditions. This approach was used at IAR with some success, on data from full scale fatigue tests of the CT-114 aircraft and is described here. Finally, as shown in the previous section, the POD model is sensitive to small sample sizes. This is addressed in more detail below.

POD Sensitivity to Small Sample Sizes: The small number of samples, which are typical of field inspection data, lead to two types of problems with POD models. First, it is possible to have a sample that is not a good representation of the actual population, giving rise to poor estimates of POD. Second, the confidence levels on the POD-flaw size relationship are often extremely broad for small sample sizes. Binomial methods, which rely on binning the data into groups based on flaw sizes, are especially affected. Curve-fit methods also give confidence levels that may be very conservative and have been shown to be sensitive to outliers [47]. The Spencer method of equation (13-12) above has been shown to be less sensitive to outliers than traditional curve fitting methods. As described in reference [43], these models were evaluated on the J85-CAN-40 data showing better results for the Spencer mode.

Test structure: A full-scale fatigue test was carried out at IAR on the CT-114 aircraft aft fuselage and empennage to five times the design life. A variety of NDT techniques were used during the test. A number of cracks were found using simple visual inspection (see example in Figure 13.35), and some were monitored periodically. These data were collected and analyzed in an attempt to better understand the process of generating POD data from field inspection results. A more detailed description of the data and analysis can be found in reference [48].

13.4.6.2 RESULTS

Using simple visual inspection, several cracks were detected in similar locations emanating from fastener holes on the port and starboard side drag cleats. While there was no prior crack growth data available for the above-mentioned locations, many of the detected cracks were monitored over time to generate crack growth rates for several locations, as shown in Figure 13.36. The data was used to

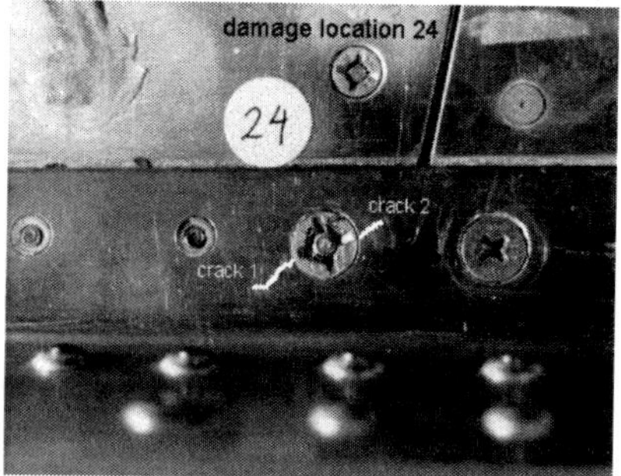

Figure 13.35 Photographs showing the CT-114 aircraft during full-scale fatigue test (left), and an example of damage location (right) and approximate size.

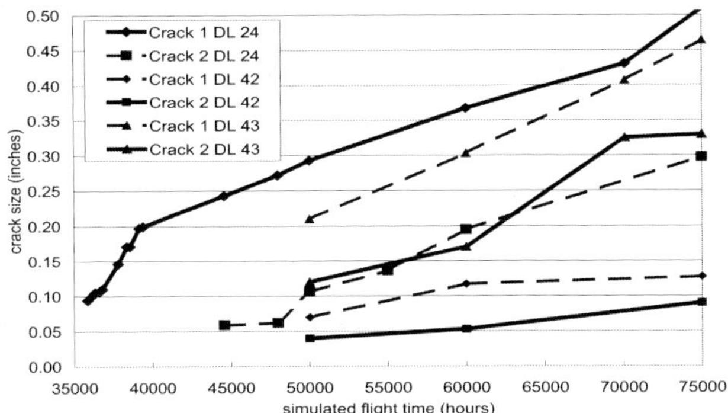

Figure 13.36 Crack growth curves generated at several locations during the CT-114 full-scale fatigue test.

estimate crack sizes at inspection times prior to the detection of each by a simple linear extrapolation. This type of extrapolation may not be suitable for large time differences from the actual first detection time and is not recommend for use in general. In this case, however, it provided a quick method to evaluate the potential usefulness of the data set for POD estimation.

A major challenge was to interpret crack growth data at several sites because after fastener removal it was found that many of the holes contained multiple cracks and some had joined together underneath the fastener. This made it impossible to determine the total crack growth rate from measurements made with the fasteners in place. Nevertheless, a few flaw sites remained with valid data. An example of an individual damage location with measured and estimated flaw sizes at inspection times is shown in Figure 13.37. This analysis was repeated at six damage locations, and the hit and miss points are plotted in Figure 13.38.

Due to the irregular nature of the small data set in Figure 13.38, a reasonable POD curve could not be constructed. However, there are a number of factors that contributed to the irregularity of the data. First, it is possible that the visual inspections carried out at the time were not very sensitive to small flaws, and the flaw sizes here were smaller than the detection level of the inspection method used. Second, the flaws may have had different detectability, caused by differences in crack closure for example, and grouping them in a single set may not have been valid for POD estimation. Finally, as it is obvious from the data, there were insufficient large flaws in the data set to achieve a meaningful POD curve. Nevertheless, this study indicated that there is a potential for use of field data to estimate POD if inspection data for all flaw sizes in the range of interest is available and a valid crack growth curve can be developed.

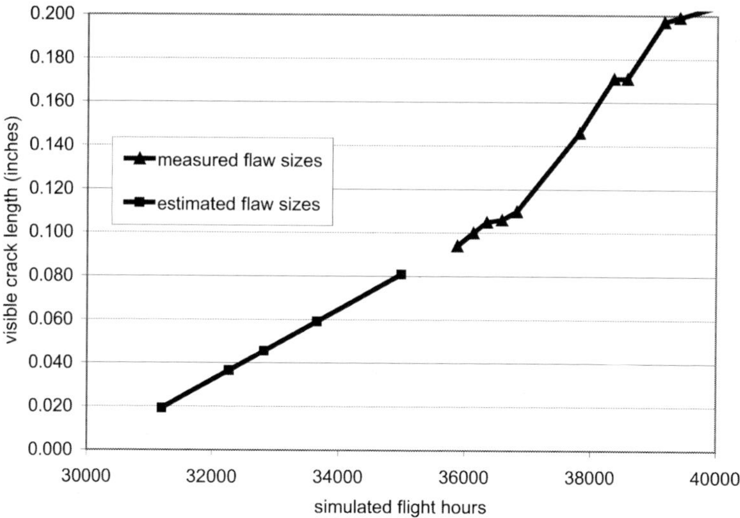

Figure 13.37 Estimated and measured flaw sizes for one damage location.

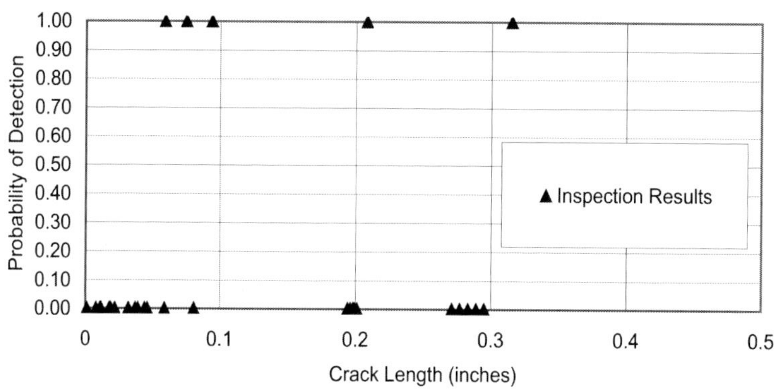

Figure 13.38 Hit/miss inspection data from the CT-114 Tutor crack growth study.

13.4.6.3 OBSERVATIONS

Estimation of probability of detection information from field data is very challenging in practice. The main problems are due to the small data sets typical of field applications, as well as uncertainty in flaw size and growth rate estimation. Despite these problems, the estimation of the POD-flaw size relationship using field inspection data remains an attractive approach as the available alternatives are very expensive. Although the methodology of developing POD from field inspection data is very simple in concept, proper application requires understanding of the sources of variability in flaw size estimation and the resulting POD curve.

13.4.6.4 Advantages and Limitations

POD estimation from field inspection data has the following advantages and limitations.

Advantages:
- Estimation of POD from field inspection data is cost-effective; however, there are major challenges.
- Does not require test samples with artificial discontinuities.
- Actual service inspection data are used, and, thus, POD results potentially will be realistic.
- Provides opportunities for improvement of inspection data recording, in-service flaw monitoring, and sizing.

Limitations:
- Very challenging to implement in practice.
- Requires good information on actual load conditions, environment, and flaw growth rate for the inspection site of interest.
- Requires adequate NDT data that often are not available.
- Requires good understanding and analysis of uncertainties involved in measurements.

13.4.7 POD Estimation using Modeling with Limited Experiments

As described in the previous sections of this chapter, empirical approaches for POD estimation may be expensive or difficult. Modeling may be used to reduce experimental measurements and obtain an estimate of POD. The present section is based on reference [49] prepared as part of work on the "Generic Eddy Current Bolt Hole POD" described earlier. Eddy current modeling was used in conjunction with experimental tests to study the potential of model-assisted POD estimation. This section focuses on the modeling aspects of eddy current inspection of lap-joint samples representing the wing skin of the CC-130 and CP-145 aircraft with the goal of reducing the amount of experimental testing, and its partial replacement with numerically simulated results in future POD projects.

13.4.7.1 Methodology

Background: As mentioned earlier, the model-assisted POD consists of two different approaches: (i) a transfer function approach that finds relational functions to convert data from one inspection situation to a similar one when testing parameters or conditions are changed, and (ii) a full model-assisted approach that employs physics-based models to complement a minimum set of experimental data for POD analysis. In this study, the first approach was employed. In a model-assisted POD estimation, the following conditions must be met prior to the use of a model: (i) The model must be validated on a small set of specimens, (ii) the model must use the same variables as the experimental study, and (iii) the model must simulate the same NDT signal features of interest.

Objectives: The objectives of the present work were to establish transfer function relationships between eddy current signals from real and simulated discontinuities, and use numerical modeling to extend the range of applications of a single study to a set of similar situations by changing input parameters or variables.

Approach: Generally, the eddy current inspection is sensitive to the material's electrical conductivity and magnetic permeability, the probe frequency, and lift-off, as well as probe-specimen coupling and other factors that directly or indirectly affect one or more of these properties. An eddy current numerical simulation software package (ECSim developed by Iowa State University and commercialized by NDE Technologies, Inc., VA.) was used to predict the eddy current signal output when there are changes in one or more of these parameters, reducing, in this way, the need for a large number of inspections.

A total of eight variables were analyzed in this work, relating to: probe characteristics (i.e., frequency, lift-off, tilt, off-center scanning); crack profile (i.e., length, depth, simultaneous variation of length and depth); as well as material electrical conductivity. However, the ECSim was not capable of taking into account the defect width as an input parameter; therefore, limiting the direct comparison between "wide" electrical-discharge-machined notches and "tight" fatigue cracks. Also, the software did not consider surface curvature and was capable of simulating only planar surfaces. However, in the case of bolt-hole inspection using a rotating eddy current probe that had a small side-mounted coil (Figure 13.39), this approximation was not limiting since the local inspection geometry was close to planar.

The material's electrical properties, probe characteristics, and other necessary information were entered into the ECSim interactive screen (Figure 13.39), and eddy current signal responses were simulated. The eddy current feature of interest (i.e., the vertical amplitude component) was plotted as each of the above-mentioned inspection parameters was changed. Also, a comparison of semi-elliptical and rectangular-shaped defects was made to provide insight in the eddy current response differences for laboratory-grown cracks and EDM notches. The transfer functions were then used to generate POD curves for real cracks from NDT data obtained on a combination of natural and laboratory induced discontinuities.

13.4.7.2 RESULTS

The following two cases will demonstrate the way transfer function modeling is used to extend the results obtained for one situation to another one.

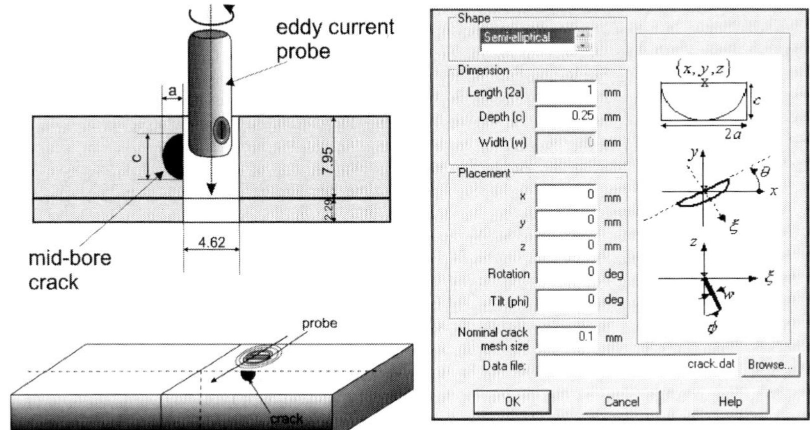

Figure 13.39 (Left): Schematic of the bolt-hole eddy current inspection and similarity to flat plate. (Right): A screen display of the ECSim software.

Case Study #1: In this case, the experimental data for a bolt-hole eddy current inspection is available for a probe frequency of 400 kHz; however, POD data is required for a 200 kHz probe. This requirement has practical implications, since lowering the driving frequency allows for a better penetration of the eddy currents in the material and consequently enables the detection of cracks located in deeper sections. Thus, in the simulation model the frequency of the probe was lowered from 400 kHz to 200 kHz, and a new set of data was generated without actually performing the experiments.

For demonstration purposes, the eddy current impedance plane diagrams generated by the simulation software for 200, 400, 600, 800, and 1000 kHz when the probe is scanning a 1 mm long, 0.25 mm deep, semi-elliptical crack, with a probe-specimen separation or lift-off of 0.1 mm is shown in Figure 13.40. Using such figures, eddy current responses at a given frequency could be related with the defect size to generate the "â vs. a" data required for POD analysis. The experimental "â" data used in this case were the amplitude of the eddy current signals (i.e., the imaginary or vertical component) obtained at 400 kHz frequency when inspecting mid-bore cracks of various depths, "a", measured by replica measurements. These were the basis of the experimental POD curve obtained for the 400 kHz probe and were also used as input parameters in the eddy current simulations.

For the same crack dimensions, the eddy current signals were simulated for 400 kHz and 200 kHz driving frequencies. Using the transfer function approach, the relationships between the signal amplitude variations with the crack depth for the two frequencies, and between the experimental and simulated eddy current amplitude data were established (Figure 13.41). In this example, it was assumed that all other testing parameters remained constant, with the exception of the driving frequency, which was lowered from 400 kHz to 200 kHz.

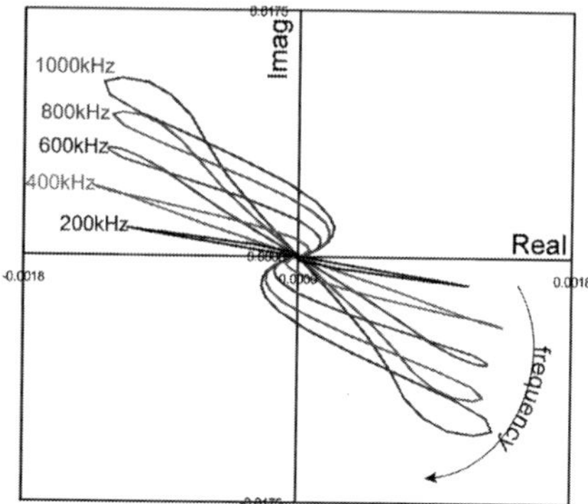

Figure 13.40 Simulation results of the eddy current signal changes with increasing driving frequency; probe 'scanning' a 1 mm-long, 0.25 mm-deep semi-elliptical crack.

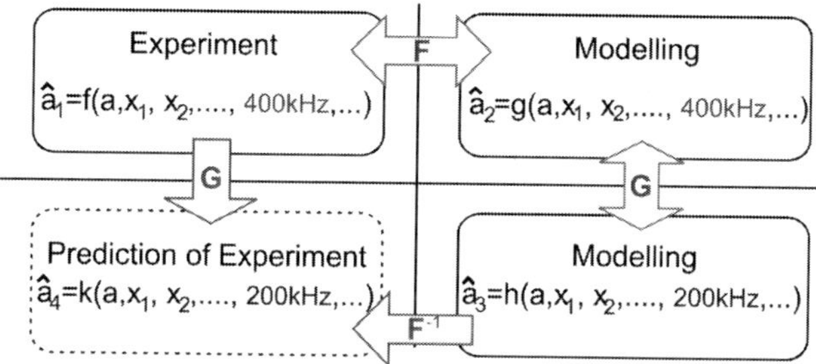

Figure 13.41 Transfer function approach for predicting eddy current results corresponding to 200 kHz excitation frequency.

As shown in Figures 13.41 and 13.42, the relation between the experimental and modelled eddy current signals at 400 kHz is established through the function **F** while the relationship between the simulated eddy current signals at 400 kHz and 200 kHz is given by function **G**. The expected experimental eddy current response for the 200 kHz driving frequency could then be found through the application of either transfer function $\mathbf{F}^{-1}(\hat{a}_3)$ or $\mathbf{G}(\hat{a}_1)$. The results of applying these functions are shown in Figure 13.43, along with the original, experimentally-collected data at 400 kHz excitation.

As evident, the two outcomes are very similar, validating the selected approach. Moreover, the predicted signal amplitudes for a driving frequency of 200 kHz verify the physical understanding expected: i.e., for shallower cracks, the 200 and 400 kHz signal responses are closer to each other because in both cases the depth of penetration of the eddy currents exceeds the depth of the defect. For deeper cracks, more eddy currents will flow under the defect for the lower frequency, resulting in less current density effect on the pick-up coils and smaller signal amplitudes as shown in Figure 13.43.

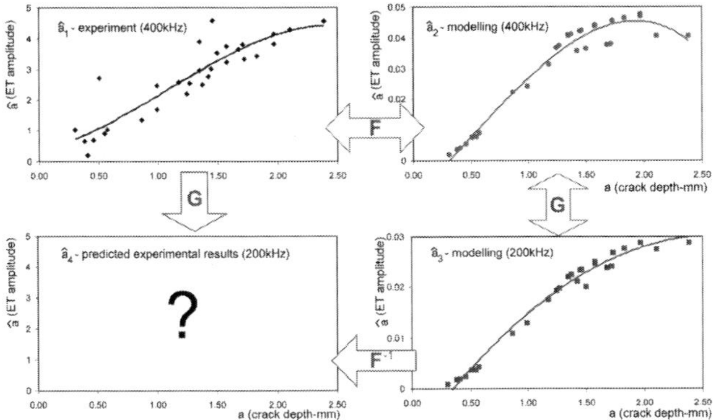

Figure 13.42 Application of the transfer function approach for a concrete situation: lowering the driving frequency from 400 kHz to 200 kHz.

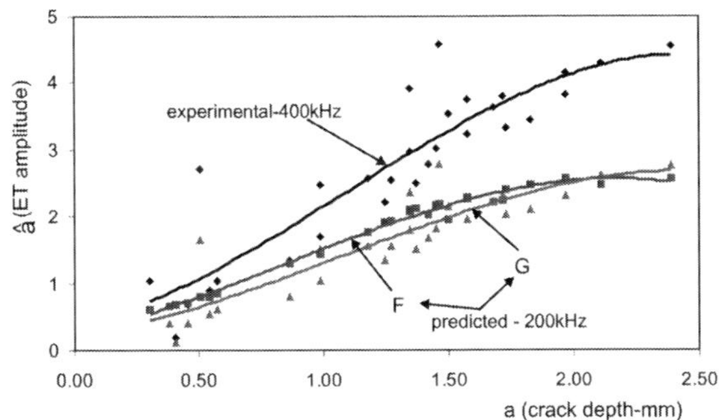

Figure 13.43 Actual eddy current inspection results at 400 kHz and predicted ones at 200 kHz, obtained through transfer functions and numerical simulations.

In order to validate the modeling predictions, bolt-hole eddy current inspections at a driving frequency of 200 kHz were performed by using the same inspection procedure as for the 400 kHz case. Only 11 cracked specimens were used for this experimental validation and were distributed in the mid-size range of depths, most critical for POD analysis. The average of the predicted eddy current amplitudes through the transfer functions **F** and **G** and the experimental results for 200 kHz driving frequency, along with the original experimental data at 400 kHz, are shown in Figure 13.44.

The similarities in amplitude and trend between modeled and experimental data at 200 kHz add extra credibility to the proposed hybrid model-assisted approach. The fitting curves in Figure 13.43 and Figure 13.44 were obtained through interpolation with polynomial functions of degrees 3 to 6 that provided the best coefficient of determination (or goodness-of-fit) value, R^2.

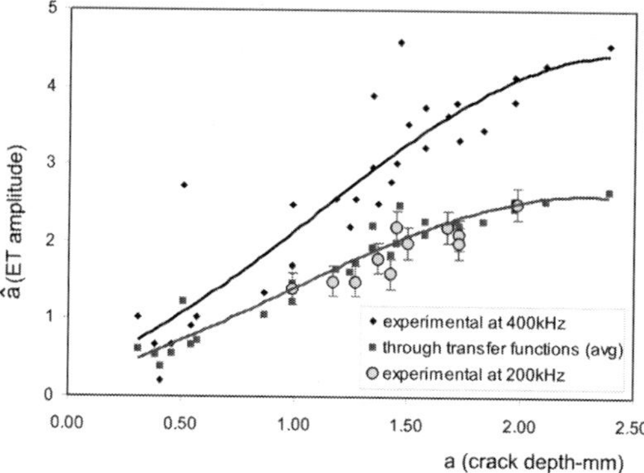

Figure 13.44 Experimental validation of the model-predicted eddy current signal amplitudes for 200 kHz driving frequency.

POD curves for 400 kHz experimental data and 200 kHz modeled inspections are plotted together in Figure 13.45. Although eddy current instrument gain and noise considerations were not taken into account, the $a_{90/95}$ value for the experimental study was found to be about 0.6 mm, while for the case of a 200 kHz driving frequency, the $a_{90/95}$ value was predicted to be about 0.5 mm, based on the data obtained through modeling and transfer functions. This result indicates that by lowering the testing frequency, one could achieve a smaller bolt hole flaw size; not a trivial outcome. However, it should be recognized that this is only a demonstration exercise, used to illustrate the inclusion of modeled inspections into POD analysis. It does not claim that this would be the case for every similar bolt hole inspection. For many of these cases, other factors would need to be taken into account, especially uncontrolled variables, such as noise or tilt.

Case Study #2: In this case, the electrical conductivity of the test part is changed due to the practical relevance it has in the reliability studies of bolt hole eddy current inspections. A change in electrical conductivity is encountered when different alloys or heat treatments are used in the manufacture of various generations of aircraft. The eddy current signatures from a 1 mm long, semi-elliptical crack are shown in Figure 13-46. The metals and their electrical conductivities used in this demonstration are shown in Table 13.5.

If the inspection procedure and signal feature for analysis, i.e., the amplitude of the vertical component, are kept the same, the results in Figure 13.46 demonstrate that changing the test piece conductivity (assuming that the material is not magnetic), does not significantly modify the imaginary component of the eddy current response. Therefore, the POD outcome would remain the same if the alloy type or the material of the structure is changed. This could have considerable implications in the portability of the POD data from one structure to another that differ only by the materials, and the cost-savings it could generate in such a case.

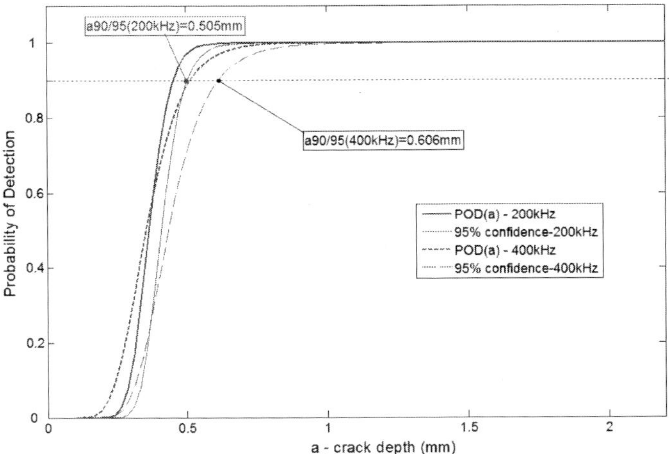

Figure 13.45 POD curves for bolt hole eddy current inspections: empirical data at 400 kHz drive frequency and model-obtained data at 200 kHz drive frequency.

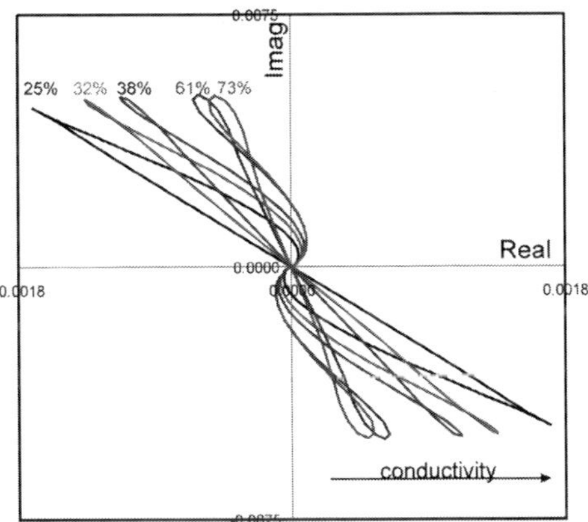

Figure 13.46 Simulation results of the eddy current signal changes with increasing material conductivity (given in percent IACS); probe 'scanning' a 1 mm long, 0.25 mm deep semi-elliptical crack.

Table 13.5 Materials and their conductivities analyzed in this section.

Material	Brass	2024 T3 Al	7075 T6 Al	2014 T3 Al	2014 T6 Al	Pure Al	Gold
Conductivity (MS/m)	14.5	17.5	18.5	19.5	22.0	35.4	42.6
Conductivity (% IACS)	25	30	32	34	38	61	73

13.4.7.3 OBSERVATIONS

The proper use of model-assisted POD enables transfer of an empirical POD to new inspection cases when changes occur in the inspection procedure or the part, potentially saving significant time and costs. In addition, the use of modeling in NDT provides valuable information to optimize inspection parameters for a specific testing situation. However, modeling cannot fully replace empirically-derived POD due to its inability to take into account environmental and human factors. Nevertheless, combining modeling with limited experiments, along with understanding of how most of the variables affect the results in a quantifiable way, model-assisted POD has the potential to greatly reduce the level of effort and cost associated with empirical POD measurements.

13.4.7.4 ADVANTAGES AND LIMITATIONS

Model-assisted POD estimation has the following advantages and limitations.

Advantages:
- Model-assisted POD estimation is cost-effective; however, the models must be first verified using limited experiments.
- Only limited specimens are required.
- Many variables can be evaluated in a timely manner, thus, improving the overall outcome
- Modeling helps in the optimization of inspection parameters and procedures.
- Data can be transferred easily to other similar situations.

Limitations:
- Requires a good model of the inspection procedure and situation at hand.
- Requires accurate data on properties of the test material, flaw characteristics, and NDT parameters for input into the model.
- Requires some experimental verification tests.

13.5 CHAPTER REFERENCES

[1]   The Detection of Fatigue Cracks by Non-destructive Testing Methods, W.D. Rummel, P.H.Todd, S.A. Frecska, S.A., and R.A. Rathke, NASA CR-2369, February 1974.

[2] Non-destructive Evaluation System Reliability Assessment, The US Department of Defence Handbook MIL-HDBK-1823, April 1999. Based on the NATO Advisory Group for Aerospace Research and Development (AGARD) Report: A Recommended Methodology for Quantifying NDE/NDI Based on Aircraft Engine Experience, C. Petrin, C.A. Annis and S.I. Vukelich, AGARD-LS-190, 1993.

[3] Non-destructive Evaluation System Reliability Assessment, The US Department of Defence Handbook MIL-HDBK-1823, April 1999.

[4] Military Standard 1783, Engine Structural Integrity Program (ENSIP), US Department of Defence, Washington, 1984.

[5] Estimate of Probability of Crack Detection from Service Difficulty, J.C. Brewer, Report Data, U.S. Department of Transportation Research and Special Programs Administration, Cambridge, DOT-VNTSC-FAA-94-4, September 1994.

[6] The Reliability and Sensitivity of NDI Techniques in Detecting LCF Cracks in Fastener Bolt Holes of Compressor Discs, N.C. Bellinger, A. Fahr, A.K. Koul, G.Gould, and P. Stoute, NRC Report: LTR-ST-1651, February 1988.

[7] NDI Reliability Study of the Detection of Damage in the CH-146 Tail Rotor Blades, D.S. Forsyth and A. Fahr, NRC Report: LTR-SMPL-2004-0010, 2004.

[8] Corrosion Detection in Aircraft Lap Joints—Proposed Approach to Development of Probability of Detection Data, J.P. Komorowski, D.S. Forsyth, D.L. Simpson and R.W. Gould, presented at the NATO Research and Technology Organization (RTO) Workshop on Airframe Inspection Reliability under Field/Depot Conditions, Brussels, RTO-MP-10, AC/323(AVT)TP/2, pp. 8-1 to 8-8, May 1998.

[9] Non-destructive Evaluation (NDE) Capabilities Data Book, W.D. Rummel, and G.A. Matzkanin, DOD Contract DLA900-90-D-0123, NTIAC, 1995.

[10] Recent Advances in Model-Assisted Probability of Detection, R.B. Thompson, L.J. Brasche, D. Forsyth, E. Lingren, P. Swindell and W. Winfree, http://ntrs.nasa.gov/archive/nasa/casi.ntrs.nasa.gov/20090025450_2009023843.pdf.

[11] Simulation-based POD Evaluation of NDI Techniques, F. Jenson, S. Mahaut, P. Calmon and C. Poidevin, CEA, LIST, F-91191 Gif-sur-Yvette, France.

[12] An Evaluation of Probability of Detection Statistics, D.S. Forsyth and A. Fahr, Proceedings of the NATO Research and Technology Organization (RTO) Meeting 10 on Airframe Inspection Reliability under Field/ Depot Conditions, NATO Research and Technology Organization (RTO Meeting Proceedings: RTO-MP-10, AC/323(AVT)TP/2, November 1998.

[13] Dvelopment of Non-destructive Inspection Probability of Detection Curves Using Field Data, D.L. Simpson, NRC Report: LTR-ST-1285, 1981.

[14] On the Independence of Multiple Inspections and the Resulting Probability of Detection, D.S. Forsyth and A. Fahr, Proceedings of the Review of Progress in Quantitative Non-destructive Evaluation, Vol. 19B, eds. D.O. Thompson and D.E. Chimenti, American Institute of Physics, New York pp. 2159–2166, 2000.

[15] Statistical Methods for Estimating Crack Detection Probabilities, Probabilistic Fracture Mechanics and Fatigue Methods: Application for Structural Design and Maintenance, A.P. Berens, and P.W. Hovey, ASTM STP 798, pp. 79–94,1983.

[16] Characterization of NDE Reliability, Review of Progress in Quantitative NDE, A.P. Berens, and P.W. Hovey Vol. 1, Plenum Press, New York, pp. 579–585, 1982.

[17] A Comparison of Probability of Detection Data Determined using Different Statistical Methods, A. Fahr, D.S. Forsyth, and M. Bullock, NRC Report: LTR-ST-1964, March 1994. Also, Statistical Functions and Computational Procedures for the POD Analysis of Hit/Miss NDI Data, M. Bullock, D. Forsyth, and A. Fahr, NRC Report : LTR-ST-1964, March 1994.

[18] One-sided Confidence Bands for Cumulative Distribution Function, R.C.H. Cheng and T.C. Iles, Technometrics, Vol.30, No.2, pp 155–159, 1988.

[19] Assessment of NDT Systems, Part I: The Relationship of True and False Detection, Material Evaluation, J.A. Swets, pp. 1294–1298, October 1983.

[20] Identifying Sources of Variation for Reliability Analysis of Field Inspections, F.W. Spencer, NATO Research and Technology Organization (RTO) Meeting Proceedings 10 on Airframe Inspection Reliability under Field/Depot Conditions, RTO-MP-10, AC/323(AVT)TP/2, pp. 11-1 to 11-8, November 1998.

[21] Assessment of NDT Systems—Part II, Indice of Performance, J.A. Swets, Material Evaluation, pp. 1299–1303, October 1983.

[22] A Unified Approach to the Model-Assisted Determination of Probability of Detection. R.B. Thompson, Materials Evaluation, 66(6): pp. 667–673, 2008.

[23] Use of Physics-Based Models of Inspection Processes to Assist in Determining Probability of Detection, R.B. Thompson, L. Brasche, J.S. Knopp, and J.C. Malas, 9th Joint FAA/DoD/NASA Conference on Aging Aircraft. Atlanta, GA., 2006.

[24] NDI Techniques for Damage Tolerance-based Life prediction of Aeroengine Turbine Disks, A. Fahr, D.S. Forsyth, and M. Bullock, NRC Report: LTR-ST-1961, February 1994.

[25] NDI Techniques for Damage Tolerance-based Life prediction of Aeroengine Turbine Disks, A. Fahr, D.S. Forsyth, and M. Bullock, NRC Report: LTR-ST-1961, February 1994.

[26] Application of Patten Recognition for Eddy Current Inspection of Aircraft Engine Components, A. Fahr, C.E. Chapman, A. Koul, A. Pelletier and D.R. Hay, Canadian Society for Non-destructive Testing (CSNDT) Journal, Vol. 14, No. 2, pp. 14–21, March 1993.

[27] Inspection of compressor Discs by Ultrasonic Leaky Waves Using an Automated C-scan System, Ultrasonic, A. Fahr, N.C. Bellinger, P. Stoute, and A.K. Koul, Review of Progress in Quantitative NDE, Plenum Press, New York, Vol. 9, 1990.

[28] Edge of Light: A New Enhanced Optical NDI Technique, D.S. Forsyth, A. Marincak, and J.P. Komorowski, Proceedings of the SPIE Conference, Scottsdale, Arizona, U.S.A, 1996.

[29] Damage Tolerance-based Life Predication of Aeroengine Compressor Discs: Part II, A Probabilistic Fracture Mechanics Approach, A.K. Koul, N.C. Bellinger and G. Gould, International Journal of Fatigue, Vol. 12, No. 5, pp. 388–396, 1990.

[30] NDI Reliability Study of the Detection of Damage in the CH-146 Tail Rotor, D.S. Forsyth, NRC Report: LTR-SMPL-2004-0010, 2004.

[31] A Study of the Reliability of Eddy Current Inspections for Cracks in the NENE 10 Engine Impeller Bore, D.S. Forsyth, M. Brothers, A. Fahr and S. Giguere, NRC Report: LTR-SMPL-2003-0064D, 2003.

[32] Eddy Current Calibration of Fatigue Cracks Using EDM Notches, S. Ross, W. Lord, NDT and E International, Vol. 30, No 5, pp. 327–327, October 1997.

[33] Comparative Responses of Cracks and Slots in Eddy Current Measurements, W. Rummel, Review of Progress in Quantitative Non-destructive Evaluation, Vol. 10A, 1991.

[34] NDE Reliability Estimation from Technique Development, D.S. Forsyth, NRC Report: LTR-SMPL-2005-0121, 2005.

[35] Development of POD From In-service NDI Data, D. S. Forsyth, A. Fahr, D.V. Leemans, and K.I. McRae, NRC Report: O-SMPL-1999-0029, 1999.

[36] Airframe Inspection Reliability Using Field Inspection Data, J.H. Heida and F.P. Grooteman, RTO Meeting Proceedings 10: Airframe Inspection Reliability under Field/Depot Conditions, pp. 5-1 to 5-5, 1998.

[37] Probability of Detection Based on Field Inspection Data, D.V. Leemans, Final Project Report for Department of National Defence (Canada) Air Vehicles Research Section, 1998.

[38] Probability of Detection Estimation for Data Sets with Rogue Points, R.W. Hyatt, G.E. Kechter, R.G. Menton, Material Evaluation, pp. 1402–1408, November 1991.

[39] The Use of Data from the CT-114 Tutor Full-Scale Test for the Development of POD", D.S. Forsyth, K. Boate, K.I. McRae, and A. Fahr, NRC-LTR-ST-2218, September 1999.

[40] Numerical Modelling as a Cost-Reduction Tool for Probability of Detection of Bolt Hole Eddy Current Testing, C. Mandache, M. Khan, A. Fahr, and M. Yanishevsky, NRC LTR-SMPL-2007-0121, 2007.

Index

About the Author

Dr. Abbas Fahr is a Senior Research Scientist in the Aerospace Portfolio of the National Research Council Canada (NRCC). He received a B.Sc. degree in Physics from Mashad University in 1972, a Master's degree in Non-destructive Testing (NDT) from Brunel University in 1975 and a Ph.D. degree in Materials Science from Imperial College of London University in 1979. He worked as a research scientist at the Materials Research Centre of the University of Missouri-Rolla, USA from 1979 to 1981 and at the Ontario Research Foundation in Mississauga, Canada from 1981 to 1984. He joined the NRCC in 1984 where he was in charge of the development of NDT technologies for aeronautical applications. Dr. Fahr is the author or co-author of over 200 technical publications and the recipient of several TTCP awards. His areas of expertise include aerospace NDT, inspection reliability assessment and materials characterization. For further information on subjects covered in this book please contact the author at: AeronauticalNDT@gmail.com.